Advances in Geographic Information Science

The ***Advances in Geographic Information Science*** series aims to: present current and emerging innovations in GIScience; describe new and robust GIScience methods for use in transdisciplinary problem solving and decision making contexts, illustrate GIScience case studies for use in teaching and training situations, analyze GIScience tools at the forefront of scientific research, and examine the future of GIScience in an expanding knowledge-based economy. The scope of the series is broad and will encompass work from all subject disciplines that develop and use an explicitly spatial perspective for analysis and problem-solving.

Kousik Das Malakar • Supriya Roy
Manish Kumar

Geospatial Technologies in Coastal Ecologies Monitoring and Management

Springer

Kousik Das Malakar
Department of Geography
GDC Memorial College (Chaudhary Bansi Lal University), BRCM Vidyagram Campus
Bahal, Haryana, India

Supriya Roy
Department of Geography
Institute of Humanities and Social Sciences, Visva-Bharati University
Santiniketan, West Bengal, India

Manish Kumar
Department of Geography
School of Basic Sciences, Central University of Haryana
Mahendragarh, Haryana, India

ISSN 1867-2434 ISSN 1867-2442 (electronic)
Advances in Geographic Information Science
ISBN 978-3-031-92016-5 ISBN 978-3-031-92017-2 (eBook)
https://doi.org/10.1007/978-3-031-92017-2

This Springer imprint is published by the registered company Springer Nature Switzerland AG
The registered company address is: Gewerbestrasse 11, 6330 Cham, Switzerland

Dedicated to young scholars in the fields of geoinformatics, coastal studies, GISciences, geography, environmental science, sustainability science, social science, and policy makers.

Preface

Coastal ecologies are dynamic and vulnerable environments shaped by natural processes and human interventions. The term "*ecologies*" refers to the interactions between organisms and their environments, encompassing biodiversity, resource distribution, and socio-ecological dynamics. It also includes the study of how living organisms interact within these ecosystems, influencing sustainability and management strategies. The increasing impacts of climate change, coastal erosion, land-use transformations, and socio-ecological vulnerabilities necessitate innovative strategies for effective coastal monitoring and management. In this context, geospatial technologies, including remote sensing (RS), geographic information systems (GIS), global navigation satellite systems (GNSS), and participatory GIS (PGIS), offer powerful tools for analyzing, interpreting, and managing coastal socio-ecological systems with enhanced precision and efficiency.

The present book, *Geospatial Technologies in Coastal Ecologies Monitoring and Management*, offers a comprehensive exploration of the theoretical foundations, practical applications, and field-based insights of geospatial methodologies in coastal research. By integrating interdisciplinary knowledge with advanced spatial technologies, this book provides a structured framework for researchers, scientists, policymakers, professionals, geographers, planners, practitioners, students, and laypeople engaged in coastal monitoring and sustainable management. Fundamentally, this book is committed to elucidating three pivotal aspects:

- Offers a comprehensive overview of geospatial technologies, including remote sensing, GIS, GPS, and participatory GIS, in a single volume
- Presents casebooks and community narratives to illustrate the application of geospatial techniques in coastal ecologies' monitoring and management
- Integrates GIScience to advance transdisciplinary solutions and inform strategic decision-making

The book is divided into 17 interrelated and critical aspect-based chapters, connected by six conjunction parts. Part I, *Fundamentals of Geospatial Technologies: Concepts and Techniques of Remote Sensing*, consists of four chapters. **Chapter 1** provides an introductory overview of remote sensing, covering radiation principles, energy sources, interaction mechanisms, heat emission, and spectral reflectance. **Chapter 2** explores sensor networks and satellite systems as tools for Earth observation, examining sensor scanning systems, resolution, satellite platforms, orbits, motion, satellite programs, and other key aspects. **Chapter 3** explores the techniques and applications of aerial imaging and photogrammetry. **Chapter 4** delves into the principles and tools for spatial analysis in geodesy and air photo interpretation within the domain of remote sensing.

Part II, *Fundamentals of Geospatial Technologies: Concepts and Techniques of Geographic Information Systems,* consists of five chapters. **Chapter 5** discusses the concepts, components, and importance of geographic information systems (GIS), along with their applications, data types, models, and sources of error. **Chapter 6** presents practical approaches to GIS data analysis, including raster and vector data analysis, data formats, and various sources of map data. **Chapter 7** provides a detailed exploration of database management systems (DBMSs), covering concepts, purposes, applications, key features, database models and languages, architecture, and design. **Chapter 8** introduces digital imaging, discussing its concepts, properties, formats, fundamental steps in image processing, preprocessing, enhancement, transformation, classification, and other related topics. **Chapter 9** examines participatory GIS (PGIS) as a solution to traditional GIS challenges and highlights the research potential of voice mapping using PGIS methodologies. Part III, *Fundamentals of Geospatial Technologies: Concepts and Techniques of GNSS and GPS*, introduces GNSS and GPS technologies through three chapters. **Chapter 10** discusses the components, functionality, interoperability, and applications of GNSS. **Chapter 11** explores the concepts, history, segments, principles, applications, accuracy, and errors associated with GPS. **Chapter 12** highlights the challenges and future directions of GPS, including its applications in everyday life, GPS literacy, differential GPS, and emerging advancements in GPS technology.

Building upon these technical foundations, Part IV, *Geospatial Solutions in Coastal Ecological Monitoring and Management: A Theoretical Exploration*, transitions into the theoretical dimensions of coastal ecological monitoring and management. It examines socio-ecological systems and integration of GIScience in coastal policy, emphasizing the need for holistic approaches to sustainability in **Chaps. 13 and 14.** Part V, *Geospatial Techniques for Monitoring and Managing Coastal Ecologies: Field Narratives, Case Studies, and Best Practices of GIScience,* presents real-world applications by showcasing case studies and community narratives from India, with a particular focus on the Bengal coast. These field-based explorations demonstrate how geospatial technologies contribute to localized coastal management solutions. In this context, **Chap. 15** compiles community narratives using PGIS, covering topics such as mangrove management and coastal stewardship, coastal management in a changing climate, disaster response, agricultural management, local knowledge in fishing ground management, and approaches to climate migration and inequality. Subsequently, **Chap. 16** presents in-depth case studies and best practices for monitoring and managing coastal ecologies using GIScience. This chapter introduces nine case studies, each offering unique insights into coastal monitoring and management, covering topics such as a century-long climatic trend analysis (casebook 1), monitoring and modeling of shoreline dynamics (casebook 2), modeling and monitoring mangrove greenness (casebook 3), land-use/land cover changes and ecological degradation (casebook 4), coastal LST and ecological changes (casebook 5), mangrove–human conflicts (casebook 6), climate adaptation strategies of traditional fishermen (casebook 7), climate gaps among climate migrants (casebook 8), and spatial aspects of coastal zone management (casebook 9), and represents the various angles of monitoring and management thoughts. These case studies illustrate diverse approaches to understanding and addressing coastal environmental challenges through GIScience applications.

Finally, Part VI, *Transforming Coastal Research: Geographic Information Science in Transdisciplinary Solutions and Strategic Decision-Making,* introduces **Chap. 17**, which provides a forward-looking perspective on the transformative role of GIScience in transdisciplinary research, strategic decision-making, and innovative coastal governance.

A defining feature of this book is its interdisciplinary scope and the application of participatory approaches. By integrating scientific methodologies with community-based knowledge systems, it underscores the significance of inclusive, data-driven policymaking. The incorporation of field narratives and case studies ensures that theoretical perspectives are substantiated with empirical evidence, making this work a valuable resource for both academic research and practical implementation. Further, the book is intended for students, researchers, and professionals in geography, environmental science, marine and coastal studies, geoinformatics, and sustainability studies. It also serves as a guide for policymakers and coastal managers seeking to incorporate geospatial insights into decision-making processes. By fostering a deeper understanding of coastal dynamics through spatial analysis, this work aims to contribute to the development of more resilient and adaptive coastal management frameworks.

In closing, we, the authors, extend our sincere gratitude to the researchers, coastal scientists, geospatial scientists, policymakers, and coastal communities whose invaluable support and knowledge have enriched this book. We hope that it serves as a catalyst for further innovation and collaboration in geospatial coastal research, ultimately contributing to more effective and sustainable coastal ecosystem management.

New Delhi, India — Kousik Das Malakar
Santiniketan, India — Supriya Roy
Mahendragarh, India — Manish Kumar

Acknowledgments

We gratefully applaud the anonymous reviewers for their informative comments, and we especially thank the coastal community of West Bengal (India) for their pioneering participation and cooperation during the field investigation.

Competing Interests The authors have no competing interests to declare that are relevant to the content of this manuscript.

Contents

About the Authors

Dr. Kousik Das Malakar is a socio-ecological geographer, GIS analyst, and researcher with a specialization in climate change and policy. He currently holds the position of Assistant Professor in the Department of Geography, GDC Memorial College (Ch. Bansi Lal University), BRCM Vidyagram Campus, Haryana. Dr. Malakar earned his Ph.D. from the Department of Geography, Central University of Haryana, India. He also possesses a master's degree in social science geography from Jawaharlal Nehru University (JNU), New Delhi, and a bachelor's degree from Vidyasagar University, Medinipur, West Bengal. His research interests encompass the climate crisis, socio-ecological systems, coastal environments and societies, sustainability, disaster studies, participatory policy planning, and the application of geospatial technologies. In his field of research, Dr. Malakar has participated in over 150 academic events and has published research papers, authored books, edited volume, edited chapters, and conference proceedings research papers in internationally reputed publications. He has received numerous certifications from the Indian Institute of Remote Sensing (IIRS) in Dehradun, the Geological Survey of India (GSI) in Hyderabad, and the National Institute of Disaster Management (NIDM) in New Delhi, recognizing his ability to engage in and acquire research ideas and knowledge. Additionally, he actively serves as a reviewer for academic journals and books, and is a member of several academic and professional organizations, including the Human Development and Capability Association (USA), Unequal World (USA), the International Society for Urban Health (USA), the International Centre of Excellence in Engineering & Technology (Portugal), and the International Centre of Excellence in Humanities, Social Sciences, and Interdisciplinary Studies (Portugal).

Ms. Supriya Roy is a GIS analyst and socio-environmental geographer. She holds a master's degree in social science geography from the Department of Geography, Institute of Humanities & Social Sciences, Visva-Bharati University, Santiniketan, and a double bachelor's degree in social science geography from the University of Gour Banga and Vidyasagar College of Education, West Bengal (India). In her field of research, she has participated in more than 100 workshops/

seminars/conferences/webinars and has contributed research monograph and articles to both national and international journals. She has been awarded multiple certifications from NIDM (New Delhi) and IIRS (Dehradun) for her active participation in research and the acquisition of knowledge and research ideas. Her research interests encompass a diverse range of topics, including coastal indigenous communities, anthropo-environmental conflicts, climate justice, climate migration, society and politics, participatory policy planning, and applications of geographic information science.

Dr. Manish Kumar has been serving as an Assistant Professor and Programme Coordinator of the M.Sc. Geoinformatics program in the Department of Geography at the School of Basic Sciences, Central University of Haryana, Mahendragarh, India, since 2020. With over 12 years of experience in teaching and research, he has authored and edited 06 books published by Springer Nature including *Climate Change and Socio-Ecological Transformation, Geographic Information Systems in Urban Planning and Management, Remote Sensing and Geographic Information Systems for Policy Decision Support, and Geoecology of Landscape Dynamics*, among others. Dr. Kumar has published more than 30 research papers in nationally and internationally recognized journals indexed in Science Citation and Scopus. He has supervised 02 Ph.D. scholars and 33 postgraduate theses. Notably, he led the Indian delegation to Kwansei Gakuin University, Japan, under the Indo-Japan Joint Research Program. He also serves as a Steering Committee Member of the International Geographical Union (IGU) Commission on Modeling Geographical Systems. His research interests include application of remote sensing and GIS in urban and regional planning, land use and land cover change, and environmental monitoring and modeling.

Abbreviations

ARVI	Atmospherically Resistant Vegetation Index
BMP	Bitmap
CAD	Computer-aided design
DBA	Database administrator
DBMS	Database management system
DCL	Data control language
DDL	Data definition language
DGPS	Differential global positioning system
DML	Data manipulation language
DSAS	Digital shoreline analysis system
EIA	Environmental impact assessment
EMR	Electromagnetic radiation
EO	Earth observation
ESA	European Space Agency
EVI	Enhanced Vegetation Index
FGD	Focus group discussion
FK	Foreign key
GEOJSON	Geographic JavaScript Object Notation
GIF	Graphics interchange format
GIS	Geographic information system
GIScience	Geographic information science
GLONASS	Global navigation satellite system
GNSS	Global navigation satellite system
GOES	Geostationary operational environmental satellite
GPS	Global positioning system
ICT	Information and communication technology
IDW	Inverse distance weighting
ISRO	Indian Space Research Organisation
JPEG	Joint Photographic Experts Group
LANDSAT	Land remote sensing satellite
LBS	Location-based service

LIDAR	Light detection and ranging
LST	Land surface temperature
LULC	Land use and land cover
MNDWI	Modified Normalized Difference Water Index
MODIS	Moderate-resolution imaging spectroradiometer
MSL	Mean sea level
MSS	Multispectral scanner
NASA	National Aeronautics and Space Administration
NDVI	Normalized Difference Vegetation Index
NDWI	Normalized Difference Moisture Index
NGO	Nongovernmental organization
NOAA	National Oceanic and Atmospheric Administration
OLI	Operational Land Imager
PGIS	Participatory geographic information system
PNG	Portable network graphics
PPGIS	Participatory public geographic information system
RADAR	Radio detection and ranging
RDBMS	Relational database management system
RS	Remote sensing
RVI	Ratio Vegetation Index
SES	Socio-ecological system
SQL	Structured query language
SVM	Support vector machine
UNECA	United Nations Economic Commission for Africa
UNESCO	United Nations Educational, Scientific and Cultural Organization
UNOOSA	United Nations Office for Outer Space Affairs
UTM	Universal Transverse Mercator
UV	Ultraviolet
WGS84	World Geodetic System 1984

Part I
Fundamentals of Geospatial Technologies: Concepts and Techniques of Remote Sensing

Chapter 1
Remote Sensing Essentials: An Introductory Overview

Remote sensing is not just about technology; it's about gaining a deeper understanding of our environment and the impact of human activity on our planet. — Jane Goodall.

Abstract Remote sensing (RS), geographic information systems (GIS), and the Global Positioning System (GPS) form the core of geospatial technology, enabling the acquisition and analysis of spatial data. This chapter introduces the fundamental concepts, history, and principles of remote sensing, a vital tool for studying the atmosphere and Earth's surface without direct contact. The chapter begins by defining remote sensing, its evolution, types, and applications, emphasizing its role in monitoring and analyzing Earth's processes. Key topics include the electromagnetic spectrum, mechanisms of radiation interaction, and the concept of spectral signatures, unique patterns of reflectance and emission that differentiate materials. Additionally, the chapter discusses components of an ideal remote sensing system, passive and active systems, and data processing stages, providing a comprehensive framework for understanding this technology. By exploring these foundational principles and critical knowledge, the chapter prepares readers to engage with advanced remote sensing applications and technological developments in subsequent sections.

Keywords Remote sensing system · Electromagnetic spectrum · Energy source · Atmospheric window · Scattering of energy

Remote sensing, as a cornerstone of geospatial technology, has transformed how we observe and analyze Earth's dynamic processes. This chapter offers a comprehensive introduction to remote sensing, tracing its evolution and highlighting its significance in studying the atmosphere and Earth's surface. Key concepts such as the electromagnetic spectrum, radiation interactions, and spectral signatures are explored, alongside an overview of system components and data processing stages.

K. D. Malakar et al., *Geospatial Technologies in Coastal Ecologies Monitoring and Management*, Advances in Geographic Information Science,
https://doi.org/10.1007/978-3-031-92017-2_1

By presenting these foundational principles, the chapter provides readers with essential knowledge to navigate advanced remote sensing applications and its role in addressing contemporary environmental and technological challenges.

Key Points of the Chapter

- Definitions and principles of remote sensing are explored.
- The historical evolution of remote sensing highlighted key milestones and advancements.
- Types and a system of remote sensing and their diverse applications are analyzed.
- Applications, benefits, and challenges of remote sensing were summarized, addressing both traditional and modern contexts.

1.1 Introduction to Remote Sensing

1.1.1 Concept

Remote sensing refers to the technology used to gather information about the atmosphere and Earth's surface from a distance. It utilizes sensors mounted on aircraft, drones, satellites, and ground-based platforms to capture data across various electromagnetic wavelengths. These sensors operate through two primary methods: active sensing, where energy is emitted and its return is measured, and passive sensing, which records natural radiation emitted or reflected by Earth's surfaces. This technology is indispensable in fields such as disaster management, urban planning, agriculture, and environmental monitoring, offering a comprehensive perspective of dynamic planetary processes. By enabling the analysis of large areas, remote sensing supports informed sustainable practices and decision-making. However, challenges such as resolution limitations, data processing demands, and atmospheric interference remain areas for further research and technological advancement. Several definitions or concepts of remote sensing are as follows:

- The National Aeronautics and Space Administration (NASA) defines "remote sensing as the process of collecting data about the Earth's surface without being in direct contact with it. This is typically done using sensors on satellites, aircraft, or other platforms".
- The United Nations Office for Outer Space Affairs (UNOOSA) conceptualizes "remote sensing as a method for obtaining information about the Earth's land, water, and atmosphere. This is achieved through sensors mounted on platforms like satellites and aircraft, which collect data in the form of images and measurements".
- The American Society for Photogrammetry and Remote Sensing (ASPRS) describes "remote sensing as the science of obtaining information about the Earth's surface without direct physical contact. This process involves detecting

and recording reflected or emitted energy, followed by the processing, analysis, and application of the gathered data".

- The European Space Agency (ESA) explains that "remote sensing entails gathering and interpreting data from a distance. This process leverages various technologies, such as satellite imagery, to observe and analyze the Earth's surface and its changes over time".

These definitions and conceptions collectively underscore the fundamental idea of remote sensing as a non-contact technique for acquiring data about the surface or environment through the use of sensors on various platforms.

1.1.2 History of Development

The history of remote sensing is a captivating narrative that stretches across centuries, marked by remarkable advancements in the twentieth and twenty-first centuries. Below is an outline of the key milestones in the evolution of remote sensing:

Seventeenth–Nineteenth Century (Early Observations) The origins of remote sensing can be traced to the early fields of cartography and astronomy, where telescopes and other optical instruments were used to observe distant objects. In the nineteenth century, the pioneering use of aerial photography emerged, with balloons and kites employed to capture images from above the Earth's surface.

World War I and II Aerial photography was instrumental in military reconnaissance during both World War I and II, driving significant advancements in remote sensing technology. In World War II, the development of radar technology for military use established a foundation for the future applications of remote sensing.

Post-World War II Era The post-war period witnessed the widespread adoption of aerial photography for diverse applications, such as natural resource management, agriculture, and land use planning.

1950s—1980s (The Space Age) The Soviet Union's launch of Sputnik 1 in 1957 signified the beginning of the space age and a landmark event in the evolution of remote sensing. In 1960, the NASA successfully launched TIROS-1, the first Earth-observing satellite, dedicated to weather monitoring. The 1960s brought further advancements with the development of sophisticated Earth-observing satellites, including the Landsat series, which provided critical data for land cover analysis and resource management.

After that the 1970s and 1980s marked a significant expansion in the use of remote sensing for environmental and scientific purposes. Applications flourished in areas such as environmental monitoring, agriculture, forestry, and geological exploration. During this period, radar remote sensing emerged as an essential tool for mapping and monitoring the Earth's surface, particularly in regions with persistent cloud cover.

1990s—Present (Modern Era) Advancements in satellite technology, particularly in high-resolution imaging and hyperspectral sensors, have significantly enhanced the capabilities of remote sensing. These technological improvements have broadened the scope of remote sensing applications, extending its use to areas such as climate monitoring, disaster management, urban planning, and other critical domains. Additionally, the advent of small satellites and unmanned aerial vehicles (UAVs or drones) has further democratized access to remote sensing technology, making it more accessible to researchers, policymakers, and industries worldwide.

The integration of remote sensing data with Geographic Information Systems (GIS) has further transformed the field, enabled advanced spatial analysis and supported informed decision-making. This synergy has allowed for more precise mapping, modelling, and monitoring of environmental and human systems, underscoring the pivotal role of remote sensing in addressing contemporary global challenges.

The history of remote sensing stands as a testament to human innovation and technological advancement. In the modern era, remote sensing has become an essential tool for comprehending and managing Earth's environment and resources, driving significant progress in scientific research, environmental stewardship, and societal development.

1.1.3 Types

Remote sensing is a versatile technology that uses passive and active methods to gather information about the atmosphere and surface (Fig. 1.1). There are methods such as:

Active Remote Sensing Active remote sensing transmits energy and measures its return after interacting with the Earth's surface or objects. The following are famous examples:

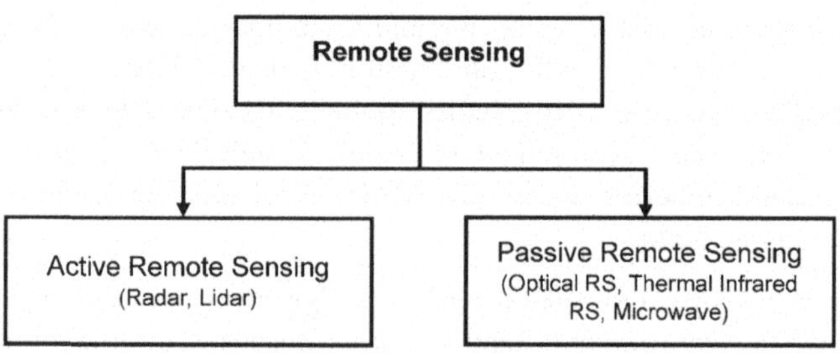

Fig. 1.1 Types of remote sensing (prepared by the authors)

- ***Radar (radio detection and ranging):*** Uses microwave or radio pulses to map terrain, monitor weather, and detect surface changes, regardless of weather.
- ***Lidar (light detection and ranging):*** Employs laser pulses for precise distance measurement, aiding in topographic mapping, forest studies, and urban planning.

Passive Remote Sensing Passive remote sensing relies on detecting and measuring natural energy emitted or reflected by objects on the Earth's surface. This method utilizes sensors that capture ambient radiation without emitting their own energy. Prominent examples include:

- ***Optical remote sensing:*** Uses visible and near-infrared light to analyze land cover and vegetation health.
- ***Thermal infrared remote sensing:*** Detects thermal radiation to measure temperature variations and identify heat anomalies.
- ***Microwave remote sensing:*** Employs long wavelengths to penetrate clouds, estimate soil moisture, monitor sea surface temperatures, and map topography.

Remote sensing techniques are widely applied in fields like environmental monitoring, agriculture, forestry, urban planning, disaster management, geology, and climate research. The choice of sensing varies based on study objectives and area characteristics, offering crucial insights into Earth's dynamic processes.

1.1.4 An Ideal Remote Sensing System

Designing an effective remote sensing system requires careful planning and several critical steps to optimize its ability to capture valuable information about the atmosphere and surface (Fig. 1.2). The following provides an overview of the essential considerations involved:

(a) ***Define objectives and requirements***

- Clearly articulate the goals and objectives of the remote sensing system, identifying the specific data required and its intended applications.
- Establish the necessary spatial and spectral resolution, define the coverage area, and determine the frequency of data acquisition to meet the system's objectives effectively.

(b) ***Select sensing platforms***

- Select suitable sensing platforms, including satellites, aircraft, drones (UAVs), or ground-based instruments, based on specific objectives and requirements.
- Evaluate factors like cost, accessibility, and the platform's ability to effectively cover the target geographic area.

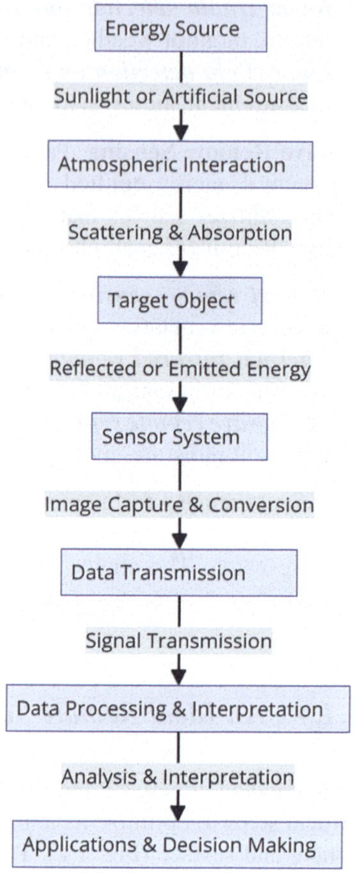

Fig. 1.2 An ideal remote sensing system (prepared by the authors)

(c) ***Choose sensors and instruments***

- Choose sensors and instruments that are compatible with the selected platform and capable of capturing the required data.
- Verify that the sensors operate in the desired spectral bands (e.g., optical, thermal, microwave) and provide the necessary spatial resolution.

(d) ***Customize sensing modes***

- Customize the sensing modes to suit specific applications, such as using active radar sensors for terrain mapping or passive optical sensors for vegetation monitoring.
- Fine-tune sensor settings, including wavelength bands and acquisition angles, to optimize data collection for each application.

(e) ***Calibrate sensors***

- Perform regular sensor calibration to ensure data accuracy and reliability, including radiometric, geometric, and atmospheric corrections.
- Establish quality control measures to monitor and sustain sensor performance over time.

(f) ***Data acquisition and processing***

- Create a data acquisition plan that takes into account factors such as orbital parameters (for satellites), flight paths (for aircraft and drones), and data acquisition schedules.
- Develop data processing workflows to transform raw sensor data into actionable information, including steps like image preprocessing, georeferencing, and atmospheric correction.

(g) ***Data storage and management***

- Establish a comprehensive data storage and management system to efficiently organize and archive collected data.
- Implement standardized data cataloguing and metadata practices to enhance data retrieval and analysis.

(h) ***Data analysis and interpretation***

- Use specialized software and algorithms for data analysis, feature extraction, and interpretation.
- Integrate artificial intelligence and machine learning techniques to automate data processing and classification.

(i) ***Data dissemination and accessibility***

- Ensure data accessibility for a broad audience, including researchers, policymakers, and the public, by providing user-friendly platforms and efficient data distribution channels.

(j) ***Quality assurance and validation***

- Regularly monitor and evaluate data quality and accuracy. Employ validation processes that incorporate ground truth data and field measurements.

(k) ***Documentation and reporting***

- Keep thorough documentation of system setup, sensor configurations, data processing procedures, and quality control measures.
- Produce regular reports and documentation to ensure transparency and reproducibility.

(l) ***Capacity building and training***

- Provide training for personnel involved in remote sensing operations and data analysis to ensure expertise and adherence to best practices.

Therefore, establishing an optimal remote sensing system necessitates meticulous planning, strategic investment in technology, and a commitment to continuous monitoring and refinement. So, customizing the system to address specific application requirements and staying informed of technological advancements are crucial for ensuring its success.

1.1.5 Applications of Remote Sensing

Remote sensing captures valuable data about the atmosphere and surface, with applications across various fields (Fig. 1.3). Here is a brief overview of key uses:

- **Weather and climate monitoring:** Weather agencies use remote sensing for forecasting, storm tracking, and climate research, while the data also enhances understanding of climate change and its effects.
- **Environmental monitoring:** Remote sensing is crucial for monitoring and assessing environmental changes, including deforestation, land dynamics, and

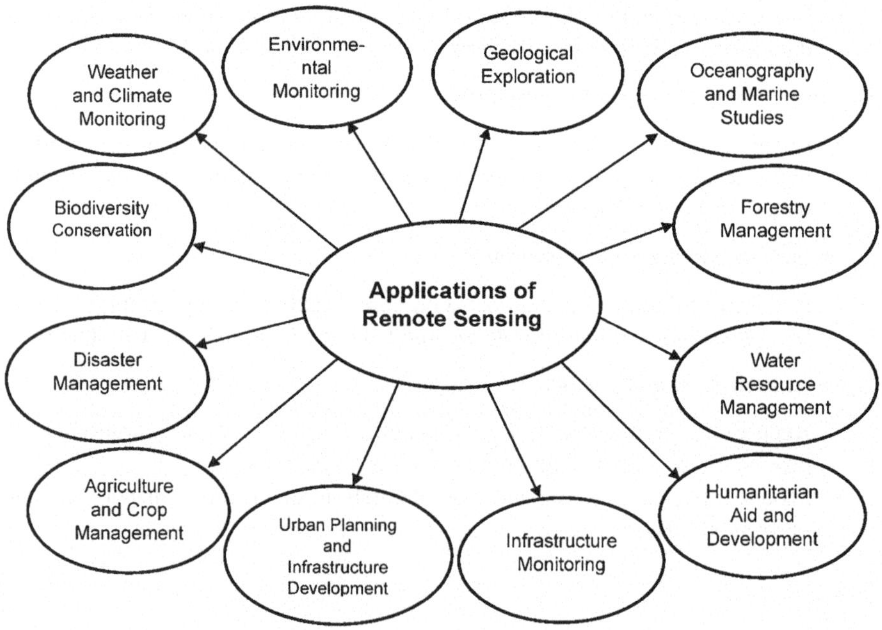

Fig. 1.3 Applications of remote sensing (prepared by the authors)

habitat loss, while also tracking phenomena such as erosion, desertification, and wetland dynamics.

- **Geological exploration:** In geology, remote sensing aids in mineral exploration, identifying geological structures, and mapping natural resources, while also supporting the study of volcanoes, landslides, and earthquake-related phenomena.
- **Oceanography and marine studies:** Remote sensing data are used to study ocean currents, sea surface temperature, and marine ecosystems, aiding in fisheries management and monitoring coastal erosion.
- **Forestry management:** Remote sensing is essential for forest inventory, deforestation detection, and assessing forest health, as well as supporting forest planning, monitoring illegal logging, and evaluating wildlife habitats.
- **Biodiversity conservation:** Conservationists utilize remote sensing to monitor wildlife habitats, track animal migrations, and assess the consequences of climate change on ecosystems, while also supporting the protection of endangered species and biodiversity hotspots.
- **Agriculture and crop management:** Farmers utilize remote sensing data for crop health assessment, yield prediction, and irrigation management, while also monitoring soil moisture, crop diseases, and pests to enable timely interventions.
- **Water resource management:** Remote sensing data are instrumental in monitoring water bodies such as lakes, rivers, and reservoirs, aiding in water quality management, pollution detection, and tracking changes in water levels.
- **Urban planning and infrastructure development:** Urban planners use remote sensing to assess urban growth, land use zoning, and transportation planning, while also monitoring infrastructure developments like roads, buildings, and utilities.
- **Disaster management:** Remote sensing supports disaster preparedness, response, and recovery by assessing damage from natural disasters such as hurricanes, floods, and earthquakes. It also aids search and rescue operations and provides real-time information during crises.
- **Infrastructure monitoring:** Remote sensing is employed for infrastructure monitoring, including the assessment of bridges, dams, pipelines, and power lines, helping to identify maintenance needs and prevent accidents.
- **Humanitarian aid and development:** Remote sensing supports humanitarian efforts by assessing the impact of conflicts, refugee movements, and food security, while also aiding in the planning and implementation of development projects in remote or underserved areas.

1.1.6 Advantage and Disadvantage of Remote Sensing

Advantages of Remote Sensing

- **Wide geographic coverage:** Remote sensing offers extensive coverage of large geographic areas, enabling the monitoring of regions that are often inaccessible by traditional methods. This capability is particularly valuable for environmental monitoring, disaster management, and land use planning.

- **Real-time or periodic data collection:** Remote sensing systems can gather data in real time or at regular intervals, providing timely information essential for decision-making. This is particularly useful for tracking processes, such as weather patterns and vegetation growth.
- **Multispectral information:** Remote sensing sensors capture data across a range of electromagnetic wavelengths, including visible light, infrared, and microwave. This multispectral data offers valuable insights into various aspects of the atmosphere and surface.
- **Consistency and reproducibility:** Data collection through remote sensing is both consistent and reproducible, facilitating comparisons over time and across different locations. This consistency is crucial for trend analysis and scientific research.
- **Cost-effectiveness:** Although the initial setup and satellite development can be costly, remote sensing is often more cost-effective in the long term, reducing the need for extensive ground-based data collection and monitoring.
- **Non-invasive:** Remote sensing enables data collection without direct contact with the area or object of interest, making it ideal for studying sensitive or hazardous environments, such as active volcanoes or disaster-affected regions.

Disadvantages of Remote Sensing

- **Environmental limitations:** Environmental factors, such as dense vegetation or water turbidity, can impact the quality of remote sensing data and reduce its effectiveness for specific applications.
- **Limited spectral and spatial resolution:** Remote sensing systems may face limitations in spatial and spectral resolution, which can hinder the detection of fine details or the differentiation of certain features. Acquiring high-resolution data can also be costly.
- **Weather dependency:** Some remote sensing methods, particularly optical sensors, are influenced by weather conditions like cloud cover, which can result in data gaps or restrict data collection during adverse weather events.
- **Data interpretation complexity:** The interpretation of remote sensing data often requires specialized knowledge and expertise. Understanding spectral signatures and performing image analysis can be challenging for individuals without technical experience.
- **Lack of ground truth:** Remote sensing data may lack the complete context for accurate interpretation. Ground truth data, which involves on-site measurements, is often necessary to validate and refine the data's accuracy.

1.2 Radiation Principles and Energy Source

1.2.1 Electromagnetic Spectrum

The electromagnetic spectrum is a broad framework that organizes all forms of electromagnetic radiation according to their wavelengths or frequencies (Fig. 1.4). It spans a wide range of energy, from very short wavelengths with high frequencies to much longer wavelengths with lower frequencies. The spectrum is essential for numerous scientific and practical applications in diverse fields, including physics, telecommunications, remote sensing, and astronomy.

1.2.2 Electromagnetic Radiation

Electromagnetic radiation (EMR) is a fundamental concept in physics that refers to the transmission of energy through space in the form of waves. It spans a wide range of wavelengths, from very short gamma rays and X-rays to visible light, microwaves, and long radio waves (Fig. 1.5). Each form of radiation possesses unique properties, and its behaviour is governed by wave-particle duality. EMR is crucial for various scientific and practical applications, such as telecommunications, medical imaging (e.g., X-rays and MRIs), astronomy (e.g., observing distant galaxies and stars), and common technologies like wireless communication and microwave ovens. A comprehensive understanding of the properties and behaviour of EMR is essential for advancing fields like GIScience, space exploration, and improving technological solutions that enhance quality of life.

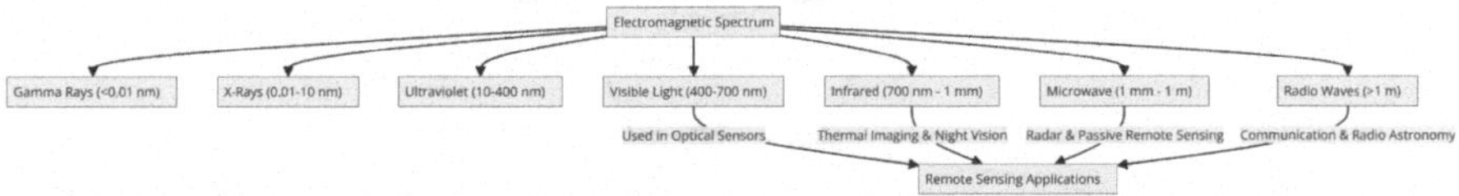

Fig. 1.4 Electromagnetic spectrum (prepared by the authors)

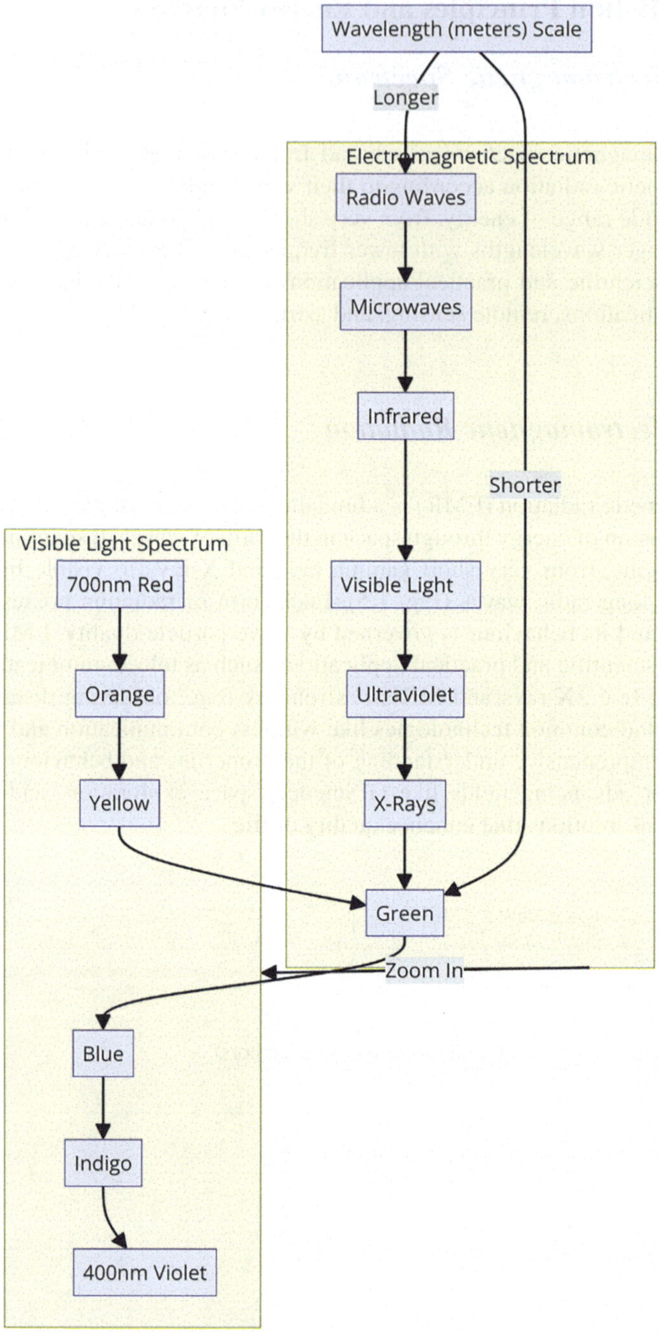

Fig. 1.5 Various spectrum of waves (prepared by the authors)

1.2.3 *Difference Between Electromagnetic Spectrum and Radiation*

In remote sensing, electromagnetic spectrum and radiation are two closely related but distinct areas of research (Fig. 1.6).

Basis	Electromagnetic spectrum	Electromagnetic radiation
Basic concept	The electromagnetic spectrum is a broad continuum of electromagnetic radiation categorized by its wavelength or frequency, ranging from extremely short wavelengths (gamma rays) to very long wavelengths (radio waves)	Electromagnetic radiation refers to energy that travels through space as electromagnetic waves. These waves travel at the speed of light and carry energy across different wavelengths and frequencies
Nature	The spectrum organizes the different types of electromagnetic radiation based on their wavelength or frequency, from radio waves at the low end to gamma rays at the high end	It consists of oscillating electric and magnetic fields that propagate through space. The energy carried by electromagnetic waves can vary in terms of frequency, wavelength, and amplitude
Characteristics	The electromagnetic spectrum includes various regions such as gamma rays, X-rays, ultraviolet radiation, visible light, infrared radiation, microwaves, and radio waves. Each type of radiation in the spectrum has distinct characteristics and is used for different applications	EMR can have a wide range of frequencies and wavelengths, and it travels at the speed of light (approximately 299,792 km/s in a vacuum)
Key point	The electromagnetic spectrum categorizes all types of electromagnetic radiation, with remote sensing utilizing specific segments (e.g., visible, infrared, microwave) to observe and analyze the atmosphere and surface	EMR refers to the actual waves or energy that is used for data collection in remote sensing
Example	The visible light we can see is just a small portion of the electromagnetic spectrum	Light, radio waves, microwaves, X-rays, and gamma rays are all forms of electromagnetic radiation

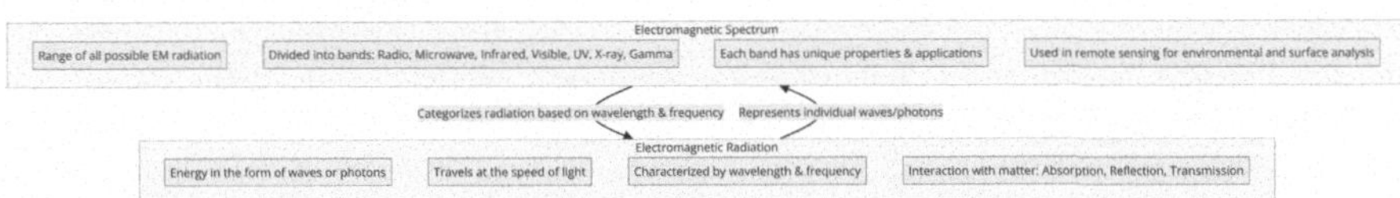

Fig. 1.6 Electromagnetic spectrum and radiation: Linkages and relations (prepared by the authors)

1.2.4 *Regions of Electromagnetic Spectrum*

The electromagnetic spectrum includes various types of radiation, each with distinct characteristics and applications in remote sensing:

- **Visible light:** This is the range of wavelengths (400–700 nm) that human eyes can perceive, responsible for the colours we see in the environment.
- **Infrared (IR) radiation:** With wavelengths between 700 nm and 1 mm, IR radiation is used in remote sensing for thermal imaging and evaluating vegetation health.
- **Microwaves:** Ranging from 1 mm to 1 m, microwaves are employed in radar systems for applications such as weather monitoring and topographic mapping.
- **Radio waves:** These have the longest wavelengths (1 m to several kilometres) and are used in radio broadcasting, GPS, and wireless communication.
- **Ultraviolet (UV) radiation:** Spanning wavelengths of 10–400 nm, UV radiation is used for sterilization and UV lamps but can be harmful to living organisms.
- **X-rays:** With wavelengths between 0.01 and 10 nm, X-rays are widely used in medical imaging and industrial inspections.
- **Gamma rays:** These have the shortest wavelengths (less than 0.01 nm) and the highest frequencies, produced in nuclear reactions and used in radiation therapy and astrophysical research.

1.2.5 *Black Body*

The concept of a black body is crucial, particularly in the analysis of thermal infrared data. A black body is a theoretical entity that absorbs all incident electromagnetic radiation and emits radiation across the entire electromagnetic spectrum, with the emission intensity varying according to its temperature. This concept is fundamental for understanding the emission and reflection of thermal energy by objects and surfaces, which has diverse applications in remote sensing (Fig. 1.7). The following provides an in-depth exploration of key aspects of black body in remote sensing:

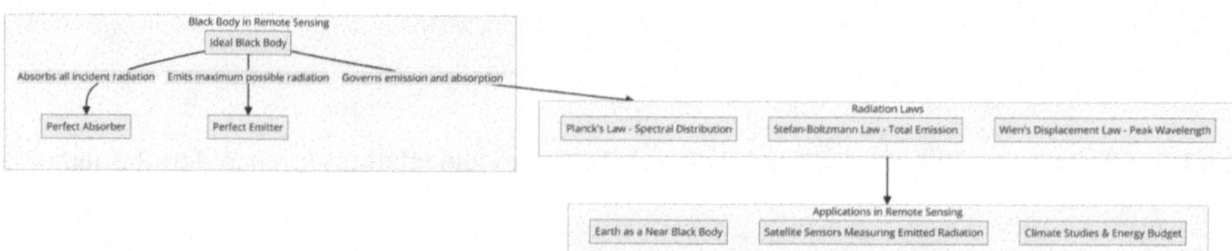

Fig. 1.7 Black body in remote sensing (prepared by the authors)

- **Perfect absorber and emitter:** A black body is an ideal absorber of all incoming radiation, reflecting none. In remote sensing, it is used as a benchmark for understanding the radiative properties of actual objects. Additionally, it emits radiation at the highest possible intensity for a given temperature, as described by Planck's law.
- **Stefan–Boltzmann law:** The total energy emitted by a black body is directly proportional to the fourth power of its temperature, as outlined in the Stefan–Boltzmann law:

$$E = \sigma T^4$$

- where, E is the emitted energy, σ is the Stefan–Boltzmann constant, and T is the temperature in Kelvin.

- **Planck's law:** This law describes the spectrum of radiation emitted by a black body at any given temperature. In remote sensing, it is essential for modelling and understanding the spectral distribution of emitted radiation, which is crucial for retrieving temperature data in thermal infrared imaging.
- **Emissivity:** In remote sensing, emissivity quantifies how closely an object's radiative behaviour aligns with that of a black body. Emissivity values range from 0 to 1, where 1 represents a perfect black body, and values less than 1 indicate objects that reflect some portion of incident radiation.
- **Use in remote sensing:** The black body model is essential for calibrating and interpreting thermal infrared data, serving as a reference for the emission characteristics of real-world objects. Remote sensing instruments, like thermal infrared radiometers, use this model to interpret surface temperature data.

In summary, the black body concept is crucial in remote sensing for sensor calibration and data interpretation, enabling accurate assessment of surface temperatures and other properties by comparing real objects to idealized radiation.

1.3 Interaction Mechanisms

1.3.1 Radiometric Aspects

Radiometric quantities are fundamental parameters in remote sensing that characterize the electromagnetic radiation (EMR) captured by sensors. These quantities offer insights into the intensity, energy, and spectral properties of the radiation, making them essential for accurate interpretation of remote sensing data (Fig. 1.8). Below are some key radiometric quantities in remote sensing:

- **Radiant flux (Φ):** Radiant flux refers to the total power or energy emitted by a source per unit of time, measured in watts (W), and in remote sensing, it represents the total energy received from a specific area or pixel.

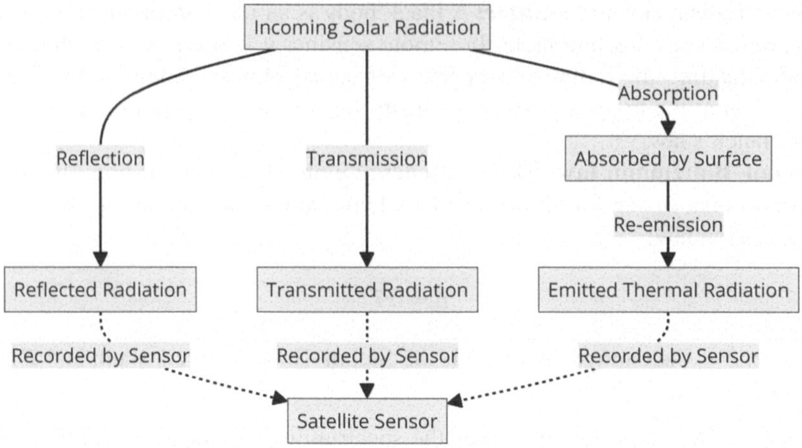

Fig. 1.8 Radiometric quantities in remote sensing (prepared by the authors)

- **Radiance (L):** Radiance is the radiant flux per unit area and per unit solid angle, quantifying the amount of radiation emitted from a surface in a specific direction, and is measured in watts per square metre per steradian ($W/(m^2 \cdot sr)$).
- **Irradiance (E):** Irradiance is the radiant flux received per unit area, typically by a sensor or detector, and is measured in watts per square metre (W/m^2), characterizing the intensity of incoming radiation.
- **Reflectance (ρ):** Reflectance is a dimensionless quantity that quantifies the ratio of reflected radiation to incident radiation, ranging from 0 (no reflection) to 1 (complete reflection), and is used to assess the amount of radiation reflected by the Earth's surface.
- **Transmittance (τ):** Transmittance is a dimensionless quantity that represents the ratio of transmitted radiation to incident radiation, characterizing the amount of radiation that passes through a medium, such as the Earth's atmosphere.
- **Emittance (ε):** Emittance refers to the radiant flux emitted by a surface per unit area. In thermal infrared remote sensing, it is crucial for estimating land surface temperature.
- **Exitance (M):** Exitance represents the total radiant flux leaving a surface per unit area, encompassing both emitted and reflected radiation from the surrounding environment.
- **Intensity (I):** Intensity measures the radiant flux per unit solid angle, quantifying the concentration of radiation in a specific direction. It is expressed in watts per steradian (W/sr).
- **Spectral radiance (Lλ):** Spectral radiance refers to radiance per unit wavelength or frequency, quantifying the radiometric properties of radiation within a specific spectral band. It is typically expressed in units such as watts per square metre per steradian per nanometre ($W/(m^2 \cdot sr \cdot nm)$).

- **Spectral irradiance (Eλ):** Spectral irradiance refers to irradiance per unit wavelength or frequency, characterizing the intensity of radiation within a specific spectral band received by a sensor. It is typically measured in watts per square metre per nanometre ($W/(m^2 \cdot nm)$).

1.3.2 Interaction with Atmosphere and Earth Surface

The interaction of electromagnetic energy with objects or features on the surface and in the atmosphere is central to the data acquisition process. These interaction mechanisms influence how remote sensing sensors detect and capture environmental information (Fig. 1.9). The key mechanisms of energy interaction in remote sensing are outlined below.

Reflection Reflection is a key interaction mechanism of electromagnetic energy in remote sensing. It occurs when electromagnetic radiation strikes a surface, causing some or all of the energy to be redirected back into space. Understanding reflection is essential for remote sensing scientists, as it enables them to interpret surface and object properties based on the way different wavelengths of radiation are reflected.

Reflection is a crucial aspect of remote sensing, especially in optical and multispectral applications. Remote sensing instruments, including satellites and airborne sensors, measure the reflected radiation across various spectral bands. Analyzing reflectance data allows for the extraction of valuable information about land cover,

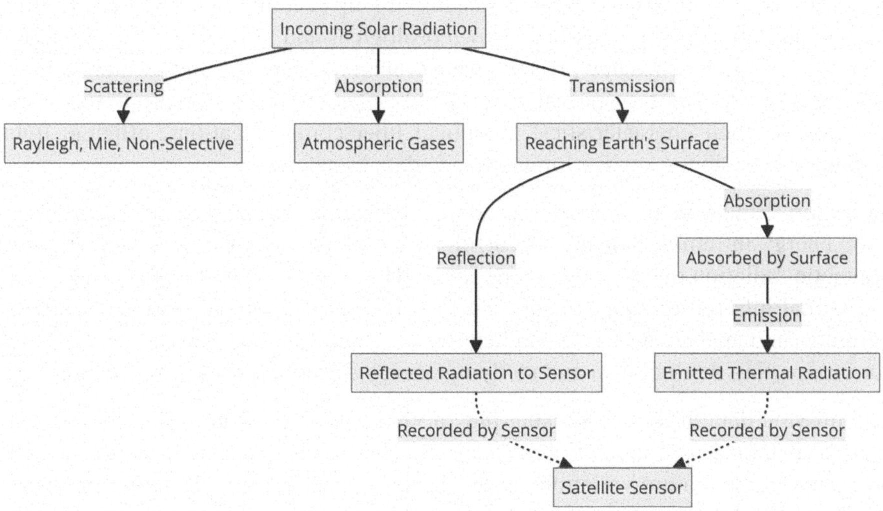

Fig. 1.9 Interaction of electromagnetic energy with the atmosphere and Earth's surface (prepared by the authors)

vegetation health, soil properties, urban features, and other environmental characteristics.

Absorption Absorption is another interaction mechanism of electromagnetic energy in remote sensing. It occurs when electromagnetic radiation penetrates a material, with a portion of the energy being absorbed by the atoms or molecules of the substance. This process converts electromagnetic energy into internal energy, typically as heat or electronic transitions. A thorough understanding of absorption is essential for accurately interpreting remote sensing data, particularly in applications related to material composition, temperature, and atmospheric studies.

Absorption plays a key role in various remote sensing applications, including mineral exploration, vegetation health assessment, atmospheric studies, thermal infrared sensing, and water monitoring. In mineral exploration, absorption bands in the infrared region help identify mineral compositions. For vegetation, changes in absorption bands indicate stress or disease. In atmospheric studies, gas absorption influences data quality and aids in correcting atmospheric effects. In thermal infrared remote sensing, surface and atmospheric absorption play a key role in temperature estimation. Absorption characteristics in the visible and near-infrared regions also help identify water bodies and evaluate water quality.

Transmission Transmission is an important interaction mechanism of electromagnetic energy in remote sensing. It occurs when electromagnetic radiation passes through a material or substance with minimal absorption or scattering. During transmission, the energy retains its original direction and wavelength as it traverses the material, emerging on the opposite side with little alteration.

Transmission of electromagnetic radiation is essential in various remote sensing applications. In atmospheric remote sensing, it helps capture data from the Earth's surface and atmosphere, supporting meteorological and climate studies. In water remote sensing, transmission properties are crucial for studying underwater features and water quality. For transparent materials like glass, transmission properties assist in assessing their characteristics. In optical fiber communications, efficient light signal transmission is key for long-distance data transfer.

Emission Emission is another important interaction mechanism of electromagnetic energy in remote sensing, occurring when objects and surfaces emit electromagnetic radiation due to their temperature. This process is particularly important in thermal infrared remote sensing, where it is used to estimate land surface temperatures and analyze heat-related phenomena. Understanding emission is essential for interpreting thermal data and studying energy dynamics in the environment.

Emission data in remote sensing play a vital role in environmental monitoring, thermal mapping, and industrial inspections. Thermal maps created from emission data reveal temperature variations in landscapes, urban areas, and natural features. Additionally, emission-based remote sensing is used to monitor heat-related environmental changes such as wildfires, volcanic activity, and soil moisture variations.

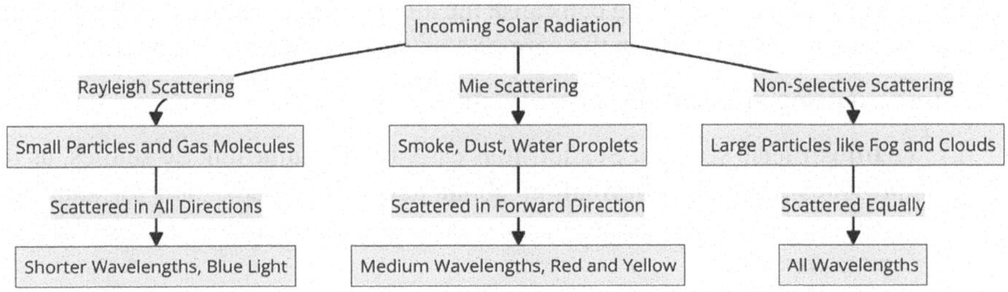

Fig. 1.10 Types of scattering (prepared by the authors)

In industrial and infrastructure inspections, emission data help identify thermal anomalies in equipment, power lines, and buildings, supporting maintenance and safety efforts.

Scattering Scattering is a key interaction mechanism of electromagnetic energy in remote sensing, occurring when electromagnetic radiation interacts with particles or objects within the atmosphere, on the surface, or in other media. This interaction causes the radiation to deviate from its original direction, influencing the intensity and spectral properties of the scattered energy (Fig. 1.10). Understanding scattering is essential for a range of remote sensing applications, including atmospheric research, optical remote sensing, and radar observations. Below is a comprehensive exploration of scattering in remote sensing:

- **Types of scattering**
  (a) **Rayleigh scattering:** It occurs when electromagnetic radiation interacts with particles significantly smaller than the wavelength of the incident radiation. It is responsible for the blue colour of the sky, as air molecules scatter short wavelengths, such as blue and violet light.
  (b) **Mie scattering:** This occurs when the particle size interacting with the radiation is comparable to the wavelength of the incident radiation. This type of scattering is commonly observed in the scattering of visible and near-infrared light by water droplets, dust, and larger atmospheric particles.
  (c) **Non-selective (non-Rayleigh):** It takes place when radiation interacts with particles much larger than the wavelength of the incident radiation. It is typically observed in the scattering of longer wavelengths, such as microwaves, by objects like buildings or terrain features.
- **Effects of scattering**
  (a) **Change in direction:** Scattering causes radiation to change direction, which can result in diffuse or redirected radiation.
  (b) **Change in spectral properties:** The scattered radiation may exhibit altered spectral properties, such as shifts in wavelength or changes in intensity.

(c) **Attenuation**: Scattering can cause the attenuation or reduction of radiation intensity, particularly in the case of Mie and non-selective scattering.

- **Remote sensing applications of scattering**

(a) **Atmospheric studies:** Scattering is essential for atmospheric studies, as it provides valuable information about aerosols, clouds, and air quality. Different types of scattering assist in analyzing particle size distribution and composition.
(b) **Optical remote sensing:** Scattering plays a key role in optical remote sensing by influencing the quality and appearance of remotely sensed imagery. It impacts contrast, visibility, and the radiometric properties of objects and features on the surface.
(c) **Radar remote sensing:** In radar remote sensing, scattering is vital for interpreting radar echoes. Radar systems utilize scattering properties to detect and characterize various targets, such as terrain features, vegetation, and precipitation.

Refraction Refraction is an interaction mechanism in remote sensing that occurs when electromagnetic waves alter their direction as they pass from one medium to another with a different refractive index. This change in direction is caused by variations in the speed of light between the two media. Refraction can distort the apparent location of objects in remote sensing imagery, especially when observing features through the Earth's atmosphere. A thorough understanding of refraction is essential for precise geospatial measurements and for correcting atmospheric effects in remote sensing data.

1.4 Heat Emission and Spectral Reflectance

1.4.1 Heat Emission

Thermal emission, often referred to as heat emission, is a key mechanism of energy interaction in remote sensing, particularly within the thermal infrared region of the electromagnetic spectrum. This process involves the emission of electromagnetic radiation by objects and surfaces due to their temperature. In the context of remote sensing, thermal emission is vital for numerous applications, most notably in estimating land surface temperature (LST) and analyzing temperature-related phenomena. The discussion below emphasizes the importance of thermal emission in remote sensing.

Thermal Infrared Emission

- Thermal emission primarily takes place in the thermal infrared region of the electromagnetic spectrum, covering wavelengths roughly between 3 and 14 μm.

- All objects and surfaces with temperatures above absolute zero (−273.15 °C or 0 Kelvin) emit thermal infrared radiation.
- The intensity and spectral properties of this emitted radiation are intrinsically linked to the temperature of the emitting object.

Estimation of Land Surface Temperature (LST)

- A key application of thermal emission in remote sensing is the estimation of land surface temperature (LST).
- Remote sensing instruments equipped with thermal infrared sensors detect the thermal radiation emitted by the Earth's surface.
- By analyzing this radiation and correcting for atmospheric influences, researchers can precisely estimate the LST across different land cover types.

Urban Heat Island (UHI) Studies

- Thermal emission data are essential for analyzing urban heat islands (UHIs), localized areas within cities that exhibit higher temperatures compared to their surrounding rural regions.
- Urban heat island (UHI) studies offer crucial insights for urban planners and policymakers, helping them tackle heat-related challenges in densely populated areas and strengthen urban climate resilience.

Industrial and Infrastructure Inspections

- Thermal infrared remote sensing is widely utilized in industrial and infrastructure inspections.
- It facilitates the detection of thermal anomalies in industrial equipment, electrical systems, power lines, and buildings, supporting maintenance, safety measures, and energy efficiency evaluations.

Environmental Monitoring and Agriculture

- Thermal emission data are critical for monitoring environmental conditions and managing agricultural resources.
- Land surface temperature (LST) data are utilized to detect temperature variations across ecosystems, urban areas, water bodies, and agricultural fields.
- LST plays a vital role in evaluating crop health, identifying stress factors, optimizing irrigation strategies, and monitoring soil moisture levels.

In summary, thermal emission in the infrared spectrum is fundamental to remote sensing, enabling land surface temperature estimation and analysis of temperature-related environmental and urban phenomena. Its applications span agriculture, urban planning, environmental monitoring, and industrial inspections, enhancing resource management and informed decision-making.

1.4.2 *Spectral Reflectance*

The spectral reflectance curve, commonly known as the spectral signature, is a foundational concept in remote sensing. It graphically depicts how an object or surface reflects electromagnetic radiation across various wavelengths of the electromagnetic spectrum. Understanding spectral signatures is crucial for identifying and characterizing materials, land cover types, and objects on the surface. The following section offers a comprehensive explanation of the spectral reflectance curve and its relevance in remote sensing.

Characteristics of Spectral Signatures

- Absorption bands: These are regions within the spectrum where reflectance significantly decreases due to the absorption of radiation by specific materials. Each material exhibits unique absorption bands at distinct wavelengths.
- Reflectance peaks: Reflectance can exhibit peaks at certain wavelengths, typically resulting from constructive interference or resonance effects, indicating a heightened response to specific wavelengths of light.
- Spectral features: Spectral signatures display distinct features that facilitate the identification of materials or land cover types. For instance, healthy vegetation generally shows high reflectance in the near-infrared (NIR) region, with lower reflectance in the visible (VIS) and shortwave infrared (SWIR) regions.

Spectral Reflectance Measurement
To construct a spectral reflectance curve, remote sensing instruments quantify the intensity of reflected or emitted radiation across a range of discrete wavelengths or narrow spectral bands. Reflectance values are generally expressed as the ratio of reflected radiation to incident (incoming) radiation, ranging from 0 (no reflection) to 1 (complete reflection), or alternatively as a percentage.

Significance of Spectral Signatures

- Spectral signatures offer a distinct and characteristic "fingerprint" of how materials interact with electromagnetic radiation.
- They are crucial for distinguishing and classifying various land cover types, such as water bodies, urban areas, vegetation, minerals, and others.
- Spectral signatures assist remote sensing scientists in interpreting imagery, mapping land features, and monitoring environmental changes.

Applications of Spectral Signatures

- Land cover classification: Remote sensing scientists utilize spectral signatures to classify and map various land cover types, such as water bodies, forests, agricultural areas, and urban environments.
- Vegetation health monitoring: Spectral signatures are employed to evaluate the health and vitality of vegetation by analyzing reflectance in specific spectral bands, notably the near-infrared (NIR) and red-edge regions.

- Water quality assessment: The spectral signature of water bodies offers key insights into water quality, enabling the detection of suspended particles, pollutants, and algae.
- Mineral exploration: The identification of minerals is frequently based on their distinct spectral signatures, particularly in the shortwave infrared (SWIR) and thermal infrared regions.

In summary, the spectral reflectance curve, or spectral signature, is essential in remote sensing for identifying materials and land cover types based on their unique responses to electromagnetic radiation. By analyzing these signatures, remote sensing scientists gain critical insights into the surface, supporting applications in environmental monitoring, mineral exploration, and land use planning.

1.4.2.1 Spectral Responses from Earth Surface

1.4.2.1.1 Spectral Signature of Vegetation

The spectral signature of vegetation is a unique and distinctive pattern of reflectance and absorption across various wavelengths of electromagnetic radiation. This signature serves as a valuable tool in remote sensing, facilitating the identification, health assessment, and classification of vegetation cover (Fig. 1.11). The following outlines the key aspects of the spectral signature of vegetation:

Low Visible (VIS) Reflectance

- Vegetation exhibits relatively low reflectance in the visible (VIS) portion of the spectrum, especially in the green region (approximately 500–600 nm).
- The reduced reflectance in the green band contributes to the characteristic green colour of vegetation, as chlorophyll absorbs much of the visible light for photosynthesis.

Near-Infrared (NIR) Reflectance

- A key characteristic of the spectral signature of healthy vegetation is its high reflectance in the near-infrared (NIR) region of the electromagnetic spectrum, typically between 700 and 1300 nm.

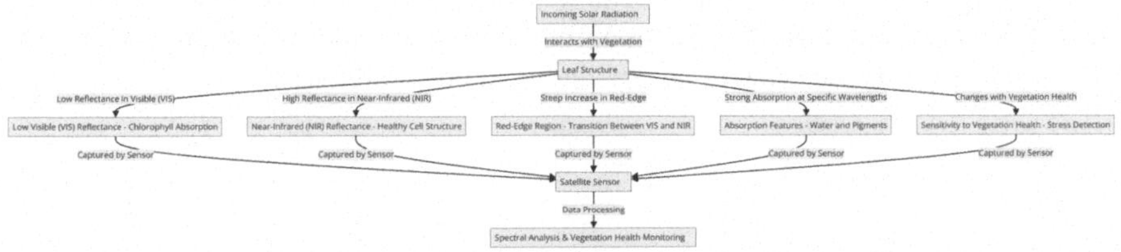

Fig. 1.11 Processes of spectral responses from earth surface: vegetation (prepared by the authors)

- This elevated reflectance is mainly attributed to the internal structure of leaves, which scatter and reflect NIR radiation.
- In NIR imagery, healthy vegetation appears significantly brighter compared to other spectral bands.

Red-Edge Region

- The red-edge region, typically spanning from 680 to 750 nm, is marked by a sharp increase in reflectance.
- This region is particularly sensitive to chlorophyll content and vegetation stress, making it a crucial indicator for assessing plant health.

Absorption Features

- The spectral signature of vegetation also exhibits absorption features in specific bands.
- Chlorophyll and other pigments absorb radiation in the blue and red regions of the spectrum for photosynthesis.
- Water absorption features, particularly in the shortwave infrared (SWIR) region, are observed due to the water content in plant tissues.

Sensitivity to Vegetation Health

- Variations in the spectral signature of vegetation can reflect its health and vitality.
- Stressed or diseased vegetation typically shows reduced NIR reflectance and altered red-edge characteristics.
- Remote sensing can detect these changes, facilitating early identification of issues in both agricultural and natural ecosystems.

Applications

- Vegetation health monitoring: Alterations in the spectral signature can be used to evaluate the health and vitality of both crops and natural vegetation.
- Land cover classification: Vegetation can be differentiated from other land cover types based on its distinct spectral signature.
- Crop Monitoring: The spectral signature aids in tracking crop growth stages and identifying stressed areas within agricultural fields.
- Forest Management: It facilitates the assessment of forest health, mapping of tree species, and monitoring of deforestation.

In summary, the spectral signature of vegetation is a unique pattern of reflectance and absorption across the electromagnetic spectrum. It is essential for identifying, classifying, and monitoring vegetation, making it a key tool in remote sensing for agriculture, forestry, ecosystem health, and land cover analysis.

1.4.2.1.2 Spectral Signature of Concrete and Asphalt

Concrete and asphalt surfaces display distinct spectral signatures in remote sensing imagery. These man-made materials exhibit unique reflectance and absorption patterns that aid in distinguishing them from natural features. Concrete generally shows a relatively uniform reflectance spectrum, with moderate to high reflectance in the visible and near-infrared (NIR) regions, with slight variations due to surface roughness and aging. It typically exhibits low reflectance in the thermal infrared region. In contrast, asphalt shows lower reflectance in the visible and NIR regions, resulting in a darker appearance, but may have increased reflectance in the shortwave infrared (SWIR) region due to its composition. Asphalt surfaces also tend to have low reflectance in the thermal infrared region.

These spectral signatures are crucial for mapping and differentiating concrete and asphalt in urban and infrastructure monitoring, such as road network analysis and maintenance evaluations (Fig. 1.12).

1.4.2.1.3 Spectral Signature of Coastal Mangrove Vegetation

Coastal mangrove vegetation exhibits a distinctive spectral signature in remote sensing imagery, reflecting its specialized adaptations to saline environments (Fig. 1.13). The following are key observations from this spectral signature:

- **Low visible (VIS) reflectance:** In the visible (VIS) portion of the spectrum, particularly in the green region (approximately 500 to 600 nm), mangrove vegetation shows lower reflectance. This is primarily because chlorophyll absorbs visible light for photosynthesis.

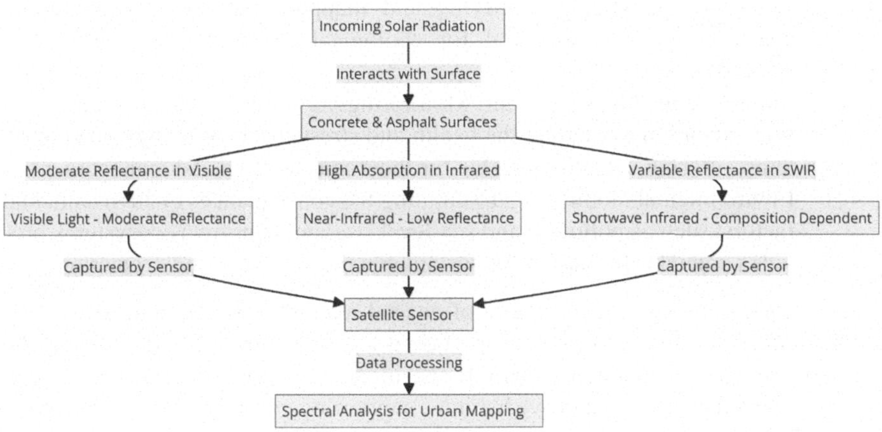

Fig. 1.12 Processes of spectral responses from earth surface: concrete and asphalt (prepared by the authors)

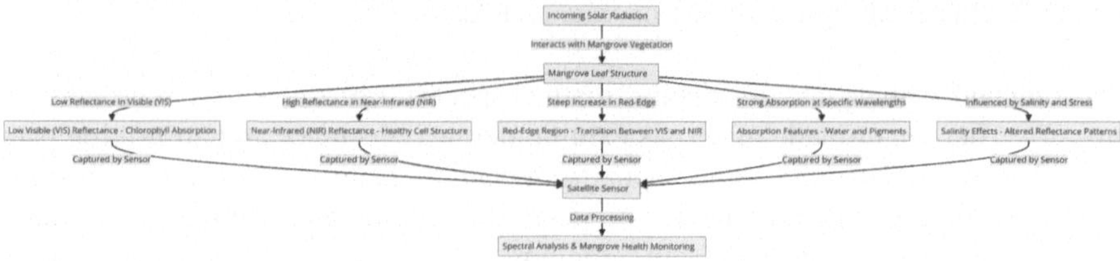

Fig. 1.13 Processes of spectral responses from earth surface: coastal mangrove vegetation (prepared by the authors)

- **Near-infrared (NIR) reflectance:** Coastal mangrove vegetation typically exhibits its high reflectance in the near-infrared (NIR) region (around 700–1300 nm), primarily due to the efficient scattering and reflection of NIR radiation by healthy mangrove leaves.
- **Red-edge region:** The spectral signature of coastal mangrove vegetation often reveals a distinct feature in the red-edge region (approximately 680–750 nm), where reflectance increases abruptly. This shift is associated with chlorophyll content and can be used to assess the health and vitality of mangrove ecosystems.
- **Absorption features:** Mangrove vegetation may exhibit absorption features in various spectral bands, linked to pigments, water content, and the unique biochemical composition of the leaves. For example, water absorption features are observed in the shortwave infrared (SWIR) region due to the water content in the leaves.
- **Salinity effects:** The spectral signature of mangrove vegetation can be influenced by the surrounding salinity levels. Increased salinity can stress the vegetation, leading to reduced NIR reflectance and altered spectral characteristics.
- **Applications:**

  (a) Mangrove mapping: Identifying and mapping mangrove areas within coastal regions is essential for ecological research and conservation initiatives.
  (b) Mangrove health assessment: Monitoring variations in the spectral signature can aid in evaluating the health and stress levels of mangrove ecosystems, offering valuable insights for their conservation and management.
  (c) Environmental monitoring: Examining the consequences of environmental factors, such as pollution and sea-level rise, on mangrove ecosystems provides crucial information for sustainable management and protection.

In summary, the spectral signature of coastal mangrove vegetation, marked by distinct reflectance and absorption properties, is key to identifying and assessing the health of these vital ecosystems. Remote sensing technology plays a critical role in studying and conserving mangrove habitats, supporting coastal ecosystem management and protection.

1.4.2.1.4 Spectral Signature of Coastal Soil

Coastal soil, found in the dynamic and saline environments along coastlines, displays distinctive spectral signatures in remote sensing imagery, reflecting its unique properties and the impact of factors such as salt content and moisture (Fig. 1.14). The following highlights the key aspects of the spectral signature of coastal soil:

- **Low visible (VIS) reflectance:** Coastal soil generally exhibits low reflectance in the visible (VIS) portion of the electromagnetic spectrum, particularly in the green region (around 500–600 nm). This low reflectance is due to the absorption of visible light by the minerals and organic matter present in the soil.
- **Near-infrared (NIR) reflectance:** In the near-infrared (NIR) region (around 700–1300 nm), coastal soil typically shows moderate to high reflectance, as soil particles effectively scatter and reflect NIR radiation.
- **Salinity effects:** The spectral signature of coastal soil is significantly influenced by salt content. Elevated salinity levels in coastal areas can alter reflectance patterns, particularly in the NIR and shortwave infrared (SWIR) regions, with salty soils often exhibiting reduced NIR reflectance.
- **Moisture content:** The moisture content of coastal soil also affects its spectral signature. Wet or saturated soils typically display higher reflectance, especially in the NIR and SWIR regions, compared to dry soils.
- **Absorption features:** Coastal soil may exhibit absorption features in specific spectral bands, related to its mineral composition. These features provide valuable information about soil types and characteristics.
- **Applications:**

  (a) Soil type mapping: Identifying and mapping various soil types and their distribution along coastlines is crucial for land use planning and agricultural applications.

  (b) Salinity assessment: Monitoring salt-affected soils aids in assessing soil health and understanding the impacts of saltwater intrusion.

  (c) Land degradation analysis: Analyzing soil erosion and degradation in coastal regions informs conservation efforts and management strategies.

  (d) Environmental studies: Investigating the influence of coastal soil properties on ecosystems, including vegetation health and water quality, provides insights into environmental dynamics.

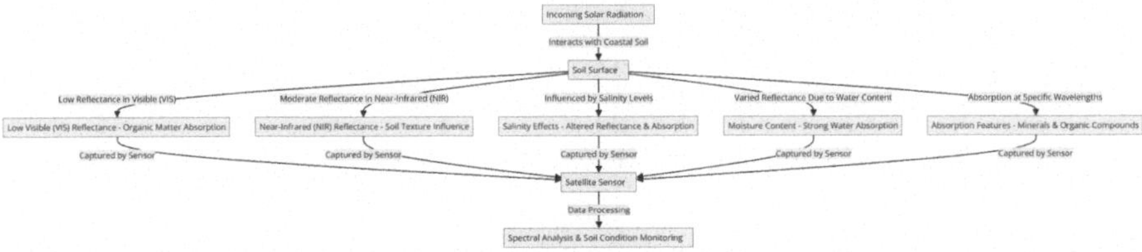

Fig. 1.14 Processes of spectral responses from earth surface: coastal soil (prepared by the authors)

In summary, the spectral signature of coastal soil, shaped by factors such as salinity, moisture content, and mineral composition, provides key insights into soil characteristics. Remote sensing technology is an essential tool for studying and monitoring these properties, supporting land management, environmental assessment, and soil conservation in coastal regions.

1.4.2.1.5 Spectral Signature of Coastal Water

The spectral signature of coastal water bodies is unique and shaped by factors such as water composition, depth, and surrounding environmental conditions. Analyzing this signature is essential for monitoring coastal ecosystems and assessing water quality (Fig. 1.15). The key characteristics of the spectral signature of coastal water are outlined below:

- **Absorption in the visible (VIS) region:** Coastal water bodies, particularly those rich in organic matter and phytoplankton, show significant absorption of visible light, especially in the blue and green wavelengths (400–600 nm). This results from waterborne particles, dissolved substances, and chlorophyll pigments.
- **Low reflectance in NIR and SWIR regions:** Coastal waters generally exhibit low reflectance in the near-infrared (NIR) and shortwave infrared (SWIR) regions (700–2500 nm) due to strong water absorption at these wavelengths. Variations may occur depending on water turbidity and its constituents.
- **Depth influence:** The spectral signature of coastal waters changes with depth. Shallow areas may exhibit distinct spectral characteristics compared to deeper waters, as increased depth enhances light absorption in the visible and NIR regions.
- **Chlorophyll and algae detection:** Chlorophyll-rich waters often display specific absorption features associated with phytoplankton and algae, making spectral signatures valuable for assessing water quality and ecosystem health.
- **Turbidity effects:** Suspended sediments or particulate matter in turbid coastal waters can increase reflectance in the visible region, limiting light penetration and altering the spectral signature.
- **Impact of salinity and temperature:** Variations in salinity and temperature affect the spectral signature, particularly in the thermal infrared region, which is sensitive to water temperature.

Fig. 1.15 Processes of spectral responses from earth surface: coastal water (Prepared by the authors)

- **Applications:**
  (a) Water quality monitoring: Assessing water quality by detecting phytoplankton blooms, pollution, and sediment transport.
  (b) Ecosystem analysis: Evaluating coastal ecosystems, such as coral reefs, seagrass beds, and mangrove forests, through water clarity and properties.
  (c) Bathymetry and depth estimation: Utilizing remote sensing to map water depth and coastal bathymetry for navigation and habitat analysis.
  (d) Environmental conservation: Monitoring the consequences of human activities, sea-level rise, climate change on coastal ecosystems to support conservation efforts.

In summary, the spectral signature of coastal water bodies, marked by high visible light absorption, varies with factors like composition, depth, turbidity, and environmental conditions. Remote sensing is essential for monitoring coastal water resources, supporting ecosystem management, conservation, and sustainable development.

1.5 Conclusion and Endnotes

This exploration of remote sensing highlights its principles, evolution, and diverse applications. Remote sensing, the science of capturing atmospheric and earth's surface data via sensors on aircraft or satellites, has revolutionized our ability to monitor and manage the planet. Its journey from early aerial photography to modern sensors reveals its vital role in meteorology, agriculture, forestry, and disaster management. Key topics included passive and active systems, optical, radar, and thermal sensors, as well as their applications in urban planning, land use mapping, and resource management. Understanding the electromagnetic spectrum and spectral signatures, unique reflectance and absorption patterns, enables the identification of materials like vegetation, mangroves, concrete, and coastal water. Further, the advantages of remote sensing, including cost-effectiveness and global reach, outweigh challenges like atmospheric interference, making it a critical tool for environmental conservation, urban planning, agriculture, and disaster response. Its advancements promise continued significance in sustainable development.

Endnotes

- Definitions and principles of remote sensing were discussed based on established literature and resources.
- The historical progression of remote sensing was summarized, highlighting key milestones and contributions.
- Types of remote sensing and their applications were derived from a wide range of sources.
- The ideal components and setup of a remote sensing system were outlined.

- Applications, advantages, and disadvantages of remote sensing were summarized, taking into account both established knowledge and contemporary challenges.
- The discussion of the electromagnetic spectrum and interaction mechanisms drew from fundamental GIScience principles and their applications in remote sensing.
- Specific spectral signatures of various materials were explained based on the known optical properties of these materials and their relevance in remote sensing contexts.

Further Readings

1. Campbell, B. A. (2002). *Radar remote sensing of planetary surfaces*. Cambridge University Press ISBN: 9780521583084.
2. Campbell, J. B., Wynne, R. H., & Thomas, V. A. (2011). *Introduction to remote sensing* (6th ed.). Guilford Press ISBN: 9781462549405.
3. Coops, N. C., & Tooke, T. R. (2017). Introduction to remote sensing. In S. Gergel & M. Turner (Eds.), *Learning landscape ecology*. Springer. https://doi.org/10.1007/978-1-4939-6374-4_1
4. Efremenko, D., & Kokhanovsky, A. (2021). Introduction to remote sensing. In *Foundations of atmospheric remote sensing*. Springer. https://doi.org/10.1007/978-3-030-66745-0_1
5. Emery, W., & Camps, A. (2017). *Introduction to satellite remote sensing: Atmosphere, ocean, land and cryosphere applications*. Elsevier. https://doi.org/10.1016/C2015-0-04517-8 ISBN: 978-0-12-809254-5.
6. Jenicka, S. (2021). Introduction to remote sensing. In *Land cover classification of remotely sensed images*. Springer. https://doi.org/10.1007/978-3-030-66595-1_1
7. Liang, S., & Wang, J. (Eds.). (2020). *Advanced remote sensing: Terrestrial information extraction and applications*. Academic Press. https://doi.org/10.1016/C2017-0-03489-4 ISBN: 978-0-12-815826-5.
8. Lillesand, T., Kiefer, R. W., & Chipman, J. (2015). *Remote sensing and image interpretation* (7th ed.). Wiley ISBN: 978-1-118-34328-9.
9. Visconti, G. (2016). Introduction to remote sensing. In *Fundamentals of physics and chemistry of the atmosphere*. Springer. https://doi.org/10.1007/978-3-319-29449-0_6
10. Weng, Q. (2012). *An introduction to contemporary remote sensing*. McGraw-Hill ISBN: 978-0-07-174012-8.

Chapter 2
Sensor Networks and Satellite Systems: Tools for Earth Observation

"We are entering a new era of planetary observation. Satellites and sensors are opening our eyes to Earth's beauty and fragility." —Al Gore.

Abstract This chapter examines the core components and functions of remote sensing systems, focusing on sensors and satellite platforms essential for Earth observation. It begins by exploring sensors, which measure electromagnetic radiation from the Earth's surface in various forms such as optical, thermal, and microwave—each tailored to specific wavelengths of the electromagnetic spectrum. The chapter discusses the advantages, resolution capabilities, and limitations of different sensor types and scanning systems. It also covers the satellite platforms that carry these sensors, including Sun-synchronous and geostationary satellites, detailing their orbits, velocities, coverage areas, and applications. The history of key satellite programmes, the Indian Remote Sensing System, and its institutions is reviewed. The chapter also highlights recent advancements, including new sensor technologies like synthetic aperture radar and hyperspectral sensors for data processing and analysis. The chapter highlights the recent advancements in sensor and satellite technology and potential future advances. This chapter explores these topics to give readers a solid understanding of the fundamental principles enabling remote sensing data collection from space.

Keywords Satellite system · Sensor technologies · Satellite motion · LANDSAT programme · Remote sensing data · Earth observation · ISRO mission

This chapter offers a detailed overview of the essential components and functions of remote sensing systems, emphasizing sensors and satellite platforms for Earth observation. It covers various sensor types, including optical, thermal, and microwave, highlighting their capabilities, advantages, and limitations. The chapter also explores satellite platforms, including Sun-synchronous and geostationary

K. D. Malakar et al., *Geospatial Technologies in Coastal Ecologies Monitoring and Management*, Advances in Geographic Information Science,
https://doi.org/10.1007/978-3-031-92017-2_2

satellites, and their roles in data collection. Additionally, it reviews the history of satellite programmes, and introduces recent advancements in sensor and satellite technologies.

Key Points of the Chapter

- Sensor types in remote sensing are discussed.
- Principles, advantages, and limitations of sensors and scanning systems are covered.
- Satellite platforms and their orbits are explored.
- The progress and strategies of various missions are highlighted.

2.1 Introduction to Sensor

2.1.1 Concept

Sensors are essential to remote sensing, allowing the observation and analysis of the Earth's surface and atmosphere from a distance. Acting as the "eyes" of remote sensing systems, these sensors detect electromagnetic radiation or other physical phenomena. They vary in type, including optical, thermal, microwave, and lidar sensors, each designed to capture specific portions of the electromagnetic spectrum. By measuring the intensity of reflected or emitted radiation, sensors generate data that is processed into images, maps, and critical information for applications such as disaster management, environmental monitoring, and agriculture. The selection of sensor type and platform (e.g., satellites, drones, or aircraft) depends on the specific objectives of the remote sensing task, with factors like spatial and spectral resolution playing a key role. Ultimately, sensors are integral to making remote sensing a powerful tool for understanding and managing Earth's systems (Fig. 2.1).

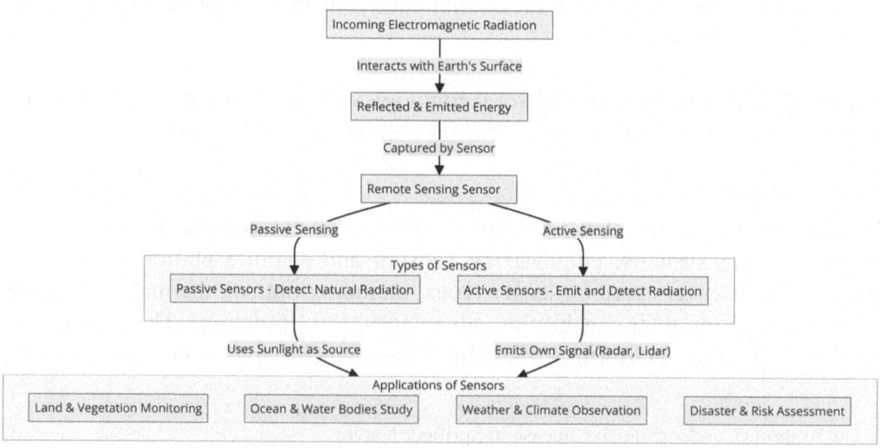

Fig. 2.1 Remote sensing sensor and its applications (prepared by the authors)

2.1.2 *Types*

Remote sensing sensors are classified based on their operational principles, the data they collect, and the portions of the electromagnetic spectrum they utilize (Fig. 2.2). The following are some common classifications of remote sensing sensors:

a) **Based on electromagnetic spectrum**

- Optical sensors: These sensors capture visible and near-infrared light, primarily utilized for colour imaging and assessing vegetation health. Example: Sentinel-2 MSI and Landsat 8 OLI.
- Thermal infrared sensors: These sensors detect heat radiation to measure land surface temperatures and identify thermal variations. Example: MODIS and ASTER.
- Microwave sensors: Radar-based sensors emit and receive microwave signals, allowing them to penetrate cloud cover and provide data on surface roughness and topography. Example: RADARSAT-2 and TerraSAR-X.

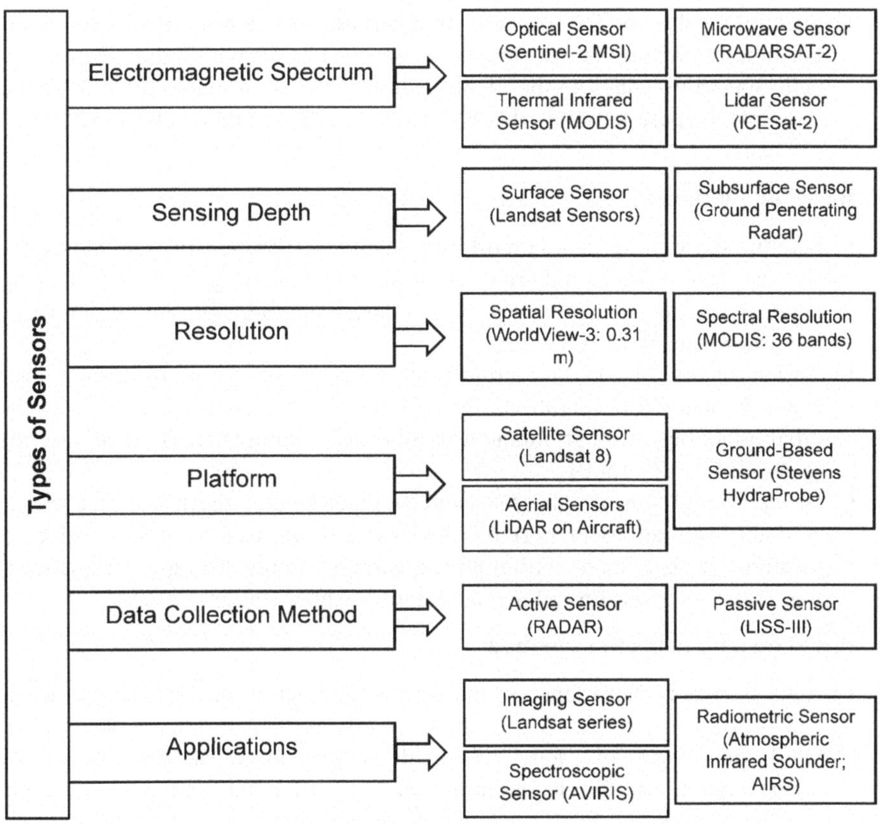

Fig. 2.2 Types of sensors (prepared by the authors)

- Lidar sensors: Lidar sensors use laser pulses to measure distances between the sensor and the Earth's surface, facilitating accurate 3D mapping and elevation analysis. Example: ICESat-2 and LiDAR on UAVs.

b) **Based on sensing depth**

- Surface sensors: These sensors capture data from the surface, offering insights into terrain, land cover, and surface features. Example: Landsat sensors (e.g., Landsat 8 OLI and WorldView-3).
- Subsurface sensors: These sensors are designed to penetrate below the Earth's surface and are used for applications such as soil moisture monitoring and geological exploration. Examples include ground penetrating radar (GPR) and soil moisture sensors like Sentinel-1 SAR.

c) **Based on resolution**

- Spatial resolution: Sensors are classified based on their spatial resolution, which refers to their capacity to capture fine or coarse details in imagery, with distinctions made between high, medium, and low resolution. Example: WorldView-3 (0.31 m) and Landsat 8 (30 m).
- Spectral resolution: This classification pertains to a sensor's ability to distinguish between various wavelengths or spectral bands within the electromagnetic spectrum, enabling the identification of different materials or features. Example: Hyperion (onboard NASA's EO-1; 220 spectral bands) and MODIS (36 bands).

d) **Based on platform**

- Satellite sensors: Mounted on orbiting satellites, these sensors provide extensive, global coverage, making them ideal for a wide range of remote sensing applications across vast geographic areas. Example: Landsat Series (e.g., Landsat 8) and WorldView-3.
- Aerial sensors: Installed on aircraft, these sensors offer high-resolution data, enabling detailed observations for regional and local-scale studies, such as urban planning or environmental monitoring. Example: LiDAR on aircraft (e.g., RIEGL VQ-1560).
- Ground-based sensors: These sensors are placed on the Earth's surface, either as fixed installations or portable devices, and are used to monitor specific locations or phenomena with high precision. Example: Soil moisture sensors (e.g., Stevens HydraProbe) and automatic weather stations (AWS).

e) **Based on data collection method**

- Active sensors: These sensors emit their own radiation, such as radar or lidar, and measure the reflected or scattered radiation that returns to the sensor. Unlike passive sensors, active sensors do not rely on external sources of radiation. Example: Radio detection and ranging (RADAR) and synthetic aperture radar (SAR).

- Passive sensors: These sensors detect naturally occurring electromagnetic radiation, such as sunlight, and measure the intensity and wavelength of the radiation that is reflected or emitted by the Earth's surface. Example: LISS-I, LISS-II, and LISS-III.

f) **Based on application**

- Imaging sensors: These sensors capture visual or multispectral images of the surface, and are frequently utilized in applications such as urban planning, land cover classification, and forestry management. Example: Landsat series (optical imaging) and WorldView-3 (multispectral imaging).
- Spectroscopic sensors: These sensors analyze the spectral signatures of materials, enabling the identification of specific substances or chemical compositions. Example: Hyperspectral imaging sensors (e.g., AVIRIS) and MODIS (moderate resolution imaging spectroradiometer).
- Radiometric sensors: Designed to measure the radiance and radiative properties of objects, these sensors are particularly valuable in atmospheric studies and climate research. Example: CERES (clouds and the Earth's radiant energy system) and atmospheric infrared sounder (AIRS).

2.1.3 Scanning System

A scanning system in remote sensing is essential for acquiring data from the surface or atmosphere and transforming it into actionable information. It serves as a core component of remote sensing platforms, including aircraft, satellites, drones, and ground-based instruments. The scanning system governs the data collection process, influencing spatial and spectral resolution, and plays a pivotal role in determining the success and efficiency of the remote sensing mission (Fig. 2.3). The following outlines the concepts of scanning systems in remote sensing in detail:

Types of Scanning Systems

- Whiskbroom scanners: These scanners utilize single or multiple detectors to sweep across the target area line by line. Commonly employed in satellite sensors, they are well-regarded for their high spatial resolution.
- Pushbroom scanners: Equipped with an array of detectors, pushbroom scanners capture an entire swath of data simultaneously. They offer exceptional spectral and spatial resolution, making them ideal for hyperspectral imaging applications.
- Cross-track scanners: These scanners operate perpendicular to the platform's direction of travel, enabling a broad field of view. They are widely used in radar and wide-swath optical sensors.
- Along-track scanners: Moving parallel to the platform's direction, along-track scanners collect data along a narrow strip. They are frequently employed in high-resolution optical sensors.

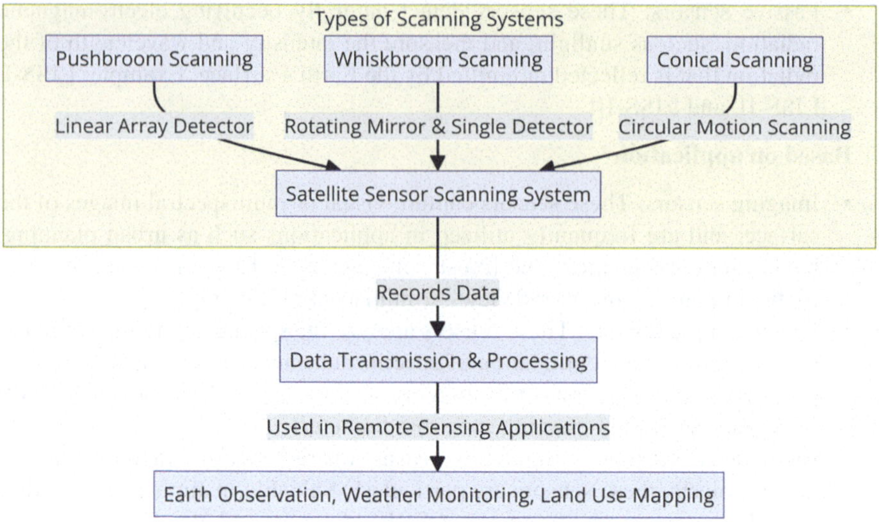

Fig. 2.3 Types of scanning systems (prepared by the authors)

- Whiskbroom–pushbroom hybrid scanners: Combining the benefits of both whiskbroom and pushbroom designs, hybrid scanners provide enhanced flexibility in data acquisition.

Scanning Patterns

- Rectilinear scan: This scanning method captures data in parallel, linear paths, ensuring systematic and comprehensive area coverage.
- Circular scan: Utilizing a rotational pattern, this method provides 360°coverage, making it ideal for applications requiring full-field observation.
- Spiral scan: In this method, data is captured in an expanding or contracting spiral pattern, offering flexibility in spatial resolution and coverage density.

Components of a Scanning System

- Sensor: The sensor serves as the core element of the scanning system, capturing data based on the application. Depending on the requirements, sensors such as optical, thermal, microwave, or lidar can be utilized to acquire specific types of information.
- Scanner mechanism: The scanner mechanism controls the movement of the sensor's line of sight, determining the scanning pattern and speed. This movement can be governed mechanically or electronically, depending on the system's design.
- Data acquisition system: This component is responsible for recording and storing the data collected by the sensor. It ensures accurate timestamping and proper organization, facilitating efficient processing and analysis.

Applications of Scanning Systems Scanning systems are widely utilized across diverse fields such as forestry, agriculture, disaster management, urban planning, environmental monitoring, and Earth sciences. They deliver critical data for applications like land cover classification, change detection, and advancing scientific research.

2.1.4 Sensor Data Nature and Characters

Sensor data represents the raw measurements obtained from devices engineered to detect and quantify physical, chemical, or biological parameters. Its nature is determined by the type of sensor, encompassing variables such as temperature, humidity, pressure, light, motion, or sound. Sensor data is typically characterized by its continuous or discrete format, unstructured nature, and high volume. Fundamental attributes include precision, accuracy, resolution, and sensitivity. Sensor data may be captured in real time or at periodic intervals, often requiring preprocessing to eliminate noise and enhance usability. This data underpins critical applications in fields such as healthcare, environmental monitoring, and smart cities, facilitating informed decision-making and automation. Table 2.1 explores the diverse nature and characteristics of satellite sensors.

2.1.5 Resolution

In remote sensing, resolution refers to the capacity of a remote sensing system to discern and record fine details and features within the collected data. It is a crucial determinant of the quality and applicability of remote sensing imagery or datasets across various domains. Resolution is multifaceted, encompassing spatial resolution, spectral resolution, temporal resolution, and radiometric resolution. Each dimension significantly influences the overall informational richness and utility of the data, collectively shaping its suitability for specific applications. The four types of resolution are discussed below:

a) **Spatial resolution**

Spatial resolution in remote sensing refers to the smallest features on the surface that can be distinguished or resolved in the captured data. It determines the level of detail visible in an image. Higher spatial resolution enables the identification of smaller objects or features, whereas lower spatial resolution provides a more generalized overview. Spatial resolution is typically expressed in units such as metres or centimetres per pixel, representing the ground area covered by a single pixel in the image (Fig. 2.4). For instance, a satellite image with a spatial resolution of 1 m signifies that each pixel corresponds to a 1 m by 1 m area on the ground.

Table 2.1 Nature and characters of sensor data

Sr. No.	Sensor	Spectral (#)	Spatial (meters)	Radiometric (bit)	Band range (μm)	Band widths (μm)	Irradiance (W m^{-2} sr^{-1} μm^{-1})	Data points (# per hectares)	Frequency of revisit (days)
A. Coarse resolution sensors									
1	MODIS (moderate resolution imaging spectroradiometer)	36/7	250, 500, 1000	12	0.62–0.67 0.84–0.876 0.459–0.479 0.545–0.565 1.23–1.25 1.63–1.65 2.11–2.16	0.05 0.36 0.02 0.02 0.02 0.02 0.05	1528.2 974.3 2053 1719.8 447.4 227.4 86.7	0.16, 0.04, 0.01 0.16, 0.04, 0.01	Daily
2	AVHRR (advanced very high-resolution radiometer)	4	1000	11	0.58–0.68 0.725–1.1 3.55–3.93 10.30–10.95 10.95–11.65	0.10 0.375 0.38 0.65 0.70	1390 1410 1510 0 0	0.01	Daily
B. Multispectral sensors									
3	Landsat-1, 2, 3, MSS	4	56 × 79	6	0.5–0.6 0.6–0.7 0.7–0.8 0.8–1.1	0.1 0.1 0.1 0.3	1970 1843 1555 1047	2.26	16
4	Landsat-4, 5, TM	7	30	8	0.45–0.52 0.52–0.60 0.63–0.69 0.76–0.90 1.55–1.74 10.4–12.5 2.08–2.35	0.07 0.80 0.60 0.14 0.19 2.10 0.25	1970 1843 1555 1047 227.1 0 80.53	11.1	16

5	Landsat-7, ETM+	8	30	8	0.45–0.52 0.52–0.60 0.63–0.69 0.50–0.75 0.75–0.90 10.0–12.5 1.75–1.55 0.52–0.90(p)	0.065 0.080 0.60 0.150 0.200 2.5 0.2 0.38	1970 1843 1555 1047 227.1 0 1368 1352.71	44.4, 11.1	16
6	Landsat-8, 9	11	30	12	0.435–0.451 0.452–0.512 0.533–0.590 0.636–0.673 0.851–0.879 1.566–1.651 10.60–11.19 11.50–12.51 2.107–2.294 0.503–0676(p) 1.363–1.384	0.016 0.06 0.057 0.037 0.028 0.085 0.59 0.01 0.187 0.173 0.021	1970 1970 1843 1555 227.1 1368 0 0 1368 1352.71 1368	44.4 11.1	16
7	ASTER	15	15, 30, 90	8	0.52–0.63 0.63–0.69 0.76–0.86 0.76–0.86 1.60–1.70 2.145–2.185 2.185–2.225 2.235–2.285 2.295–2.365 2.360–2.430	0.11 0.06 0.1 0.1 0.1 0.04 0.04 0.05 0.07 0.07	1846.9 1546 1117.6 1117.6 232.5 80.32 74.96 69.20 59.82 57.32	44.4, 11.1, 1.23	16

(continued)

Table 2.1 (continued)

Sr. No.	Sensor	Spectral (#)	Spatial (meters)	Radiometric (bit)	Band range (μm)	Band widths (μm)	Irradiance (W m^{-2} sr^{-1} μm^{-1})	Data points (# per hectares)	Frequency of revisit (days)
8	IRS-1	5	23.5	8	0.52–0.59 0.62–0.68 0.77–0.86 1.55–1.70 0.5–0.75(p)	0.07 0.06 0.09 1.15 0.25	1852.1 1577.38 1096.7 240.4 1603.9	8.1	16
9	IRS-1C	5	23.5	8	0.52–0.59 0.62–0.68 0.77–0.86 1.55–1.70 0.5–0.75(p)	0.07 0.06 0.09 0.15 0.25	1851.1 1583.8 1102.5 240.4 1627.1	18.1	16
10	IRS-P6-AWiFs	4	56	10	0.52–0.59 0.62–0.68 0.77–0.86 1.55–1.70	0.07 0.06 0.09 0.15	1857.7 1556.4 1082.4 239.84	3.19	16
C. Hyper-spatial sensors									
11	IKONOS	4	1–4	11	0.445–0.516 0.506–0.595 0.632–0.698 0.757–0.853	0.71 0.89 0.66 0.96	1930.9 1854.8 1156.5 1156.9	10,000, 625	5
12	QuickBird	4	0.61–2.44	11	0.45–0.52 0.52–0.60 0.63–0.69 0.76–0.89	0.07 0.08 0.06 0.13	1381.79 1924.59 1843.08 1574.77	14,872, 625	5

13	RapidEye A–E	5	6.5	12	0.44–0.51 0.52–0.59 0.63–0.68 0.69–0.73 0.77–0.89	0.07 0.07 0.05 0.04 0.12	1979.33 1752.33 1499.18 1343.67 1039.88	236.7	1–2
14	WorldView	1	0.55	11	0.45–0.51	0.06	1996.77	40,000	1.7–5.9
D. Hyperspectral sensors									
15	ALI (advanced land imager)	10	30	12	0.048–0.69(p) 0.433–0.453 0.450–0.515 0.425–0.605 0.633–0.690 0.775–0.805 0.845–0.890 1.200–1.300 1.550–1.750 1.080–2.350	0.64 0.20 0.65 0.80 0.57 0.30 0.45 1.00 2.00 2.70	1747.86 1849.5 1985.07 1732.18 1485.23 1134.29 948.36 439.61 223.39 78.07	11.1	16

Source: NASA Terra; United States Geological Survey (USGS); The European Space Agency; Satellite Imaging Corporation; and Indian Space Research Organisation (ISRO)

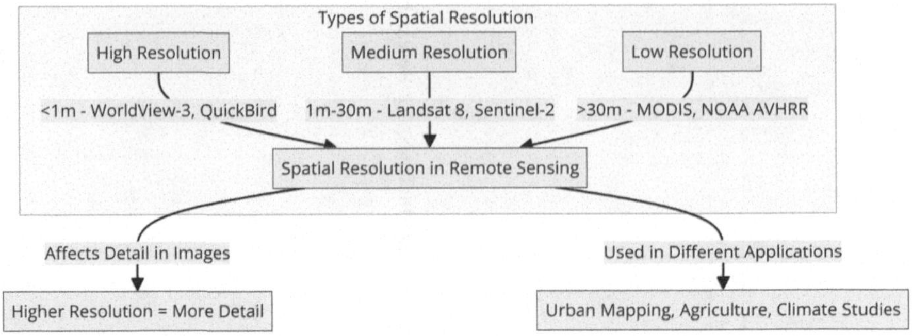

Fig. 2.4 Types of spatial resolution (prepared by the authors)

Example

Satellite	Sensor	Spatial resolution (mts.)
IRS 1C	Panchromatic (PAN)	6
Landsat	Thematic mapper (TM)	30
SPOT	Panchromatic (PAN)	10

b) **Spectral resolution**

Spectral resolution describes a remote sensing sensor's ability to detect and differentiate between wavelengths or spectral bands in the electromagnetic spectrum, determining the number and width of bands available for analysis. Sensors with high spectral resolution can capture a large number of narrow spectral bands, facilitating detailed spectral analysis and enabling the precise identification of specific materials or vegetation types. Conversely, sensors with low spectral resolution have broader spectral bands, providing less detailed spectral information (Fig. 2.5).

Example

Landsat TM band(s)	Range (μm)	Spectral resolution (mts.)
Band 1 (B)	0.45–0.52	30
Band 2 (G)	0.52–0.60	30
Band 3 (R)	0.63–0.69	30
Band 4 (NIR)	0.76–0.90	30
Band 5 (SWIR)	1.55–1.75	30
Band 6 (TIR)	10.40–12.50	120
Band 7 (SWIR/MIR)	2.08–2.35	30

c) **Temporal resolution**

Temporal resolution refers to the frequency at which a remote sensing system revisits or acquires data for the same geographic area over a given period. It plays a vital role in monitoring dynamic processes such as vegetation growth, land cover changes, and disaster management (Fig. 2.6). Sensors with high temporal resolution

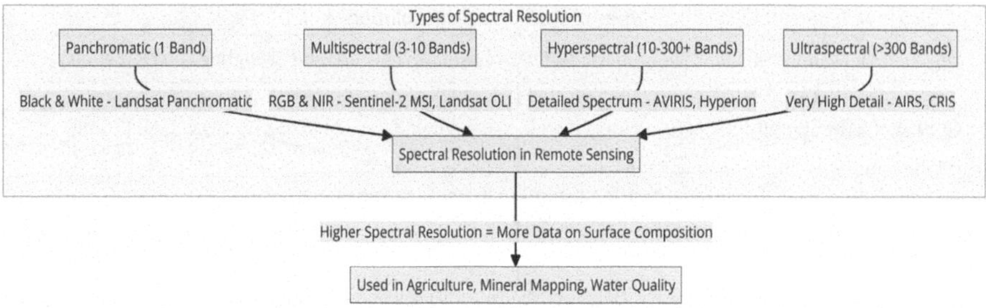

Fig. 2.5 Types of spectral resolution (prepared by the authors)

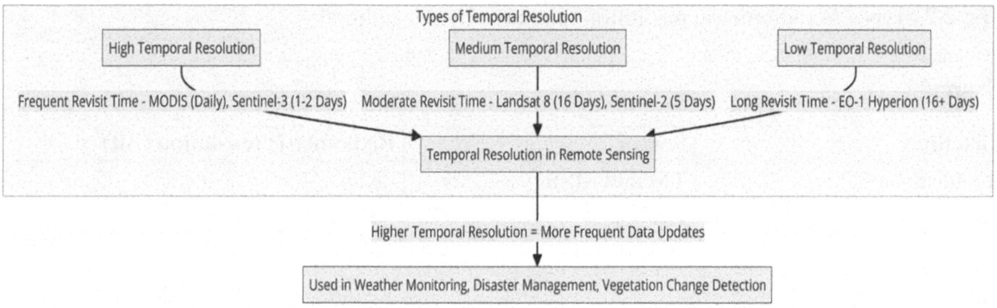

Fig. 2.6 Types of temporal resolution (prepared by the authors)

capture data at shorter intervals, enabling the observation of rapid changes, whereas sensors with low temporal resolution revisit areas less frequently, making them more suitable for monitoring slower-evolving phenomena.

Example

Satellite	Temporal resolution (days)
Landsat 4 and 5	16
IRS 1A and 1B	18
SPOT 1, 2, and 3	26

d) **Radiometric resolution**

Radiometric resolution refers to a sensor's capability to distinguish between varying levels of brightness or radiance within the collected data. It reflects the sensor's sensitivity to subtle variations in reflected or emitted energy. Sensors with high radiometric resolution can detect minute differences in reflectance or radiance, making them particularly useful for applications such as land cover classification. In contrast, sensors with low radiometric resolution may be less effective in capturing fine radiometric details (Fig. 2.7).

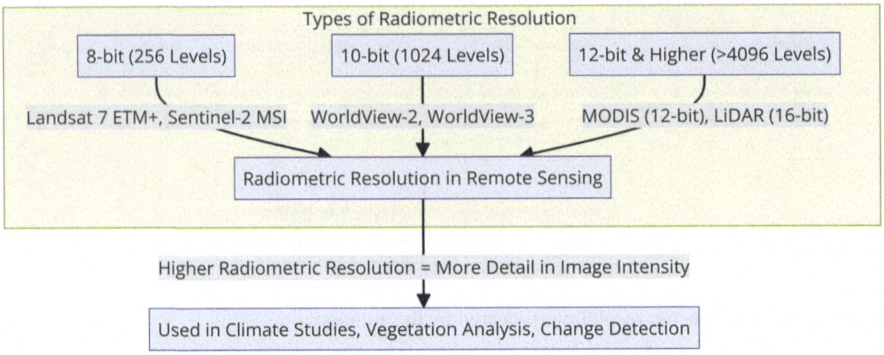

Fig. 2.7 Types of radiometric resolution (prepared by the authors)

Example

Satellite	Sensor	Radiometric resolution (Bit)
Landsat 4 and 5	TM and MSS	8
SPOT	HRV	8
EO-1	HYPERION	16

In summary, resolution is a key factor in remote sensing that affects data quality and usability. Choosing the right resolution parameters for specific applications is essential for deriving meaningful insights from imagery and datasets.

2.1.6 Different Sensors

Remote sensing utilizes a range of sensors to gather data about the atmosphere and surface. These sensors are designed to capture specific portions of the electromagnetic spectrum and are classified based on their operating principles and applications. Below are some of the common types of sensors employed in remote sensing:

Optical Sensors

a. Multispectral sensors: These sensors collect data across multiple narrow bands within the visible and near-infrared spectrum. They are widely used for applications such as vegetation health assessment, land cover classification, and agricultural monitoring.

   Example: Landsat series of satellites, which are used for vegetation monitoring, land cover mapping, and environmental management.

b. Panchromatic sensors: Capturing data in a single, broad spectral band, these sensors offer high spatial resolution but limited spectral detail. They are ideal for detailed mapping of land features.

   Example: IKONOS and WorldView satellites are examples of panchromatic sensors.

c. Hyperspectral sensors: Hyperspectral sensors capture data across multiple narrow and continuous spectral bands, allowing for detailed spectral analysis. They are commonly applied in mineral exploration, environmental monitoring, and crop analysis.

Example: AVIRIS (airborne visible/infrared imaging spectrometer), used for mineral exploration and environmental monitoring.

Thermal Infrared Sensors

a. Thermal infrared sensors: These sensors detect infrared radiation emitted by objects as a result of their temperature. They are utilized for monitoring land surface temperatures, identifying heat anomalies, and analyzing energy balance.

Example: The MODIS on NASA's Terra and Aqua satellites incorporates thermal bands to monitor land surface temperatures and detect wildfires.

Microwave Sensors

a. Radar (radio detection and ranging): Radar sensors emit microwave pulses and measure the time taken for the signals to return after reflecting off objects. Effective in all weather conditions, radar is widely used for applications such as topographic mapping, soil moisture estimation, and flood monitoring.

Example: The European Space Agency's Sentinel-1 mission is equipped with C-band radar sensors to monitor changes in the surface, including ground deformation, deforestation, and ocean wind patterns.

b. Passive microwave sensors: These sensors detect natural microwave radiation emitted by the surface and atmosphere. They are primarily used for soil moisture estimation and atmospheric profiling.

Example: The soil moisture active passive (SMAP) satellite integrates both passive and active microwave sensors to measure soil moisture content, supporting weather forecasting and agricultural applications.

Radiometers

a. Radiometers: Radiometers measure the intensity of electromagnetic radiation across different spectral bands. They are commonly used in atmospheric research, climate monitoring, and satellite-based Earth observations.

Example: The atmospheric infrared sounder (AIRS) on NASA's Aqua satellite measures temperature and humidity profiles in the atmosphere, providing valuable data for weather forecasting and climate research.

Lidar (Light Detection and Ranging) Sensors

b. Lidar sensors: Lidar sensors use laser pulses to measure the distance between the sensor and the surface. They are extensively used in topographic mapping, vegetation structure analysis, and urban planning.

Example: NASA's Ice, Cloud, and Land Elevation Satellite-2 (ICESat-2) utilizes lidar technology to measure changes in the elevation of ice sheets, glaciers, and forests.

Ultraviolet and X-ray Sensors

a. Ultraviolet and X-ray sensors: These sensors detect radiation in the ultraviolet and X-ray regions of the electromagnetic spectrum. They are employed in specialized applications such as solar observations, atmospheric ozone monitoring, and mineral exploration.

   Example: The Solar and Heliospheric Observatory (SOHO) employs ultraviolet and X-ray sensors to investigate the Sun's activities, such as solar flares and coronal mass ejections.

2.2 Satellite System

2.2.1 Concept

Satellite systems are central to modern Geographic Information Sciences (GISciences), facilitating geospatial data acquisition and analysis. These systems, which include Earth-observing satellite constellations, global navigation satellite systems (GNSS) such as GPS, and communication satellites, are crucial for capturing high-resolution imagery, providing precise positioning data, and enabling communication for GIS applications. They generate vast amounts of geospatial data essential for mapping, monitoring, and analyzing the Earth's surface. GIS professionals and researchers leverage this data to create accurate maps, track land-use and land cover changes, monitor environmental shifts, assess natural disasters, and enhance navigation and location-based services. The integration of satellite systems into GISciences has transformed the understanding and management of spatial data, making it an invaluable tool in fields like urban planning, environmental management, agriculture, and disaster response (Fig. 2.8).

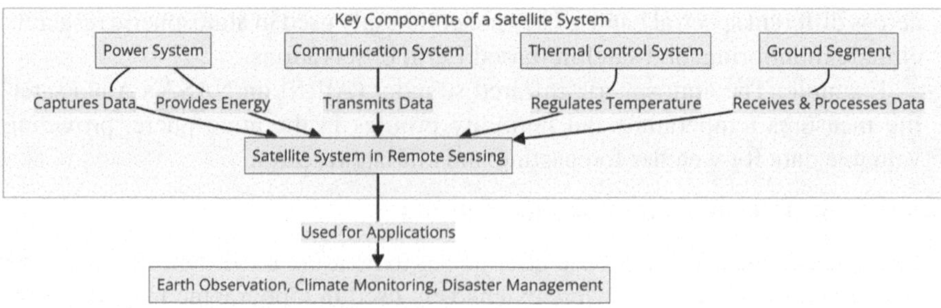

Fig. 2.8 Key components of a satellite system (prepared by the authors)

2.2.2 *Types*

Satellite systems can be categorized into various types based on their functions, orbits, and applications (Fig. 2.9). The major classifications of satellite systems are as follows:

Based on Orbits

a. Geostationary satellites: These satellites orbit the Earth at a speed matching the planet's rotation, allowing them to remain fixed relative to a specific point on the Earth's surface. They are primarily used for communication, weather monitor-

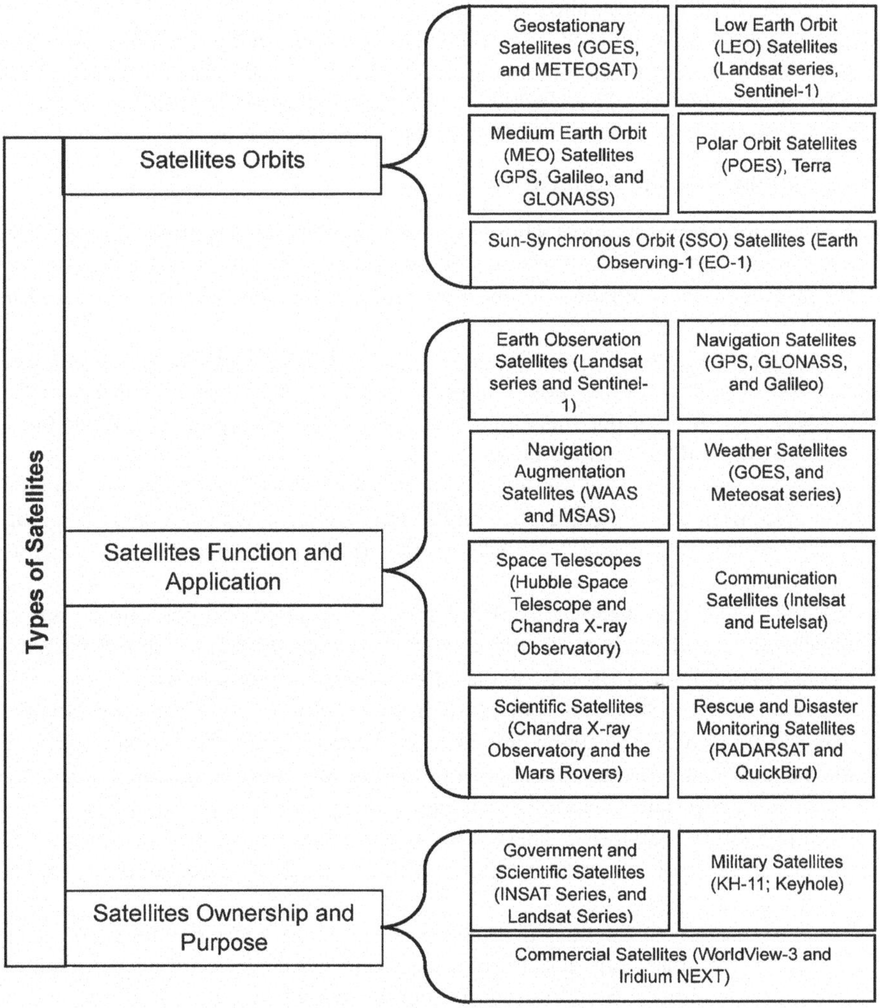

Fig. 2.9 Types of satellite system (prepared by the authors)

ing, and broadcasting. Example: GOES (geostationary operational environmental satellite), and METEOSAT.

b. Low Earth orbit (LEO) satellites: Orbiting at altitudes ranging from 160 to 2000 km above the Earth's surface, LEO satellites are employed for Earth observation, remote sensing, and scientific research. Example: Landsat series, Sentinel-1, and WorldView.
c. Medium Earth orbit (MEO) satellites: These satellites orbit at altitudes between 8000 and 20,000 km and are primarily used for navigation and positioning systems. Example: Global Positioning System (GPS), Galileo, and GLONASS.
d. Polar orbit satellites: These satellites travel in a north-south direction, passing over the Earth's poles. They are primarily used for Earth observation, climate monitoring, and environmental studies. Example: Polar-orbiting environmental satellites (POES), Terra, and Aquarius.
e. Sun-synchronous orbit (SSO) satellites: Designed to synchronize with the Sun's position, SSO satellites maintain consistent lighting conditions during each pass over a particular area. They are frequently employed in remote sensing missions. Example: Earth Observing-1 (EO-1) and Copernicus Sentinel-3.

Based on Function and Application

a. Earth observation satellites: These satellites capture data regarding the surface, atmosphere, and oceans. They are essential for applications such as disaster management, environmental monitoring, and urban planning. Example: Landsat series and Sentinel-1.
b. Navigation satellites: Satellite systems such as GPS, GLONASS, Galileo, and BeiDou provide accurate positioning, navigation, and timing information for applications in aviation, maritime, and terrestrial navigation. Example: GPS, GLONASS, Galileo, and BeiDou.
c. Navigation augmentation satellites: Satellites like WAAS (wide area augmentation system) enhance the accuracy and reliability of signals from navigation satellite systems. Example: WAAS and MSAS (multifunctional satellite augmentation system).
d. Weather satellites: Weather satellites monitor atmospheric conditions, gather meteorological data, and provide weather forecasts. Example: GOES (geostationary operational environmental satellite) and the Meteosat series.
e. Space telescopes: Space telescopes, such as the Hubble Space Telescope and the James Webb Space Telescope, are used for astronomical observations, studying distant celestial objects, and exploring cosmic phenomena. Example: Hubble Space Telescope and Chandra X-ray Observatory.
f. Communication satellites: These satellites facilitate telecommunications and broadcasting services, enabling global voice, data, and video communication. Example: Intelsat and Eutelsat.
g. Scientific satellites: Dedicated to scientific research, these satellites focus on areas such as space physics, Earth's magnetic field, and cosmic phenomena. Example: Chandra X-ray Observatory and the Mars Rovers.

h. Rescue and disaster monitoring satellites: Equipped for rapid response, these satellites provide real-time data for disaster assessment and aid in relief efforts. Example: RADARSAT and QuickBird.

Based on Ownership and Purpose

a. Government and scientific satellites: Operated by governmental bodies or scientific institutions, these satellites support research, environmental monitoring, and national security initiatives. Example: INSAT Series, Landsat Series, and Sentinel Series.
b. Military satellites: Designed for defence and security purposes, these satellites facilitate communication, reconnaissance, and navigation for military operations. Example: KH-11 (Keyhole) by USA, Kosmos Series by Russia, and Cartosat-2 Series by India.
c. Commercial satellites: These satellites are owned and managed by private enterprises and serve commercial purposes such as telecommunications, remote sensing, and Earth observation. Example: WorldView-3 and Iridium NEXT.

2.2.3 Platform

Remote sensing platforms are vehicles or structures that carry instruments like sensors and cameras, facilitating data collection on the Earth's surface and atmosphere. These platforms are tailored to operate in diverse environments and at varying altitudes to meet specific remote sensing objectives. The following are the various types of platforms (Fig. 2.10) utilized in remote sensing systems:

- **Satellites**
    a. Earth-observing satellites: These satellites, including polar-orbiting and geostationary types, gather data on the surface, atmosphere, and oceans. Examples include Landsat, MODIS, and GOES.

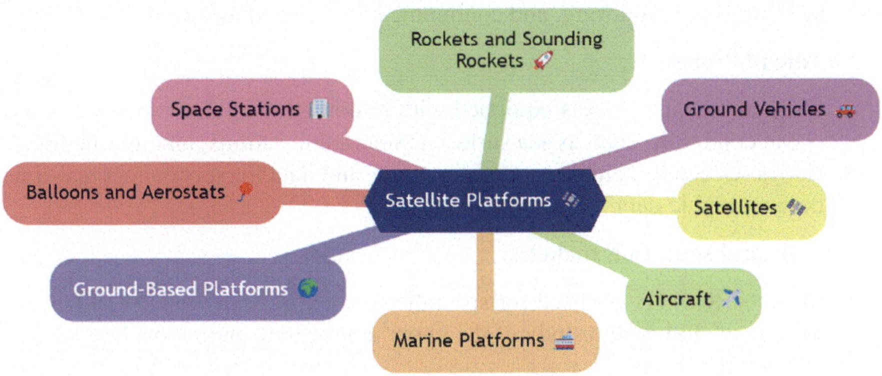

Fig. 2.10 Various satellite platforms (prepared by the authors)

b. Communication satellites: Primarily for telecommunications, satellites like Intelsat and Inmarsat also support remote sensing through transponders for communication and data relay.

- **Aircraft**

a. Manned aircraft: Piloted airplanes equipped with remote sensing instruments capture high-resolution imagery for applications like aerial photography, land-use mapping, and disaster assessment.
b. Unmanned aerial vehicles (UAVs or drones): UAVs are widely used for high-resolution aerial imagery, particularly in agriculture, forestry, and environmental monitoring.

- **Ground-based platforms**

a. Terrestrial sensors: Ground-based sensors, either fixed or mobile, collect data on local weather, air quality, and other environmental parameters.
b. Weather balloons: Instrumented balloons equipped with sensors collect meteorological data at various altitudes.

- **Ground vehicles**

a. Mobile mapping systems: Vehicles equipped with remote sensing instruments, such as LiDAR and cameras, capture geospatial data for applications like road mapping and infrastructure monitoring.

- **Space stations**

a. International space station (ISS): The ISS supports scientific research and Earth observation, including remote sensing experiments and microgravity studies.

- **Balloons and aerostats**

a. Weather balloons: High-altitude balloons equipped with instruments measure atmospheric conditions, providing data for weather forecasting and research.
b. Tethered aerostats: Tethered balloons or aerostats with sensors are used for surveillance, monitoring, and communication in fixed locations.

- **Marine platforms**

a. Ships and boats: Vessels equipped with remote sensing instruments monitor ocean conditions such as sea surface temperature, salinity, and marine life.
b. Buoys: Oceanographic buoys with sensors and data loggers collect real-time oceanographic data.

- **Rockets and sounding rockets**

a. Rockets: Rockets are used for specialized scientific experiments and observations, including the study of the upper atmosphere and ionosphere.

2.2.4 *Satellite Orbit*

Satellite orbit refers to the path or trajectory a satellite follows as it orbits a celestial body, such as Earth, the Moon, or another planet or star. Understanding satellite orbits is essential in space science and technology, as it influences a satellite's position, movement, and coverage area. Following is a comprehensive overview of satellite orbits:

Characteristics of Satellite Orbits

a. Altitude: The altitude of a satellite is the distance between it and the surface, and it influences the satellite's function. Common altitude categories include:
   - Low Earth orbit (LEO): Orbiting between 160 and 2000 km, LEO satellites are used for Earth observation, remote sensing, and communication, with the International Space Station (ISS) as an example.
   - Medium Earth orbit (MEO): Orbiting at 8000–20,000 km, MEO satellites are used for navigation, such as GPS and Galileo constellations.
   - Geostationary orbit (GEO): At approximately 36,000 km above the equator, GEO satellites remain fixed relative to Earth's surface, ideal for communication, broadcasting, and weather monitoring.

b. Inclination: The inclination of a satellite's orbit is the angle between its orbital plane and Earth's equator. Satellites may have polar orbits (90° inclination), equatorial orbits (0° inclination), or intermediate inclinations, impacting their coverage area and observable regions.
c. Eccentricity: Eccentricity measures how much a satellite's orbit deviates from a perfect circle. Higher eccentricity results in an elongated orbit, affecting the satellite's distance from Earth and its speed variation during the orbit.

Types of Satellite Orbits

a) Geostationary orbit (GEO): GEO satellites remain fixed relative to a specific point on the Earth's equator, matching Earth's rotational speed. This provides continuous coverage, making them ideal for communication, broadcasting, and weather observation.
b) Polar orbit: Polar orbiting satellites pass over Earth's poles in a north-south direction, offering global coverage over time. These orbits are suitable for Earth observation, environmental monitoring, and scientific research.
c) Sun-synchronous orbit (SSO): SSO satellites maintain a consistent angle to the Sun during their orbit, ensuring uniform lighting conditions. This makes them ideal for remote sensing and Earth observation.
d) Molniya orbit: Molniya orbits are highly eccentric with a high inclination (around 63.4°), providing extended coverage over high-latitude regions, especially the polar areas, for communication purposes.

e) Heliocentric orbit: Heliocentric orbits revolve around the Sun, as opposed to Earth. Space telescopes like the Hubble Space Telescope are placed in heliocentric orbits to study distant celestial objects.

Applications of Satellite Orbits

a. Earth observation: Polar and Sun-synchronous orbiting satellites gather data for Earth observation, including weather monitoring, environmental assessments, and disaster management.
b. Navigation: MEO satellites, such as GPS, Galileo, and BeiDou, offer accurate positioning and navigation services for aviation, maritime, and land-based applications.
c. Communication: GEO satellites are primarily used for global communication, television broadcasting, and Internet services, providing continuous coverage over specific regions.
d. Military and surveillance: Certain satellites are placed in specialized orbits for military and surveillance purposes, including reconnaissance and intelligence collection.
e. Scientific research: Various satellites in different orbits are used for scientific research, including studies in astronomy, space physics, and atmospheric sciences.

2.2.5 *Motion*

The motion of a satellite system pertains to the trajectory and movement of satellites as they orbit celestial bodies, such as the Earth. A comprehensive understanding of this motion is essential for the effective functioning of satellites across diverse applications, including communication, navigation, Earth observation, and space exploration. The following outlines the key aspects of satellite system motion:

Orbital Motion

- Satellites traverse a curved trajectory around celestial bodies as a result of gravitational attraction, a path referred to as an orbit.
- The motion of satellites is dictated by Newton's laws of motion and the universal law of gravitation, which collectively describe the forces and dynamics influencing orbital behaviour.
- Satellites are subject to two principal forces: gravitational attraction, which pulls them towards the celestial body, and centrifugal force, which counterbalances this pull due to their orbital velocity.

Kepler's Laws of Planetary Motion

Johannes Kepler established three fundamental laws of planetary motion, which are equally applicable to the motion of satellites:

- Kepler's first law (law of ellipses): The trajectory of a satellite is an elliptical orbit, with the celestial body positioned at one of the focal points.

- Kepler's second law (law of equal areas): A line connecting the satellite to the celestial body sweeps out equal areas in equal time intervals. This law accounts for the variation in a satellite's speed, as it moves faster when closer to the celestial body (perihelion) and slower when farther away (aphelion).
- Kepler's third law (law of harmonic proportions): The square of a satellite's orbital period is proportional to the cube of the semi-major axis of its orbit. This law establishes a relationship between the orbital period and the satellite's average distance from the celestial body.

Applications

The motion of satellite systems is utilized across diverse applications, including telecommunications, navigation (e.g., GPS), weather forecasting, Earth observation, scientific investigations, and space exploration.

In summary, satellite motion, governed by the principles of physics and celestial mechanics, is critical for the effective design, deployment, and operation of satellites, enabling their diverse societal and scientific contributions.

2.2.6 Location of Space, and GIScience-Related Law

The precise location of a place on surface is a cornerstone of remote sensing, enabling accurate mapping, monitoring, and analysis of Earth's features and phenomena. Several laws and principles underpinning spatial positioning and remote sensing are integral to this field, including:

a) **Keplar's law and its application in remote sensing**

Johannes Kepler, in the early seventeenth century, formulated three fundamental laws of planetary motion, known as Kepler's laws. These laws describe the orbits of planets and other celestial objects around the Sun, forming the basis of modern celestial mechanics and our understanding of planetary motion. The laws are as follows:

Kepler's First Law—The Law of Ellipses

- Kepler's first law states that the orbit of a planet around the Sun is an ellipse, with the Sun positioned at one of its two foci.
- An ellipse, a flattened and elongated circle, is defined by two focal points. In the context of planetary motion, the Sun occupies one focus, while the other remains empty.
- The closest point in the orbit to the Sun is termed the perihelion, whereas the farthest point is known as the aphelion.
- This law supplanted the earlier notion of circular orbits proposed by Ptolemy and Copernicus, providing a more accurate representation of planetary motion.

Kepler's Second Law—The Law of Equal Areas

- Kepler's second law states that a line connecting a planet to the Sun sweeps out equal areas in equal intervals of time.
- This means that as a planet moves along its elliptical orbit, it travels faster when closer to the Sun (perihelion) and slower when farther away (aphelion), ensuring equal areas are covered in the same time span.
- The law highlights the variation in a planet's orbital speed, providing key insights into the dynamics of planetary motion.

Kepler's Third Law—The Law of Harmonic Proportions

- Kepler's Third Law establishes a relationship between a planet's orbital period (the time required to complete one orbit) and its average distance from the Sun, represented by the semi-major axis.
- Mathematically, the law states that the square of a planet's orbital period is directly proportional to the cube of its semi-major axis.
- This principle enables astronomers to determine the relative distances of planets in the solar system based on their orbital periods, demonstrating that planets farther from the Sun have longer orbital periods.

Application of Kepler's Laws in Remote Sensing

- **Orbital mechanics and satellite design:** Kepler's laws are essential for designing and operating Earth-observing satellites. By applying these laws, engineers can accurately calculate orbital parameters such as altitude, inclination, and orbital period to meet specific remote sensing goals. For instance, satellites in Sun-synchronous orbits are designed to maintain a constant angle relative to the Sun, ensuring consistent lighting conditions for remote sensing over a given area.
- **Orbit prediction and satellite tracking:** Remote sensing and Earth observation satellites depend on precise orbital predictions to ensure timely passes over designated areas of interest. Kepler's laws, combined with the satellite's initial conditions, enable accurate forecasting of satellite positions and orbits, which is crucial for mission planning and data collection scheduling.
- **Accurate georeferencing:** Georeferencing, the process of assigning geographic coordinates to remote sensing data, depends on a precise understanding of satellite orbits. The satellite's position and velocity at the time of data capture, determined using Kepler's laws, are crucial for accurately georeferencing and aligning remote sensing images and data.
- **Earth observation and climate monitoring:** Remote sensing satellites in polar orbits, as described by Kepler's laws, offer a global view of Earth's surface, atmosphere, and oceans. These satellites are essential for applications such as climate monitoring, environmental assessment, disaster management, and the study of long-term environmental changes.
- **Accurate positioning for navigation:** Satellite navigation systems, such as GPS (Global Positioning System), are grounded in Kepler's laws and are vital for accurate geopositioning on Earth. GPS operates through a network of satellites, with their precise positions and orbits determined using principles derived from Kepler's laws.

b) **Newton's gravitational law and its application in remote sensing**

Newton's law of universal gravitation, formulated by Sir Isaac Newton in the seventeenth century, describes the gravitational force between two masses. This law plays a crucial role in remote sensing, especially in analyzing satellite orbits and ensuring precise satellite positioning for Earth observation. Below is an overview of its applications in remote sensing:

Newton's Law of Universal Gravitation

Newton's law of universal gravitation states that every point mass exerts an attractive force on every other point mass, which is proportional to the product of their masses and inversely proportional to the square of the distance between them. The law is mathematically expressed as:

$$F = G\frac{m_1 m_2}{rr^2}$$

where, F is the gravitational force between the two masses; G is the gravitational constant; m_1 and m_2 are the masses of the two objects; and r is the distance between the centers of the two masses.

Application of Newton's Law in Remote Sensing

- **Satellite orbits and trajectories:** Newton's law is crucial for understanding the motion of satellites in orbit around Earth. It is employed to calculate the gravitational force acting on a satellite and to determine the centripetal force necessary to sustain a stable orbit. Remote sensing satellites rely on these calculations to maintain their intended orbits and altitudes, ensuring consistent coverage of target regions for Earth observation and data collection.
- **Orbital perturbations:** While Kepler's laws offer an idealized model of orbital motion, real-world satellite orbits are influenced by various perturbing forces, such as the gravitational effects of the Moon, Sun, and other celestial bodies. Newton's Law is used to calculate and model these perturbations, enabling engineers and mission planners to predict and correct deviations from the intended orbital path.
- **Satellite positioning and navigation:** Precise knowledge of gravitational forces, as outlined by Newton's law, is crucial for accurate satellite positioning and navigation systems. Systems like the Global Positioning System (GPS) depend on the gravitational interactions between satellites and Earth to determine the exact positions of receivers on the Earth's surface. The law is integral in calculating the gravitational potential field, which underpins GPS functionality.
- **Gravity field and geoid modelling:** Remote sensing instruments, such as satellite-based gravimeters and altimeters, measure variations in Earth's gravitational field. Newton's law is essential for modelling the gravity field and geoid, which is critical for applications like geodesy, sea-level monitoring, and precise elevation measurements.
- **Gravity recovery and climate experiment (GRACE):** The GRACE satellite mission used precise measurements of gravitational anomalies to track changes

in Earth's water and ice distribution, offering valuable data for studying climate change and managing water resources.

c) **Inverse square law and its application in remote sensing**

The inverse square law is a fundamental principle in remote sensing, which asserts that the intensity of electromagnetic radiation decreases with the square of the distance from the source. In this context, the law is crucial for understanding how radiation strength diminishes as it travels from its source to the sensor or target.

Application of Inverse Square Law in Remote Sensing

- **Radiometric corrections:** When remote sensing data is acquired, the sensor measures the intensity of electromagnetic radiation. To ensure accurate comparisons across different measurements, radiometric corrections are applied to account for variations in radiation intensity caused by changes in the distance from the sensor to the target. The inverse square law is used to adjust for this variation, enabling consistent radiometric analysis.
- **Atmospheric corrections:** Remote sensing data often requires atmospheric corrections due to the scattering and absorption of radiation by the atmosphere. The inverse square law aids in estimating the pathlength of radiation through the atmosphere, facilitating corrections for atmospheric effects on the data.
- **Calibration:** The inverse square law is vital for calibrating remote sensing instruments. By understanding how radiation intensity diminishes with distance, scientists and engineers can calibrate sensors to ensure accurate measurements at varying distances from the target.

d) **Planck's law and its application in remote sensing**

Planck's law is a fundamental concept in remote sensing and physics, which characterizes the spectral radiance (brightness) of electromagnetic radiation emitted by a blackbody at a specific temperature.

Application of Planck's Law in Remote Sensing

- **Thermal remote sensing:** Planck's law is essential for interpreting thermal emissions from Earth's surface. It enables scientists and researchers to estimate the temperature of various features, such as land, water bodies, and buildings, based on the thermal radiation they emit.
- **Temperature monitoring:** In thermal infrared remote sensing, sensors capture radiation in specific wavelength bands. Planck's law assists in converting these radiation measurements into temperature values, allowing for the monitoring of temperature variations across the surface and atmosphere.
- **Urban heat island analysis:** Planck's law is valuable for studying urban heat islands, enabling the assessment of temperature variations in urban areas and helping urban planners and policymakers address heat-related challenges.
- **Climate science:** Planck's law aids climate studies by providing data on surface temperature trends and changes, helping researchers evaluate climate impacts in different regions.

- **Environmental studies:** Planck's law is applied in environmental monitoring, including agriculture and forestry, to assess land surface temperatures, identify areas of vegetation stress, and track ecosystem health.
- **Disaster Management:** In disaster management, such as wildfire monitoring and volcanic activity analysis, Planck's law helps estimate temperatures in hazardous areas, supporting decision-making and mitigation strategies.

e) **Stefan–Boltzmann law and its application in remote sensing**

The Stefan-Boltzmann law is a fundamental principle in remote sensing and physics, which defines the relationship between the total energy radiated by a blackbody (an idealized emitter) and its temperature.

Application of Stefan–Boltzmann Law in Remote Sensing

- **Thermal remote sensing:** The Stefan–Boltzmann law is applied in thermal remote sensing to estimate the total energy emitted by objects on surface and in the atmosphere. It enables scientists and researchers to convert thermal radiation measurements into temperature values.
- **Temperature monitoring:** In thermal infrared remote sensing, sensors measure the radiant flux from objects. By applying the Stefan–Boltzmann law, professionals can calculate the temperature of features like land, water bodies, and industrial facilities based on the total energy radiated.
- **Energy efficiency and building analysis:** In urban and infrastructure studies, the law aids in evaluating the thermal properties of buildings, assessing heat loss, and optimizing energy efficiency.
- **Environmental studies:** The law is used in environmental monitoring, agriculture, and forestry to assess land surface temperatures, identify areas of ecological interest, and monitor vegetation stress.
- **Climate science:** The Stefan–Boltzmann law plays a crucial role in climate studies by estimating the Earth's radiative energy balance. It helps researchers understand how temperature changes impact thermal radiation emission, contributing to climate modelling and predictions.
- **Disaster management:** The law is valuable in disaster management, particularly for assessing temperatures in potentially hazardous areas, such as wildfire hotspots, volcanic regions, and industrial sites.

f) **Wein's displacement law and its application in remote sensing**

Wein's displacement law, developed by Wilhelm Wien in the late nineteenth century, defines the relationship between the temperature of a blackbody radiator and the wavelength at which it emits the highest intensity of radiation.

Application of Wein's Displacement Law in Remote Sensing

- **Thermal remote sensing:** Wein's displacement law is applied in thermal remote sensing to determine the temperature of objects based on the peak wavelength of their thermal emissions, with hotter objects emitting shorter wavelengths and cooler ones emitting longer wavelengths.

- **Temperature estimation:** Remote sensing instruments that capture thermal infrared data analyze the peak wavelength of spectral radiance to estimate the temperature of land, water bodies, buildings, and other surface features.
- **Thermal mapping:** The law aids in creating thermal maps for applications in agriculture, forestry, and land management, helping to monitor temperature variations for informed decision-making.
- **Industrial and environmental monitoring:** The law helps identify temperature anomalies, hotspots, and potential issues in both industrial facilities and natural environments, aiding in monitoring and assessment.
- **Fire detection:** In wildfire monitoring, the law helps quantify fire temperatures and identify affected areas by analyzing thermal data.
- **Climate studies:** Wein's displacement law contributes to understanding temperature distribution and variations on Earth's surface, supporting climate models and assessments.

g) **GIScience (others)-related law**

Kirchhoff's Law of Thermal Radiation Kirchhoff's law states that materials that are good absorbers of radiation are also good emitters. It is crucial for understanding the absorption and emission of thermal radiation by Earth's surface objects.

Beer–Lambert Law The Beer–Lambert law explains how the concentration of an absorbing substance in a medium affects the absorption of light or radiation. In remote sensing, it is used to estimate the concentration of substances in the atmosphere, water, and vegetation.

Snell's Law Snell's law describes how light changes direction when passing between media with different refractive indices. It is important for understanding how light bends when entering Earth's atmosphere, impacting remote sensing observations.

Fresnel's Equations Fresnel's equations describe the reflection and transmission of light at the interface between two different media. They help in understanding how light interacts with surfaces, including the generation of polarized light.

Norton–Beer Law The Norton–Beer law describes the attenuation of electromagnetic radiation as it travels through a medium, often used to model atmospheric absorption effects on remote sensing data.

Kirchhoff's Law of Thermal Emission This law states that the emissivity of an object at a given wavelength equals its absorptivity at the same wavelength, helping relate reflectance and emissivity in thermal remote sensing.

Lambert's Cosine Law Lambert's cosine law explains how irradiance (light intensity) on a surface varies with the angle of incidence. It is used to correct remote sensing data for changes in sun angle and surface orientation.

2.3 Satellite Programmes: History and Programmes

2.3.1 *History of Satellite/Observation Programmes*

Satellite programmes have significantly advanced space exploration, communication, Earth observation, scientific research, and military applications. These programmes exemplify the integration of cutting-edge technology, engineering, and global cooperation. The following is a detailed overview of various satellite programmes:

Early Satellite Programmes (1950s–1960s)

- **Sputnik (Soviet Union, 1957):** Sputnik 1, launched by the Soviet Union, was the first artificial satellite, marking the beginning of the space age.
- **Explorer (USA, 1958):** In response to Sputnik, the United States launched the Explorer programme, with Explorer 1 becoming the first American satellite and making notable scientific discoveries.
- **Luna (Soviet Union, 1959):** The Luna programme, focused on lunar exploration, achieved a milestone with Luna 2, the first human-made object to reach the Moon.

Human Spaceflight Programmes (1960s–Present)

- **Mercury (USA):** The Mercury programme (1961–1963) was the United States' first human spaceflight initiative.
- **Vostok (Soviet Union):** The Vostok program launched Yuri Gagarin into orbit in 1961, making him the first human to journey into space.
- **Apollo (USA):** The Apollo programme (1961–1972) culminated in the historic moon landings, with Apollo 11's Neil Armstrong and Buzz Aldrin becoming the first humans to set foot on the Moon.
- **Intelsat (International, 1965–present):** The Intelsat network, managed by the International Telecommunications Satellite Organization, provides global satellite communication services.
- **Landsat (USA, 1972–present):** The Landsat programme has been a critical source of Earth observation data, supporting environmental monitoring and resource management.
- **GPS (USA, 1978–present):** The Global Positioning System (GPS) offers precise global positioning and timing information.
- **GLONASS (Russia, 1982–present):** GLONASS provides navigation and positioning services, primarily within Russia.
- **INSAT Series (India, 1983–present):** The INSAT satellites provide services in telecommunications, broadcasting, meteorology, and disaster management.
- **IRS Series (India, 1988–present):** The IRS satellites support Earth observation and resource management, with applications in agriculture, forestry, and land use.
- **Hubble Space Telescope (USA, 1990–present):** The Hubble Space Telescope has transformed our understanding of the universe with its deep space observations.

- **Iridium (USA, 1998–present):** The Iridium satellite constellation delivers worldwide mobile satellite communication services.
- **International Space Station (ISS):** The ISS (1998–present) is a symbol of global collaboration, supporting crewed missions and scientific research from multiple nations.
- **Globalstar (USA, 1999–present):** Globalstar's satellite network enables voice and data communication on a global scale.
- **Shenzhou (China):** China's Shenzhou programme (2003–present) has facilitated crewed space missions and space station development.
- **Galileo (European Union, 2016–present):** Galileo is Europe's global navigation satellite system (GNSS), enhancing global navigation capabilities.
- **SpaceX Starlink (USA, 2018–present):** SpaceX's Starlink project aims to provide global broadband internet coverage through its satellite constellation.
- **OneWeb (International, 2019–present):** OneWeb's satellite network seeks to offer global Internet access, connecting underserved regions.
- **BeiDou (China, 2020–present):** The BeiDou system provides global navigation services, developed and operated by China.

2.3.2 Satellite Programmes

Satellite programmes are organized efforts focused on the design, construction, launch, and operation of satellites for diverse purposes. These programmes have greatly advanced science, technology, communication, navigation, and national security. The following are some notable satellite programmes:

a) **CORONA satellite Programme (1960–1972)**

The CORONA satellite programme, launched by the United States in the early 1960s, was among the first major reconnaissance satellite initiatives. Overseen by the U.S. Air Force with assistance from the Central Intelligence Agency (CIA), the programme involved a series of photographic reconnaissance satellites designed to gather intelligence during the Cold War. These satellites were equipped with film-based cameras to capture high-resolution images of the surface.

- **Purpose:** The primary objective of the CORONA satellites was military reconnaissance, focusing on monitoring Soviet activities. They provided critical intelligence by capturing images of military installations, missile sites, and other strategic locations.
- **Technological contributions:** CORONA satellites were pioneers in remote sensing and satellite photography, driving significant advancements in optical imaging systems and satellite-based intelligence collection.
- **Legacy:** Though declassified in the 1990s, the success of the CORONA programme laid the foundation for contemporary reconnaissance satellites and advanced Earth observation technologies.

b) **LANDSAT satellite programme (1972–present)**

The LANDSAT programme, launched by NASA in partnership with the U.S. Geological Survey (USGS), is one of the longest-operating Earth observation satellite programmes. It has been instrumental in providing vital data for land-use planning, environmental monitoring, agriculture, forestry, and resource management.

- **Purpose:** The primary goal of LANDSAT satellites is to monitor Earth's surface and analyze changes in land cover and use, including tracking deforestation, urbanization, agricultural trends, and the impact of natural disasters.
- **Technological contributions:** Equipped with multispectral imaging sensors, LANDSAT satellites offer high-resolution imagery across various spectral bands, aiding scientific research in environmental change, biodiversity conservation, and climate studies.
- **Legacy:** Over five decades, the LANDSAT programme has significantly advanced geospatial data analysis and Earth observation science. It continues to contribute to a deeper understanding of environmental dynamics, supporting applications from resource management to disaster response.
- **Key Milestones:**
  - ***LANDSAT 1 (1972):*** Launched in 1972, LANDSAT 1 marked the start of systematic Earth observation through satellite imagery.
  - ***LANDSAT 4 (1982):*** The 1982 launch of LANDSAT 4 introduced the Thematic Mapper (TM) sensor, enhancing data quality by offering seven spectral bands with improved spatial resolution.
  - ***LANDSAT 7 (1999):*** Launched in 1999, LANDSAT 7 featured the Enhanced Thematic Mapper Plus (ETM+) sensor, providing higher resolution and better data quality. However, a failure in the scan line corrector limited data collection in certain areas.
  - ***LANDSAT 8 (2013):*** With the introduction of the Operational Land Imager (OLI) and Thermal Infrared Sensor (TIRS), LANDSAT 8, launched in 2013, provided enhanced imagery and an additional thermal infrared band.
  - ***LANDSAT 9 (2021):*** Launched in 2021, LANDSAT 9 continues the legacy of its predecessors, equipped with similar sensors to LANDSAT 8 to maintain continuity in Earth observation data.

- **Applications and impact**
  - ***Environmental monitoring:*** LANDSAT's high-resolution imagery enables the tracking of global environmental changes such as deforestation, urbanization, and land degradation, allowing researchers to monitor shifts in land use over time.
  - ***Climate change studies:*** LANDSAT plays a key role in studying climate change, providing data on sea-level rise, glacier retreat, and other environmental effects.

- *Water resources:* LANDSAT helps track changes in water bodies and monitor climate change impacts, including shifts in water availability and droughts.
- *Agriculture:* The programme is used to monitor crop health, irrigation patterns, and the effects of pests and diseases on crops, as well as assess land suitability for farming.
- *Disaster management:* LANDSAT data supports disaster response by offering real-time information on areas affected by events like floods, earthquakes, and wildfires, assisting in damage assessment and recovery efforts.
- *Urban planning and development:* The satellite provides data on urban sprawl, land use, and infrastructure, aiding in sustainable resource management and urban planning.

- **Scientific contributions and global impact:** The LANDSAT programme has significantly advanced our understanding of surface by providing consistent, long-term data for studying environmental changes. Its contributions have been crucial in resource management, environmental conservation, and supporting sustainable development, improving our capacity to address environmental challenges.

c) **SPOT satellite programme (1986–present)**

The SPOT programme, developed by the French Space Agency (CNES), focuses on high-resolution Earth observation to support applications such as agriculture, urban planning, and environmental monitoring.

- **Purpose:** SPOT satellites are designed to capture detailed images of surface for mapping, resource management, and environmental monitoring. They are widely used in fields like urban planning, forestry, agriculture, and disaster management.
- **Technological contributions:** Equipped with multispectral and panchromatic imaging systems, SPOT satellites have advanced remote sensing technology, enabling the detection of changes in vegetation, urbanization, and land use.
- **Legacy:** The SPOT programme has played a vital role in mapping and monitoring both natural and human-made features, providing crucial imagery for disaster management, land-use planning, and environmental conservation.
- **Key milestones:**

  - *SPOT 1 (1986):* The inaugural satellite in the series, SPOT 1, delivered 20 m resolution imagery and became the first to offer commercial Earth observation services, marking a significant advancement in high-resolution remote sensing.
  - *SPOT 4 (2002):* Launched in 2002, SPOT 4 enhanced imaging capabilities with the addition of the Vegetation (VGT) instrument, enabling global vegetation monitoring.
  - *SPOT 5 (2002):* Also launched in 2002, SPOT 5 provided even higher resolution imagery (2.5 m) and enabled stereo imaging, which proved essential for topographic mapping.

  - ***SPOT 6 (2012):*** SPOT 6, launched in 2012, introduced 1.5 m resolution imagery and complemented earlier satellites with more frequent revisits of key areas.
  - ***SPOT 7 (2014):*** Launched in 2014, SPOT 7 continued the 1.5 m resolution standard of SPOT 6, enhancing imaging capacity and extending revisit capabilities to ensure global data continuity.

- **Applications and impact**
  - ***Environmental monitoring:*** SPOT data tracks land-use changes, deforestation, and climate impacts such as desertification and glacial retreat, aiding environmental risk assessments and conservation efforts.
  - ***Forestry and land cover mapping:*** SPOT contributes to mapping forests, monitoring deforestation, and assessing vegetation health, supporting sustainable forestry management and ecosystem monitoring.
  - ***Marine and coastal studies:*** SPOT data is used to monitor water bodies, track shoreline changes, assess water quality, and study climate change impacts on coastal ecosystems.
  - ***Disaster management:*** SPOT imagery is crucial for disaster monitoring and response, providing timely data on areas affected by floods, wildfires, and earthquakes to assess damage and plan recovery.
  - ***Defence and security:*** SPOT satellites provide detailed imagery for surveillance, border control, and military planning, making them valuable tools for intelligence and security agencies.
  - ***Urban and regional planning:*** The high-resolution imagery from SPOT satellites aids in land-use management, urban planning, infrastructure development, and evaluating the environmental impact of urbanization.

- **Scientific contributions and global impact:** The SPOT satellite programme has contributed to scientific research by providing high-resolution imagery for monitoring environmental changes, urbanization, and natural disasters. It supports global studies on climate change and habitat loss, as well as initiatives like the Global Forest Observation Initiative. And the SPOT data holds commercial value, aiding sectors like agriculture, urban planning, defence, and conservation. Its frequent, high-quality imagery enables informed decision-making across governments, businesses, and NGOs.

2.4 Indian Remote Sensing System and Institutions

2.4.1 History of Indian Remote Sensing System

The Indian Remote Sensing (IRS) system, spearheaded by the Indian Space Research Organisation (ISRO), has undergone remarkable advancements since its inception, becoming a cornerstone in Earth observation, environmental monitoring, resource management, and disaster mitigation.

- **Early foundations (1970s):** India's remote sensing initiatives began in the 1970s with the establishment of the Space Applications Centre (SAC) and the Indian National Committee for Space Research (INCOSPAR) under the visionary leadership of Dr. Vikram Sarabhai. The launch of the Aryabhata satellite in 1975 marked India's initial step into space-based observation, though its capabilities were limited to scientific experiments rather than Earth observation.
- **Launch of the IRS programme (1980s):** The formal introduction of the IRS programme in the 1980s signified India's commitment to leveraging satellite technology for resource management and national development. The programme targeted key applications such as agriculture, water resources, forestry, land use, urban planning, and disaster management. The launch of IRS-1A in 1988 marked India's entry into operational remote sensing with sophisticated sensors for Earth observation.
- **Expansion and innovation (1990s–2000s):** The IRS programme expanded during the 1990s and early 2000s with the deployment of IRS-1B (1991), IRS-1C (1995), and IRS-1D (1997), which introduced improved spatial resolution and advanced sensors. These satellites facilitated significant contributions to agricultural monitoring, resource mapping, and urban planning. Integration of remote sensing data with Geographic Information Systems (GIS) further enhanced decision-making processes in multiple domains.
- **Technological advancements (2010s):** The 2010s witnessed substantial progress with the launch of high-resolution satellites, including IRS-P6 (2003), Cartosat-1 (2005), and Cartosat-2 (2007). These satellites supported precise urban and rural planning, forestry, disaster management, and environmental monitoring. Remote sensing data became integral to national policies and sustainable development strategies.
- **Modern innovations and the future (post-2010s):** Recent advancements include the launch of RISAT (radar imaging satellite) and Cartosat-3, featuring cutting-edge technologies such as synthetic aperture radar (SAR) for all-weather imaging. These satellites have enhanced the accuracy of monitoring environmental and agricultural parameters, even in challenging conditions, ensuring data continuity and reliability.

Today, the IRS programme is vital to India's national development, offering invaluable data for agriculture, forestry, urban planning, disaster response, and environmental conservation. As a pillar of ISRO's space mission, the IRS system not only addresses domestic needs but also contributes to global Earth system sciences, sustainable development, and climate research.

2.4.2 Indian Institution (ISRO) and Mission

2.4.2.1 Indian Space Research Organisation

The Indian Space Research Organisation (ISRO), founded in 1969, serves as India's leading space agency under the Department of Space, Government of India. Established under the visionary leadership of Dr. Vikram Sarabhai, often hailed as the father of India's space programme, ISRO has evolved into a global pioneer in space science and technology. Over the years, it has showcased its expertise in designing, developing, and launching satellites and space exploration missions, playing a pivotal role in advancing scientific innovation and contributing significantly to national development.

Historical Overview

ISRO's origins date back to the establishment of the Indian National Committee for Space Research (INCOSPAR) in 1962, led by Dr. Vikram Sarabhai. He envisioned space technology as a tool for national development, focusing on applications like communication, resource management, weather forecasting, and education. Formally founded in 1969, ISRO succeeded INCOSPAR and came under the Department of Space in 1972. Its first major achievement was the launch of Aryabhata, India's inaugural satellite, in 1975. This marked the beginning of India's journey in satellite technology and set the foundation for ISRO's future landmark missions, establishing the country as a key player in space exploration.

Mission and Vision

ISRO's mission is centered on leveraging space technology for India's socio-economic progress while advancing space science and planetary exploration. Its vision focuses on addressing critical challenges like poverty alleviation, resource management, and disaster mitigation, alongside establishing India as a global leader in space exploration. The primary objectives of ISRO include:

- **Satellite development:** Designing and deploying satellites for communication, navigation, remote sensing, and Earth observation.
- **Space exploration:** Undertaking scientific missions to study planets, the Moon, and other celestial bodies.
- **Launch vehicle innovation:** Developing indigenous launch vehicles to deploy satellites into various orbits.
- **National development:** Offering technological solutions for agriculture, education, healthcare, disaster management, and urban planning.
- **International collaboration:** Partnering with global space agencies and providing cost-effective satellite launch services.

Major Achievements and Contributions

a) **Satellite programmes**

- Indian remote sensing (IRS) satellites: Launched in 1988 with IRS-1A as the first operational satellite, this series supports resource management, environmental monitoring, and disaster mitigation efforts.
- INSAT/GSAT Series: These satellites are dedicated to communication, broadcasting, and meteorology, playing a critical role in telecommunication and weather forecasting since the INSAT programme's inception in 1982.
- Navigation satellites (NavIC): The Navigation with Indian Constellation (NavIC) system delivers precise positioning and timing services, essential for navigation, disaster management, and emergency response.

b) **Launch vehicle development**

- SLV (satellite launch vehicle): India's first experimental launch vehicle, it marked a significant milestone with the successful deployment of the Rohini satellite in 1980.
- PSLV (polar satellite launch vehicle): Known as ISRO's workhorse, the PSLV is celebrated for its reliability and versatility, having successfully launched landmark missions such as Chandrayaan-1 and the Mars Orbiter Mission (Mangalyaan).
- GSLV (geosynchronous satellite launch vehicle): Featuring a cryogenic upper stage, the GSLV is pivotal for placing heavier satellites into geosynchronous orbit.
- GSLV Mk III (LVM-3): A heavy-lift launch vehicle, it supports advanced missions like Chandrayaan-2 and the forthcoming Gaganyaan human spaceflight programme.

c) **Space exploration missions**

- Chandrayaan programme:
  - Chandrayaan-1 (2008): India's first lunar mission, which confirmed the presence of water molecules on the Moon.
  - Chandrayaan-2 (2019): Included an orbiter, lander, and rover, showcasing India's technological advancements despite challenges during the landing phase.
  - Chandrayaan-3 (2023): Achieved a historic milestone by successfully landing on the lunar south pole, making India the first nation to do so.
- Mars orbiter mission (Mangalyaan) (2013): Demonstrated India's prowess in cost-effective space exploration, becoming the first country to achieve Mars orbit insertion on its maiden attempt.
- Aditya-L1 (2023): A solar mission aimed at studying the Sun's corona and understanding its influence on space weather.

Impact and Global Recognition
ISRO's achievements have significantly advanced India's scientific, economic, and societal progress. Its cost-effective innovations have earned global recognition, establishing India as a major player in the global space economy and inspiring other developing nations to embrace space technology.

2.4.2.2 India and Satellite/Observation Programmes

India, through its Indian Space Research Organisation (ISRO), has achieved remarkable advancements in satellite technology and space exploration. Its satellite programmes encompass diverse areas such as Earth observation, communication, navigation, and planetary exploration. A concise/major overview of these programmes is provided below:

- **Aryabhata (1975):** India's first satellite, launched by the Soviet Union, marked the nation's entry into space exploration, facilitating research in solar physics and aeronomy.
- **Rohini Satellite Series (1980–1994):** India's first series of indigenously developed satellites for Earth observation, primarily used for weather monitoring, agricultural planning, and resource management.
- **INSAT Series (1983–present):** The Indian National Satellite System (INSAT) consists of multipurpose satellites that provide services in telecommunications, television broadcasting, meteorology, and disaster warning.
- **IRS Series (1988–present):** The Indian remote sensing (IRS) satellites are dedicated to Earth observation, environmental monitoring, and resource management, including agriculture, forestry, and land use.
- **GSAT Series (2001–present):** Geostationary satellites focused on enhancing communication, broadcasting, and broadband services, improving connectivity across India.
- **Cartosat Series (2005–present):** High-resolution Earth observation satellites used in urban planning, infrastructure development, and disaster management.
- **NavIC (2016–present):** The Navigation with Indian Constellation (NavIC) is India's regional satellite navigation system, offering precise positioning and timing services within India and surrounding regions.
- **Mangalyaan (Mars Orbiter Mission, 2013):** India's Mars mission, which made it the first Asian country to reach Mars orbit, providing valuable data on the Martian atmosphere and surface.

2.5 Conclusion and Endnotes

In the field of sensor and satellite systems, significant advancements have transformed sectors like Earth observation, environmental monitoring, telecommunications, and national security. These systems have greatly expanded our understanding of the planet and beyond, providing essential data for scientific research, disaster management, and daily applications. This chapter has explored the evolution of satellite programmes, from the early days of Sputnik to modern Earth-observing constellations, highlighting advancements in sensor technology that have improved precision, sensitivity, and versatility. We have also discussed the vital role of international collaboration in satellite development and operation. Looking ahead, emerging technologies like small satellites, machine learning, and advanced sensors promise to enhance these systems further. In conclusion, sensor and satellite systems are crucial to tackling global challenges and exploring space. To fully harness their potential, ongoing innovation, collaboration, and regulation are essential for the benefit of humanity.

Endnotes

- Sensor and satellite systems have revolutionized Earth observation, environmental monitoring, telecommunications, and national security.

- These systems provide invaluable data for scientific research, disaster management, and everyday applications.
- The evolution of satellite programmes from Sputnik to modern constellations showcases significant advancements in sensor technology.
- International collaboration plays a crucial role in the development and operation of satellite programmes.
- Emerging technologies like small satellites, machine learning, and advanced sensors hold great potential for the future of satellite systems.

Further Readings

1. Chuvieco, E. (Ed.). (2007). *Earth observation of global change: The role of satellite remote sensing in monitoring the global environment.* Springer . ISBN: 978-1-4020-6357-2. https://doi.org/10.1007/978-1-4020-6358-9
2. Laignel, B., Vignudelli, S., Almar, R., et al. (2023). Observation of the coastal areas, estuaries and deltas from space. *Surveys in Geophysics, 44*, 1309–1356. https://doi.org/10.1007/s10712-022-09757-6
3. McCarthy, M. J., Colna, K. E., El-Mezayen, M. M., et al. (2017). Satellite remote sensing for coastal management: A review of successful applications. *Environmental Management, 60*, 323–339. https://doi.org/10.1007/s00267-017-0880-x
4. Melet, A., Teatini, P., Le Cozannet, G., et al. (2020). Earth observations for monitoring marine coastal hazards and their drivers. *Surveys in Geophysics, 41*, 1489–1534. https://doi.org/10.1007/s10712-020-09594-5
5. Morton, Y. J., van Diggelen, F., Spilker, J. J., Jr., Parkinson, B. W., Lo, S., & Gao, G. (Eds.). (2021). *Position, navigation, and timing technologies in the twenty-first century: Integrated satellite navigation, sensor systems, and civil applications.* The Institute of Electrical and Electronics Engineers, Wiley. ISBN: 9781119458418. https://doi.org/10.1002/9781119458449
6. Qu, J. J., Gao, W., Kafatos, M., Murphy, R. E., & Salomonson, V. V. (Eds.). (2006). *Earth science satellite remote sensing: Vol.1: Science and instruments.* Springer . ISBN: 978-3-540-35606-6. https://doi.org/10.1007/978-3-540-37293-6
7. Siegel, H., Gerth, M., & Ohde, T. (2002). Remote sensing applications in coastal zone management. In G. Schernewski & U. Schiewer (Eds.), *Baltic coastal ecosystems. Central and eastern European development studies.* Springer. https://doi.org/10.1007/978-3-662-04769-9_21
8. Strong, A. (1989). Greater global warming revealed by satellite-derived sea-surface-temperature trends. *Nature, 338*, 642–645. https://doi.org/10.1038/338642a0
9. Victorov, S. (1996). *Regional satellite oceanography.* CRC Press (Taylor and Francis Group) ISBN 9780748402731.
10. Wahner, A., Schoeberl, M., Brune, W. H., Wahner, A., & Schoeberl, M. (Eds.). (2006). *Observing Systems for Atmospheric Composition: Satellite, aircraft, sensor web and ground-based observational methods and strategies.* Springer . ISBN: 978-0-387-30719-0. https://doi.org/10.1007/978-0-387-35848-2
11. Yu, Z., Hao, Z., Xu, Z., et al. (2024). Satellite observations of coastal upwelling in the northern Arafura Sea. *Journal of Oceanology and Limnology, 42*, 361–376. https://doi.org/10.1007/s00343-023-2390-4

Chapter 3
Aerial Imaging and Photogrammetry: Techniques and Applications

> *"In the age of digital information, the aerial photograph has become the symbol of both insight and intrusion."*
> — *Susan Sontag.*

Abstract This chapter delves into the principles, techniques, and applications of aerial photography and photogrammetry, two interrelated fields integral to remote sensing and geospatial analysis. Together, they enable the capture of high-resolution imagery from airborne platforms and the extraction of precise spatial information. The chapter begins by exploring the concept, history, and methodologies of aerial photography capturing images from elevated positions, typically via aircraft or unmanned aerial vehicles. It discusses aerial photography systems, cameras, and their applications, alongside their advantages, such as detailed imagery, flexible coverage, and specialized data acquisition, as well as their limitations. Subsequently, the chapter examines photogrammetry, a method for deriving accurate measurements and 3D spatial data from aerial photographs. It outlines the process of constructing photogrammetric approaches, along with the types, scales, and measurement techniques essential for producing reliable spatial information. The practical applications of aerial photography and photogrammetry are also explored across fields such as topographic mapping, land surveying, urban planning, environmental monitoring, and infrastructure development. Their role in disaster management, resource exploration, and archaeological research is emphasized, highlighting the value of aerial imagery in delivering up to date, detailed for informed decision-making. In conclusion, this chapter offers a comprehensive overview of aerial photography and photogrammetry, examining their technological advancements and multidisciplinary applications. By addressing these topics, it equips readers with a robust understanding of their capabilities and potential in remote sensing and geospatial analysis.

Keywords Aerial photography system · Aerial camera · Environmental monitoring · Photogrammetry · Scale determination

K. D. Malakar et al., *Geospatial Technologies in Coastal Ecologies Monitoring and Management*, Advances in Geographic Information Science,
https://doi.org/10.1007/978-3-031-92017-2_3

This chapter introduces the foundational principles and practical applications of aerial photography and photogrammetry, pivotal tools in remote sensing and geospatial analysis. It examines how high-resolution imagery captured from airborne platforms, such as aircraft and drones, is transformed into precise spatial data through photogrammetry. Readers will explore the history, methodologies, and technological advancements of these fields, alongside their multidisciplinary applications in topographic mapping, urban planning, environmental monitoring, and more. By providing insights into their techniques and value in decision-making, this chapter offers a concise yet comprehensive understanding of these transformative geospatial technologies.

Key Points of the Chapter
- Concept and history of aerial photography.
- Various components of aerial camera are discussed.
- Photogrammetry integrates principles and applications for precise spatial analysis.
- Its versatility addresses challenges in environmental monitoring, agriculture, and urban planning.

3.1 Aerial Photography

3.1.1 Concept of Aerial Photography

Aerial photography, the practice of capturing images from elevated perspectives using aircraft, drones, or satellites, has become an indispensable tool across various fields (Fig. 3.1). Originally developed for basic imaging, it now supports applications in cartography, geography, and urban planning by aiding in map creation, landscape analysis, and infrastructure monitoring. In agriculture, it helps optimize yields by tracking crop health, while in environmental science, it monitors ecological changes and conservation efforts. Beyond these, aerial photography enriches art and media, offering unique visual narratives. The rise of drones has made this technology more accessible, revolutionizing industries and fostering creativity.

The principles of aerial photography serve as fundamental guidelines for capturing images from elevated platforms such as aircraft, drones, or satellites. These principles are crucial for ensuring the quality, precision, and reliability of aerial imagery across diverse applications. Key principles include:

- **Overlap and coverage:** Overlapping images are captured to ensure complete area coverage and facilitate seamless mosaicking during post-processing.
- **Scale and resolution:** Scale links image distances to ground distances, while resolution determines detail; both depend on platform altitude and camera specifications.
- **Ground control points (GCPs):** GCPs with known coordinates align aerial images to real-world geography, ensuring accurate georeferencing and mapping.

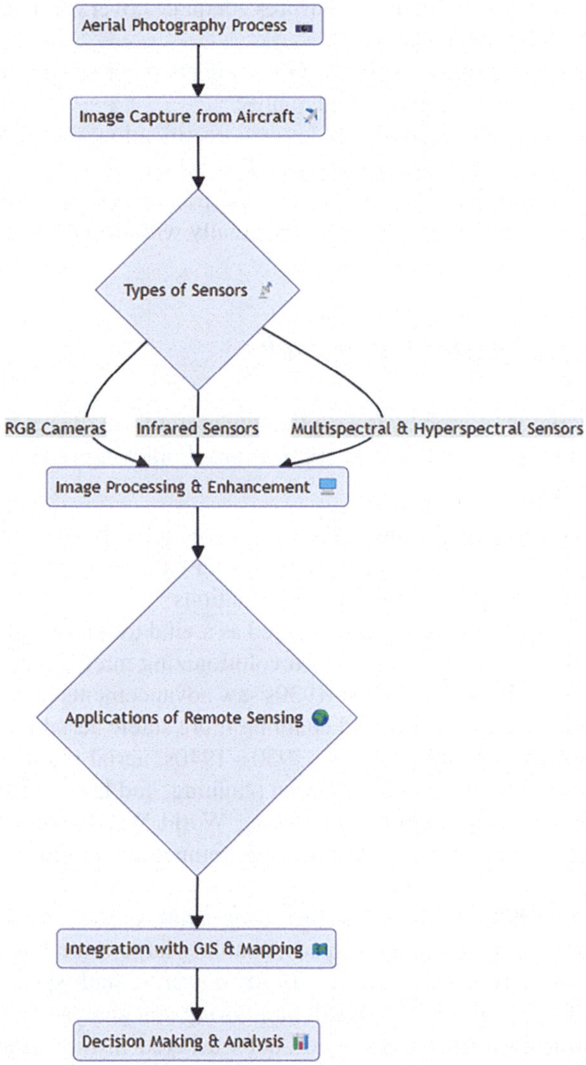

Fig. 3.1 Concept of aerial photography (prepared by the authors)

- **Sun angle and shadows:** Sun position influences image quality, with low angles creating long shadows and high angles reducing contrast and colour.
- **Weather conditions:** Clear skies optimize image sharpness, while fog or haze reduces visibility and quality.
- **Camera calibration:** Precise calibration corrects lens distortions, enabling accurate photogrammetric measurements.
- **Camera orientation:** Inertial navigation systems (INS) record camera orientation, ensuring precise positioning of images.

- **Flight planning:** Careful planning ensures adequate coverage, overlap, and navigation using GPS technology.
- **Ground sampling distance (GSD):** GSD reflects pixel spacing on the ground and determines image resolution and quality.
- **Image processing:** Processing includes orthorectification, mosaicking, and georeferencing to enhance image accuracy and usability.
- **Regulatory compliance:** Adherence to legal frameworks and airspace regulations is critical for aerial photography, especially with drones.

3.1.2 History of Aerial Photography

The history of aerial photography is a captivating narrative of over a century, defined by technological progress and diverse applications. Following is a brief overview of its key milestones:

- **Nineteenth-century beginnings:** French photographer Nadar captured the first aerial photograph in 1858 using a tethered hot air balloon, marking the early days of aerial imaging despite technological limitations.
- **World War I:** Aerial photography emerged as a vital tool for military reconnaissance, with aircraft-mounted cameras revolutionizing intelligence gathering.
- **Interwar Innovations:** The 1920s–1930s saw advancements in cameras and aircraft, improving image quality and enabling more stable aerial platforms.
- **Mapping and cartography:** By the 1930s–1940s, aerial photography became essential for topographic mapping, urban planning, and land-use studies.
- **Post-war civilian applications:** Following World War II, aerial photography expanded into fields like agriculture, environmental monitoring, and urban development.
- **Satellite era (1950s–1960s):** Satellite programmes, such as CORONA and Landsat, shifted aerial imaging to space, enabling large-scale Earth observation.
- Remote sensing advances: From the 1970s onwards, multispectral and hyperspectral imaging revolutionized Earth monitoring and analysis.
- **Digital revolution and drones:** The 2000s ushered in high-resolution digital cameras and drones, democratizing aerial photography and broadening its applications.
- **Art and recreation:** Aerial photography has gained artistic significance, with drone-captured images offering unique perspectives for media, documentaries, and social sharing.

The history of aerial photography highlights its transformation from a military and cartographic instrument into a versatile technology with applications spanning agriculture, environmental science, urban planning, and art. Ongoing advancements in hardware and software continue to redefine its potential, offering innovative ways to explore and comprehend our world.

3.1.3 *Classification of Aerial Photography*

Aerial photography can be categorized based on various criteria, such as the platform used, camera orientation, altitude, purpose, and spectrum type. Categories include:

- **Platform type**
  - (a) Manned aerial photography: Captured from aircraft, helicopters, or balloons, this traditional method has long been used for mapping and reconnaissance.
  - (b) Drone photography: UAVs or drones provide flexible, cost-effective aerial imaging for diverse applications, including film-making, agriculture, and environmental studies.
- **Altitude**
  - (a) Low-level aerial photography: Captured below 500 feet, this method focuses on specific objects or areas, often using drones.
  - (b) High-level aerial photography: Taken above 500 feet, this type is used for mapping, land surveys, and large-area coverage.
- **Temporal resolution**
  - (a) Single-time aerial photography: Captured during a single flight, this approach provides a snapshot of the area at a specific moment, often used in land surveys.
  - (b) Repeat or time-series aerial photography: Involves multiple flights over time to track landscape changes, useful for applications such as environmental monitoring, disaster assessment, and construction progress.
- **Purpose and application**
  - (a) Oblique aerial photography: Captured at an angle, this method provides a three-dimensional perspective of the subject, frequently applied in artistic photography, urban planning, and real estate.
  - (b) Vertical aerial photography: Images are taken directly downwards, offering top-down views of the terrain, commonly used in cartography, land-use planning, and environmental monitoring.
- **Imaging technology**
  - (a) Analog aerial photography: Traditionally used film-based cameras for aerial imaging, though still employed in specialized applications.
  - (b) Digital aerial photography: Predominantly used today, digital cameras provide higher resolution, faster data transfer, and more efficient post-processing.

- **Specialized techniques**
  (a) Infrared aerial photography: Utilizes infrared-sensitive cameras to capture images outside the visible spectrum, aiding in applications like crop health monitoring and detecting vegetation stress in environmental studies.
  (b) Multispectral and hyperspectral aerial photography: Involves capturing images across multiple spectral bands, allowing for detailed analysis of features such as soil composition, vegetation health, and mineral deposits.

3.1.4 An Aerial Photography System

An aerial photography system is a sophisticated arrangement of equipment, technology, and processes designed to capture high-resolution images of the surface from an elevated perspective, typically using aircraft, drones, or satellites. The system comprises various components and elements essential for its operation:

- **Platform:** The platform refers to the vehicle used for aerial photography, such as a manned aircraft, helicopter, drone (UAV), or satellite. Platform selection is based on factors like altitude requirements, payload capacity, and mission goals.
- **Camera:** The camera is the core imaging device, specially designed for high-resolution, distortion-free capture. Features may include gyroscopic stabilization, multiple lenses, and remote triggering mechanisms.
- **Lenses and filters:** Specific lenses (wide-angle, zoom, or specialized) and filters (polarizers or infrared) are used to enhance image quality, depending on the application.
- **Image sensors:** Digital image sensors (CCD or CMOS) in modern aerial cameras convert light into digital data, enabling high-resolution images and compatibility with various spectral bands for applications like multispectral imaging.
- **GPS and IMU:** Equipped with Global Positioning System (GPS) and inertial measurement units (IMU), the system record's precise location and orientation data, crucial for georeferencing and mapping accuracy.
- **Control systems:** Advanced control systems, such as autopilots in drones, manage flight stability and ensure precise image capture.
- **Communication equipment:** Real-time communication systems, especially in drones, allow operators to monitor and control platforms, receiving live video feeds.
- **Data storage:** High-capacity storage devices (e.g., SSDs or memory cards) store images and data, which are later processed and analyzed.
- **Ground control stations:** For drone systems, ground control stations (GCS) manage flight, camera settings, and telemetry data.
- **Software and image processing tools:** Post-processing tools enable image stitching, georeferencing, orthorectification, and analysis for diverse applications.
- **Safety and redundancy systems:** Safety features like fail-safes and redundancy systems ensure safe operations, especially in drones and manned aircraft.

- **Payloads:** Additional sensors or equipment, such as LiDAR scanners, thermal cameras, or multispectral sensors, may be included to meet specific mission objectives.

An aerial photography system (Fig. 3.2) integrates these components and technologies to capture accurate, high-resolution images for diverse applications, including cartography, environmental monitoring, scientific research, and artistic expression. The effective integration and maintenance of these elements are crucial for ensuring the system's reliability and optimal performance.

3.1.5 Application of Aerial Photography

Aerial photography has wide-ranging applications across various industries and fields, offering unique perspectives and valuable data. The following (Fig. 3.3) highlights its varied applications:

- **Mapping and cartography:** Aerial photography is vital for producing accurate maps, topographic surveys, and GIS applications, facilitating land-use planning, urban development, and infrastructure design.
- **Environmental monitoring**
  (a) Forest management: Helps evaluate forest health, monitor deforestation, and plan conservation strategies.
  (b) Wetland assessment: Tracks ecosystem changes, water quality, and wildlife habitats in wetland areas.
  (c) Disaster management: Provides critical data for damage assessment and emergency response during natural disasters like floods, hurricanes, and wildfires.

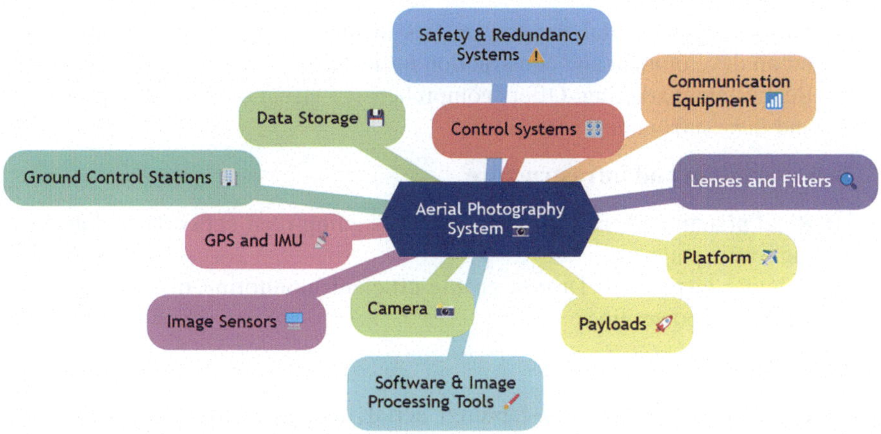

Fig. 3.2 An aerial photography system (prepared by the authors)

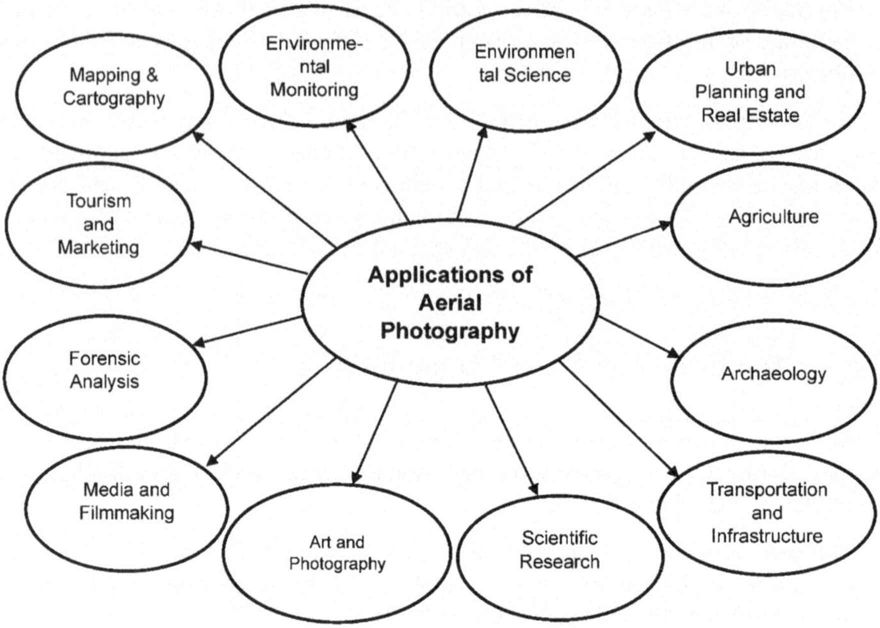

Fig. 3.3 Applications of aerial photography (prepared by the authors)

- **Environmental science**
  (a) Coastal monitoring: Tracks coastal erosion, shoreline changes, and impacts of sea-level rise.
  (b) Ecosystem research: Provides data for studying ecosystems, wildlife, and habitat conservation.
- **Urban planning and real estate**
  (a) Site selection: Aerial imagery assists in identifying suitable locations for urban development and construction projects.
  (b) Real estate marketing: Offers comprehensive visual overviews of properties and their surroundings for buyers.
- **Transportation and infrastructure**
  (a) Road planning: Aerial views support designing road networks and evaluating traffic patterns.
  (b) Pipeline inspection: Drones enable efficient monitoring of pipelines, railways, and power lines.
- **Agriculture**
  (a) Crop monitoring: Drone-captured aerial images assist in assessing crop health, detecting diseases, and optimizing irrigation and fertilization.

(b) Precision agriculture: Aerial data supports resource-efficient precision farming by enabling targeted interventions.

- **Archaeology:** Drones and aerial photography help uncover and document archaeological sites, including buried structures that are invisible from ground level.
- **Scientific research:** Facilitates studies in disciplines like geography, geology, oceanography, and climate science, contributing valuable data and insights.
- **Tourism and marketing:** Promotes tourist destinations by showcasing scenic landscapes, landmarks, and attractions through compelling aerial visuals.
- **Forensic analysis:** Aerial images are utilized in crime scene reconstruction and evidence collection during investigations.
- **Art and photography:** Aerial photography serves as a creative medium to capture breathtaking views of landscapes and urban settings.
- **Media and filmmaking:** Aerial shots enhance storytelling in movies, documentaries, and commercials by offering dramatic and unique perspectives.

3.1.6 Advantage and Disadvantage of Aerial Photography

Advantages of Aerial Photography

- Comprehensive coverage: It enables extensive coverage of large areas, making it essential for mapping, land-use planning, and environmental monitoring.
- Unique perspective: Aerial photography offers an elevated vantage point, uncovering patterns and details that are often obscured at ground level.
- High resolution: Modern aerial sensors deliver high-resolution imagery, ensuring precise and detailed data capture.
- Analytical utility: Aerial imagery provides critical datasets for analysis across disciplines like agriculture, environmental science, and urban development.
- Efficiency: Aerial surveys are time-efficient, making them ideal for urgent tasks such as disaster assessment or monitoring construction progress.
- Real-time monitoring: Drones equipped with live video capabilities enable real-time observation and immediate decision-making during operations.
- Enhanced safety: By removing the need for on-site access in remote or hazardous areas, aerial photography improves safety for surveyors and researchers.
- Affordability and accessibility: The advent of drones has made aerial photography more affordable and accessible, even for smaller-scale projects.

Disadvantages of Aerial Photography

- Weather dependence: Aerial photography is highly reliant on favourable weather conditions. Adverse elements like rain, fog, or strong winds can disrupt or delay operations.

- Cost: Manned aerial photography incurs significant costs due to the need for aircraft, specialized equipment, and skilled personnel. While drones are generally more affordable, initial investments can still be substantial.
- Regulations: The use of aerial platforms, particularly drones, is subject to complex regulatory requirements and permits, which can be time-intensive to secure.
- Altitude limitations: Aerial photography is constrained by altitude, with higher elevations often compromising image resolution and detail.
- Data processing: Post-capture processing, including tasks like georeferencing, orthorectification, and image stitching, can be labour-intensive and time-consuming.
- Technical expertise: Successful operation of aerial photography systems requires specialized training and expertise to ensure safety and effectiveness.
- Privacy concerns: Capturing aerial images may inadvertently infringe on privacy, raising ethical and legal issues, especially when individuals or private properties are photographed without consent.

3.2 Aerial Camera

3.2.1 Basic of Aerial Camera

Aerial cameras are specialized devices designed to capture high-quality images from elevated platforms like aircraft or drones. Unlike standard cameras, they are built to meet the unique demands of overhead imaging. Stability is a key feature, with gyroscopic stabilization systems minimizing vibrations to produce sharp images. These cameras are equipped with lenses optimized for aerial photography, such as wide-angle or zoom lenses, and allow for remote adjustments in real-time during flight.

Integrated GPS and inertial measurement units (IMU) record precise location and orientation, ensuring accurate georeferencing. Modern aerial cameras also utilize digital sensors, such as CCD or CMOS, for high-resolution imagery, seamless data transfer, and compatibility with specialized applications like multispectral imaging. These features collectively ensure the production of stable, clear, and georeferenced images, essential for mapping, analysis, and diverse applications.

3.2.2 Components of Aerial Camera

The components of an aerial camera are essential to its design and functionality, enabling it to capture high-quality images from elevated platforms such as aircraft or drones. These components can be categorized into various types:

- **Lens:** The lens is a pivotal component of an aerial camera, focusing light onto the image sensor. Specialized lenses, such as wide-angle or zoom lenses, enable precise capture of diverse scenes.
- **Image sensor:** Modern aerial cameras employ digital sensors like charge-coupled device (CCD) or complementary metal-oxide-semiconductor (CMOS), which convert light into digital data, offering high resolution and sensitivity for detailed imagery.
- **Shutter mechanism:** The shutter controls exposure time, ensuring images are neither overexposed nor underexposed, thus capturing scenes with optimal light balance.
- **Gyroscopic stabilization:** To counteract vibrations and movement from the platform, gyroscopic stabilization systems ensure sharp and stable image quality.
- **GPS and IMU:** Integrated Global Positioning System (GPS) and inertial measurement unit (IMU) record precise location and orientation during image capture, essential for georeferencing and mapping.
- **Camera control system:** Remote control systems allow operators to adjust settings like exposure, aperture, and focus in real time during flight.
- **Data storage:** High-capacity storage devices, such as SSDs or memory cards, store large volumes of high-resolution images during aerial missions.
- **Communication interfaces:** Equipped with Ethernet, Wi-Fi, or specialized connections, these interfaces facilitate data transfer and enable remote operation.
- **Power supply:** A robust power source, usually through batteries or external connections, ensures uninterrupted operation during missions.
- **Lens filters:** Filters, such as polarizers or infrared, optimize image quality and enable the capture of specific spectral data based on application needs.
- **Lens hoods and sun shields:** These accessories reduce lens flare and minimize the effects of harsh sunlight, improving image clarity.
- **Housing and enclosure:** Protective enclosures shield the camera from environmental elements like moisture, dust, and temperature fluctuations.
- **Mounting hardware:** Secure mounting hardware stabilizes the camera on platforms like drones or fixed-wing aircraft, ensuring consistent and reliable image capture.

These components work together to enable aerial cameras to capture high-resolution, georeferenced images for various applications, such as mapping, environmental monitoring, and agriculture. Their integration ensures the camera's functionality, allowing it to overcome the challenges of aerial photography while delivering accurate and reliable results.

3.2.3 Types of Aerial Camera

Aerial cameras are available in various types, each designed to meet specific applications and requirements, such as:

- **Metric cameras**
  (a) Large-format metric cameras: Primarily utilized in photogrammetry, these cameras are designed for applications such as mapping, surveying, and geospatial analysis. They produce high-resolution imagery with precise geometric accuracy, making them ideal for generating detailed maps and three-dimensional models.
- **Non-metric cameras**
  (a) Panoramic cameras: Panoramic cameras capture wide-angle or 360°views in a single shot, making them suitable for immersive photography, virtual tours, and surveillance applications.
  (b) Oblique cameras: These cameras capture images at angles, offering a three-dimensional perspective of the subject. They are commonly employed in urban planning, real estate, and architectural photography to creatively display buildings and landscapes.
- **Multispectral and hyperspectral cameras**
  (a) Multispectral cameras: Multispectral aerial cameras capture images across multiple spectral bands beyond the visible spectrum, such as infrared and ultraviolet. These cameras are essential for applications like crop health monitoring in agriculture and vegetation analysis in environmental studies.
  (b) Hyperspectral cameras: Hyperspectral cameras capture images across numerous narrows, contiguous spectral bands, enabling detailed spectral analysis. They are primarily used in mineral exploration, remote sensing, and vegetation classification.
- **Thermal infrared (IR) cameras**
  (a) Thermal IR cameras: Thermal infrared (IR) cameras detect heat emitted by objects, capturing thermal radiation. They are utilized in diverse applications, including building inspections, search and rescue operations, and environmental monitoring, such as tracking wildlife or assessing wildfires.
- **Digital cameras**
  (a) Consumer-grade digital cameras: When mounted on drones or aircraft, consumer-grade digital cameras are employed for various applications, including real estate photography, environmental monitoring, and artistic projects. These cameras are cost-effective and offer versatility for a wide range of uses.
- **LiDAR cameras**
  (a) LiDAR cameras: LiDAR (light detection and ranging) cameras integrate laser sensors with imaging technology to capture 3D point cloud data. These cameras are essential for applications such as terrain modelling, forestry management, and autonomous vehicle navigation.

- **Scientific cameras**

  (a) Scientific cameras: Designed for research applications, scientific cameras are used in fields like atmospheric studies, geological surveys, and environmental research. These cameras typically feature specialized sensors and advanced data collection capabilities tailored to meet the specific needs of scientific investigations.

3.3 Photogrammetry

3.3.1 Concept of Photogrammetry

Photogrammetry is a sophisticated technique that utilizes the principles of geometry and photography to derive accurate three-dimensional data from two-dimensional images. This process involves the precise measurement of objects, landscapes, and terrain features by analyzing photographs captured from various angles. Central to photogrammetry is the principle of triangulation, where overlapping images are used to calculate spatial positions of points. By identifying corresponding points across these images and measuring their relative positions, photogrammetry enables the creation of detailed 3D models, maps, and measurements of real-world environments. The technique has widespread applications in fields such as cartography, topographic mapping, urban planning, architecture, archaeology, environmental monitoring, and 3D modelling for industries like film and gaming. With advancements in technology, including the use of drones and high-resolution cameras, photogrammetry has become increasingly accessible and precise, transforming industries reliant on accurate spatial data. Photogrammetry is defined by several key characteristics and attributes, which shape its nature and application (Fig. 3.4). These include:

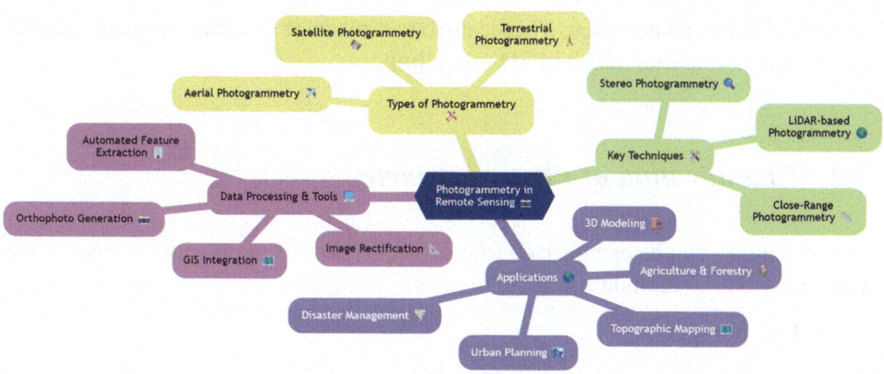

Fig. 3.4 Concept of photogrammetry in remote sensing system (prepared by the authors)

- **Triangulation principle:** The core of photogrammetry is triangulation, where overlapping images are used to calculate the spatial positions of points through geometric calculations.
- **Non-contact measurement:** Photogrammetry is a non-contact technique, allowing analysis of objects or terrain without physical interaction, making it ideal for sensitive or hard-to-reach subjects.
- **Precision and accuracy:** Known for high precision, photogrammetry can achieve sub-centimetre accuracy depending on the equipment and methods used.
- **Use of imagery:** This technique relies on analyzing images captured from various platforms, including satellites, drones, and ground-based cameras, from different perspectives.
- **Versatility:** It is applicable across diverse fields, such as cartography, engineering, and environmental science, enabling tasks like map creation, land monitoring, and 3D reconstruction.
- **Remote sensing:** Photogrammetry is often combined with remote sensing technologies like multispectral imaging to capture data beyond the visible spectrum.
- **Georeferencing:** For integration with GIS, photogrammetry data must be georeferenced, assigning real-world geographic coordinates to the measured points.
- **Digital transformation:** Digital advancements have enhanced photogrammetry, enabling faster, more accurate processing with digital cameras and software.
- **Aerial and terrestrial applications:** It can be conducted from both aerial (e.g., drones, aircraft) and terrestrial (ground-based) perspectives, expanding its usability.
- **Data integration:** Photogrammetry frequently integrates data from GPS, LiDAR, and multispectral sensors to enhance spatial accuracy and analysis.
- **Data visualization:** The results are often presented as 3D models, point clouds, or DEMs, simplifying the interpretation of complex spatial information.
- **Efficiency and cost-effectiveness:** It offers a cost-effective method for capturing detailed spatial data over large areas, reducing the need for extensive fieldwork.

In summary, photogrammetry is defined by its non-contact, image-based measurement method, high precision, versatility across various industries, and integration with digital technologies. These attributes make it essential for applications requiring precise spatial data and 3D modelling.

3.3.2 Classification of Photogrammetry

Photogrammetry can be categorized into several types based on various criteria and applications. These include:

- **Based on perspective**

  (a) Aerial photogrammetry: Aerial photogrammetry involves capturing images from elevated platforms, such as aircraft or drones. It is commonly applied in mapping, land surveying, and environmental monitoring.

(b) Terrestrial photogrammetry: Terrestrial photogrammetry is performed from ground level and is employed for tasks such as architectural documentation, crime scene reconstruction, and cultural heritage preservation.

- **Based on data source**

(a) Close-range photogrammetry: Close-range photogrammetry captures objects or scenes from a short distance and is used in applications such as industrial measurement, product design, and forensics.
(b) Aerial photogrammetry: Aerial photogrammetry captures images from elevated positions, typically for large-scale mapping and environmental monitoring.

- **Based on perspective geometry**

(a) Metric photogrammetry: Metric photogrammetry is focused on achieving precise measurements and generating accurate 3D data. It is commonly applied in cartography, land surveying, and engineering.
(b) Non-metric (oblique) photogrammetry: Non-metric photogrammetry is used for applications that do not require exact measurements but benefit from a 3D perspective. It is widely utilized in fields such as architecture, archaeology, and urban planning.

- **Based on image acquisition**

(a) Single-frame photogrammetry: Single-frame photogrammetry utilizes individual images to generate 3D models or measurements, ideal for applications where a single image provides adequate information.
(b) Multi-frame (stereophotogrammetry): Multi-frame photogrammetry captures multiple overlapping images from different angles to facilitate 3D reconstruction. Stereophotogrammetry, a key method in this category, enhances spatial accuracy.

- **Based on image orientation**

(a) Direct photogrammetry: Direct photogrammetry calculates the camera orientation and the relative positions of features directly from images, often utilizing ground control points for accuracy.
(b) Analytical photogrammetry: Analytical photogrammetry uses mathematical models and instruments to determine camera orientation and 3D coordinates, providing high precision for mapping and engineering applications.

- **Based on application**

(a) Topographic photogrammetry: Topographic photogrammetry is used to generate topographic maps, assist in land-use planning, and model terrain features.
(b) Engineering photogrammetry: Engineering photogrammetry supports infrastructure design, construction monitoring, and detailed building documentation.

(c) Forensic photogrammetry: Forensic photogrammetry aids in crime scene reconstruction, accident analysis, and evidence preservation.
(d) Cultural heritage photogrammetry: Cultural heritage photogrammetry is employed to document and preserve historical sites, artifacts, and monuments.

3.3.3 Scales and Measurements of Photogrammetry

In photogrammetry, precision and accuracy are critical to ensuring reliable results. The integration of scale factors and measurements guarantees that photogrammetric data is both visually informative and mathematically sound. These principles form the foundation for the application of photogrammetry across diverse fields, including engineering, construction, environmental monitoring, and cultural heritage preservation. The following concepts are central to this process:

Scales in Photogrammetry

- Scale factor: The scale factor represents the relationship between real-world dimensions and their corresponding image dimensions, often expressed as a ratio (e.g., 1:10,000). It is crucial for accurate measurements in photogrammetry.
- Large scale vs. small scale: Large-scale photogrammetry captures detailed measurements of smaller areas, ideal for urban planning and engineering. Small-scale photogrammetry covers larger areas with less detail, suited for regional mapping and environmental monitoring.
- Orthophotos: Orthophotos are corrected aerial or satellite images that remove distortions and maintain a consistent scale, making them valuable for mapping and integrating with other geospatial data.

Example
Question: Focal length of a camera is 6 inches and flying height above the flat terrain is 24,000 ft. Now calculate the scale of the photograph.

Calculation:
We know, scale of a photograph (R.F.) = *Focal length / Flying height.*
Here, Focal length = 6 inches
Flying height = 24, 000 ft or (24, 000 × 12) 2, 88, 000 inches [1 ft =12 inches]

$$\text{Therefore, R.F.} = 6/2{,}88{,}000 = 1/48{,}000$$

So, scale of the photograph (R.F.) = 1:48000

Measurements in Photogrammetry

- Length measurements: Photogrammetry enables precise distance measurements between points on the ground or objects in images, using the scale factor to convert to real-world units (e.g., meters or feet).

- Area measurements: By analyzing object dimensions and applying the scale factor, photogrammetry provides accurate area calculations, essential for applications like land-use planning and agriculture.
- Height and elevation: Photogrammetry can determine the height or elevation of objects or terrain features through multiple images from different angles, enabling accurate height measurements for buildings, trees, and terrain changes.
- Volume calculations: Photogrammetry enables volume calculations, such as for stockpiles or landfills, by comparing images over time or using stereo pairs to measure volume changes.
- 3D models: Photogrammetry creates 3D models of objects, structures, or landscapes, capturing spatial information in three dimensions for detailed visualization and analysis.
- Georeferencing: Georeferencing assigns geographic coordinates (latitude and longitude) to points, images, or maps, ensuring integration with other geospatial data in a common coordinate system.
- Control points: Control points are accurately measured reference points on the ground, used to establish scale, orientation, and georeferencing in photogrammetric projects.
- Ground control points (GCPs): Ground control points are specific reference points placed on the ground and identified in images, critical for accurate georeferencing by linking image coordinates to real-world locations.
- Quality assessment: Quality assessment in photogrammetry involves evaluating measurement accuracy, precision of control points, scale factor reliability, and identifying potential errors or distortions.

3.3.4 Applications of Photogrammetry

Photogrammetry is widely applicable across numerous industries and fields, owing to its capacity to generate accurate 3D models and measurements from images. The following outlines its key applications:

- **Cartography and mapping:** Essential for creating topographic maps, digital elevation models (DEMs), and orthophotos used in geographic information systems (GIS).
- **Wildlife and environmental research:** Enables wildlife tracking and habitat condition assessment for conservation and ecological studies.
- **Environmental monitoring:** Helps monitor changes like coastline erosion, land subsidence, and forest health, supporting conservation efforts.
- **Mining and quarrying:** Facilitates volume calculations, slope stability assessments, and excavation planning to enhance operational safety and efficiency.
- **Disaster management:** Vital for rapid assessment and mapping of disaster zones, aiding response and recovery efforts.
- **Mining exploration:** Assists in geological analysis, identifying mineral deposits, and planning drilling operations for resource extraction.

- **Urban planning and development:** Used to assess land use, plan infrastructure, and simulate urban growth for informed city planning.
- **Architecture and building documentation:** Used by architects to document buildings and create 3D models for restoration and renovation projects.
- **Archaeology:** Supports the documentation and reconstruction of archaeological sites and artifacts, aiding preservation and research.
- **Forestry:** Aids in assessing tree growth, timber volumes, and forest health for sustainable forest management.
- **Agriculture:** Assists in crop monitoring, soil analysis, and yield prediction through data captured by drones with photogrammetric sensors.
- **Civil engineering:** Supports site planning, road design, and infrastructure development by providing accurate 3D models and measurements, reducing time and costs in surveying.
- **Film and video production:** Creates realistic 3D models and environments for visual storytelling and virtual worlds.
- **Search and rescue:** Provides detailed terrain models to improve search strategies in challenging environments.
- **Infrastructure inspection:** Used to inspect and assess infrastructure, facilitating maintenance and repair planning.

3.4 Conclusion and Endnotes

This chapter explores the principles, techniques, and applications of aerial photography and photogrammetry, essential tools in remote sensing and geospatial analysis. It begins with an overview of aerial photography, including its history, methods, systems, and applications, highlighting its advantages like detailed imagery and flexible coverage, as well as its limitations. The discussion then shifts to photogrammetry, focusing on its role in extracting accurate measurements and 3D spatial data from aerial photographs. Key topics include photogrammetric approaches, scales, and measurement techniques, which are crucial for reliable spatial information. Applications in fields such as topographic mapping, urban planning, environmental monitoring, and disaster management are also discussed, emphasizing their value in providing detailed, up-to-date data for decision-making. In summary, the chapter provides a concise overview of aerial photography and photogrammetry, their technological advancements, and their diverse applications, offering a solid foundation in these key areas of remote sensing and geospatial analysis.

Endnotes

- Photogrammetry combines principles and applications, offering a comprehensive technique for precise spatial analysis.
- Its adaptability to digital technologies, such as drones and cameras, ensures continued relevance in modern applications.
- Versatility across fields like environmental monitoring, agriculture, heritage preservation, and urban planning highlights its significance in solving complex challenges.

- Control points and georeferencing enhance accuracy, making photogrammetry a reliable tool for various industries.
- Transformative applications in filmmaking, mining, and conservation drive innovation across diverse sectors.

Further Readings

1. Aber, J. S., Marzolff, I., Ries, J., & Aber, S. E. W. (2019). *Small-format aerial photography and UAS imagery: Principles, techniques and geoscience applications*. Academic Press (Elsevier) ISBN: 9780128129425.
2. DeWitt, B. A., & Wolf, P. R. (2000). *Elements of photogrammetry with applications in GIS*. McGraw-Hill Education ISBN: 9780071761123.
3. Falkner, E. (1995). *Aerial mapping: Methods and applications*. CRC Press, (Taylor and Francis Group) ISBN: 9781566701037.
4. Linder, W. (2016). *Digital photogrammetry: A practical course*. Springer . ISBN: 978-3-662-50462-8. https://doi.org/10.1007/978-3-662-50463-5
5. Morgan, D., & Falkner, E. (2001). *Aerial mapping: Methods and applications* (2nd ed.). CRC Press, (Taylor and Francis Group) ISBN 9780367578725.
6. Paine, D. P., & Kiser, J. D. (2012). *Aerial photography and image interpretation*. Wiley ISBN:9780470879382. https://doi.org/10.1002/9781118110997
7. Sajinkumar, K. S., & Oommen, T. (2018). Photogrammetry. In P. T. Bobrowsky & B. Marker (Eds.), *Encyclopedia of engineering geology. Encyclopedia of earth sciences series*. Springer. https://doi.org/10.1007/978-3-319-73568-9_221
8. Schmidt, D., & Kühn, F. (2007). Aerial photography. In *Environmental geology*. Springer. https://doi.org/10.1007/978-3-540-74671-3_3
9. Schmidt, D., Herms, P., Steinbach, A., & Kühn, F. (2007). Photogrammetry. In *Environmental geology*. Springer. https://doi.org/10.1007/978-3-540-74671-3_4
10. Trorey, L. G. (2013). *Handbook of aerial mapping and photogrammetry*. Cambridge University Press ISBN: 9781107623309.

Chapter 4
Geodesy and Air Photo Interpretation: Principles and Tools for Spatial Analysis

> *"To map is to create a model of reality; to interpret that map is to unveil the hidden patterns that shape our understanding of space." — J.B. Harley.*

Abstract The chapter "Geodesy and Air Photo Interpretation" provides a thorough introduction to the ideas and applications of geodesy and air photo interpretation. Geodesy is concerned with the measurement and comprehension of the Earth's shape, size, and gravitational field, whereas air photo interpretation is concerned with the analysis and extraction of information from aerial photos. The chapter begins with a discussion of geodesy and its significance in understanding the Earth's geodetic reference system. There is information about the ellipsoid, geoid, datum, coordinate system, reference, great circle, spherical angle, spherical triangle, spherical trigonometry, world geodetic system, and Everest spheroid. The next part explores the principles and practices of air photo interpretation. It demonstrates how aerial photographs captured by planes or unmanned aerial vehicles can be utilized to collect spatial data. In addition, it dives into the concept, types, history, and various stereoscopic equipment. The chapter also looks at the various applications of geodesy and air photo interpretation. It highlights the value of topographic mapping, land surveying, and cartography. It also emphasizes their value in environmental monitoring, natural resource management, and other sectors. The importance of geodesy and air photo interpretation in comprehending landforms, plant patterns, land-use changes, and other spatial characteristics is emphasized in the chapter. The chapter finishes by emphasizing the significance of precision, accuracy, and critical thinking in geodesy and air photo interpretation.

Keywords Air photo interpretation · Coordinate system · World geodetic system · Geodesy · Stereoscopic

K. D. Malakar et al., *Geospatial Technologies in Coastal Ecologies Monitoring and Management*, Advances in Geographic Information Science,
https://doi.org/10.1007/978-3-031-92017-2_4

This chapter provides a comprehensive introduction to two foundational geospatial disciplines, such as geodesy and air photo interpretation. Geodesy, focused on understanding Earth's shape, size, and gravitational field, is explored in depth, covering essential concepts such as the ellipsoid, datum, coordinate systems, and spherical trigonometry. The chapter also delves into the principles and practices of air photo interpretation, examining how aerial photographs can be analyzed to extract spatial data. With applications ranging from topographic mapping to environmental monitoring, the chapter highlights the critical role of precision and accuracy in both fields for effective geospatial analysis.

Key Points of the Chapter

- Geodesy enables precise spatial referencing by measuring Earth's shape, size, and gravity field.
- Covers key geodetic concepts: ellipsoid, datum, coordinate systems, spherical geometry, and others.
- Describe air photo interpretation extracts spatial data from aerial images, supporting mapping and surveying.
- Explains how stereoscopic techniques enhance aerial image analysis for accurate spatial data extraction.
- Highlights how geodesy and air photo interpretation support mapping, environmental monitoring, and resource management.

4.1 Geodesy

4.1.1 Concept of Geodesy

Geodesy is a scientific discipline dedicated to measuring and comprehending the Earth's shape, size, and gravitational field. It serves as a cornerstone for various domains, including surveying, cartography, navigation, and geophysics. By integrating mathematical, physical, and observational methods, geodesy enables precise determination of point positions on the Earth's surface and variations in its gravitational field. A central concept in geodesy is the geoid, a theoretical surface that represents the gravitational field and corresponds to mean sea level. Geodesists extensively utilize satellite technologies, such as the Global Positioning System (GPS), to achieve highly accurate position measurements. Geodetic data underpin a wide range of activities, from infrastructure development and natural resource management to forecasting sea-level changes and analyzing tectonic movements. In an increasingly interconnected world, geodesy continues to gain prominence as a foundational geospatial science, driving numerous applications that influence daily life (Fig. 4.1). The inherently interdisciplinary and multifaceted nature of geodesy encompasses several critical aspects:

- **Measurement and observation:** Geodesy focuses on precise measurement and observation of the Earth's surface, encompassing the determination of coordi-

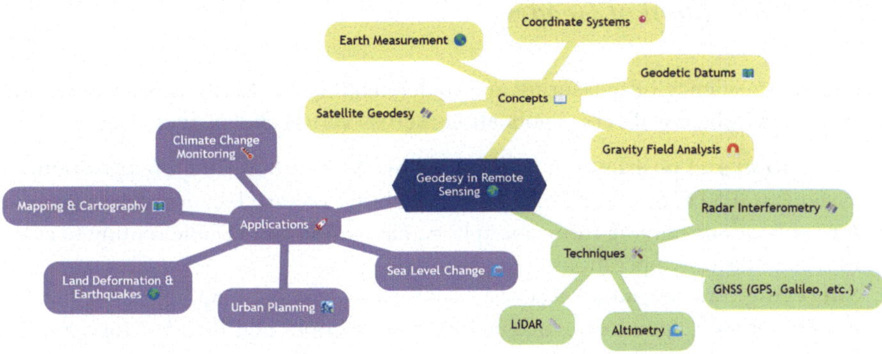

Fig. 4.1 Concept of geodesy in remote sensing system (prepared by the authors)

nates, distances, angles, and elevations. Advanced tools and techniques, such as satellites and geodetic instruments, are often employed to achieve this precision.

- **Earth's shape and size:** A central focus of geodesy is understanding the shape and size, including the determination of the geoid, which represents mean sea level, and identifying deviations from a perfect sphere, such as the flattening at the poles.
- **Gravitational field:** Geodesy explores the gravitational field by measuring its variations and anomalies. These measurements offer insights into the mass distribution and support geophysical research.
- **Global positioning:** Technologies like GPS have revolutionized geodesy, enabling precise location determination on the surface. These advancements support applications in navigation, mapping, and surveying.
- **Reference Systems:** The discipline establishes reference systems and datums that provide the foundation for all spatial measurements and mapping, ensuring consistency and accuracy in geospatial data.
- **Interdisciplinary nature:** Geodesy integrates insights and methodologies from various disciplines, including astronomy, geophysics, mathematics, and computer science, to develop advanced models, methods, and tools for Earth measurement.
- **Dynamic Earth:** Geodesy accounts for the Earth's dynamic nature, incorporating phenomena like plate tectonics, sea-level changes, and geodynamic processes, which influence the accuracy of measurements over time.
- **Applications:** Beyond scientific research, geodesy has wide-ranging practical applications in civil engineering, construction, land surveying, cartography, and environmental monitoring. It is essential for infrastructure development and disaster management.

In summary, geodesy is a multidisciplinary science integrating mathematics, physics, and technology to study the Earth's shape, size, and gravitational field, with wide-ranging applications that significantly influence modern society.

4.1.2 Application of Geodesy

Geodesy, the science of measuring and understanding the Earth's shape, size, and gravitational field, has diverse applications across various fields, including:

- **Navigation and positioning:** Geodesy forms the backbone of global positioning systems (GPS) and satellite navigation, enabling precise location determination for applications ranging from smartphone navigation and vehicle routing to aviation and maritime operations.
- **Surveying and mapping:** Accurate land surveying and mapping, essential for urban planning, land-use management, construction, and detailed topographic representation, rely on geodesy.
- **Satellite orbit determination:** Geodesy ensures the precision of satellite orbits, vital for communication, Earth observation, and weather forecasting.
- **Astronomy and space exploration:** Geodesy enhances celestial object positioning in astronomy and assists spacecraft navigation and planetary landing site selection in space exploration.
- **Environmental monitoring:** Geodesy tracks icebergs, urban land subsidence, and geological or environmental changes, aiding in environmental conservation and monitoring efforts.
- **Climate change research:** Geodetic techniques provide critical data on polar ice cap and glacier dynamics, contributing to studies on global sea-level rise and climate change impacts.
- **Natural resource management:** In agriculture, forestry, and mining, geodesy supports sustainable management by monitoring land subsidence, soil erosion, and land-use changes.
- **Sea-level monitoring:** Geodesy is crucial for tracking sea-level changes and assessing their impacts on coastal areas, aiding climate adaptation strategies for coastal communities.
- **Geophysical studies:** Geodesy contributes to understanding Earth's internal structure and gravitational variations, providing valuable insights for geophysical research.
- **Flood modelling:** Elevation data from geodesy is indispensable for accurate floodplain mapping, disaster management, and early warning systems.
- **Infrastructure development:** It ensures precision in constructing and maintaining infrastructure such as bridges, roads, dams, and tunnels.

In essence, geodesy encompasses a wide range of applications, enhancing our understanding of the Earth and its dynamic processes while serving as a cornerstone for various scientific, engineering, and societal advancements.

4.2 Some Aspects of Geodesy

4.2.1 Ellipsoid

In geodesy, an ellipsoid, also referred to as a reference ellipsoid or spheroid, is a mathematical model used to approximate the Earth's shape. The Earth is not a perfect sphere but an oblate spheroid, characterized by a slight flattening at the poles and a bulging at the equator due to its rotation (Fig. 4.2). To account for this deviation and facilitate geodetic calculations, geodesists employ ellipsoids. The following are key aspects of ellipsoids in geodesy:

- **Mathematical representation:** An ellipsoid is mathematically defined by three key parameters: the semi-major axis (a), the semi-minor axis (b), and the flattening (f). The semi-major axis represents the distance from the Earth's centre to the equator, while the semi-minor axis is the distance from the Earth's centre to a point on the surface through the poles. Flattening quantifies the deviation of the ellipsoid from a perfect sphere.
- **Variety of ellipsoids:** Different ellipsoids are used in geodesy to best approximate the Earth's shape in specific regions or countries. These regional ellipsoids provide more accurate representations of local curvature variations.
- **Reference Ellipsoids:** Standard reference ellipsoids, such as the World Geodetic System (WGS84) and the Geodetic Reference System (GRS80), are widely utilized for global geodetic applications, including GPS and precise geospatial measurements.
- **Geodetic coordinates:** Ellipsoids are used to define geodetic coordinates—latitude, longitude, and height (or elevation) above the ellipsoid's surface—critical for mapping, navigation, and surveying.
- **Geodetic datums:** A geodetic datum combines an ellipsoid with a specific reference point on the surface, serving as the basis for spatial data. Common datums like WGS84 and NAD83 establish consistent reference frames for global positioning and mapping.

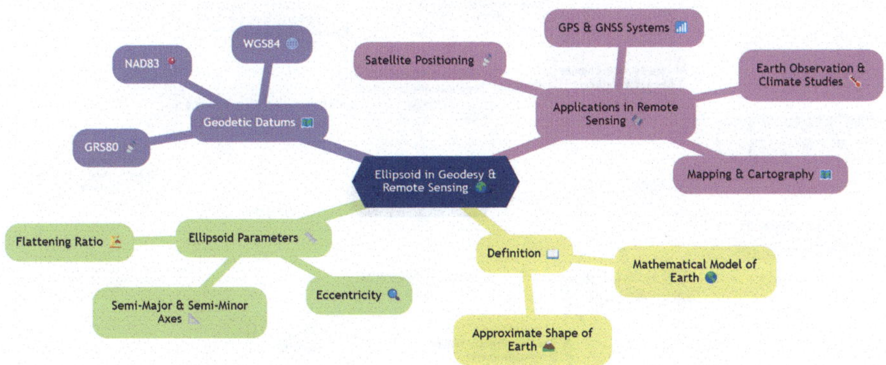

Fig. 4.2 Concept of ellipsoid in geodesy and remote sensing (prepared by the authors)

- **Geoid relationship:** The geoid represents mean sea level, shaped by the Earth's gravitational field. Geodesy relates the ellipsoid to the geoid to account for gravitational variations, enabling precise positioning and height measurements.
- **Applications:** Ellipsoids are foundational to cartography, GPS navigation, land surveying, and geospatial mapping. They ensure accurate representation of coordinates and distances on maps and in navigation systems, supporting reliable location-based services.

In summary, ellipsoids are fundamental to geodesy, serving as mathematical models for approximating the Earth's shape and enabling precise geospatial measurements and mapping at both global and local scales. They underpin modern navigation systems and form the basis of contemporary geodetic practices.

4.2.2 Geoid

In geodesy, the geoid is a fundamental reference surface that represents mean sea level and serves as the basis for measuring elevations on the surface. It is an irregular, hypothetical surface corresponding to the equipotential gravity field, reflecting variations in gravitational pull due to the uneven distribution of mass within the planet (Fig. 4.3). The following are key aspects of the geoid in geodesy:

- **Representation of mean sea level:** The geoid serves as a reference surface representing the average sea level around the globe, influenced by Earth's gravity. It is essential for defining elevations and measuring heights relative to or above/below mean sea level.
- **Irregular shape:** Unlike a simple geometric shape such as an ellipsoid or sphere, the geoid is irregular and undulating. Its elevation varies due to the uneven distribution of Earth's mass, resulting in a gravitational field that is not uniform.

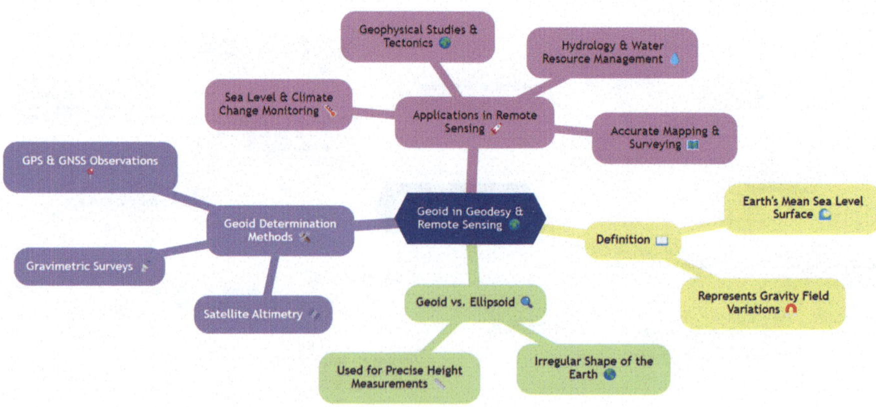

Fig. 4.3 Concept of geoid in geodesy and remote sensing (prepared by the authors)

- **Geodetic datums:** In geodesy, the geoid is used in conjunction with an ellipsoid to establish geodetic datums, which serve as reference systems for geospatial measurements. A datum defines the relationship between the ellipsoid and the geoid at a specific reference point.
- **Height above geoid:** When determining elevations or heights above the geoid, geodesists account for gravitational field variations. This ensures precise elevation measurements relative to mean sea level, crucial for applications in construction, surveying, and mapping.
- **Geoid models:** Geoid models are mathematical representations of the geoid's shape and undulations. These models are continually refined through satellite data from missions such as GRACE (Gravity Recovery and Climate Experiment) and GOCE (Gravity Field and Steady-State Ocean Circulation Explorer) to provide accurate geoid information.
- **Practical applications:** The geoid plays a vital role in a range of applications, including civil engineering, aviation, sea-level monitoring, and the study of ocean currents and tides. Accurate height measurements, facilitated by the geoid, are crucial for tasks such as flood risk assessment and infrastructure design.
- **Geoid separation:** "Geoid separation" or "geoid height" refers to the vertical offset between the geoid and the ellipsoid at a given location. This separation value varies across the Earth's surface.

In summary, the geoid is a complex, irregular reference surface that represents mean sea level and incorporates variations in Earth's gravity field. It is essential in geodesy for precise elevation measurements, offering a consistent reference for geospatial data and supporting numerous applications that depend on accurate elevation information.

4.2.3 Datum

In geodesy, a datum is a critical reference framework used to define positions and measurements on the Earth's surface. It provides the foundation for geospatial data, ensuring that coordinates, distances, and heights are accurately referenced and consistent (Fig. 4.4). The following are key concepts about datums in geodesy:

- **Concept:** A geodetic datum consists of a specific ellipsoid (a mathematical model of the Earth's shape) and a reference point on the Earth's surface. This reference point establishes the origin and orientation for the coordinate system.
- **Coordinate system:** A datum includes a coordinate system that typically specifies latitude, longitude, and often elevation (height above or below a reference surface, such as the geoid). These coordinates are used to define positions on the Earth's surface.
- **Horizontal datum:** The horizontal datum defines latitude and longitude coordinates, playing a crucial role in mapping, navigation, and spatial referencing.

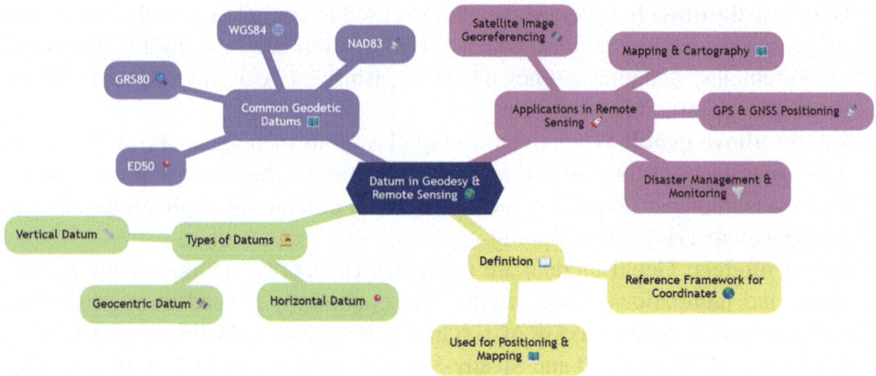

Fig. 4.4 Concept of datum in geodesy and remote sensing (prepared by the authors)

Common examples include NAD83 (North American Datum 1983) and WGS84 (World Geodetic System 1984).

- **Vertical datum:** The vertical datum defines elevations or heights relative to a reference surface, such as the geoid or mean sea level. It is essential for precise elevation measurements in applications like construction and flood modelling.
- **Transformation parameters:** When converting coordinates between different datums, transformation parameters are used. These parameters account for variations in ellipsoid size, orientation, and the reference point's position.
- **Local datums:** In addition to global or national datums, local datums are established for smaller regions or specific projects requiring high accuracy. These datums may consider local geoid variations or other regional factors.
- **Datum evolution:** Datums evolve over time as geodetic techniques advance and our understanding of Earth's shape improves. Regular updates are made to maintain accuracy and ensure consistency with modern measurements.
- **Geodetic infrastructure:** Datums are integral to a country's geodetic infrastructure, forming the foundation for mapping, surveying, and navigation systems. They enable the integration of diverse geospatial data and ensure the consistency and reliability of geospatial information.
- **Global datums:** Global datums, such as WGS84, are widely used in global positioning systems (GPS) and international applications. They provide a common reference for global navigation and geospatial data exchange.

In summary, a datum in geodesy is a reference framework that encompasses an ellipsoid, a reference point, and a coordinate system. It is crucial for accurately defining positions and measurements on the surface, ensuring consistency in geospatial data across a wide range of applications, including mapping, navigation, construction, and environmental monitoring.

4.2.4 Coordinate System

In geodesy, a coordinate system is a fundamental framework used to represent and define locations on the surface. It offers a standardized method for specifying geographic positions, which is essential for navigation, mapping, surveying, and various other geospatial applications (Fig. 4.5). The following are key points regarding coordinate systems in geodesy:

- **Latitude and longitude:** The geographic coordinate system, based on latitude and longitude, is the most widely used in geodesy. Latitude measures the north–south position, ranging from −90° (South Pole) to +90° (North Pole), while longitude measures the east–west position, ranging from −180° to +180°, with the Prime Meridian at 0°.
- **Degrees, minutes, and seconds:** Latitude and longitude are often expressed in degrees, minutes, and seconds (DMS) for ease of communication. For example, the coordinates of New York City are approximately 40° 42′ N and 74° 0′ W.
- **Decimal degrees:** To facilitate calculations and data exchange, coordinates are frequently represented in decimal degrees (DD), where fractions of a degree are expressed as decimals. For example, the Eiffel Tower's coordinates are approximately 48.858844° N, 2.294351° E.
- **Coordinate grids:** Alternative coordinate systems, such as the Universal Transverse Mercator (UTM) and State Plane Coordinate Systems (SPCS), use grid-based systems, typically in metres or feet. These systems divide the Earth's surface into zones, providing precise local measurements.
- **Height or elevation:** In addition to horizontal coordinates, a complete geodetic coordinate system includes a vertical component to define height or elevation relative to a reference surface, such as the geoid or mean sea level.
- **Datum and reference frame:** A coordinate system is defined by its associated datum and reference frame. The datum specifies the ellipsoid (the model of

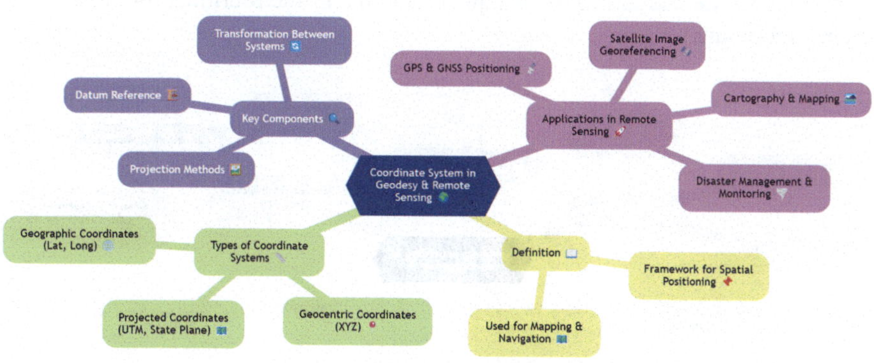

Fig. 4.5 Concept of coordinate systems in geodesy and remote sensing (prepared by the authors)

Earth's shape) and the reference point, with different datums used for different regions or applications.
- **Geodetic transformations:** When working with data from different coordinate systems or datums, geodetic transformations are necessary to accurately convert coordinates. These transformations account for variations in the reference ellipsoid, orientation, and reference point position.
- **Global Positioning Systems (GPS):** GPS technology utilizes coordinate systems to determine precise locations on Earth's surface, calculating latitude, longitude, and elevation by triangulating signals from multiple satellites.
- **Cartesian coordinates:** For specialized applications, such as engineering and construction, Cartesian coordinate systems (x, y, z) are used within local grids to simplify calculations.

In summary, coordinate systems in geodesy are vital for defining locations on the Earth's surface, ensuring accurate positioning and mapping across various fields such as navigation, surveying, GIS, and remote sensing. The selection of a coordinate system and datum depends on the application's requirements and the region of interest.

4.2.5 *Reference*

In geodesy, a reference is a crucial concept that provides a fixed point, surface, or framework for defining and comparing measurements and coordinates. References are essential for maintaining accuracy, consistency, and precision in geospatial positioning and measurements (Fig. 4.6). The following are key aspects of references in geodesy:

- **Horizontal reference:** Geodesy uses horizontal references, such as latitude and longitude, to define the positions of points on the surface. These coordinates describe locations relative to the equator and the Prime Meridian (or other reference meridians).

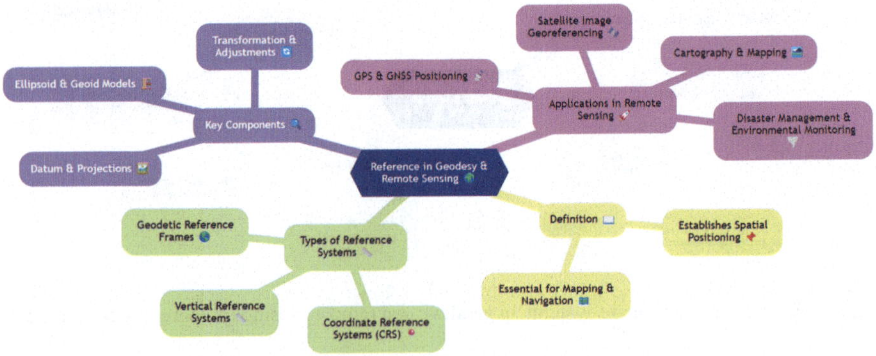

Fig. 4.6 Concept of reference in geodesy and remote sensing (prepared by the authors)

- **Vertical reference:** In addition to horizontal references, geodesy utilizes vertical references to determine elevations or heights above or below a specific reference surface. The geoid, representing mean sea level, is a common vertical reference, along with benchmarks or tidal gauges.
- **Geodetic datums:** Geodetic datums are comprehensive reference systems that integrate both horizontal and vertical references. These datums specify the ellipsoid, reference point, and coordinate system orientation, providing a consistent framework for accurate geospatial measurements and mapping.
- **Datum transformations:** When working with geospatial data across regions or different datums, datum transformations are required to align coordinates. These transformations account for variations in ellipsoid parameters and reference points between datums.
- **National and global datums:** Countries often establish national datums for precise local mapping and surveying. Globally, commonly used horizontal datums include WGS84 and NAD83, which serve as standard references for applications like GPS.
- **Local references:** For specific projects or regions, local references may be established to accommodate regional geodetic conditions or to provide high-precision control, often based on custom datums or grid systems.
- **Accuracy and precision:** The selection of a reference is crucial for ensuring the accuracy and precision of geospatial data. A well-defined reference minimizes errors and discrepancies in systems like GIS, surveying, and mapping.
- **GPS and satellite navigation:** The Global Positioning System (GPS) relies on a reference frame based on WGS84 to calculate precise locations on Earth's surface. GPS accuracy is contingent on the stability and accuracy of this reference frame.

In summary, references in geodesy provide the essential framework for geospatial measurements and positioning. They encompass both horizontal and vertical components and are defined by geodetic datums that establish the Earth's shape, orientation, and reference points. These references ensure the accuracy, consistency, and compatibility of coordinates, elevations, and measurements across diverse geospatial applications.

4.2.6 Great Circle

In geodesy, the concept of a great circle is fundamental to understanding the geometry of the surface. A great circle is the largest possible circle that can be drawn on a sphere or an ellipsoid, representing the shortest path between two points on the surface (Fig. 4.7). The following are key aspects of great circles in the context of geodesy:

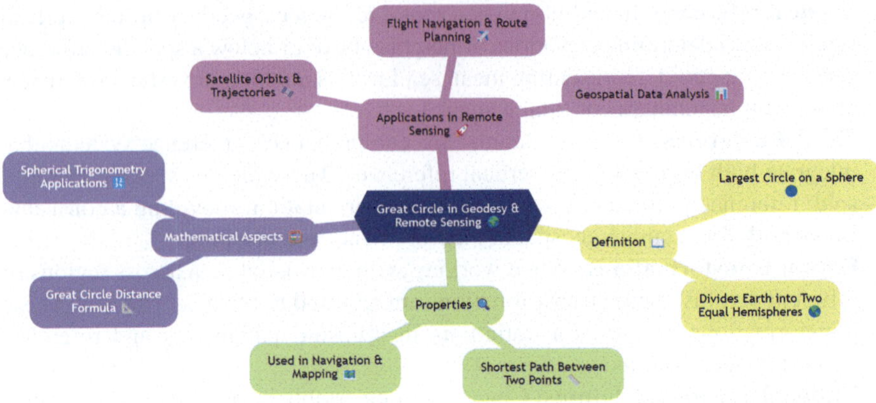

Fig. 4.7 Concept of great circle in geodesy and remote sensing (prepared by the authors)

- **Concept:** A great circle is formed when a plane passes through the centre of a sphere or ellipsoid, intersecting its surface. This circle divides the sphere or ellipsoid into two equal halves.
- **Examples:** The equator is a prominent example of a great circle. Similarly, lines of longitude (meridians) that run from the North Pole to the South Pole are great circles. Any pair of antipodal points defines a great circle.
- **Special characteristics**

  a) Great circles have the largest possible circumference of any circle that can be drawn on a sphere or ellipsoid.
  b) The centre of a great circle coincides with the centre of the sphere or ellipsoid.
  c) The shortest path between two points on a sphere or ellipsoid follows the arc of a great circle, making it vital for navigation and route planning.

- **Applications:** Great circles are essential in geodesy for calculating distances and azimuths (directions) between locations. They also play a critical role in cartography and geodetic computations.
- **Great circle routes:** Great circles are used in navigation and aviation to determine the shortest routes between two locations on Earth's surface. These routes reduce travel distance, time, and fuel consumption.
- **Latitude and longitude:** A great circle route does not follow a constant direction relative to Earth's surface but changes as it intersects with meridians (lines of longitude). This behaviour results from the curved geometry of the surface.
- **Great circles on an ellipsoid:** Since the Earth is not a perfect sphere but an ellipsoid, great circles on the Earth are approximated using an ellipsoidal model. This model accounts for the Earth's flattening at the poles and bulging at the equator, improving the accuracy of geodetic calculations.

In summary, great circles are a crucial concept in geodesy, serving as the basis for calculating the shortest distance between two points on the surface. Their practical applications in navigation, aviation, and geospatial analysis make them indispensable for precise distance and direction computations, while their geometric properties enhance our understanding of Earth's surface.

4.2.7 *Spherical Angle*

In geodesy, a spherical angle is a core concept utilized to measure the angle formed between two intersecting lines or planes on the surface of a sphere or an ellipsoid. It holds significant importance in geodetic computations, especially in determining distances, directions, and positional relationships on the Earth's surface (Fig. 4.8). The following discussion highlights the spherical angles in the context of geodesy:

- **Concept:** A spherical angle is the angle formed at the intersection of two arcs (great circles) on the surface of a sphere or ellipsoid. It is expressed in angular units such as degrees or radians.
- **Great circles:** To define a spherical angle, three points are required on the sphere or ellipsoid: the apex (the vertex of the angle) and two points that determine the arcs (great circles) forming the angle.
- **Angular units:** Spherical angles are typically measured in degrees (°) or radians (rad). In geodesy, degrees are the most commonly used unit, with a full circle measuring 360° or 2π radians.
- **Properties of spherical angles**

  a) A complete circle on a sphere or ellipsoid corresponds to 360° (or 2π radians).

  b) Spherical angles are positive when measured counterclockwise from the initial arc to the terminal arc.

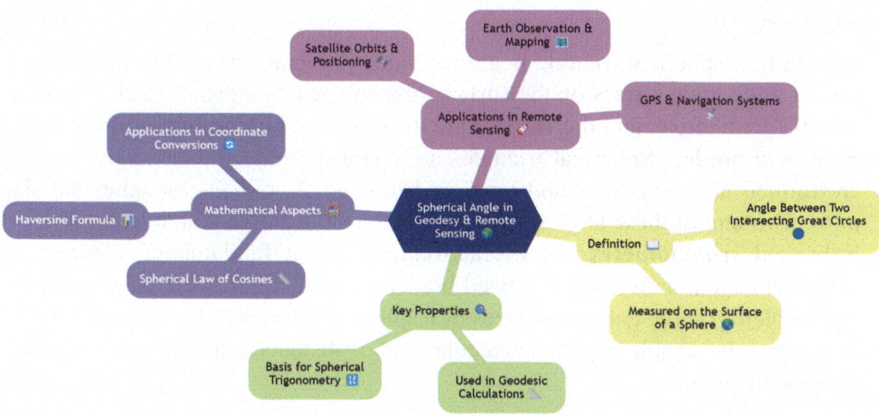

Fig. 4.8 Concept of spherical angle in geodesy and remote sensing (prepared by the authors)

c) In a spherical triangle (formed by the intersection of three great circles), the sum of the interior angles exceeds 180°.

- **Applications:** Spherical angles have numerous applications in geodesy, including:

  a) Calculating distances and directions between locations on Earth's surface.
  b) Determining positions using spherical trigonometry.
  c) Defining geodetic coordinate systems, such as latitude and longitude.
  d) Performing navigational calculations for aviation and maritime activities.

- **Spherical trigonometry:** This specialized branch of trigonometry is dedicated to solving problems involving spherical surfaces. It uses spherical angles and their relationships to enable precise geodetic calculations.
- **Ellipsoidal considerations:** While spherical angles are vital for computations on a spherical surface, geodetic calculations often use ellipsoidal models that better represent the Earth's actual shape. Ellipsoidal angles account for the Earth's flattening and improve accuracy in geodetic computations.

In summary, spherical angles are essential in geodesy for measuring angles on the surface of a sphere or ellipsoid. They form the basis for geodetic calculations, facilitating the accurate determination of distances, directions, and positions on the Earth's surface, while also underpinning geodetic coordinate systems such as latitude and longitude.

4.2.8 Spherical Triangle

In geodesy, a spherical triangle is a fundamental geometric construct utilized to address and resolve challenges in navigation, surveying, and geodetic computations on the surface of a sphere or an ellipsoid (a more accurate model of the Earth's shape) (Fig. 4.9). The following are the key aspects of spherical triangles in the context of geodesy:

- **Concept:** A spherical triangle is a three-sided polygon formed by the intersection of three great circles on the surface of a sphere or ellipsoid. Each side is an arc of a great circle, and the vertices are the points of intersection of these arcs.
- **Spherical angles:** Spherical triangles are characterized by their angles, which are formed at the vertices and measured in degrees, radians, or other angular units. The sum of these angles always exceeds 180°, unlike planar triangles.
- **Spherical trigonometry:** Spherical triangles are a fundamental element of spherical trigonometry, a specialized branch of mathematics designed for solving problems on the curved surfaces of a sphere or ellipsoid. It plays a vital role in geodesy by enabling precise calculations of distances, directions, and positions on the surface.

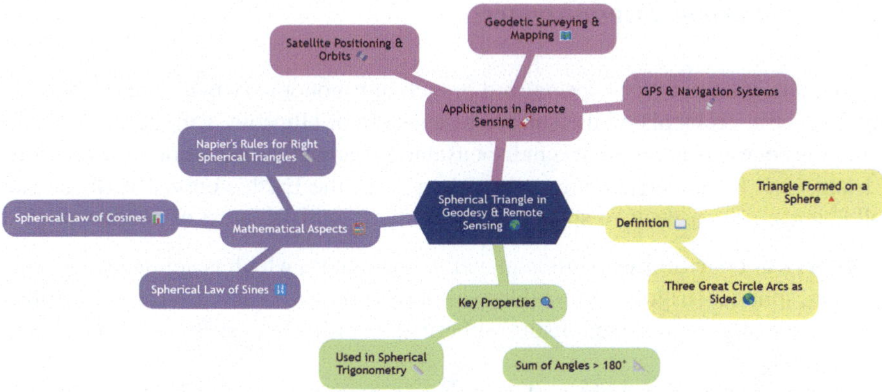

Fig. 4.9 Concept of spherical triangle in geodesy and remote sensing (prepared by the authors)

- **Types of spherical triangles**

  a) Spherical right triangles: Triangles with one 90° angle.
  b) Spherical oblique triangles: Triangles without any right angles.
  c) Spherical equilateral triangles: Triangles with three equal sides and three equal angles.

- **Applications**

  a) Navigation: Calculating optimal routes and distances for air and sea travel.
  b) Surveying: Determining distances and positions between points on the Earth's surface.
  c) Geodetic computations: Solving problems related to azimuths, distances, and coordinates.
  d) Astronomy: Understanding the positions and motions of celestial objects relative to the observer's location on Earth.

- **Law of sines and cosines:** Spherical trigonometry uses specialized formulas, such as the law of sines and the law of cosines, to establish relationships between the sides and angles of a spherical triangle, aiding in precise problem-solving.
- **Ellipsoidal consideration:** Although spherical triangles provide a useful approximation for many calculations, geodesy often requires ellipsoidal models to account for the Earth's true shape. In such cases, ellipsoidal trigonometry is employed to address deviations from a perfect sphere.

In summary, spherical triangles are fundamental in geodesy, serving as a critical tool for addressing various geodetic challenges related to navigation, surveying, and positioning on the Earth's surface. They form the foundation of spherical trigonometry, facilitating precise calculations and measurements essential to the discipline.

4.2.9 Spherical Trigonometry

Spherical trigonometry is a specialized branch of trigonometry that focuses on solving geometric problems on the surface of a sphere or ellipsoid, with particular relevance to geodesy (Fig. 4.10). It is indispensable for geodetic calculations and provides a framework for understanding and working with the Earth's curved surface. The following are key aspects of spherical trigonometry in the context of geodesy:

- **Spherical coordinates:** Unlike planar trigonometry, which is applied to flat surfaces, spherical trigonometry addresses problems on the curved surface. Positions are typically represented using spherical coordinates, such as latitude and longitude.
- **Spherical triangles:** Spherical triangles are essential in spherical trigonometry, formed by the intersection of three great circles on the surface of a sphere or ellipsoid. These triangles are characterized by their angles, sides, and the interrelationships among them.
- **Spherical trigonometric functions:** Spherical trigonometry introduces specialized trigonometric functions, such as the haversine, and versine. These functions are used to calculate various parameters of spherical triangles, including distances, azimuths, and angles.
- **Law of sines:** The law of sines for spherical triangles links the lengths of the sides to the sines of the opposite angles. This law, while analogous to the planar law of sines, is specifically adapted for the spherical surface.
- **Law of cosines:** The law of cosines for spherical triangles relates the cosines of the sides to the cosines of the corresponding angles. Similar to the planar version, this law is used to determine missing angles or sides within a spherical triangle.
- **Applications in geodesy:** Spherical trigonometry plays a pivotal role in geodesy, with applications including:

  a) Navigation: Determining courses, bearings, and distances for maritime and aerial navigation.

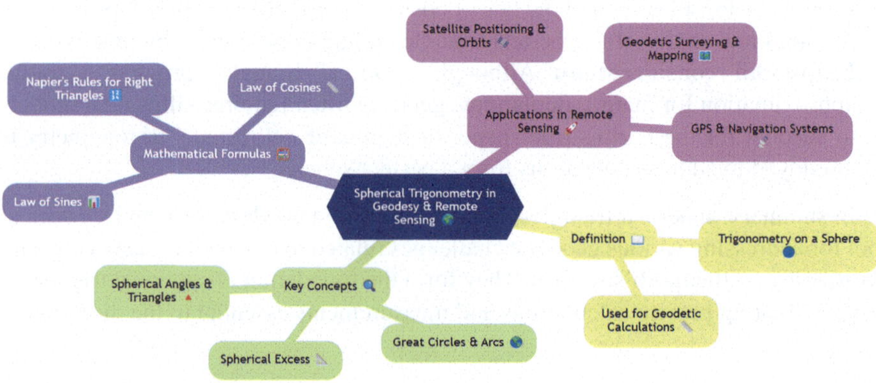

Fig. 4.10 Concept of spherical trigonometry in geodesy and remote sensing (prepared by the authors)

b) Surveying: Calculating positions and distances between survey points on the Earth's surface.
c) Geodetic computations: Solving problems related to positions, azimuths, and distances.
d) Celestial navigation: Analyzing celestial objects' positions and their relationship to the observer's location on Earth.

- **Ellipsoidal trigonometry:** While spherical trigonometry provides a useful approximation, geodesy often requires ellipsoidal models to account for the Earth's true shape. Ellipsoidal trigonometry is employed to correct for the deviations from a perfect sphere.
- **Practical tools:** In geodesy, specialized software and calculators equipped with spherical trigonometric functions are essential for performing complex geodetic computations, enabling professionals to make precise measurements and calculations for various applications.

In summary, spherical trigonometry is a specialized branch of trigonometry designed for the curved surface. It plays a crucial role in geodesy, enabling the solution of complex problems in navigation, surveying, and geodetic computations, especially when using spherical or ellipsoidal models of the Earth.

4.2.10 World Geodetic System

The World Geodetic System (WGS) is a global geodetic reference system widely adopted for accurate positioning, navigation, and geospatial data exchange. It is essential in geodesy and plays a pivotal role in the functioning of global positioning systems (GPS) and satellite-based navigation (Fig. 4.11). The following discussion outlines the fundamental aspects of the World Geodetic System (WGS) in the context of geodesy:

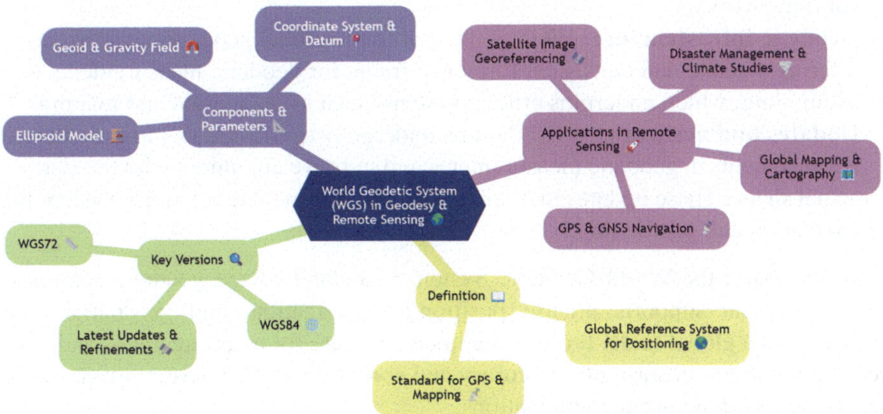

Fig. 4.11 Concept of World Geodetic System in geodesy and remote sensing (prepared by the authors)

- **Concept:** The World Geodetic System (WGS) is a geodetic reference framework that defines the Earth's shape, orientation, and coordinate system. It comprises a reference ellipsoid (a mathematical model of the Earth's shape), a reference point, and a coordinate system based on latitude, longitude, and ellipsoidal height.
- **WGS datums:** Various versions of the World Geodetic System have been developed over time, each identified as a WGS datum. Commonly used versions include WGS72, WGS84, and more recent iterations such as WGS84(G1762) and WGS84(G1674). These datums are periodically updated to enhance accuracy and ensure better alignment with the Earth's measurements.
- **WGS84:** Introduced in 1984, WGS84 is one of the most widely utilized datums globally. It serves as the reference frame for the GPS system and is crucial for satellite-based navigation, mapping, and geospatial data exchange. WGS84 is defined by a specific reference ellipsoid and reference point.
- **Coordinate system:** WGS employs a geographic coordinate system based on latitude and longitude, where latitude indicates north–south positions and longitude represents east–west positions. Elevations can also be included as ellipsoidal height. Coordinates are typically expressed in degrees, minutes, and seconds or decimal degrees.
- **GPS and satellite navigation:** WGS84 compatibility is vital for the accuracy and interoperability of GPS devices and satellite navigation systems. It ensures consistency between GPS receivers and satellites in their reference frames, facilitating precise positioning and navigation.
- **Geospatial data exchange:** WGS serves as the standard reference system for geospatial data exchange across multiple fields, including cartography, Geographic Information Systems (GIS), and remote sensing. It enables seamless integration of data collected from various locations and instruments by providing a common reference framework.
- **Coordinate transformations:** In cases involving local or regional coordinate systems, geodesists often perform coordinate transformations to align these systems with the WGS datum, ensuring consistency and compatibility across different frameworks.
- **Geodetic infrastructure:** WGS is integral to the global geodetic infrastructure, offering a stable and consistent reference frame for geodetic measurements and positioning, which underpins critical systems such as navigation and mapping.
- **Updates and accuracy:** WGS datums undergo periodic updates to incorporate advancements in geodetic measurements and improve alignment with the Earth's actual shape. These updates maintain WGS as a dependable reference system for navigation and positioning.

In summary, the World Geodetic System is a standardized geodetic reference framework that supports precise positioning, navigation, and geospatial data exchange on a global scale. It offers a unified structure for mapping, GPS, and geodetic applications, ensuring the accuracy and consistency of geospatial information across various domains and applications.

4.2.11 Everest Spheroid

The Everest spheroid, also referred to as the Indian 1956 spheroid or the Everest 1830 spheroid, is a geodetic reference ellipsoid predominantly used in India and surrounding regions for mapping, surveying, and geodetic applications (Fig. 4.12). The following are key aspects of the Everest spheroid in the context of geodesy:

- **Origin and history:** The Everest spheroid, introduced in 1830, was named after Sir George Everest, the British Surveyor General of India. It was specifically created to serve as the reference ellipsoid for geodetic surveys conducted in India during the British colonial era.
- **Ellipsoid parameters:** The Everest spheroid is defined by key parameters, including the semi-major axis (a), semi-minor axis (b), and flattening (f). These values were specifically tailored to optimize the spheroid's fit for the geodetic characteristics of the Indian subcontinent.
- **Regional application:** Although the Everest spheroid is recognized internationally, it was primarily used within the Indian subcontinent. It provided a more accurate representation of the Earth's shape in this region, accounting for the complex topography and geodetic challenges of the Indian terrain.
- **Updates and revisions:** Over time, advancements in geodetic measurement techniques led to multiple revisions of the Everest spheroid, aiming to improve its accuracy and better align it with modern geodetic standards.
- **Transition to WGS84:** In the latter half of the twentieth century, India, along with many other countries, adopted the World Geodetic System 1984 (WGS84) as the standard geodetic reference system. WGS84 offers high precision and compatibility with satellite-based positioning systems, such as GPS.
- **Legacy use:** While WGS84 has become the global standard for positioning and navigation, the Everest spheroid remains prevalent in older maps, surveys, and historical geodetic data in the Indian subcontinent. For contemporary applications and global system interoperability, however, WGS84 is more commonly employed.

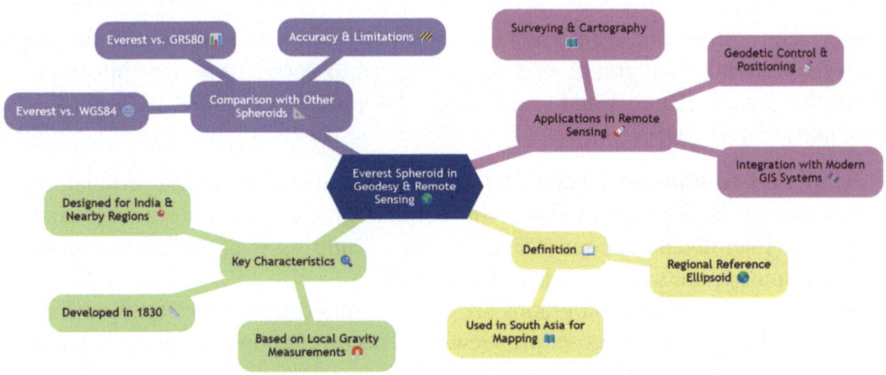

Fig. 4.12 Concept of Everest spheroid in geodesy and remote sensing (prepared by the authors)

- **Local adjustments:** In certain cases, regional adjustments are made to align measurements and coordinate systems with the Everest spheroid when working with legacy data or projects specific to India and neighbouring regions.

In summary, the Everest spheroid is a historically significant geodetic reference ellipsoid primarily used in India and neighbouring regions. While it played a key role in geodetic surveys during the colonial period, it has largely been supplanted by global systems such as WGS84 for contemporary positioning, navigation, and geospatial applications.

4.3 Air Photo Interpretation

4.3.1 Concept of Air Photo Interpretation

Air photo interpretation is a crucial technique in remote sensing and geospatial analysis, involving the systematic examination and analysis of aerial photographs or satellite imagery to extract valuable information about the surface features and conditions. This process relies on the observer's ability to recognize and interpret patterns, shapes, and spectral characteristics present in the images. Through this skilled analysis, it becomes possible to identify landforms, vegetation types, urban development, geological features, and environmental changes such as deforestation or land degradation. Air photo interpretation is crucial for applications in agriculture, urban planning, forestry, disaster management, and environmental monitoring. It enhances geospatial technologies like GIS and remote sensing by providing visual and contextual insights into the Earth's surface, making it an invaluable tool for decision-making and resource management.

4.3.2 Principles of Air Photo Interpretation

The principle of air photo interpretation is based on the systematic and thorough analysis of aerial photographs or satellite images to derive important information about the surface features and conditions. This process is guided by several key principles (Fig. 4.13):

- **Scale and resolution:** Understanding the scale of the imagery is crucial, as it dictates the level of detail visible in the photograph. Higher-resolution images provide finer details but cover smaller areas, requiring interpretation to be done at the appropriate scale for accurate analysis.
- **Stereoscopy:** Aerial photographs are often captured from two slightly different angles, allowing for stereoscopic viewing. This three-dimensional perspective is

essential for identifying elevation, depth, and surface relief, with stereo pairs being particularly useful in topographic mapping.
- **Tonal variation:** Variations in brightness or tonal patterns within the imagery can offer valuable insights into land cover, land use, and natural features. Recognizing tonal differences is vital for the identification of objects and environmental phenomena.
- **Texture and pattern:** The arrangement of features and their repetitive patterns can provide critical information regarding land cover, vegetation types, and even human-made structures. Different textures and patterns may indicate agricultural fields, urban areas, or natural landscapes.
- **Shape and size:** Analyzing the shapes and sizes of features helps distinguish between different objects. Geometric shapes may indicate human-made structures, while natural features typically display more irregular shapes.
- **Shadow and relief:** Shadows cast by objects can reveal details about their height, shape, and orientation. Understanding the angle and direction of sunlight is essential for interpreting these features accurately.
- **Contextual information:** The broader geographical and ecological context is important when making informed judgements during interpretation. Knowing the surrounding environment aids in accurately identifying and classifying objects.

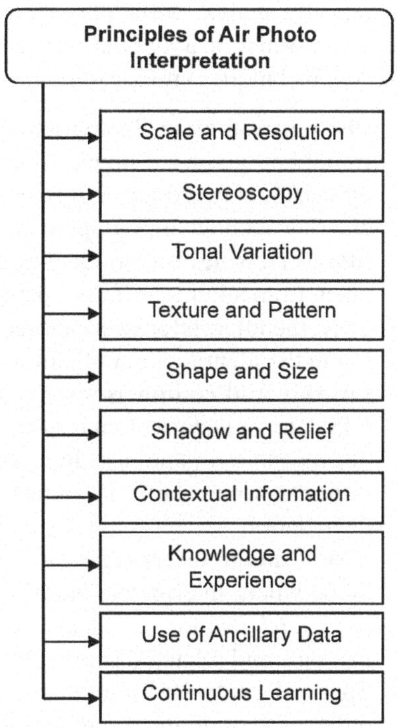

Fig. 4.13 Principles of air photo interpretation (prepared by the authors)

- **Knowledge and Experience:** A solid foundation in disciplines such as geography, geology, and ecology are essential for accurate interpretation. Expertise and experience significantly enhance the ability to recognize and interpret features correctly.
- **Use of ancillary data:** Additional sources of data, such as maps, ground truth information, and multispectral imagery, can be integrated with aerial photographs to improve the accuracy of interpretation.
- **Continuous learning:** Interpretation is a skill that improves with practice. Geospatial professionals continuously refine their abilities by studying new imagery, participating in training programmes, and staying updated on technological advancements.

In summary, air photo interpretation involves analyzing imagery based on factors like scale, stereoscopy, and tonal variation. Expertise is crucial for accurate interpretation, making it vital for environmental assessment, land-use planning, and disaster management.

4.3.3 Techniques of Air Photo Interpretation

Air photo interpretation techniques refer to the methods and strategies used to systematically analyze aerial photographs or satellite images in order to extract meaningful information about the surface features and conditions. The following outlines the key techniques employed in air photo interpretation:

- **Photogrammetry:** Photogrammetry is the science of extracting measurements from photographs, utilizing stereo pairs of aerial images to determine the three-dimensional coordinates of points on the surface. This technique is essential for creating accurate topographic maps and obtaining elevation data.
- **Stereo viewing:** Stereo viewing involves examining pairs of aerial photographs taken from slightly different perspectives. By viewing these images stereoscopically, the interpreter gains a three-dimensional understanding of the terrain, aiding in the identification of elevation, depth, and surface relief.
- **Overlay and comparisons:** Overlaying aerial photographs with existing maps, GIS data, or historical images allows interpreters to identify temporal changes or discrepancies in land use and land cover. Comparisons between images taken at different times help in detecting urban expansion, deforestation, or land degradation.
- **Visual image interpretation:** Visual image interpretation is the primary technique where interpreters analyze photographs for patterns, shapes, tonal variations, and textures to identify and classify features. Expertise in geography, geology, and related disciplines is crucial for making accurate interpretations.
- **Spectral analysis:** For multispectral or hyperspectral imagery, spectral analysis involves examining the reflectance or emission patterns across various wave-

lengths of light. This technique is useful for identifying vegetation types, mineral deposits, and environmental conditions.

- **Pattern recognition:** Pattern recognition entails identifying recurring shapes, arrangements, or configurations of features on the surface. Recognizable patterns can indicate agricultural fields, urban infrastructure, road networks, or natural features.
- **Shadow analysis:** Shadow analysis involves considering the shadows cast by objects in aerial photographs to infer their height, shape, and orientation. By understanding the angle and direction of sunlight, interpreters can assess relief and identify features like buildings or mountains.
- **Digital techniques of air photo interpretation:** Digital techniques in air photo interpretation have significantly advanced remote sensing and geospatial analysis. High-resolution satellite and aerial imagery, coupled with powerful computing tools, have enhanced the efficiency and accuracy of interpretation. These techniques utilize Geographic Information Systems (GIS), remote sensing software, and specialized algorithms for tasks like image classification, object recognition, and change detection. The integration of multiple data sources, such as multispectral imagery, LiDAR data, and ground-based measurements, enables comprehensive analysis. Digital methods have broadened the scope of interpretation, allowing for faster, more precise, and versatile applications in areas like urban development, land-use planning, and environmental monitoring.
- **Feature extraction:** Feature extraction techniques employ computer algorithms to automatically detect and delineate specific objects or land cover types within an image. This approach is particularly effective for large-scale mapping and classification tasks.
- **Geospatial software:** Geospatial software, including Geographic Information Systems (GIS) and remote sensing tools, supports the analysis of aerial photographs by offering capabilities for image enhancement, geometric correction, and data integration. These tools improve the efficiency and accuracy of interpretation.
- **Training and expertise:** Air photo interpretation requires specialized training and expertise. Professionals in the field often undergo formal education and continuous learning to enhance their skills and remain current with technological advancements.

In summary, air photo interpretation techniques include methods such as visual analysis, stereo viewing, spectral analysis, and feature extraction. The choice of technique depends on the imagery type, analysis objectives, and the interpreter's expertise. These methods collectively enable the extraction of valuable geospatial data, supporting applications in environmental monitoring, land management, and urban planning.

4.3.4 An Air Photo Interpretation System

An air photo interpretation system is a structured collection of tools, methods, and techniques employed by geospatial professionals to analyze and extract meaningful information from aerial photographs or satellite imagery. These systems are designed to optimize the interpretation process, ensuring both efficiency and accuracy in geospatial analysis (Fig. 4.14). So, the air photo interpretation system includes the following components and features:

- **Geospatial software:** Central to the system is the integration of geospatial software, including Geographic Information Systems (GIS) and remote sensing tools. These platforms support image analysis, enhancement, georeferencing, and the integration of additional geospatial data layers.
- **Imagery acquisition:** The system begins with the acquisition of high-resolution aerial photographs or satellite images, typically sourced from specialized providers or government agencies, serving as the primary data for interpretation.
- **Stereo viewing equipment:** To interpret stereoscopic images, the system may include stereoscopes or software that simulates stereoscopic viewing, enabling three-dimensional perception and terrain analysis.
- **Digital image processing:** Key processes such as image classification, change detection, and feature extraction are integral to the system, with automation improving efficiency and reducing human error.
- **Reference data:** The system may incorporate reference data such as ground truth information, existing maps, and historical imagery to validate interpretations and enhance accuracy.
- **Training and expertise:** Skilled personnel, trained in geospatial analysis, remote sensing, and cartography, are essential for accurate interpretation and optimal use of the system.
- **Data management:** Effective data management capabilities are included to organize and store large volumes of imagery and associated geospatial data layers.
- **Integration with GPS:** Many systems are integrated with Global Positioning System (GPS) technology, ensuring precise georeferencing and positioning for accurate feature location on the Earth's surface.
- **Visualization tools:** Advanced visualization tools enable the display and analysis of images in conjunction with other geospatial data layers, assisting with contextual understanding and informed decision-making.
- **Customization:** Interpretation systems can be customized to cater to the specific needs of various applications, such as urban planning, environmental monitoring, agriculture, and disaster response.
- **Quality control:** Quality control measures, including peer reviews and validation, are integrated to ensure the accuracy and reliability of the interpreted data.
- **Report generation:** The system often includes features for generating reports or maps that summarize the interpreted data for communication and decision-making purposes.

Fig. 4.14 An air photo interpretation system (prepared by the authors)

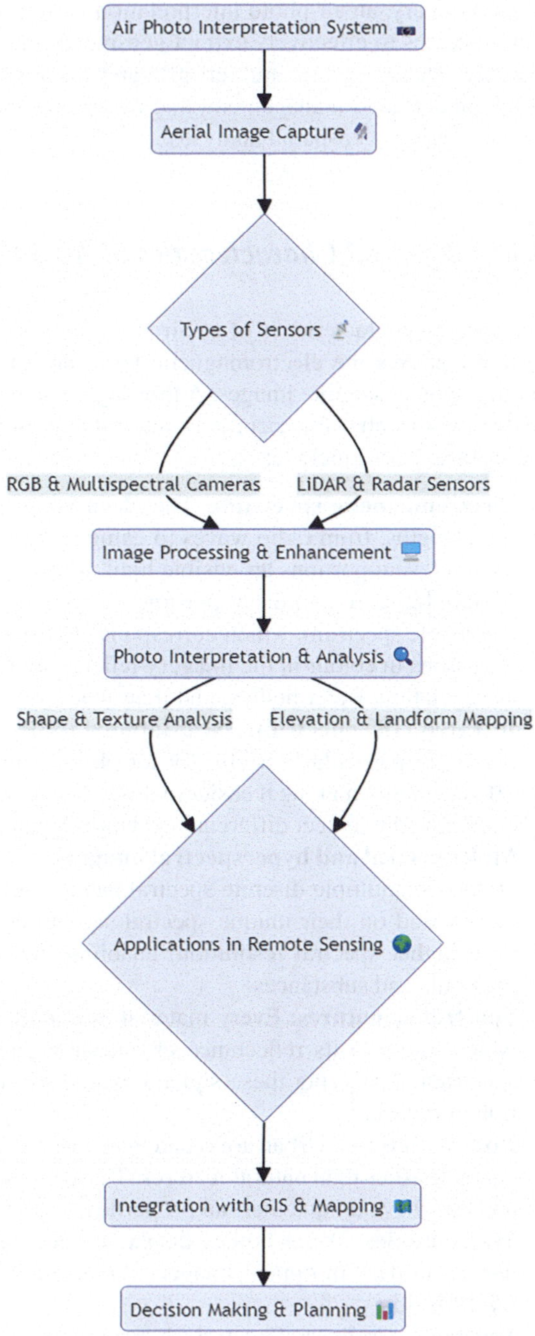

In summary, an air photo interpretation system integrates technology, expertise, and processes to effectively extract geospatial information from aerial and satellite imagery. These systems are versatile and essential for applications such as urban development, land management, and natural resource monitoring, offering critical data for various geospatial analyses.

4.3.5 Spectral Characteristics of Air Photo Interpretation

The spectral characteristics of an air photo interpretation system pertain to how the system analyses the electromagnetic spectrum to extract information from aerial photographs or satellite images. A thorough understanding of these spectral characteristics is essential for accurately interpreting and classifying various features on the surface. These include:

- **Electromagnetic spectrum:** The electromagnetic spectrum covers a range of wavelengths, from radio waves to gamma rays. For air photo interpretation, the most relevant portions are visible light, infrared, and ultraviolet radiation.
- **Visible light:** A significant portion of aerial and satellite imagery is captured in the visible spectrum, which corresponds to the colours visible to the human eye. Variations in colour in the imagery reflect differences in surface properties, such as vegetation, water bodies, and man-made structures.
- **Infrared (IR) spectrum:** Near-infrared (NIR) and shortwave infrared (SWIR) bands are particularly useful for air photo interpretation. NIR strongly reflects off vegetation, making it easier to differentiate between healthy and stressed vegetation, and to detect different vegetation types.
- **Multispectral and hyperspectral imaging:** Multispectral sensors acquire imagery across multiple discrete spectral bands, facilitating the identification of features based on their unique spectral signatures. Hyperspectral sensors provide even higher spectral resolution, enabling the detection of a broader range of materials and substances.
- **Spectral signatures:** Every material or feature has a unique spectral signature, which refers to its reflectance or emission pattern across the electromagnetic spectrum. Analyzing these signatures helps in the identification and classification of objects.
- **Built structures:** Urban areas and man-made structures exhibit different spectral characteristics than natural features. These areas tend to reflect more visible light and may have unique thermal properties.
- **Water bodies:** Water bodies display distinct spectral characteristics, typically appearing dark in optical imagery due to their low reflectance in the NIR and SWIR bands.
- **Vegetation indices:** Spectral characteristics are used to calculate vegetation indices, such as the Normalized Difference Vegetation Index (NDVI), which assesses the health and density of vegetation by comparing reflectance in the red and NIR bands.

- **Change detection:** Spectral characteristics are key in change detection analyses. By comparing spectral signatures of the same location over time, changes such as deforestation, urban growth, or crop development can be tracked.

In summary, the spectral characteristics of an air photo interpretation system are crucial for identifying and classifying surface features. Analyzing reflectance and emission patterns across various wavelengths enables interpreters to distinguish materials, land cover types, and environmental conditions, making it an invaluable tool for agriculture, forestry, environmental monitoring, and land-use planning.

4.4 Stereoscopic

4.4.1 Concept of Stereoscope

A stereoscope is an optical device that allows the viewing of two nearly identical images or photographs, creating a three-dimensional (3D) perception of the scene. This principle is based on stereopsis, the human brain's ability to merge slightly different images from each eye to perceive depth and volume. In the case of a stereoscope, two images, often aerial or satellite photographs, are presented separately to each eye. When viewed through the lenses, the brain combines the images, generating a sense of depth and spatial relationships within the scene. Stereoscopes have been widely utilized in fields such as geology, cartography, and aerial photo interpretation to extract information about terrain elevation, topography, and 3D structures. While traditional handheld stereoscopes are still in use, modern digital technologies now enable virtual stereoscopic viewing, allowing for 3D analysis without physical instruments.

4.4.2 History of Stereoscope

The history of the stereoscope is a compelling narrative that spans several centuries, highlighting humanity's enduring fascination with three-dimensional imagery. The following outlines significant milestones in the evolution of the stereoscope:

- **Early concepts (seventeenth–eighteenth century):** The foundation of stereoscopy, the ability to perceive depth from two slightly offset images, emerged in the seventeenth century. Various inventors and scientists, including Sir Charles Wheatstone and Sir David Brewster, explored this concept, leading to the development of early stereoscopic devices in the nineteenth century.
- **The Brewster stereoscope (1849):** Sir David Brewster, a Scottish scientist, created the Brewster stereoscope, also known as the lenticular stereoscope. This

device utilized a lenticular lens to direct two images to each eye, producing a 3D effect, and represented a major advancement in stereoscopic technology.
- **The Wheatstone reflective stereoscope (1838):** Sir Charles Wheatstone introduced the reflective stereoscope, which employed mirrors to direct the images, offering a more compact and practical solution for viewing stereographic images. It became popular for the viewing of stereographic drawings and photographs.
- **Stereographic photography (1850s):** The invention of stereographic photography, involving pairs of photographs taken from slightly different angles, revolutionized stereoscopy. Photographers, such as Oliver Wendell Holmes Sr., began capturing stereoscopic images of landscapes, urban scenes, and daily life.
- **Stereoscope cards (stereoviews):** In the late nineteenth century, stereoscope cards, or stereoviews, became widely popular. These cards, featuring paired images mounted side by side, were sold as both entertainment and educational tools. They offered scenes of travel, historical events, and natural wonders.
- **The stereopticon (magic lantern):** A variation of the stereoscope, the stereopticon, was developed to project 3D images onto a screen. It became a popular device for public lectures, exhibitions, and entertainment, making 3D imagery accessible to larger audiences.
- **Revival in the twentieth century:** Stereoscopes and stereographic imagery experienced a revival in the twentieth century, notably through the View-Master, a popular toy that utilized circular reels for displaying 3D images, which became a household name.
- **Digital and virtual reality:** Advances in digital technology have brought stereoscopic imagery into the realm of virtual reality (VR). VR headsets and digital stereoscopes use advanced optics, sensors, and digital displays to provide immersive 3D experiences, widely applied in gaming, education, and simulations.

The history of the stereoscope illustrates humanity's enduring pursuit of capturing and experiencing depth and realism in three dimensions. From its early optical innovations to the advent of modern digital and virtual reality technologies, stereoscopy has continually evolved, making significant contributions to entertainment, education, and visual storytelling.

4.4.3 Types of Stereoscopes

Stereoscopes are optical devices that allow for the viewing of two separate but nearly identical images, producing a three-dimensional (3D) effect. Various types of stereoscopes exist, each designed for specific purposes and applications. They are:

- **Handheld stereoscope:** The traditional stereoscope, consisting of two eyepieces and lenses or prisms, enables viewers to merge separate images into a three-dimensional (3D) perception. It is commonly used for viewing printed stereographic cards or photographs.

- **Mirror stereoscope:** Also referred to as the Holmes stereoscope, this version uses mirrors to direct images towards the viewer's eyes. It features a compact design and is typically used for viewing stereographic cards or printed materials.
- **Pocket stereoscope:** A small, portable device designed for easy carrying, the pocket stereoscope is ideal for quick 3D viewing of stereographic cards or photos, often used in fieldwork.
- **Viewer stereoscope:** Commonly found in public spaces such as museums and tourist attractions, viewer stereoscopes are often coin-operated or mechanically operated, allowing multiple viewers to simultaneously experience 3D imagery.
- **Virtual reality (VR) stereoscope:** Leveraging advancements in technology, VR headsets offer an immersive 3D experience by tracking head movements and using sophisticated optics. These devices are used for gaming, virtual tours, and simulations.
- **Digital stereoscope:** Digital stereoscopes display separate images to each eye via digital screens, simulating the 3D effect. These can be standalone devices or integrated into computers and smartphones, commonly used for 3D movies, gaming, and virtual reality applications.
- **View-Master:** A popular toy that uses circular reels with pairs of images to create 3D views, the View-Master allows users to experience 3D scenes or educational content by looking through the viewer.
- **Auto-stereoscope**: Known as glasses-free 3D displays, auto-stereoscopes use lenticular lenses, parallax barriers, or similar technologies to display 3D images without the need for special glasses.

4.4.4 Instruments of Stereoscope

Stereo-plotting is an essential technique in photogrammetry and geospatial mapping, enabling the transformation of stereo pairs of aerial photographs into precise three-dimensional (3D) models of the surface. To carry out stereo-plotting effectively, a range of instruments and tools are utilized. The following provides an overview of the primary tools/instruments used in this process:

- **Stereoscope:** A fundamental instrument in stereo-plotting, the stereoscope allows the operator to view two overlapping images simultaneously, creating the illusion of 3D depth. Variants include:

  a) Mirror stereoscope: Utilizes mirrors to direct light from the two images into the viewer's eyes.
  b) Lens stereoscope: Employs lenses to separate the images for each eye.
  c) Digital stereoscope: Modern versions use electronic displays and optics to simulate the stereoscopic effect.

- **Planimeter:** A device used to measure the area of features or objects on a map or photograph, assisting in the quantification of size and shape to facilitate accurate map creation.
- **Stereo plotter:** A specialized instrument designed for stereo-plotting, integrating the stereoscope with additional features like measuring systems and control panels, enabling the operator to measure distances, angles, and elevations in the 3D model.
- **Parallax bar:** A ruler-like tool with crosshairs used to measure horizontal parallax between corresponding points in the stereo images, which is essential for determining distances and elevations.
- **Floating markers:** Specialized markers placed on the stereoscope or digital display to help the operator mark and measure points of interest in the stereo pair. These markers come in various shapes and sizes for different applications.
- **Tablet digitizer:** In digital stereo-plotting, a tablet digitizer is used to interact with digital stereoscopic images, allowing the operator to place markers, measure distances, and manipulate the 3D model using a stylus or other input devices.
- **Digital photogrammetric workstation (DPW):** Modern stereo-plotting involves digital workstations equipped with specialized software that provides an interface for viewing and manipulating stereo images and 3D models, supporting precise measurement and feature extraction.
- **Precision measuring instruments:** Tools like callipers, micrometres, and digital rulers ensure accurate measurements of distances, angles, and elevations, which are crucial for precision in stereo-plotting.
- **Digital elevation models (DEMs):** DEMs, generated through stereo-plotting, provide detailed elevation data of the surface, useful for applications such as terrain analysis and flood modelling.
- **Geographic Information Systems (GIS):** GIS software is often integrated with stereo-plotting tools to support georeferencing, feature extraction, and data integration, aiding in the creation of geospatial databases and maps.
- **Global Positioning System (GPS):** GPS technology is used to accurately georeferenced stereo imagery and resulting 3D models, ensuring alignment with real-world coordinates.

In brief, the instruments used in stereo-plotting are essential for transforming stereo pairs of aerial photographs into precise 3D representations of the Earth's surface. Ranging from traditional stereoscopes to advanced digital workstations, these tools facilitate accurate measurement, feature extraction, and geospatial mapping, establishing stereo-plotting as a vital technique in disciplines such as cartography, geology, urban planning, and environmental management.

4.5 Conclusion and Endnotes

In conclusion, this discussion has explored key topics related to geodesy, air photo interpretation, stereoscopes, and stereo-plotting. Geodesy, the study of Earth's shape and size, is fundamental to accurate mapping and navigation, with concepts like the geoid, ellipsoid, and datum supporting precise positioning. On the other hand, air photo interpretation involves analyzing aerial and satellite images to extract geospatial data, relying on scale, stereo viewing, and spectral analysis, with digital tools like GIS enhancing its accuracy. And the discussion about stereoscopes, which create 3D perceptions from stereo pairs, plays a crucial role in interpreting topographic and geological imagery. Finally, stereo-plotting, which converts stereo pairs into accurate 3D models, relies on a variety of instruments and is essential for creating precise maps and digital elevation models.

Through the inclusion of endnotes, we recognize the rich historical context, scientific principles, and technological innovations that form the foundation of these geospatial fields. The tools and concepts discussed have extensive applications, spanning urban planning, environmental monitoring, disaster management, and navigation. The ongoing integration of digital technologies is driving further advancements in geodesy, air photo interpretation, and stereo-plotting, solidifying their essential role in the contemporary understanding and application of Earth's spatial data.

Further Readings

1. Alomar-Garau, G., & Grimalt-Gelabert, M. (2021). Air photo interpretation for spatial analysis of heritage agrarian structures in Mediterranean settings as sea-breezes proxy-data. Application to the island of Mallorca. *Remote Sensing, 13*(22), 4652. https://doi.org/10.3390/rs13224652
2. Bhatta, B. (2021). *Remote sensing and GIS* (3rd ed.). Oxford University Press ISBN: 978-0199496648.
3. Hofmann-Wellenhof, B., & Moritz, H. (2006). *Physical geodesy*. Springer Vienna. ISBN: 978-3-211-33544-4. https://doi.org/10.1007/978-3-211-33545-1
4. Seeber, G. (2003). *Satellite geodesy: Foundations, methods, applications* (2nd ed.). Walter de Gruyter ISBN: 9783110175493.
5. Strang, G., & Borre, K. (1997). *Linear algebra, geodesy, and GPS*. Wellesley College ISBN: 9780961408862.
6. Strozzi, T., Ambrosi, C., & Raetzo, H. (2013). Interpretation of aerial photographs and satellite SAR interferometry for the inventory of landslides. *Remote Sensing, 5*(5), 2554–2570. https://doi.org/10.3390/rs5052554
7. Tempfli, K., Kerle, N., Huurneman, G. C., & Janssen, L. L. F. (Eds.). (2009). *Principles of remote sensing: An introductory textbook*. The International Institute for Geo-Information Science and Earth Observation (ITC) ISBN: 978-90-6164-270-1.
8. Turner, S. (Ed.). (1987). *Applied geodesy: Global positioning system — Networks — Particle accelerators — Mathematical geodesy*. Springer . ISBN: 978-3-540-18219-1. https://doi.org/10.1007/BFb0010105
9. Vanicek, P., & Krakiwsky, E. J. (1986). *Geodesy: The concepts*. Elsevier Science Publisher ISBN: 0444877754.
10. Xu, G. (Ed.). (2010). *Sciences of geodesy – I: Advances and future directions*. Springer . ISBN: 978-3-642-11740-4. https://doi.org/10.1007/978-3-642-11741-1

Part II
Fundamentals of Geospatial Technologies: Concepts and Techniques of Geographic Information System

Chapter 5
Geographic Information Systems: The Basics and Beyond

"Geographic Information Systems empower us to see the world as it is, as it was, and as it could be, all at once."— Roger Tomlinson.

Abstract GIS is a powerful technology for capturing, organizing, analyzing, and visualizing spatial data, enabling users to gain valuable insights and make informed decisions. This chapter provides an in-depth exploration of the concept, history, scope, components, capabilities, and applications of Geographic Information Systems (GIS). It begins by introducing GIS and its significance in spatial data management, illustrating how spatial data (such as satellite images, maps, and survey data) is combined with attribute data (such as population statistics or land-use information). The chapter emphasizes data types and models that facilitate the investigation of data relationships, patterns, and trends. It also addresses system limitations and errors within GIS. Furthermore, the chapter explores the diverse applications of GIS across disciplines, including urban planning, environmental management, transportation, public health, and natural resource management. It highlights the pivotal role of GIS in decision-making, resource allocation, and problem-solving by providing geographical context and analytical capabilities. The chapter concludes by discussing recent advancements and future trends in GIS, along with its interdisciplinary connections to fields such as environmental science and ecology, agriculture and forestry, tourism and hospitality, and various other domains.

Keywords Geographic information systems · GIS data model · Error in GIS · Spatial and non-spatial data · Open-source GIS software

K. D. Malakar et al., *Geospatial Technologies in Coastal Ecologies Monitoring and Management*, Advances in Geographic Information Science,
https://doi.org/10.1007/978-3-031-92017-2_5

This chapter delves into the transformative potential of Geographic Information Systems (GIS) as a tool for managing, analyzing, and visualizing spatial data. It provides a comprehensive overview of GIS, including its concept, historical development, components, and applications. Readers will gain insights into how GIS integrates spatial and attribute data to uncover patterns, trends, and relationships. The chapter addresses system limitations and explores GIS applications across diverse fields, such as environmental management, urban planning, and public health. Concluding with recent advancements and interdisciplinary connections, it underscores the critical role of GIS in informed decision-making and problem-solving.

Key Points of the Chapter

- Basic concept of GIS and its applications.
- GIS integrates spatial and attribute data for insightful analysis.
- Key components include data types, models, and analytical tools.
- GIS supports diverse fields like urban planning and environmental management.

5.1 Introduction to Geographic Information Systems (GIS)

5.1.1 Concept of GIS

Geographic Information Systems (GIS) is a transformative and multifaceted technology that fundamentally reshapes the way spatial data is analyzed, interpreted, and visualized. At its core, GIS provides a comprehensive framework for collecting, managing, and analyzing geographic data, supporting informed decision-making in fields such as urban planning, environmental management, emergency response, and marketing. By integrating various layers of data, including maps, satellite imagery, and location-specific information, GIS facilitates a holistic understanding of geographic phenomena. This technology enables users to identify patterns, relationships, and trends within spatial datasets, enhancing both decision-making and problem-solving capabilities. GIS is widely applied across industries such as agriculture, healthcare, transportation, and real estate, positioning it as a vital tool in contemporary society. Its capacity to deliver actionable insights, optimize resource allocation, and support evidence-based planning highlights its critical role in an increasingly interconnected world (Fig. 5.1). Thus, the discussion examined several definitions and concepts of GIS, as outlined below:

- The United States Geological Survey (USGS) describes "GIS as a computer-based system designed to capture, store, analyze, and manage data associated with specific locations on the Earth's surface. GIS facilitates the visualization, interpretation, and analysis of data through various formats, including maps, reports, and charts, enabling users to uncover relationships, patterns, and trends".

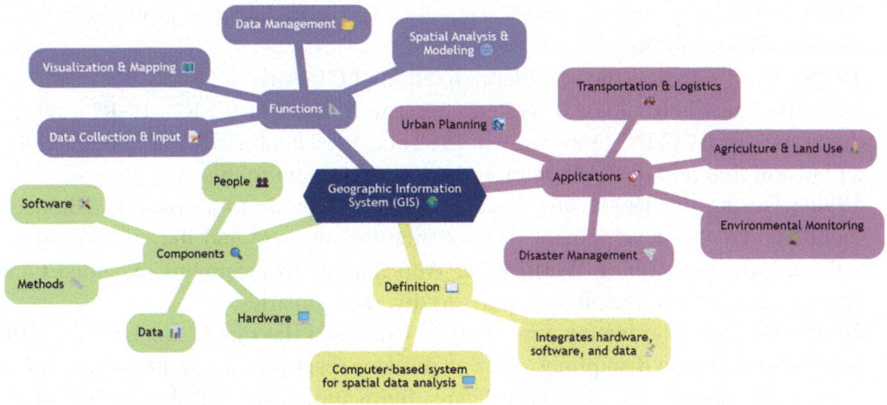

Fig. 5.1 Concept of Geographic Information Systems (prepared by the authors)

- ESRI (Environmental Systems Research Institute) conceptualizes "GIS as a framework for the collection, management, and analysis of data. Grounded in the principles of geography, GIS integrates various types of data, examining spatial locations and organizing layers of information into visual representations, such as maps and 3D scenes".
- The United Nations Economic Commission for Africa (UNECA) describes "GIS as a robust tool for capturing, storing, analyzing, and visualizing spatial data. This capability supports informed decision-making and enhances the management of resources, infrastructure, and environmental challenges".
- The University of California, Berkeley explains that "GIS as a computer-based system that enables users to integrate spatial data, such as maps and images, with corresponding attribute data, facilitating tasks such as mapping, visualization, spatial analysis, and querying"

5.1.2 History of GIS

The history of Geographic Information Systems (GIS) is a compelling narrative that dates back to the early 1960s, marking the beginning of its evolution as a transformative technology.

- **Early foundations:** The origins of spatial data analysis can be traced back to the nineteenth century when cartographers and mapmakers began overlaying maps to derive meaningful insights. However, the true development of GIS began with the advent of computer technology.
- **1960s–1970s:** The early 1960s marked the advent of computer-based mapping systems. Roger Tomlinson's development of the "Canada Geographic Information System" (CGIS) in 1960 is widely regarded as the first true GIS. Focused on

land-use planning and resource management, CGIS laid the groundwork for modern GIS technology.

- **1980s:** The 1980s saw the commercialization of GIS, driven largely by organizations like the Environmental Systems Research Institute (ESRI). ESRI's introduction of the ARC/INFO software in 1982 became a landmark event, establishing a platform that remains influential in GIS development to this day.
- **1990s:** During the 1990s, GIS became more accessible to a broader audience, facilitated by the development of user-friendly software and the integration of GIS with the burgeoning World Wide Web. These advancements extended GIS applications beyond technical experts to diverse industries.
- **2000s–present:** The twenty-first century witnessed the rapid expansion of GIS applications across disciplines such as disaster response, urban planning, environmental management, and healthcare. Innovations like web-based GIS platforms and mobile applications, exemplified by services like Google Maps and location-based technologies, have made GIS an essential part of daily life.
- **Contemporary trends:** GIS is continually advancing with the integration of big data, artificial intelligence, and the Internet of Things (IoT). It plays a crucial role in tackling global challenges like climate change, public health crises, and sustainable urban development, reinforcing its significance in shaping the future.

The history of GIS reflects its remarkable evolution from a specialized analytical tool to a pervasive technology that drives decision-making across diverse industries and applications. It continues to serve as an indispensable resource for understanding and analyzing the complex spatial relationships that shape our world.

5.1.3 Components of GIS

Geographic Information Systems (GIS) comprise several interconnected components that collaboratively enable the capture, storage, analysis, and visualization of spatial data (Fig. 5.2). The key components of GIS are outlined below:

- **Hardware:** The hardware component encompasses physical devices such as computers, servers, GPS devices, scanners, and other equipment essential for collecting, processing, and storing geographic data. High-performance computing systems are critical for managing the complex calculations and large datasets involved in GIS analysis.
- **Software:** GIS software provides the tools and applications necessary for managing, analyzing, and visualizing spatial data. Prominent examples include ESRI's ArcGIS, QGIS, and open-source platforms like GRASS GIS. These tools enable users to create maps, conduct spatial analysis, and manage geodatabases effectively.
- **Data:** Data forms the foundation of GIS and includes a wide range of information, such as maps, satellite imagery, GPS coordinates, and attribute data. Geographic data is typically categorized as vector data (e.g., points, lines, polygons)

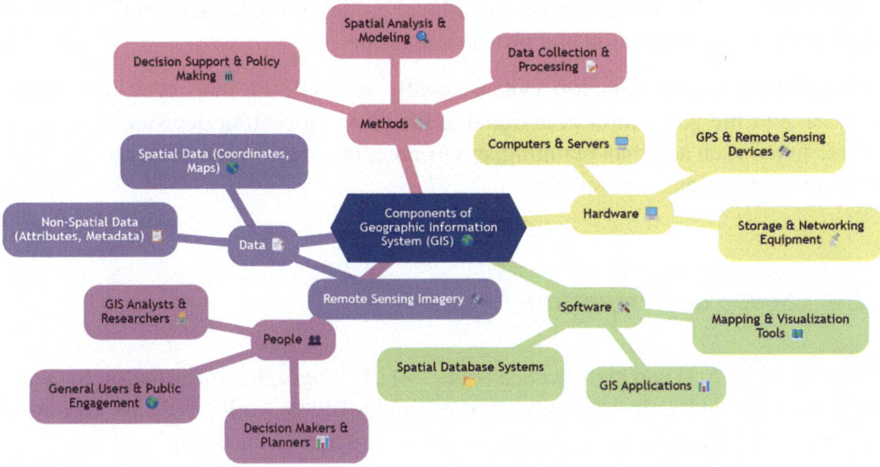

Fig. 5.2 Components of Geographic Information Systems (prepared by the authors)

and raster data (e.g., grid-based cells). Accurate, high-quality, and up-to-date data are essential for reliable analysis and informed decision-making.

- **People:** Skilled professionals, including GIS analysts, cartographers, and data scientists, are indispensable to GIS operations. These experts design and implement GIS projects, manage and analyze data, and interpret results to support decision-making across various sectors.
- **Methods/procedures:** The methodologies and procedures in GIS define the workflows for collecting, processing, analyzing, and presenting spatial data. This includes data modelling, geoprocessing techniques, and established best practices to ensure effective and accurate spatial analysis.
- **Networks/communication:** Effective dissemination and sharing of geographic information require robust communication infrastructure. The Internet, intranet systems, and cloud platforms facilitate the distribution of GIS data and results to stakeholders, enabling collaboration across organizations and users.
- **Standards:** Established standards ensure data consistency and interoperability within GIS. Guidelines provided by organizations such as the Open Geospatial Consortium (OGC) and widely used data formats like Shapefiles and GeoJSON enable seamless data exchange between different GIS platforms and software.
- **Quality control:** Ensuring the accuracy and reliability of GIS data is critical. Quality control processes, such as data validation, error correction, and metadata creation, are essential for maintaining the integrity of GIS datasets and analyses.
- **Training and education:** The effective use of GIS requires ongoing training and education. Programmes and courses tailored to GIS software, data management, and analytical techniques are widely available to build the expertise needed for advanced GIS applications.
- **Legal and ethical considerations:** GIS projects often involve sensitive data and raise concerns about privacy and intellectual property. Adhering to legal regula-

tions, such as data privacy laws, and ethical guidelines is fundamental to responsible GIS implementation and project execution.

These components function synergistically to facilitate map creation, spatial analysis, and the generation of actionable insights, supporting decision-making in diverse fields such as urban planning, environmental conservation, emergency management, and business operations.

5.1.4 *Scope and Importance*

The scope (Fig. 5.3) and significance (Fig. 5.4) of Geographic Information Systems (GIS) are vast and ever-growing in today's data-driven world. With its continual evolution, GIS has become an essential tool across numerous disciplines.

Fig. 5.3 Scope of GIS (prepared by the authors)

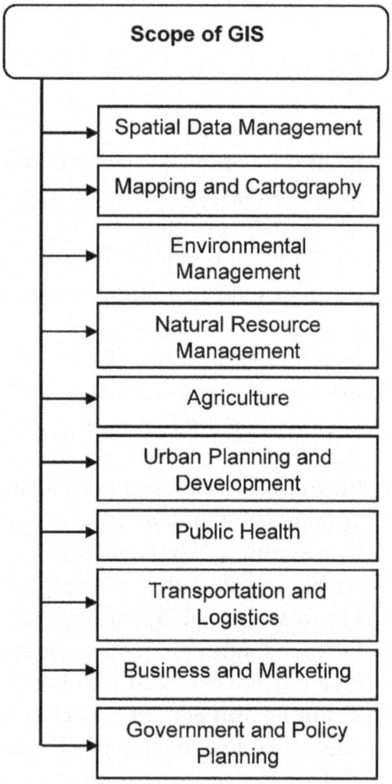

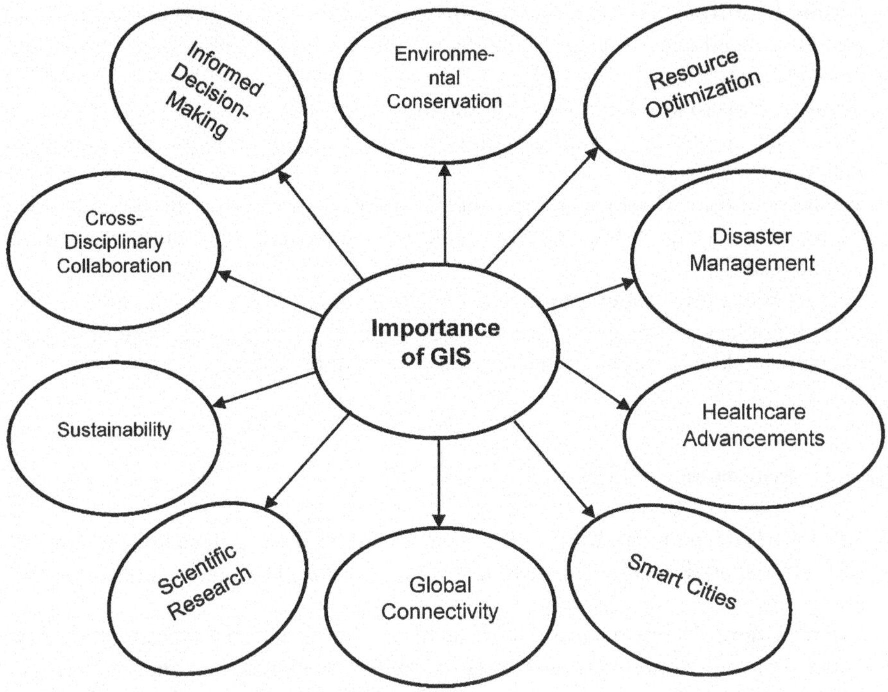

Fig. 5.4 Importance of GIS (prepared by the authors)

5.1.4.1 Scope of GIS

- **Spatial data management:** GIS provides a structured system for storing, organizing, and retrieving geographic information, including maps, satellite imagery, and attribute data, to support analysis and decision-making.
- **Mapping and cartography:** GIS is integral to cartography, enabling the creation of accurate, accessible maps that present complex data in visual formats.
- **Environmental management:** GIS is essential for monitoring ecosystems, managing natural resources, and supporting disaster response and mitigation efforts.
- **Natural resource management:** GIS plays a critical role in managing natural resources, supporting exploration, extraction, and environmental assessments in industries like mining and oil.
- **Agriculture:** GIS supports precision farming by optimizing crop management, soil analysis, and irrigation, improving yields while minimizing environmental impact.
- **Urban planning and development:** GIS is widely used in urban planning to manage land use, infrastructure, and zoning regulations, supporting analyses of population growth and environmental impacts.

- **Public health:** GIS aids in tracking disease patterns, managing epidemics, and planning healthcare services by identifying high-risk areas and optimizing resource allocation.
- **Transportation and logistics:** GIS enhances route optimization, traffic management, and logistics planning, contributing to more efficient transportation networks.
- **Business and marketing:** GIS is valuable for market analysis, site selection, and customer profiling, aiding decision-making in industries like retail, real estate, and advertising.
- **Government and policy planning:** Governments use GIS for land administration, boundary delineation, and disaster response, supporting evidence-based decision-making in areas such as taxation and public services.

5.1.4.2 Importance of GIS

- **Informed decision-making:** GIS supports data-driven decision-making by revealing spatial relationships, trends, and patterns, leading to more effective choices across various sectors.
- **Environmental conservation:** GIS aids in monitoring and managing ecosystems, supporting conservation efforts for forests, wetlands, and endangered species, and contributing to climate change mitigation strategies.
- **Resource optimization:** GIS enables businesses and organizations to optimize resource allocation, improving efficiency and reducing costs in areas such as field team deployment, inventory management, and infrastructure planning.
- **Disaster management:** GIS plays a crucial role in disaster preparedness, response, and recovery, assisting with evacuation planning, damage assessment, and resource allocation during natural disasters.
- **Healthcare advancements:** GIS supports epidemiology by tracking disease outbreaks and optimizing vaccine distribution, enhancing public health strategies and emergency responses.
- **Smart cities:** GIS is pivotal in the development of smart cities, facilitating efficient infrastructure management, traffic control, and public service delivery.
- **Global connectivity:** GIS technology connects people globally, enabling location-based services and navigation systems, as seen in platforms like Google Maps and GPS devices.
- **Cross-disciplinary collaboration:** GIS promotes interdisciplinary collaboration, facilitating research and problem-solving across diverse fields.
- **Sustainability:** GIS fosters sustainability by assessing environmental impacts, optimizing resource use, and supporting conservation efforts.
- **Scientific research:** GIS is a valuable tool for researchers across diverse fields, including ecology, geology, archaeology, and climatology, providing spatial analysis capabilities and visualizing complex data.

5.1.5 Key Differences with GIS

5.1.5.1 Remote Sensing (RS) and GIS

Remote sensing and Geographic Information Systems (GIS) are interconnected technologies that work synergistically to enhance spatial data analysis and management. The relationship (Fig. 5.5) between remote sensing and GIS is as follows:

- **Data source:** Remote sensing serves as a primary source of spatial data for GIS. Imagery from satellites or aerial platforms is imported into GIS as layers, forming the foundation of various GIS applications.
- **Enhanced data:** Remote sensing enriches GIS by providing high-resolution, up-to-date information about the Earth's surface. For instance, satellite imagery offers detailed data on land use and land cover, essential for urban planning, agriculture, and environmental monitoring.
- **Spatial analysis:** GIS enhances remote sensing by enabling advanced spatial analyses, such as detecting land cover changes, calculating vegetation indices, or assessing the impacts of natural disasters.
- **Visualization:** GIS transforms raw remote sensing imagery into informative maps and graphics, making it easier for researchers and decision-makers to communicate findings effectively.
- **Integration:** Combining remote sensing with GIS creates a powerful analytical framework. GIS integrates multiple remote sensing datasets with other geographic data, such as demographic or infrastructure information, to provide a holistic view of a region.
- **Environmental monitoring:** Remote sensing facilitates continuous monitoring of Earth's surface, while GIS analyses and interprets temporal data to track environmental changes, such as deforestation, crop health, or ecosystem shifts.

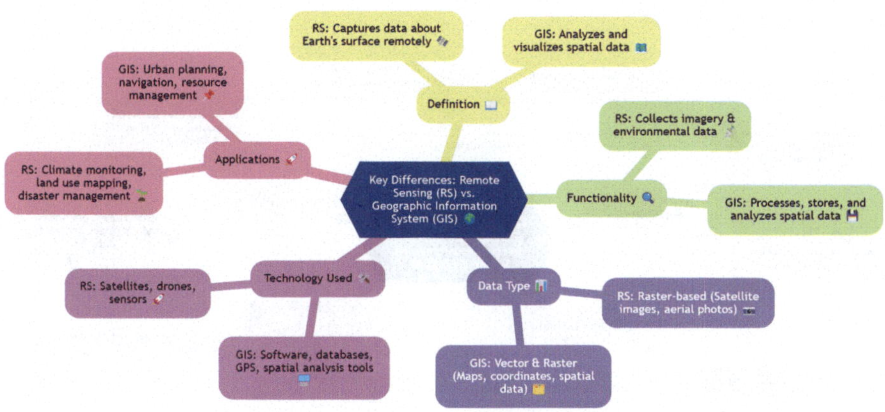

Fig. 5.5 Remote sensing (RS) and Geographic Information Systems (GIS) (prepared by the authors)

- **Emergency response:** Remote sensing provides rapid data on disaster-affected areas, while GIS supports response coordination by mapping affected zones, allocating resources, and evaluating impacts.

5.1.5.2 Artificial Intelligence (AI) and GIS

The interplay between artificial intelligence (AI) and Geographic Information Systems (GIS) creates a dynamic synergy that significantly enhances spatial data analysis and informed decision-making. The relationship (Fig. 5.6) between GIS and AI can be outlined as follows:

- **Data analysis:** AI algorithms enable the rapid processing of large spatial datasets, uncovering hidden patterns and trends that traditional methods might overlook. This capability is invaluable for GIS applications requiring in-depth data analysis.
- **Predictive modelling:** AI enhances GIS by enabling the creation of predictive models to forecast urban expansion, disease outbreaks, or environmental changes. Machine learning algorithms utilize historical data to generate accurate predictions of future events or trends.
- **Image recognition:** AI-powered image recognition automates the analysis of satellite and aerial imagery within GIS. This includes identifying objects, land cover types, and temporal changes, significantly reducing manual analysis efforts.
- **Natural language processing (NLP):** NLP techniques enable GIS to extract geographic information from unstructured text sources like reports and social media posts, facilitating the visualization and analysis of location-based data.
- **Spatial pattern recognition:** AI algorithms excel at detecting complex spatial patterns, including anomalies, clusters, and trends in geographic data, enhancing the depth and accuracy of GIS analysis.

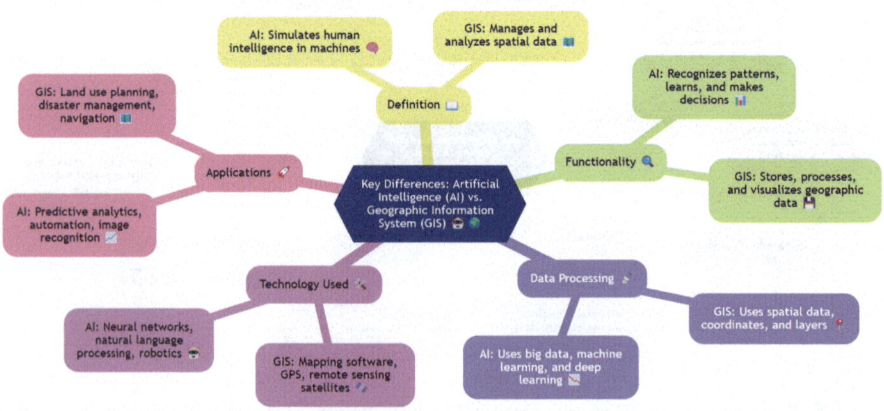

Fig. 5.6 Artificial intelligence (AI) and Geographic Information Systems (GIS) (prepared by the authors)

- **Autonomous systems:** AI powers autonomous systems like self-driving cars and drones, which rely on GIS for navigation and spatial awareness. GIS data is used in real time for decision-making and route optimization.
- **Data fusion:** AI facilitates the integration of diverse data sources, including GIS, IoT devices, and sensors, to build comprehensive spatial models. This is particularly valuable in applications such as smart cities and environmental monitoring.
- **Enhanced decision support:** AI-driven GIS tools provide actionable insights and strategic recommendations derived from complex spatial data, aiding informed decision-making in urban planning, disaster management, and resource allocation.

5.1.5.3 GIS and Participatory Geographic Information Systems (PGIS)

The relationship between Geographic Information Systems (GIS) and Participatory Geographic Information Systems (PGIS) is grounded in their shared objective of leveraging spatial data and technology to facilitate collaborative decision-making and promote community engagement. Consequently, the connection (Fig. 5.7) between GIS and PGIS:

- **Community engagement:** While GIS is typically expert-driven, PGIS emphasizes active community involvement. It promotes local knowledge sharing and decision-making by those directly impacted by geographic issues.
- **Community empowerment:** By providing communities with access to spatial tools and data, PGIS empowers them to analyze, visualize, and advocate for their needs, fostering self-reliance.

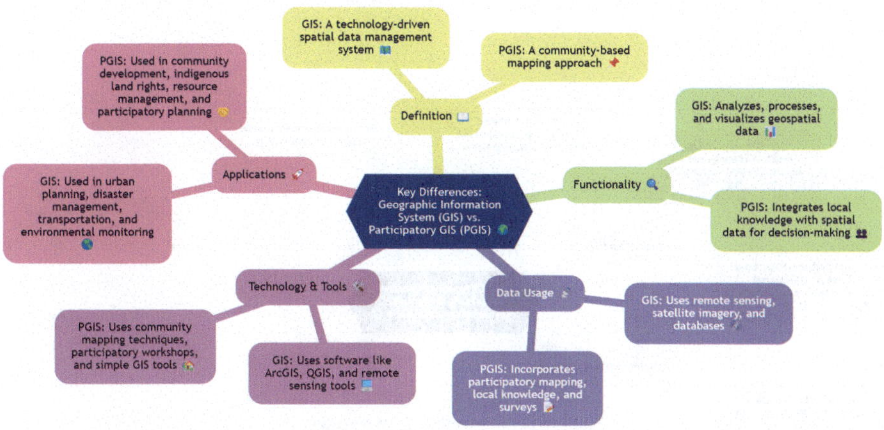

Fig. 5.7 Geographic Information Systems (GIS) and Participatory Geographic Information Systems (PGIS) (prepared by the authors)

- **Data collection:** GIS data is usually gathered by professionals, whereas PGIS involves community members in the data collection process using tools such as GPS devices, smartphones, and participatory mapping techniques.
- **Local knowledge:** PGIS values and integrates local and traditional knowledge, acknowledging that community members often possess unique insights into their environment that enhance GIS data.
- **Collaborative decision-making:** PGIS facilitates collaborative decision-making by involving community members in spatial planning, allowing them to voice concerns and contribute, resulting in more inclusive and equitable outcomes.
- **Environmental justice:** PGIS supports marginalized communities by helping them document environmental changes, communicate their impacts, and advocate for their rights, addressing issues of environmental justice.
- **Resource management:** PGIS is particularly valuable in natural resource management, enabling communities to map resources, document land use, and propose sustainable solutions tailored to their needs.
- **Cultural preservation:** PGIS aids in mapping and preserving cultural heritage sites and practices, protecting indigenous knowledge and traditions for future generations.

5.1.5.4 Manual System and GIS

A manual system relies on paper maps, charts, and physical records to manage and analyze geographic information. Tasks such as map creation, data entry, and spatial analysis are carried out manually, often using tools like compasses and rulers. While this method laid the groundwork for modern spatial analysis, it is labour-intensive, error-prone, and restricts the complexity and scale of analyses that can be performed. Consequently, the relationship (Fig. 5.8) between manual systems and GIS is as follows:

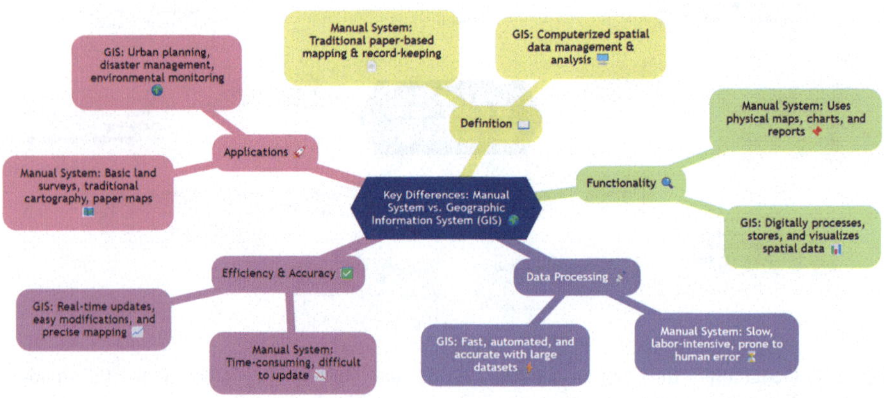

Fig. 5.8 Manual system and Geographic Information Systems (GIS) (prepared by the authors)

- **Efficiency:** GIS significantly enhances efficiency by enabling faster data entry, retrieval, and analysis, saving both time and resources compared to manual systems.
- **Accuracy:** Manual systems are susceptible to human errors such as measurement inaccuracies and transcription mistakes. GIS reduces these errors through automated processes and digital data capture, ensuring more precise results.
- **Complex analysis:** GIS supports advanced spatial analyses, including spatial modelling, overlay analysis, interpolation, and predictive modelling, which are impractical with manual methods, offering deeper insights for decision-making.
- **Data integration:** GIS integrates diverse data sources, such as satellite imagery, GPS data, and socio-economic information, providing a comprehensive understanding of geographic phenomena that manual systems cannot achieve.
- **Visualization:** GIS delivers dynamic, interactive visualizations that allow users to explore and interpret data effectively, unlike static manual maps that lack real-time updates or interactivity.
- **Data management:** GIS centralizes spatial data storage in digital databases, simplifying maintenance, updates, and sharing, whereas manual systems rely on cumbersome physical record-keeping.
- **Remote access:** GIS enables remote access to spatial data over networks, facilitating collaboration and decision-making across distances, unlike manual systems constrained by physical records.
- **Data queries:** GIS allows users to execute complex queries and extract specific information from extensive datasets rapidly, a capability that manual systems, with their labour-intensive processes, lack.
- **Environmental impact:** Transitioning to GIS reduces paper usage and waste, promoting environmental sustainability compared to resource-heavy manual systems.

5.1.6 Applications of GIS

Geographic Information Systems (GIS) are applied across numerous fields and industries, owing to their capacity to capture, store, manage, analyze, and visualize spatial data. The following discussion highlights the various applications of GIS (Fig. 5.9):

- **Environmental management:** GIS is essential for monitoring and managing environmental resources. It is used to track deforestation, assess biodiversity, manage watersheds, and model the impacts of climate change.
- **Natural resource management:** In industries such as forestry, agriculture, and mining, GIS supports land management, crop yield analysis, and resource exploration, promoting sustainable resource use.

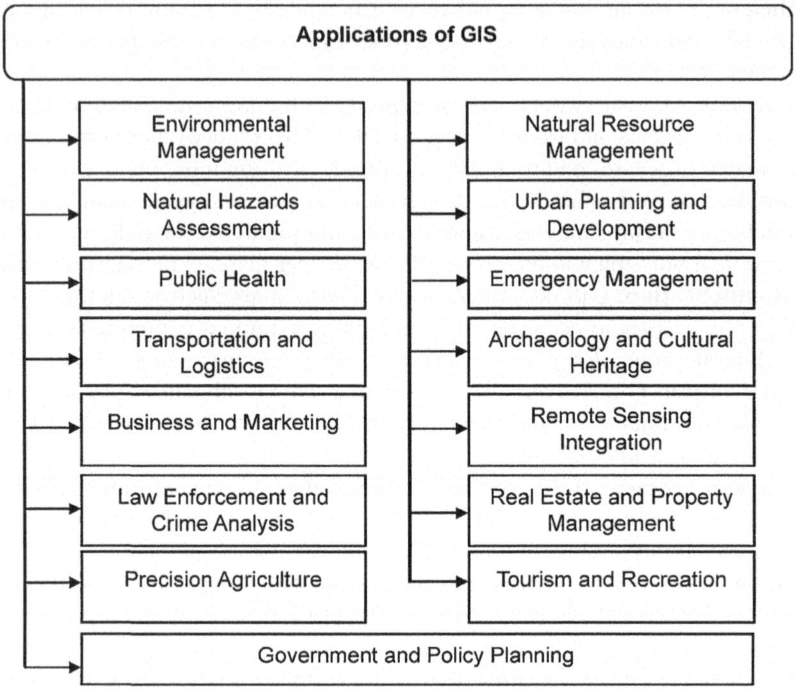

Fig. 5.9 Application of Geographic Information Systems (GIS) (prepared by the authors)

- **Natural hazards assessment:** GIS helps assess risks associated with natural hazards such as floods, earthquakes, and landslides, supporting the creation of hazard maps and informing land-use decisions in vulnerable areas.
- **Urban planning and development:** GIS plays a crucial role in urban planning by facilitating land-use analysis, infrastructure design, and transportation planning. It aids cities in optimizing resource allocation and planning for sustainable development.
- **Public health:** GIS is instrumental in tracking disease outbreaks, analyzing health trends, and planning healthcare services. It plays a critical role in epidemiological investigations and resource allocation during health crises.
- **Emergency management:** GIS supports disaster preparedness, response, and recovery, facilitating evacuation planning, damage assessment, and resource allocation during natural disasters and other emergencies.
- **Transportation and logistics:** GIS optimizes route planning, traffic management, and logistics operations, and is integral to navigation systems and public transportation planning.
- **Business and marketing:** GIS aids businesses in market analysis, site selection, and customer profiling, enabling targeted advertising and optimizing supply chain management.

- **Archaeology and cultural heritage:** GIS is used by archaeologists to analyze historical sites, manage excavation data, and document cultural heritage, assisting in the preservation and study of cultural artifacts.
- **Remote sensing integration:** GIS integrates with remote sensing technologies (e.g., satellite imagery and drones) for applications like disaster assessment, land cover classification, and environmental monitoring.
- **Law enforcement and crime analysis:** GIS is used by law enforcement for crime mapping, hotspot analysis, and resource deployment, enhancing public safety and operational efficiency.
- **Precision agriculture:** GIS supports precision farming by optimizing irrigation, fertilization, and pest control, improving crop yields while minimizing environmental impact.
- **Tourism and recreation:** GIS enhances the tourist experience by providing interactive maps, location-based services, and information on points of interest.
- **Real estate and property management:** GIS is applied in real estate for property valuation, land parcel identification, and location-based decision-making in property development.
- **Government and policy planning:** GIS assists governments in land administration, electoral boundary delineation, and evidence-based policymaking in areas such as taxation, zoning, and public services.

The diverse applications of GIS highlight its broad utility and significance across various sectors, enhancing decision-making, resource optimization, and our understanding of complex, interconnected systems. As technological advancements continue, the scope of GIS applications is expected to grow, fostering innovation and improving efficiency across multiple industries.

5.1.7 Today's GIS and Future Research

Modern Geographic Information Systems (GIS) have undergone significant evolution from their origins, continually transforming how spatial data is collected, managed, analyzed, and visualized.

5.1.7.1 Today's GIS

- **Technological advancements:** Contemporary GIS is defined by enhanced computing capabilities, cloud-based solutions, and intuitive software interfaces. These innovations have significantly broadened the accessibility of GIS, enabled the management of big data and facilitated intricate spatial analyses.
- **Web-based GIS:** Web GIS applications, such as Google Maps and ArcGIS Online, have become increasingly widespread, providing users with remote

access to geographic data via Internet connectivity. This development has expanded GIS's reach and user base.
- **Integration with remote sensing:** The integration of GIS with remote sensing enables the analysis of high-resolution satellite imagery and aerial photography, revolutionizing fields like environmental monitoring and disaster management.
- **Mobile GIS:** Mobile GIS applications have taken the technology into the field, allowing for real-time data collection, GPS tracking, and field mapping. This is particularly valuable in sectors like agriculture, forestry, and utilities.
- **Data sharing and collaboration:** GIS promotes the exchange of spatial data and encourages collaboration among organizations and stakeholders. Through standardized data formats and open data initiatives, GIS fosters greater interoperability and facilitates the sharing of geographic information.

5.1.7.2 Future Research Directions

- **Real-time GIS:** The incorporation of real-time data from sources such as IoT sensors and social media platforms will enable dynamic GIS applications, improving the effectiveness of areas like emergency response and traffic management.
- **AI and machine learning integration:** The integration of AI and machine learning into GIS is expected to enhance capabilities in predictive modelling, object detection within imagery, and automated feature extraction, significantly advancing spatial analysis.
- **3D and indoor mapping:** The expansion of GIS into three-dimensional and indoor mapping is critical for the development of smart cities, urban planning, and navigation within complex indoor environments like airports and shopping malls.
- **Big data analytics:** As GIS continues to handle the vast volumes of data generated globally, developing efficient data storage, processing, and analytical techniques will be crucial to manage the growing scale of spatial data.
- **Privacy and ethics:** With the increasing collection and analysis of location-based data, addressing concerns related to privacy and ethical implications of data usage and surveillance will become essential for responsible GIS practices.
- **Climate change and sustainability:** GIS will play an integral role in monitoring and mitigating the impacts of climate change, with research focusing on modelling environmental changes and their consequences on communities.
- **Health GIS:** The ongoing global health challenges, including pandemics, will drive advancements in health GIS, particularly in areas such as disease mapping, vaccine distribution, and healthcare resource allocation.
- **Crisis management:** Continued research will refine GIS applications for disaster response and recovery, enhancing tools for resource allocation, damage assessment, and evacuation planning in crisis situations.

- **AI-assisted decision support:** The development of AI-driven decision support systems that leverage GIS will be pivotal for industries such as public administration, transportation, and urban planning, enhancing decision-making processes.
- **Ethnic and social inclusion:** Future GIS research will prioritize ensuring that technology is inclusive and equitable, addressing concerns of social justice and the fair distribution of GIS benefits across all communities.

In conclusion, contemporary Geographic Information Systems (GIS) represent a dynamic and continuously evolving field with a broad range of applications across diverse industries. Ongoing research and technological advancements are poised to further enhance GIS capabilities, reinforcing its role as a crucial tool for tackling complex spatial challenges and deepening our understanding of the world.

5.2 GIS Data Types and Models

5.2.1 GIS Data Types

Geographic Information Systems (GIS) utilize diverse types of data (Fig. 5.10), each playing a distinct role in supporting spatial analysis and visualization.

Vector Data

Vector data serves as one of the fundamental data models in Geographic Information Systems (GIS) for representing spatial features (Fig. 5.11). It utilizes geometric shapes like points, lines, and polygons to accurately model real-world phenomena. Renowned for its precision, vector data is extensively employed in applications that demand detailed spatial representation, including urban planning, environmental analysis, and transportation mapping.

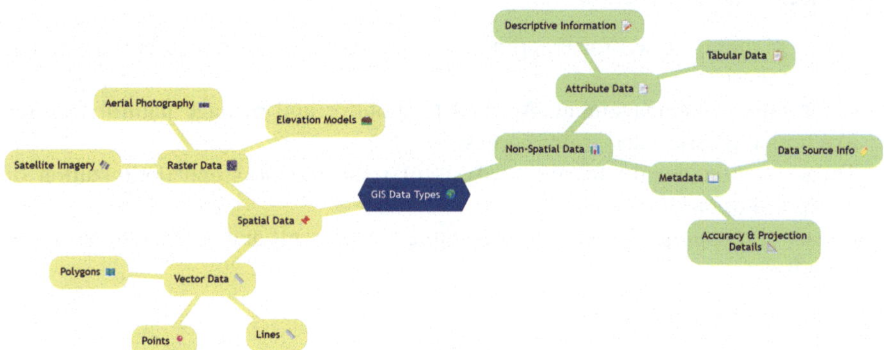

Fig. 5.10 GIS data types (prepared by the authors)

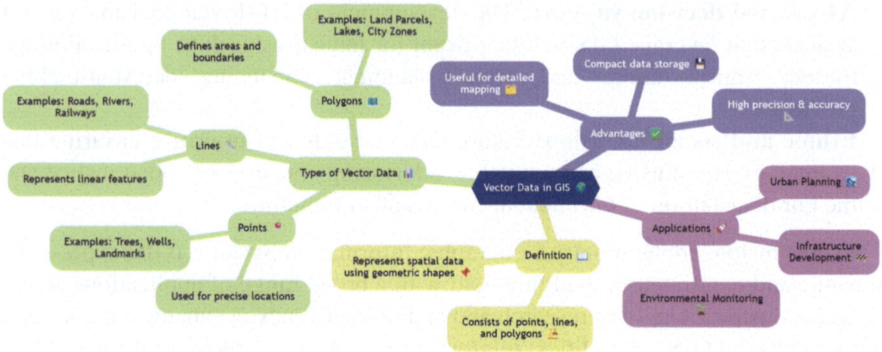

Fig. 5.11 GIS data type: vector data (prepared by the authors)

Characteristics of Vector Data

- Attribute data: Vector features are linked to attribute data stored in a database, allowing for comprehensive analysis of both spatial and non-spatial properties.
- Precision: Vector data provides a high level of accuracy in representing geographic features, making it superior to raster data for detailed spatial modelling.
- Topology: Vector data preserves spatial relationships such as adjacency, connectivity, and containment, which are essential for advanced spatial analysis.

Types of Vector Data

- Point data: Represents discrete locations on the Earth's surface. Examples: Wells, streetlights, or landmarks.
- Line data: Represents linear features with a defined start and end point. Examples: Roads, rivers, or pipelines.
- Polygon data: Represents closed areas defined by a series of connected lines. Examples: Land parcels, lakes, or administrative boundaries.

Applications of Vector Data

- Urban planning: Used for mapping infrastructure such as roads, buildings, and zoning districts.
- Environmental management: Applied to monitor and manage natural features like protected areas and water bodies.
- Transportation: Facilitates the analysis of road networks and optimization of transportation routes.
- Cadastral mapping: Essential for depicting land parcels and delineating property boundaries.

Advantages of Vector Data

- High accuracy: Ideal for applications demanding precise geometric representation, such as cadastral mapping.

- Storage efficiency: Requires less storage space than raster data for representing discrete spatial features.
- Scalability: Can be resized or scaled without compromising data quality.
- Advanced analysis: Enables complex spatial operations, including buffering, overlay analysis, and network analysis.

Limitations of Vector Data

- Algorithmic complexity: Vector data analysis requires more sophisticated algorithms compared to raster data, increasing computational demands.
- Time-intensive processing: Tasks such as topology editing and maintaining geometric relationships can be time-consuming.
- Challenges in visualizing continuous data: Representing continuous phenomena, such as elevation or temperature, is less intuitive with vector data compared to raster formats.

Raster Data

Raster data is a fundamental data model in Geographic Information Systems (GIS) used to represent spatial information (Fig. 5.12). It consists of a grid of cells, or pixels, where each cell holds a value corresponding to a specific attribute, such as elevation, temperature, or land cover. This data model is especially effective for representing continuous spatial phenomena that vary across a landscape.

Characteristics of Raster Data

- Grid structure: Raster data is arranged in a grid format consisting of rows and columns, with each cell representing a specific geographic location.
- Resolution: The resolution of raster data is determined by the size of each cell. Smaller cells provide finer detail and higher resolution, while larger cells reduce detail but require less storage space.

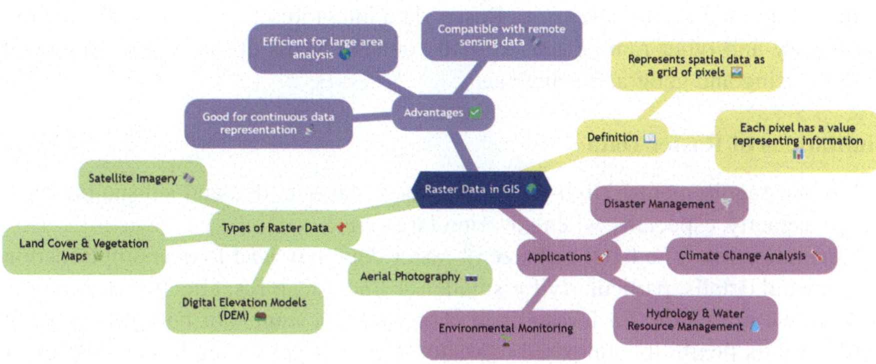

Fig. 5.12 GIS data type: raster data (prepared by the authors)

- Attribute values: Each cell in a raster dataset holds a single value, representing either a physical measurement (e.g., elevation) or a categorical attribute (e.g., forest, urban).
- Extent: Raster data covers continuous geographic areas, making it particularly suitable for modelling spatial phenomena such as temperature, rainfall, and vegetation patterns.

Applications of Raster Data

- Environmental analysis: Raster data is instrumental in modelling climate change, monitoring vegetation growth, and assessing land degradation over time.
- Hydrological modelling: It plays a crucial role in analyzing watersheds, modelling flood scenarios, and studying groundwater flow dynamics.
- Land use and land cover mapping: Satellite-derived raster data is extensively used for classifying and monitoring changes in land use and land cover patterns.
- Disaster management: Raster data is essential in mapping and analyzing natural hazards, including floods, forest fires, and earthquakes, aiding in preparedness and response efforts.
- Terrain analysis: Digital elevation models (DEMs) are commonly employed for terrain analysis, including the generation of slope, aspect, and contour maps to assess landscape characteristics.

Advantages of Raster Data

- Representation of continuous data: Raster data is well-suited for representing continuous variables, such as temperature, elevation, and soil moisture, where values change gradually across space.
- Simplicity: The grid-based structure of raster data is simple and efficient, making it easy to process using computational techniques.
- Analysis: Raster data supports a broad spectrum of spatial analyses, including surface analysis, interpolation, and raster algebra, facilitating complex environmental and geographic studies.
- Integration with remote sensing: Raster data integrates seamlessly with satellite imagery and other remote sensing data, enhancing its utility in environmental monitoring and geographic analysis.

Limitations of Raster Data

- Storage requirements: High-resolution raster datasets demand substantial storage capacity, especially when covering large areas.
- Generalization: The fixed cell size of raster data may lead to oversimplification of spatial details, particularly for smaller or linear features, affecting accuracy.
- Attribute limitation: Each raster cell is limited to a single attribute value, which reduces its flexibility compared to vector data in representing more complex or multiple attributes.

- Resolution dependency: The effectiveness of raster data is heavily reliant on its resolution, and lower resolution may not be suitable for applications that require high precision or detailed analysis.

5.2.2 *GIS Data Models*

Geographic Information Systems (GIS) utilize a variety of data models to effectively organize, structure, and manage spatial information (Fig. 5.13). These models determine how geographic features and their associated attributes are represented and interrelated within the GIS framework. The following discussion highlights the major data models used in GIS:

Vector Data Model

- Representation: The vector data model represents geographic features as distinct geometric shapes, including points, lines, and polygons (Fig. 5.14).
- Attributes: Each vector feature, such as a point symbolizing a city, is linked to an attribute table that stores non-spatial information, such as the city's name or population.
- Examples: The vector data model is commonly used to represent features like roads, administrative boundaries, land parcels, and individual trees.

Raster Data Model

- Representation: The raster data model partitions geographic areas into a uniform grid of cells or pixels, each representing a specific location (Fig. 5.15).
- Attributes: Each cell is assigned a value corresponding to a particular attribute, such as elevation, temperature, or land cover type.
- Examples: Raster data models are widely applied in digital elevation models (DEMs), satellite imagery, and land cover classification.

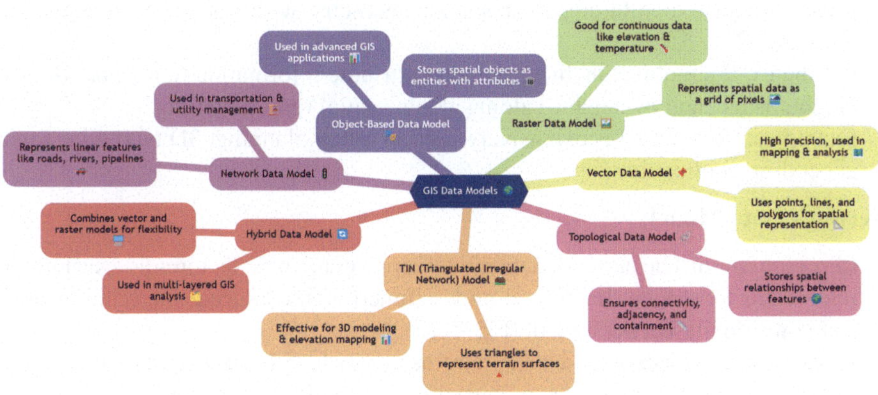

Fig. 5.13 GIS data models (prepared by the authors)

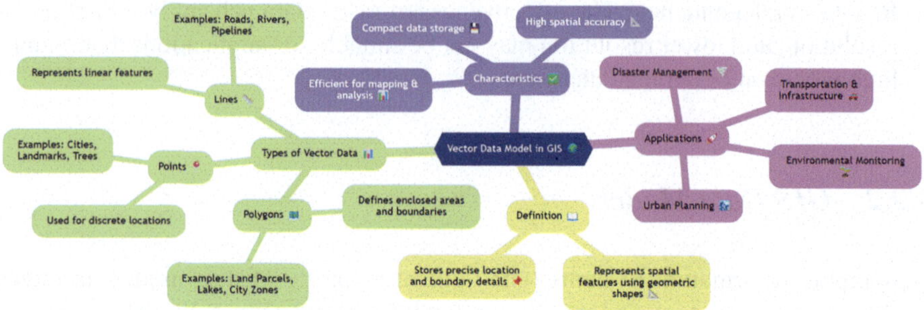

Fig. 5.14 Vector data models (prepared by the authors)

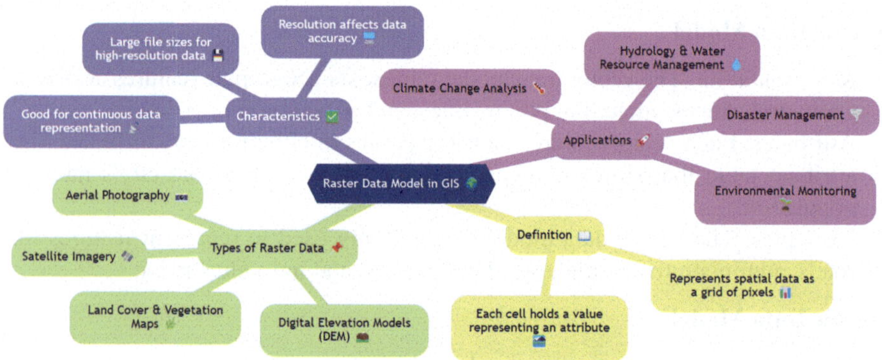

Fig. 5.15 Raster data models (prepared by the authors)

Object-Based Data Model

- Representation: The object-based data model represents geographic features as distinct objects, combining their spatial geometry with associated non-spatial attributes (Fig. 5.16).
- Attributes: Each object includes detailed attribute information (e.g., a building's height, usage, and ownership) alongside its spatial representation.
- Examples: This model is extensively used in urban planning, 3D modelling, and geocoding applications.

Network Data Model

- Representation: The network data model represents linear features as a system of interconnected edges (links) and nodes (junctions), capturing the relationships and connectivity between them (Fig. 5.17).
- Attributes: Both edges and nodes are associated with attributes such as road names, lengths, travel speeds, or capacities, enabling detailed analysis.

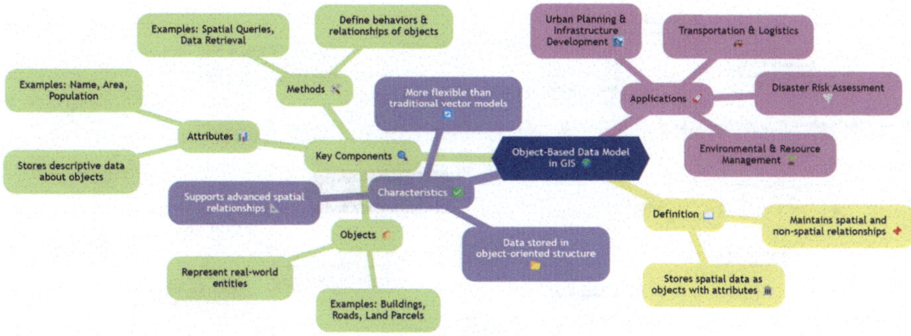

Fig. 5.16 Object-based data model (prepared by the authors)

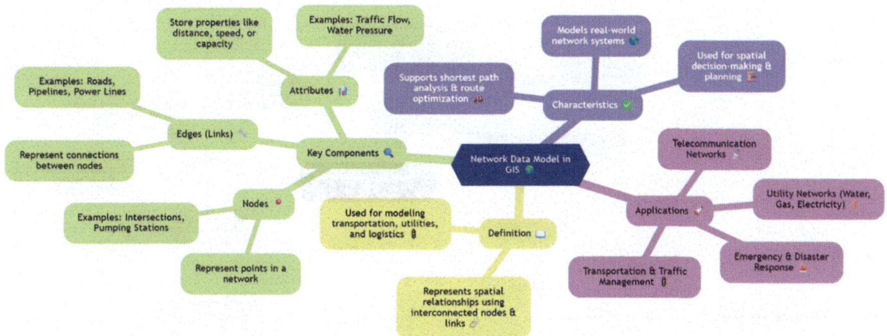

Fig. 5.17 Network data model (prepared by the authors)

- Examples: Network data models are widely used in transportation systems (e.g., road networks, railways), utility networks (e.g., pipelines, power grids), and routing or network analysis.

Topological Data Model

- Representation: The topological data model defines spatial relationships between features by representing connections, adjacencies, and containment, ensuring accurate spatial relationships (Fig. 5.18).
- Attributes: Explicitly stores topological relationships, such as shared boundaries, connected lines, and enclosed areas, enabling efficient spatial analysis and error detection.
- Examples: Topological data models are utilized in GIS for tasks like map overlay, network analysis, and ensuring data integrity in applications such as cadastral mapping and land-use planning.

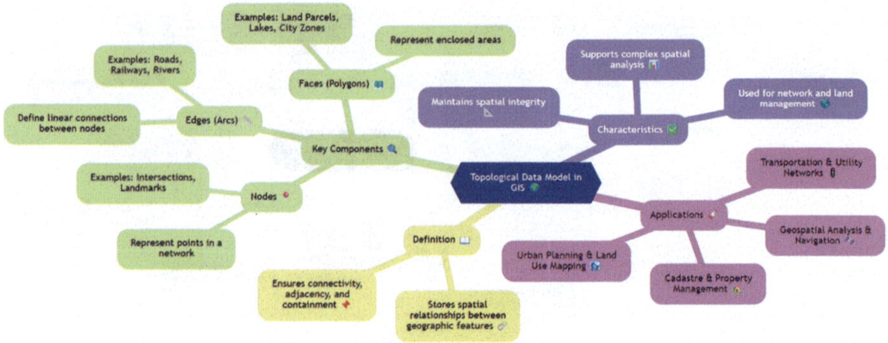

Fig. 5.18 Topological data model (prepared by the authors)

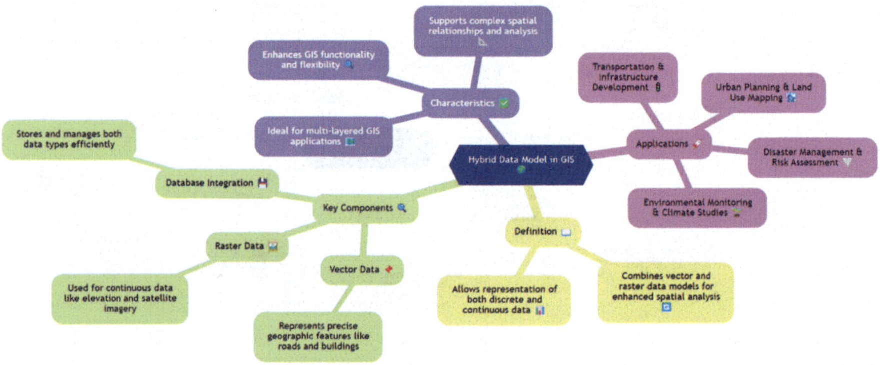

Fig. 5.19 Hybrid data model (prepared by the authors)

Hybrid Data Model

- Representation: Hybrid data models integrate elements of both vector and raster data models, combining the strengths of each to represent geographic features and spatial phenomena (Fig. 5.19).
- Attributes: These models enable the representation of vector-like discrete features (e.g., roads, boundaries) within a raster framework, allowing for seamless integration and flexibility in spatial analysis.
- Examples: Hybrid data models are commonly applied in land-use planning, where vector boundaries are superimposed on raster imagery for enhanced analysis and visualization.

TIN (Triangulated Irregular Network) Data Model

- Representation: TIN (triangulated irregular network) data models represent surfaces by partitioning geographic areas into a network of interconnected, non-overlapping triangles (Fig. 5.20).

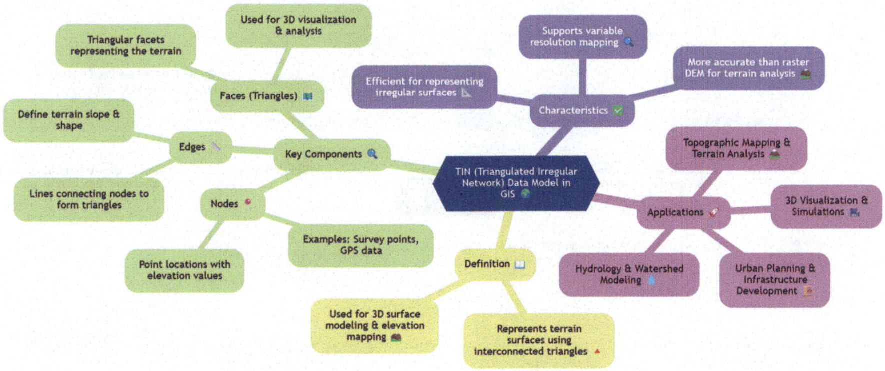

Fig. 5.20 Triangulated irregular network data model (prepared by the authors)

- Attributes: Each triangle is assigned elevation or other surface attributes, making it suitable for accurately modelling terrain and 3D surfaces.
- Examples: TIN models are widely utilized in terrain analysis, hydrological studies, and 3D landscape visualization, such as slope and contour mapping.

5.3 Error in GIS

Error in Geographic Information Systems (GIS) refers to inaccuracies in spatial data or analysis results, arising from sources like data acquisition, processing, or modelling. Acquisition errors may result from imprecise GPS measurements or outdated maps, while processing errors can occur during digitization or georeferencing. Model-based errors stem from simplifications or assumptions in analytical models. Such errors can significantly impact precision-critical applications like land surveying or navigation. To mitigate these issues, GIS professionals use techniques such as error propagation modelling and accuracy assessment, ensuring reliable spatial data and informed decision-making (Fig. 5.21).

5.3.1 Conceptual Error in GIS

Conceptual errors in Geographic Information Systems (GIS) arise from the misinterpretation or misunderstanding of geographic concepts, models, or spatial relationships applied in GIS analysis. Unlike errors related to data accuracy, these issues stem from how spatial information is conceptualized and utilized by GIS users. Such errors can significantly affect the reliability and validity of GIS outcomes (Fig. 5.22). Common causes of conceptual errors include:

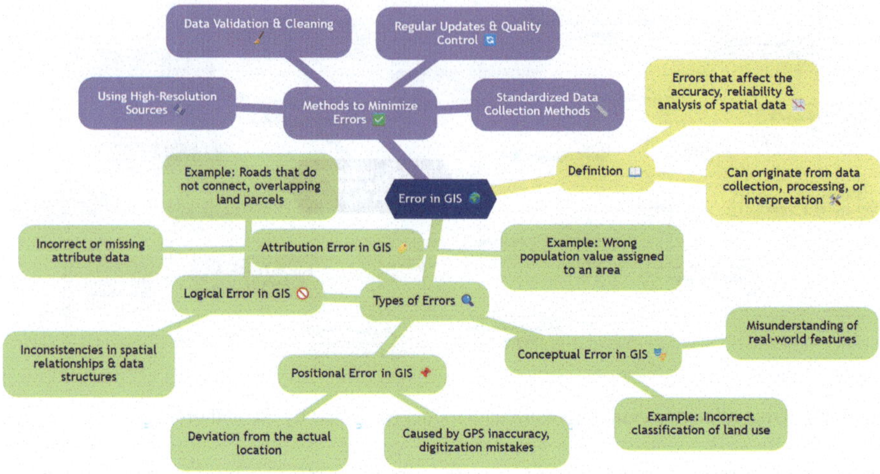

Fig. 5.21 Error in GIS (prepared by the authors)

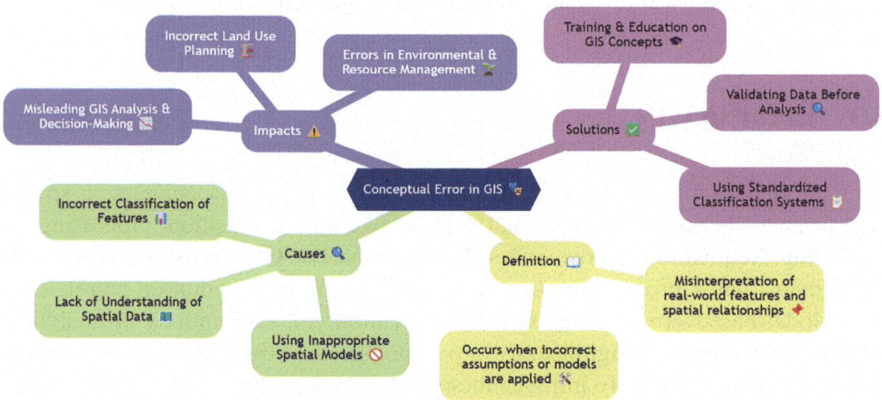

Fig. 5.22 Conceptual error in GIS (prepared by the authors)

- Misconceptions: Users may hold incorrect assumptions about the geographic phenomena being analyzed, resulting in flawed interpretations or conclusions.
- Modelling Errors: Errors may occur when users apply inappropriate or overly simplified models to represent complex spatial processes.
- Spatial Scale: Misunderstanding or neglecting the scale at which data is collected and analyzed can lead to errors, as geographic phenomena often differ at various scales.
- Data Integration: Merging datasets with incompatible conceptual frameworks can cause errors in spatial analysis and interpretation.
- Semantic Errors: Mis-defining or misassigning meanings to geographic attributes and relationships can lead to inaccurate outcomes.

Addressing conceptual errors in GIS requires enhancing user training and education to foster greater spatial literacy, ensuring users comprehend the fundamental geographic principles and theories relevant to their analyses. Additionally, it involves regularly reviewing and refining the conceptual frameworks applied in GIS projects. By minimizing conceptual errors, GIS professionals can enhance the accuracy and reliability of spatial analyses and decision-making processes.

5.3.2 Positional Error in GIS

Positional error in Geographic Information Systems (GIS) pertains to inaccuracies or deviations in the geographic coordinates or locations of spatial data relative to their true, real-world positions. This type of error is primarily linked to the measurement and representation of geographic features (Fig. 5.23). Positional error can arise from various factors:

- Data collection: Positional inaccuracies can arise from errors in data collection methods, such as GPS measurements or manual digitization. GPS devices may have limited accuracy, particularly in environments like dense forests or urban canyons.
- Map projections: Distortions are inevitable when representing the Earth's curved surface on flat maps, resulting in positional errors. The selection of map projection and coordinate system can affect the extent and distribution of these errors.
- Datum transformations: Variations in geographic datums, which serve as reference models for the Earth's shape, can lead to positional discrepancies during data transformations. Datum shifts are often required when integrating data from different sources or altering coordinate systems.
- Data resolution: Coarse-resolution data may fail to accurately represent fine-scale geographic features, leading to positional inaccuracies in these features.

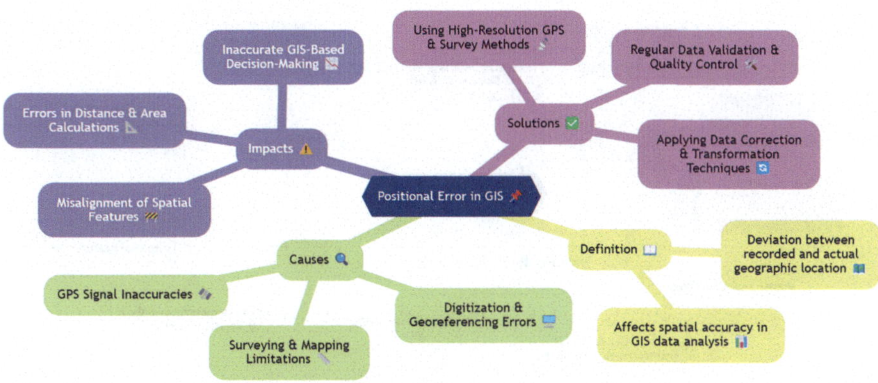

Fig. 5.23 Positional error in GIS (prepared by the authors)

- Georeferencing and registration: Errors in georeferencing can occur when aligning scanned or paper maps with spatial data. Misregistration of data can result in positional discrepancies between datasets.
- Data editing and updates: Manual edits or updates to spatial data, if not carefully validated, can introduce positional errors. Maintaining data integrity during editing processes is essential.
- Temporal changes: Geographic features may shift over time due to natural or human-induced changes. Failing to account for these changes can lead to positional inaccuracies when using historical data.

Mitigating positional errors requires the use of accurate data collection methods, the careful selection of suitable map projections and coordinate systems, the application of precise georeferencing techniques, and the regular validation and updating of spatial datasets. GIS professionals must remain mindful of positional errors and account for them in spatial analysis and decision-making to maintain the reliability and accuracy of GIS outcomes.

5.3.3 *Attribution Error in GIS*

Attribution error in Geographic Information Systems (GIS) refers to inaccuracies or inconsistencies in the attribute data linked to spatial features. While positional errors concern the geographic location of features, attribution errors focus on the accuracy, completeness, and consistency of the descriptive information associated with those features (Fig. 5.24). These errors may occur due to the following reasons:

- Data inaccuracy: Errors can occur when attribute data is outdated or incorrect, such as inaccuracies in population figures, land-use classifications, or property values due to outdated sources or mistakes in data entry.

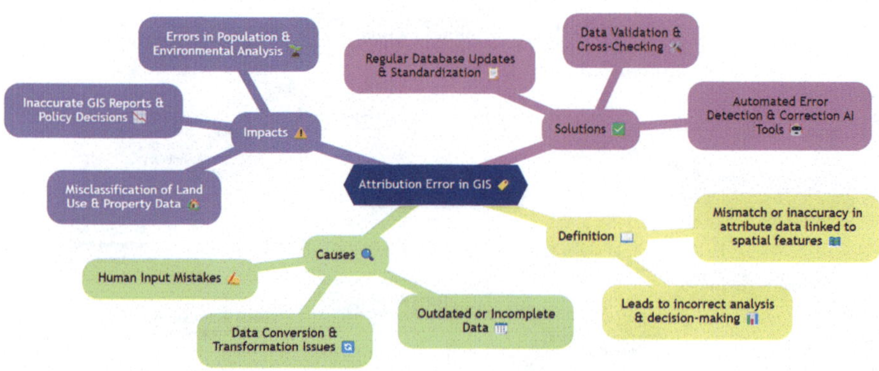

Fig. 5.24 Attribution error in GIS (prepared by the authors)

- Missing data: The absence of crucial attribute information, such as ownership details or demographic data, can compromise the integrity of GIS analysis.
- Inconsistencies: Variations in naming conventions or measurement units can introduce inconsistencies in attribute data. For instance, discrepancies in place names or the use of both metric and imperial units can result in errors.
- Data entry errors: Human errors, such as typographical mistakes, omissions, or incorrect values during data entry, can lead to attribution errors. These errors may not be immediately evident but can significantly impact analysis outcomes.
- Temporal changes: Changes in attribute data over time can lead to inaccuracies if historical and current data are not properly managed and synchronized, affecting trend analysis and decision-making.
- Metadata issues: Inadequate or inconsistent metadata, which describes the data's source, quality, and usage, can cause difficulties in understanding and appropriately utilizing attribute data.
- Data integration: Combining datasets from various sources with differing levels of quality may result in attribute errors. Incompatible or mismatched attributes can distort analysis results.

To mitigate attribution errors in GIS, professionals should implement rigorous data validation and verification processes. Regular updates to data sources and thorough metadata documentation are essential. Standardized data models, data dictionaries, and quality control protocols can enhance the accuracy and reliability of GIS analysis and decision-making.

5.3.4 Logical Error in GIS

Logical errors in Geographic Information Systems (GIS) occur due to inconsistencies, contradictions, or flawed reasoning within the data or analysis processes. Unlike positional or attribution errors, which concern data accuracy and completeness, logical errors pertain to the internal consistency and validity of GIS operations and workflows (Fig. 5.25). The following provides an overview of logical errors in GIS:

- Topology errors: These occur when spatial relationships between geographic features violate defined topological rules, such as gaps or overlaps between polygons, undershoots or overshoots in lines, or disconnected road networks. Such errors can hinder spatial analysis, including network routing or overlay operations.
- Data integrity issues: Logical errors can arise from problems with data integrity, such as duplicate records, inconsistent attribute values, or invalid data types. These issues can compromise the accuracy of queries, calculations, and analyses in GIS.

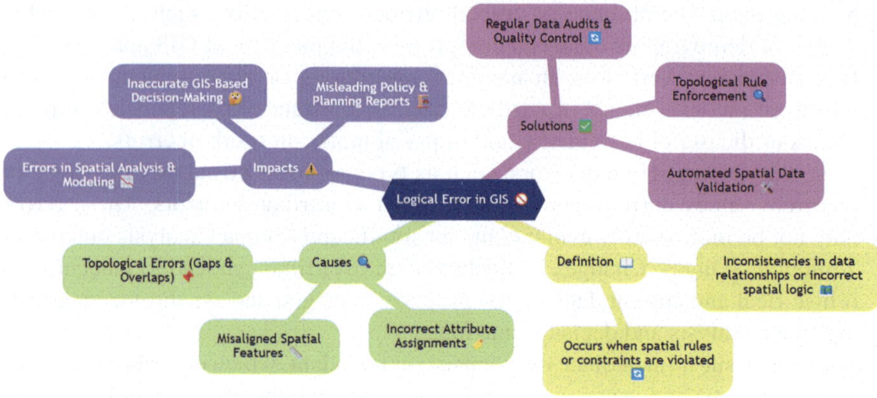

Fig. 5.25 Logical error in GIS (prepared by the authors)

- Inconsistent data models: Conflicts between data models, formats, or schemas in different parts of a GIS project can lead to logical errors. Incompatible models during data integration can cause misalignment and errors.
- Incorrect geoprocessing workflows: Errors occur when users design geoprocessing workflows that do not align with the intended spatial operations or logical conditions. This can result in incorrect analysis outcomes.
- Conflicting data definitions: Ambiguous or poorly defined geographic features or attributes can lead to logical errors. Misinterpretation or misapplication of data may arise from insufficient documentation or unclear definitions.
- Improper spatial analysis techniques: Using unsuitable spatial analysis methods or algorithms for certain tasks can result in logical errors. For instance, applying a nearest-neighbours analysis to non-point data may produce invalid results.

Addressing logical errors in GIS requires careful planning, adherence to data standards, thorough documentation, and ongoing quality control. Users should validate spatial relationships, apply robust data validation techniques, and establish consistent workflows to minimize errors and ensure the integrity of GIS analyses.

5.4 GIS and Other Disciplines

Geographic Information Systems (GIS) is a versatile, interdisciplinary tool with significant impact across various fields. Its capability to capture, manage, analyze, and visualize spatial data makes it an essential resource for professionals in diverse domains. The following is a comprehensive discussion on the relationship between GIS and several key disciplines.

Environmental science and ecology	Ecosystem analysis: GIS assists environmental scientists in mapping land cover, tracking habitat changes, and monitoring species distribution, supporting conservation efforts by identifying key protection areas. Natural resource management: GIS aids in managing resources like forests, water bodies, and wildlife, modelling resource dynamics and predicting environmental impacts to promote sustainability. Climate change modelling: GIS supports climate change research by analyzing trends in temperature, sea-level rise, and carbon emissions, assisting in planning for adaptation and mitigation strategies.
Agriculture and forestry	Precision agriculture: GIS enhances agricultural practices by analyzing soil data, monitoring crop health, and optimizing irrigation and fertilization through precision techniques. Forestry management: GIS aids forestry professionals in forest inventory, fire risk assessment, and promoting sustainable timber harvesting practices.
Geology and earth sciences	Geological mapping: GIS supports geologists in generating detailed geological maps, modelling subsurface structures, and evaluating geological hazards. Natural hazard assessment: GIS is essential for analyzing natural hazards such as volcanoes, earthquakes, and landslides by mapping vulnerable areas and assessing risk.
Energy and utilities	Infrastructure management: GIS is used by energy and utility companies for asset management, network analysis, and maintenance planning.
Urban planning and architecture	City planning: Urban planners leverage GIS to analyze urban growth, traffic patterns, and land use, supporting sustainable city design, optimizing transportation networks, and managing urban sprawl. Architectural design: Architects employ GIS for site suitability analysis, viewshed studies, and contextual design, enabling informed decisions on building placement and urban layout.
Social sciences	Demographic analysis: Sociologists and demographers utilize GIS to examine population trends, migration patterns, and social disparities through spatial data visualization.
Public health	Epidemiology: Public health professionals use GIS to monitor disease outbreaks, analyze health disparities, and optimize healthcare resource allocation, enhancing disease surveillance and response strategies. Healthcare access: GIS identifies underserved areas with limited healthcare access, aiding in the development of more equitable healthcare systems.
Education	Geospatial education: GIS is incorporated into educational programmes to develop spatial reasoning and analytical abilities, promoting geospatial literacy and preparing students for diverse career opportunities.
Criminal justice and law enforcement	Crime mapping: GIS supports law enforcement by mapping crime locations, conducting hotspot analysis, and aiding in predictive policing. This enhances resource allocation and crime prevention efforts. Emergency response: GIS plays a vital role in emergency management, facilitating real-time incident tracking, resource distribution, and evacuation planning.

Military and defence	Geospatial intelligence (GEOINT): GIS is integral to military operations, supporting terrain analysis, reconnaissance, mission planning, and target tracking.
Tourism and hospitality	Destination planning: The tourism industry utilizes GIS for destination management, site selection, and marketing. It helps identify key tourist attractions and optimize travel routes.
Transportation and logistics	Logistics planning: GIS enhances transportation and supply chain logistics by optimizing routes, managing fleets, and supporting location-based decision-making. Public transit: GIS aids public transit systems by analyzing ridership patterns, optimizing bus routes, and improving overall accessibility.
Business and marketing	Location intelligence: Businesses use GIS for market analysis, site selection, and location-based marketing. It aids retailers in identifying target demographics and optimizing store locations for maximum impact.

5.5 Conclusion and Endnotes

In conclusion, Geographic Information Systems (GIS) are a versatile and interdisciplinary tool that has significantly influenced a broad array of fields and disciplines. As highlighted throughout this chapter, GIS has become a vital component in sectors such as environmental science, urban planning, public health, agriculture, geology, transportation, criminal justice, business, social sciences, education, military, energy, tourism, and beyond. By enabling the capture, management, analysis, and visualization of spatial data, GIS facilitates informed decision-making and problem-solving in an increasingly complex and interconnected world. In essence, GIS transcends disciplinary boundaries, deepening our understanding of spatial relationships and facilitating more informed decision-making across multiple sectors. Its ongoing development and integration hold the potential to further transform research, planning, and decision-making in the future.

Further Readings

1. Antenucci, J., Brown, K., Croswell, P., Kevany, M., & Archer, H. (1991). *Geographic information systems: A guide to the technology*. Springer ISBN: 978-0-412-99351-0.
2. Chor Pang Lo, C., & Yeung, A. K. W. (2016). *Concepts and techniques of geographic information systems*. Pearson Education ISBN: 978-9332581883.
3. Fischer, M. M., & Nijkamp, P. (2011). *Geographic information systems, spatial modelling and policy evaluation*. Springer . ISBN: 978-3-642-77502-4. https://doi.org/10.1007/978-3-642-77500-0
4. Greenberg, J. D., Logsdon, M. G., & Franklin, J. F. (2002). Introduction to geographic information systems (GIS). In S. E. Gergel & M. G. Turner (Eds.), *Learning landscape ecology*. Springer. https://doi.org/10.1007/0-387-21613-8_3

5. Grueau, C., Laurini, R., & Ragia, L. (2023). *Geographical information systems theory, applications and management* (Communications in computer and information science). Springer . ISBN: 978-3-031-44111-0. https://doi.org/10.1007/978-3-031-44112-7
6. Hanchette, C. L. (2003). Geographic information systems. In P. W. O'Carroll, L. H. Ripp, W. A. Yasnoff, M. E. Ward, & E. L. Martin (Eds.), *Public health informatics and information systems. Health informatics*. Springer. https://doi.org/10.1007/0-387-22745-8_21
7. Huisman, O., & de By, R. A. (Eds.). (2009). *Principles of Geographic Information Systems: An introductory textbook*. The International Institute for Geo-Information Science and Earth Observation (ITC) ISBN: 978-90-6164-269-5.
8. Kresse, W., & Danko, D. (2012). *Springer handbook of geographic information* (Springer handbooks) (2nd ed.). Springer . ISBN: 978-3-030-53124-9. https://doi.org/10.1007/978-3-030-53125-6
9. Zhang, Y., & Drake, W. (2014). Geographic information systems (GIS). In A. C. Michalos (Ed.), *Encyclopedia of quality of life and Well-being research*. Springer. https://doi.org/10.1007/978-94-007-0753-5_1159
10. Zhu, X. (2024). *Geographical information systems: A practical approach* (2nd ed.). Routledge, (Taylor and Francis Group) ISBN 9781032380469.

Chapter 6
Practical Approaches to GIS Data Analysis

"The true value of GIS lies in its ability to answer real-world questions and solve everyday challenges". — Jack Dangermond.

Abstract This chapter explored the practical approaches of GIS data analysis, focusing on methods and tools for extracting insights from geographical data to support decision-making. It began by highlighting the significance of GIS data analysis and outlining processes such as data approaches, practical practices, formats, and integration to ensure accuracy and reliability. Techniques for analyzing raster and vector data were discussed, alongside the use of geoprocessing tools and spatial manipulation methods like buffering, clipping, and interpolation. The chapter also examined diverse data sources, including analogue maps, satellite imagery, GPS, and agency reports, and emphasized the importance of geoprocessing workflows, model builders, and scripting for automating and streamlining analysis. Emerging trends such as integrating remote sensing and big data analytics with PGIS were explored, along with the challenges of ensuring data privacy, ethics, and transparency. In conclusion, the chapter provided a comprehensive overview of GIS methodologies, offering practical insights into data preparation, analysis, visualization, and decision support while addressing key issues and advancements in the field. Therefore, the chapter emphasizes the importance of clear communication as well as the issues and emerging developments in GIS data analysis. The chapter investigates these concerns in order to give readers practical knowledge and skills for efficiently studying geographical data with GIS tools and procedures.

Keywords GIS data analysis · Map data · GIS data format · Map manipulation · GIS tools

K. D. Malakar et al., *Geospatial Technologies in Coastal Ecologies Monitoring and Management*, Advances in Geographic Information Science,
https://doi.org/10.1007/978-3-031-92017-2_6

This chapter provides a comprehensive guide to the practical applications of GIS data analysis, emphasizing its role in deriving meaningful insights and supporting decision-making. It covers essential methodologies, including approaches, practical practices, formats, and integration to ensure accuracy and reliability. Key topics such as geoprocessing tools, spatial manipulation methods, and automated workflows are explored. The chapter also addresses emerging trends, including the integration of remote sensing and big data analytics, while highlighting critical ethical considerations. Readers will gain practical skills and insights for effective geospatial analysis using GIS tools and processes.

Key Points of the Chapter

- Concept, approaches, importance of GIS data analysis and its various formats.
- Practical practices of GIS (raster and vector) data analysis.
- Various dimensions of map data and the process of input map data.

6.1 GIS Data Analysis

6.1.1 Concept of GIS Data Analysis

Geographic Information System (GIS) data analysis is a powerful tool for interpreting and visualizing spatial data to support informed decision-making. By integrating geographical information such as maps, satellite imagery, and databases with analytical techniques, GIS addresses diverse challenges across fields like environmental management, urban planning, disaster response, and business strategy. It enables the examination of spatial patterns, relationships, and trends, allowing users to identify correlations, optimize resources, assess risks, and formulate policies. Key steps in GIS analysis include data collection and preprocessing to ensure quality, applying spatial techniques like buffering and overlay to extract insights, and visualizing results through maps and graphs. These findings guide decision-making and actions, often in iterative processes that refine analysis and ensure reproducibility. Collaboration and result-sharing further enhance GIS applications, making it an essential tool in tackling spatial challenges and shaping our understanding of the world.

GIS data analysis encompasses multiple dimensions, each with a distinct purpose, contributing to a holistic understanding of spatial data. These dimensions include:

- **Spatial analysis:** Examines spatial relationships, patterns, and distributions through operations like buffering, proximity analysis, and spatial interpolation, identifying clusters, hotspots, and trends.
- **Attribute analysis:** Focuses on non-spatial attributes of geographic features using statistical analysis, data aggregation, and classification to derive insights.

- **Temporal analysis:** Incorporates the time dimension to study changes over time, crucial for monitoring phenomena such as climate change, traffic patterns, and population growth.
- **Network analysis:** Explores connectivity and accessibility within spatial networks like roads or utilities, aiding in route optimization, service area analysis, and network modelling.
- **3D and 4D analysis:** Extends GIS to include elevation (3D) and time (4D), supporting applications like urban planning, terrain modelling, and temporal change tracking.
- **Remote sensing analysis:** Integrates satellite imagery and aerial data for tasks such as vegetation health monitoring, land cover classification, and disaster assessment.
- **Geostatistics:** Combines statistical methods with spatial data to predict values in unsampled locations, useful for resource estimation and pollution mapping.
- **Decision support and modelling:** Creates spatial models and scenarios to evaluate the potential impacts of decisions on environments, societies, and infrastructure.
- **Spatial query and visualization:** Provides tools for querying and visualizing spatial data through maps, charts, and dashboards to communicate insights effectively.
- **Geospatial big data analysis:** Addresses the challenges of managing and analyzing large geospatial datasets from IoT devices and sensors, extracting actionable information.

Collectively, these dimensions empower GIS professionals to derive profound insights into spatial phenomena, address complex challenges, and make well-informed decisions across diverse fields, including environmental management, urban planning, disaster response, and business analytics.

6.1.2 Approaches of GIS Data Analysis

GIS data analysis employs diverse approaches (Fig. 6.1), each designed to address specific spatial challenges and objectives. These include:

- **Descriptive analysis:** Summarizes spatial data using statistics, tabulation, and visualizations (e.g., histograms, pie charts) to understand the distribution and characteristics of geographic features.
- **Spatial query and selection:** Retrieves specific data based on spatial or attribute criteria, enabling targeted data extraction and filtering.
- **Spatial modelling:** Simulates real-world phenomena, such as land-use changes or disease spread, through mathematical or computational models based on spatial variables.
- **Multi-criteria decision analysis:** Combines spatial data with weighted criteria to support complex decisions in site selection, resource allocation, and planning.

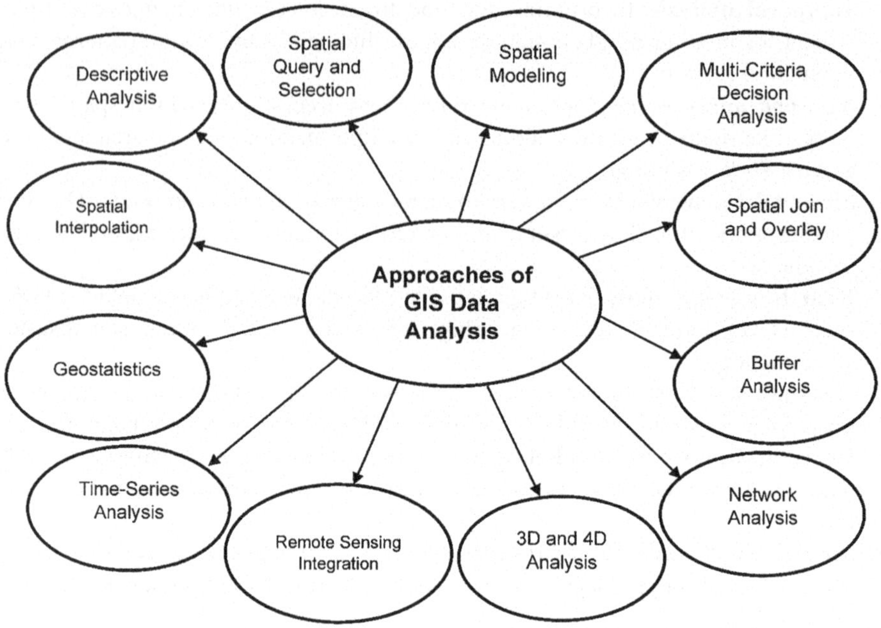

Fig. 6.1 Approaches of GIS data analysis (prepared by the authors)

- **Spatial join and overlay:** Combines datasets through operations like intersection and union to identify overlaps, relationships, or new patterns between spatial features.
- **Buffer analysis:** Creates proximity zones around geographic features for applications like service area analysis, hazard assessment, and environmental impact studies.
- **Network analysis:** Models transportation or utility networks to optimize routes, define service areas, and evaluate connectivity and accessibility.
- **Spatial interpolation:** Estimates values at unsampled locations using methods like kriging or inverse distance weighting, useful for creating elevation models or pollution maps.
- **Geostatistics:** Applies statistical methods to account for spatial autocorrelation, predicting values at unsampled locations and assessing data uncertainty.
- **Time-series analysis:** Incorporates temporal data to study changes and trends over time, crucial for climate monitoring, urban growth analysis, and disaster tracking.
- **3D and 4D analysis:** Adds elevation (3D) and time (4D) dimensions for tasks like terrain modelling, 3D visualization, and monitoring temporal changes in space.
- **Remote sensing integration:** Integrates GIS with satellite imagery and aerial data for applications like land cover classification, vegetation health analysis, and environmental monitoring.

6.1.3 Importance of GIS Data Analysis

The significance of GIS data analysis in today's data-driven world is unparalleled. It serves as a vital tool across diverse domains, profoundly influencing decision-making, problem-solving, and our comprehension of spatial phenomena. These include:

- **Natural resource management:** GIS supports forestry, mining, and agriculture through land-use planning, yield prediction, and inventory management.
- **Environmental management:** GIS monitors natural resources, tracks environmental changes, and assesses human impact, supporting sustainable practices and climate change mitigation.
- **Disaster response:** GIS facilitates real-time monitoring, risk assessment, and evacuation planning, enhancing preparedness and minimizing disaster impacts.
- **Environmental awareness:** GIS visualizes data to promote education, foster public awareness, and encourage community engagement on environmental issues.
- **Informed decision-making:** GIS provides spatial insights and a comprehensive perspective, aiding decisions in urban planning, disaster management, and resource allocation.
- **Resource optimization:** Spatial analysis enables efficient resource use in agriculture, logistics, and infrastructure, leading to cost savings and higher efficiency.
- **Public health:** GIS supports disease tracking, epidemiology, and healthcare planning, proving essential in managing outbreaks and resource allocation.
- **Archaeology:** GIS maps historical sites, manages cultural heritage, and analyzes spatial relationships in archaeological research.
- **Urban planning and transportation:** GIS optimizes transportation systems, traffic management, and city planning, driving smart city initiatives and reducing congestion.
- **Business intelligence:** GIS aids in market analysis, site selection, and supply chain management, enhancing business strategies and customer targeting.
- **Policy development:** Governments use GIS for evidence-based policy-making, implementation monitoring, and impact assessment on society and the environment.
- **Scientific research:** GIS advances research in geology, ecology, and other disciplines by enabling spatial modelling, visualization, and hypothesis testing.

In conclusion, GIS data analysis plays a pivotal role in tackling complex spatial challenges, enhancing operational efficiency, and supporting informed decision-making across various sectors. Its capacity to integrate and analyze spatial and non-spatial data establishes it as an essential tool for understanding and managing our world more effectively.

6.2 Practical Practices of Raster Data Analysis

Practical raster data analysis follows a systematic approach to extract valuable insights from gridded datasets. It begins with data acquisition and preprocessing to ensure proper formatting, georeferencing, and error-free datasets. This is followed by data exploration to examine characteristics like spatial resolution, scale, and extent. Common techniques include reclassification, resampling, and mosaicking to prepare data for analysis. Raster algebra is a key component, enabling mathematical operations and the combination of multiple raster layers to derive new insights. Spatial statistics and modelling further uncover patterns and relationships within the data, while visualization tools like colour maps, heatmaps, and contour plots aid in interpretation. Effective documentation and metadata management ensure reproducibility and facilitate data sharing. Collaboration and interdisciplinary approaches maximize the utility of raster data analysis across fields such as environmental science, geology, remote sensing, and agriculture. Key operations in raster data analysis include local, neighbourhood, zonal, and distance operations, which are fundamental for processing and deriving meaningful information from gridded datasets (Fig. 6.2).

6.2.1 Local Operations

Practical raster data analysis frequently involves the use of local operations, fundamental techniques for processing and manipulating individual cells or pixels within a raster dataset (Fig. 6.3). These operations focus on the value of a specific pixel and its immediate neighbours, making them crucial for various spatial tasks. Key aspects of local operations in raster data analysis include:

- **Filtering and smoothing:** Local operations can apply filters or smoothing techniques to reduce noise in raster data, improving its quality for further analysis. For example, a moving window filter can remove outliers or refine elevation data.

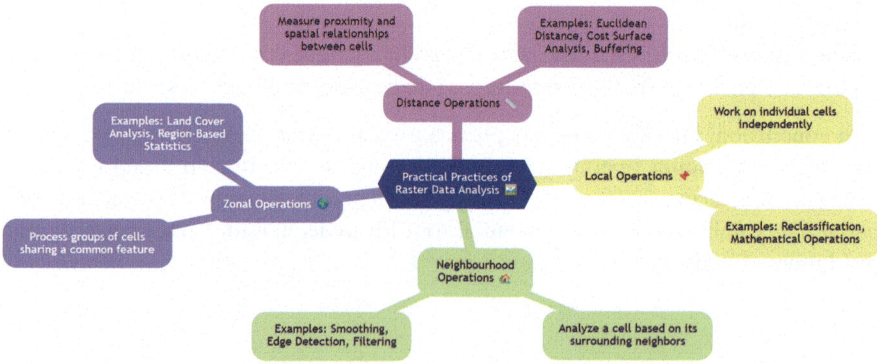

Fig. 6.2 Practical practices of raster data analysis (prepared by the authors)

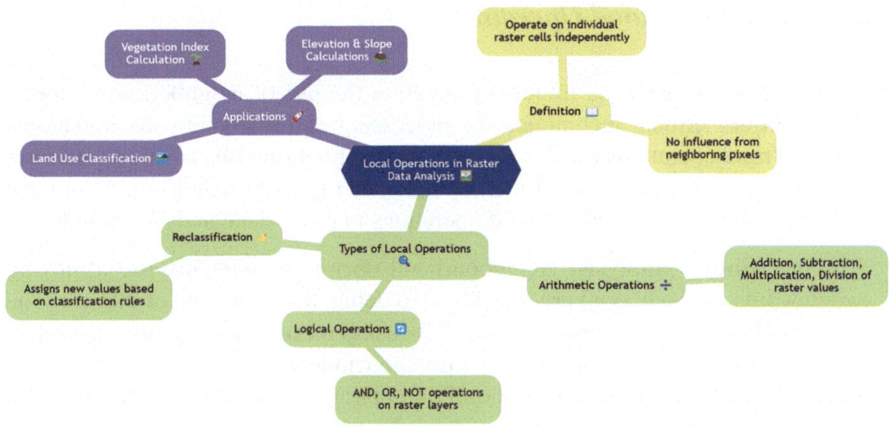

Fig. 6.3 Local operations in practices of raster data analysis (prepared by the authors)

- **Edge detection:** Local operations can detect edges and boundaries within raster datasets, crucial for tasks like feature extraction or object detection in remote sensing imagery.
- **Mathematical transformations:** Raster data can undergo mathematical transformations, such as convolution or Fourier analysis, using local operations to highlight specific spatial patterns or features.
- **Local statistics:** Local operations compute statistics, like mean, median, or standard deviation, within a moving window, helping to characterize local variability and trends.
- **Neighbourhood analysis:** Analyzing neighbouring cell values through local operations identifies clusters, hotspots, or spatial patterns, useful in fields like epidemiology and ecology.
- **Local reclassification:** Local operations enable reclassification of raster values based on specific conditions or rules within the neighbourhood, facilitating data aggregation or thematic mapping.
- **Cost surface analysis:** In network analysis, local operations calculate cost surfaces, representing the cost or impedance of traveling through a raster dataset, considering cell values.
- **Local Modelling:** Local operations support the development of models where each cell's value is influenced by its neighbours, enabling tasks like spatial autocorrelation analysis or cellular automata modelling.

Practitioners of raster data analysis employ these local operations alongside Geographic Information System (GIS) software or specialized raster processing tools to derive meaningful insights from raster datasets. These techniques are crucial for a variety of applications, including terrain analysis, image processing, environmental modelling, and geospatial decision-making.

6.2.2 *Neighbourhood Operations*

Practical raster data analysis frequently involves the use of neighbourhood operations, which are critical techniques for processing and analyzing raster datasets (Fig. 6.4). These operations examine a group of neighbouring pixels or cells within a defined radius or shape and perform computations or transformations on these local groups. Key aspects of neighbourhood operations in raster data analysis include:

- **Moving window analysis:** Neighbourhood operations typically use a moving window (also known as a kernel or filter) to define the local area of interest. The window slides across the raster dataset, focusing on one cell at a time, performing operations based on the values within the window.
- **Filtering and smoothing:** Neighbourhood operations are commonly employed for filtering and smoothing raster data. Techniques such as mean filtering, Gaussian smoothing, and median filtering help reduce noise and clarify patterns in the data.
- **Edge detection:** Neighbourhood operations are essential for edge detection and feature extraction, identifying abrupt changes in pixel values and highlighting object boundaries or edges in the raster dataset.
- **Local statistics:** These operations compute statistics, such as mean, standard deviation, or variance, within the neighbourhood, aiding in the characterization of local variability and value distribution.
- **Focal statistics:** Focal statistics, a type of neighbourhood operation, calculate statistics for each cell based on its neighbours' values. This is valuable for tasks like terrain analysis, where slope, aspect, and curvature are derived.
- **Local reclassification:** Neighbourhood operations can reclassify raster values based on specific conditions or rules within the local area, facilitating thematic mapping or categorization of raster datasets.

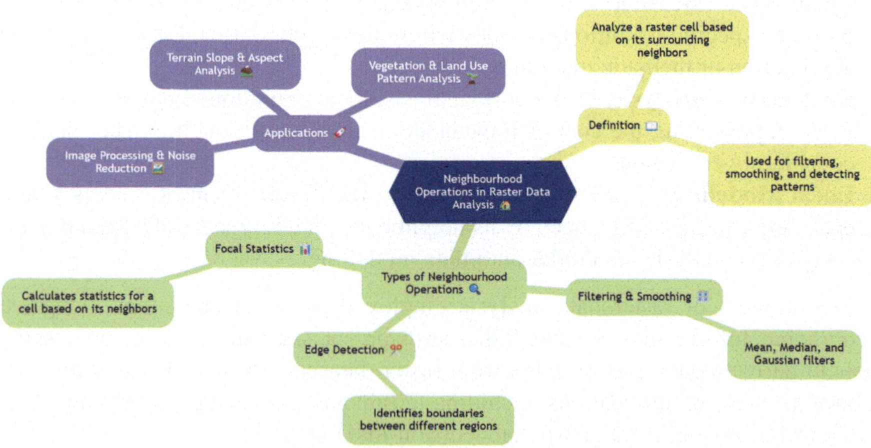

Fig. 6.4 Neighbourhood operations in practices of raster data analysis (prepared by the authors)

- **Cost surface analysis:** In network analysis, neighbourhood operations help calculate cost surfaces, representing the cost or impedance of traveling between locations in the raster dataset, considering nearby cells.
- **Spatial filtering:** Neighbourhood operations apply spatial filters to enhance particular features in raster data. For example, anisotropic diffusion filters can enhance linear features in images.
- **Cellular automata modelling:** Neighbourhood operations play a crucial role in cellular automata models, where a cell's state depends on the states of its neighbouring cells. This is applied in ecological modelling, urban growth modelling, and other simulations.

Practitioners of raster data analysis utilize neighbourhood operations with Geographic Information System (GIS) software or specialized raster processing tools to extract meaningful insights, reduce noise, and derive valuable information from raster datasets. These techniques are widely applied across various fields, including remote sensing, geospatial analysis, image processing, and spatial modelling.

6.2.3 Zonal Operations

Practical raster data analysis frequently involves the use of zonal operations, which are key techniques for analyzing and summarizing raster data within specified zones or regions of interest (Fig. 6.5). These zones can be defined by vector features (e.g., polygons) or determined by raster values. Key aspects of zonal operations in raster data analysis include:

- **Zone definition:** Zonal operations begin by defining regions of interest, which may represent administrative boundaries, land parcels, or study areas. Zones can

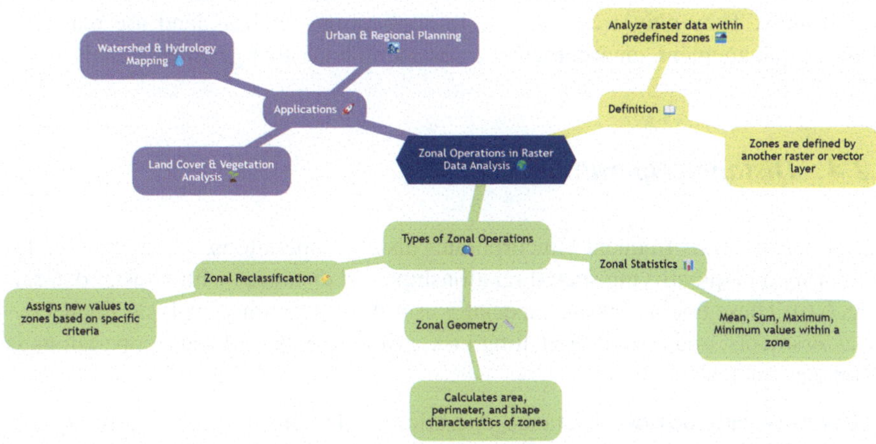

Fig. 6.5 Zonal operations in practices of raster data analysis (prepared by the authors)

be delineated using vector polygons or by specifying raster values that define different zones.

- **Summarization:** Zonal operations summarize raster data within each defined zone, calculating common statistics such as sum, mean, median, maximum, minimum, and standard deviation. This provides a compact representation of the data for each zone.
- **Area weighting:** When summarizing raster data, zonal operations often incorporate the area of each zone, ensuring accurate statistics, especially when zone sizes vary. Weighted calculations account for the influence of zone size on the results.
- **Zonal histograms:** Zonal operations can generate histograms that display the distribution of raster values within each zone, offering insights into data variability within the zones.
- **Zonal functions:** Some zonal operations involve more advanced functions, such as determining the majority class within a zone, identifying the zone with the highest or lowest value, or locating the maximum or minimum value within each zone.
- **Zonal overlay:** Zonal operations may involve overlaying summarized data onto the original zones, resulting in new raster datasets that represent the summarized information, useful for thematic mapping and visualization.
- **Spatial join:** Zonal operations resemble spatial join operations, integrating vector and raster data to combine information from both data types.
- **Applications:** Zonal operations are used in various fields, including agriculture (for analyzing crop yields in fields), ecology (to study habitat characteristics), hydrology (to assess watershed properties), and urban planning (for analyzing land use in zones).

Zonal operations are implemented using GIS software and specialized raster analysis tools. These operations allow spatial analysts, researchers, and decision-makers to derive valuable insights and statistics from raster data within defined spatial units. They play a crucial role in applications such as land-use planning, resource management, environmental assessment, and other geospatial analyses.

6.2.4 Distance Operations

Raster data analysis frequently incorporates distance operations, which are vital for measuring and quantifying spatial relationships and distances within raster datasets (Fig. 6.6). These operations are especially useful for proximity analysis, connectivity assessments, and spatial modelling tasks. Key aspects of distance operations in raster data analysis include:

- **Distance calculation:** Distance operations calculate the distance from each raster cell to a reference point or feature, using metrics like Euclidean or Manhattan distance, based on the specific application.

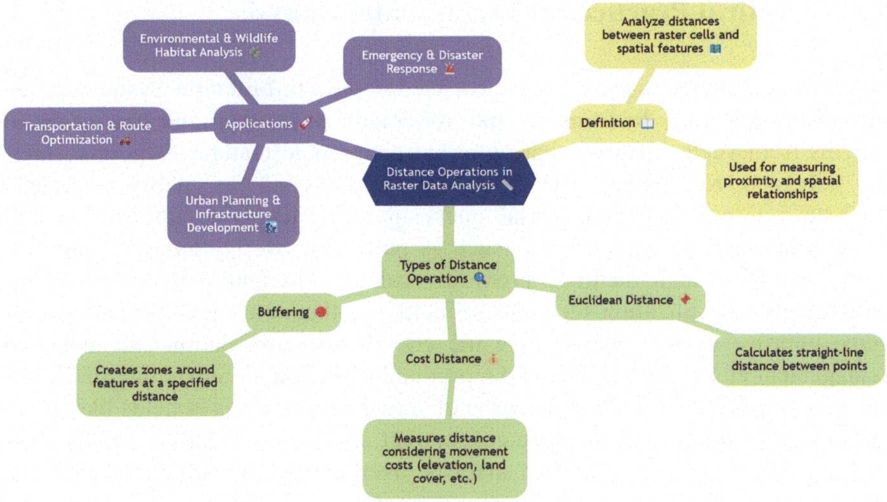

Fig. 6.6 Distance operations in practices of raster data analysis (prepared by the authors)

- **Cost surfaces:** These operations generate cost surfaces that represent travel impedance, considering factors such as terrain or land use. They enable more realistic distance calculations for tasks like least-cost path analysis.
- **Proximity analysis:** Distance operations quantify relationships based on proximity, such as identifying cells within a buffer distance or finding the nearest neighbour for each cell.
- **Accessibility analysis:** These operations assess access to resources or services by calculating distances or travel times, supporting decisions in service area optimization and transportation planning.
- **Connectivity analysis:** In network analysis, distance operations help determine connectivity between cells or nodes, crucial for shortest path analysis and service area assessment.
- **Spatial modelling:** Distance operations define spatial relationships and constraints in models, including those for disease spread, habitat suitability, or urban growth.
- **Buffer analysis:** Distance operations create buffers around features of interest, used in environmental assessments, land-use planning, and site selection.
- **Visualization:** Distance maps are visualized using colour-coded maps or contour plots, highlighting varying distances or costs across a study area.

Distance operations are implemented through Geographic Information System (GIS) software and specialized raster analysis tools. These techniques are crucial for addressing various spatial challenges and are widely used in fields such as transportation planning, epidemiology, and urban planning.

6.3 Practical Practices of Vector Data Analysis

Vector data analysis is a core aspect of Geographic Information Systems (GIS) and spatial analysis, focusing on the systematic processing and extraction of insights from vector datasets that represent geographic features as points, lines, and polygons. Key practices involve data preprocessing for quality assurance, attribute manipulation, and spatial querying to retrieve relevant information. Geoprocessing tools support tasks such as buffering, overlay analysis, and network analysis, which facilitate proximity calculations and spatial relationship assessments. Visualization techniques, including thematic mapping and spatial joins, improve data presentation, while effective documentation and metadata management ensure traceability and reproducibility. Collaborative, interdisciplinary approaches enhance the application of vector data analysis in fields like urban planning, environmental management, market analysis, and public health. This analysis typically involves buffer analysis, overlay operations, pattern analysis, and map manipulation (Fig. 6.7).

6.3.1 Buffer Aspects

Vector data analysis frequently employs buffers, which are essential tools for spatial analysis within Geographic Information Systems (GIS). Buffers are generated around specific geographic features, such as points, lines, or polygons, and serve various important functions (Fig. 6.8):

- **Proximity analysis:** Buffers are primarily employed to assess the spatial relationship between features. For instance, a buffer can be applied around a school (point feature) to identify residential parcels (polygon features) within a designated distance, supporting school planning or bus route optimization.

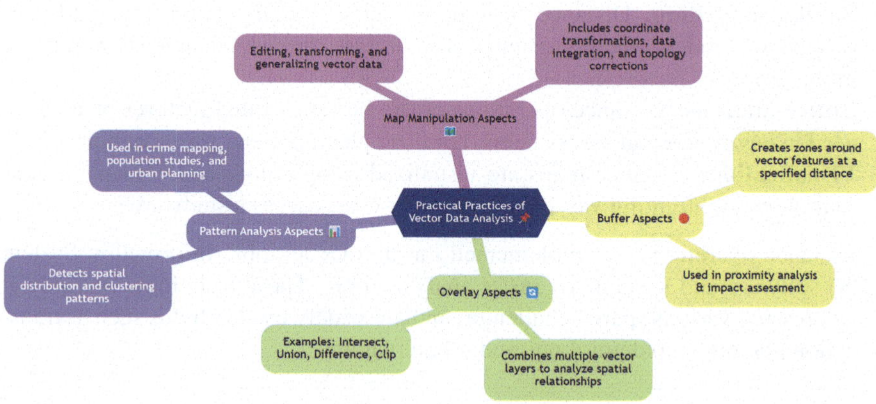

Fig. 6.7 Practical practices of vector data analysis (prepared by the authors)

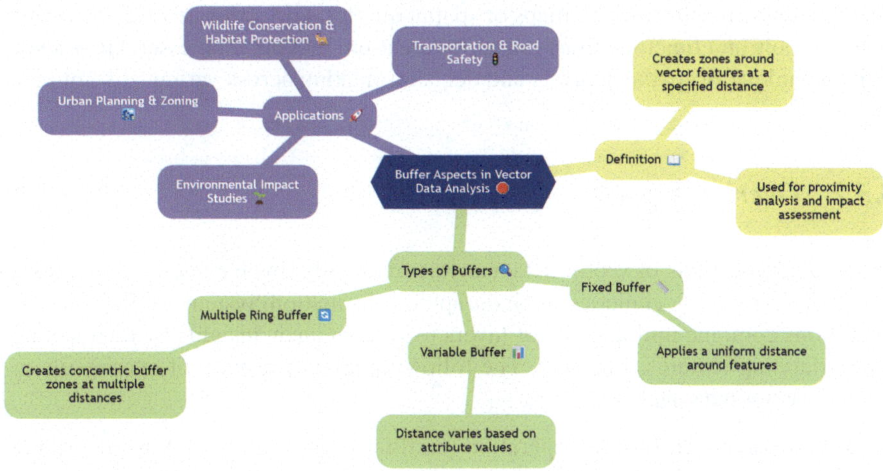

Fig. 6.8 Buffer aspects in vector data analysis (prepared by the authors)

- **Environmental impact assessment:** Buffers are critical for assessing the potential environmental impact of developments or activities. By creating buffers around sensitive areas like wetlands or riverbanks, analysts can pinpoint regions requiring protection or regulation to mitigate environmental damage.
- **Zoning and land-use planning:** Buffers are frequently utilized in zoning regulations to define setbacks. For example, a buffer around commercial properties may be required to separate them from residential zones, maintaining neighbourhood integrity.
- **Emergency response planning:** Buffers are used in emergency management to delineate affected areas during incidents like chemical spills or natural disasters. Buffer analysis aids in evacuation planning and resource distribution.
- **Wildlife conservation:** Buffers are essential in establishing protected zones around critical wildlife habitats, nesting sites, or migration corridors, thereby supporting biodiversity conservation.
- **Infrastructure and facility location:** Buffers play a significant role in determining the optimal locations for essential facilities such as hospitals, fire stations, and power plants. By considering buffer zones, planners ensure accessibility to services for the population.
- **Market analysis:** In business, buffers are applied to evaluate market coverage. For example, a retail store can analyze the population within a specific distance to assess potential customer reach.
- **Public health:** Buffers are employed to analyze public health data, such as the distribution of healthcare facilities or the spread of diseases. Buffer analysis helps identify areas with limited healthcare access.

Practical applications of buffer analysis involve selecting suitable buffer distances, accounting for the spatial reference and coordinate system of the dataset, and visualizing the results through maps or spatial queries. GIS software offers a diverse array of tools and functionalities for buffer creation, making it an essential and versatile technique in spatial analysis and decision-making across various disciplines.

6.3.2 *Overlay Aspects*

Practical applications of vector data analysis frequently involve overlay operations, which are essential techniques in Geographic Information Systems (GIS). Overlay operations combine multiple vector layers to generate new insights by intersecting, or overlapping features (Fig. 6.9). The following are key aspects of overlay operations in vector data analysis:

- **Data integration:** Overlay operations facilitate the integration of diverse spatial datasets, such as land parcels, roads, and zoning boundaries, enabling analysts to explore relationships between different geographic features.
- **Spatial queries:** Overlay supports spatial queries by identifying shared areas or intersections between features across multiple layers. For instance, it helps identify parcels located within specific land-use zones.
- **Union and intersection:** Union operations combine features from multiple layers, preserving all geometries, while intersection operations focus on areas where features from different layers overlap.
- **Difference and symmetric difference:** Difference operations identify areas unique to one layer, excluding common areas with another, while symmetric difference operations highlight areas exclusive to each layer, excluding their intersection.

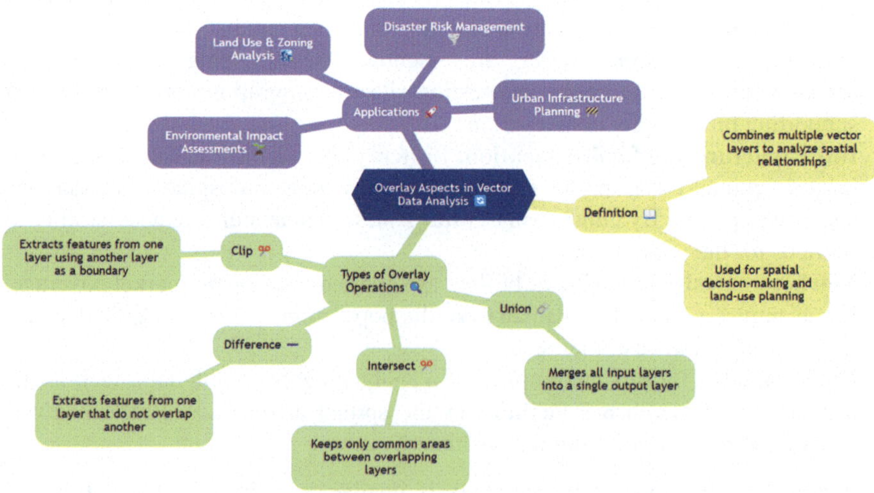

Fig. 6.9 Overlay aspects in vector data analysis (prepared by the authors)

- **Clipping and clipping masks:** Overlay allows for the creation of clipped datasets by defining a region of interest (mask) and extracting only features that fall within it, making it ideal for focusing on specific geographic regions.
- **Topology checking:** Overlay operations assist in identifying and resolving topological errors, such as gaps and overlaps, ensuring data integrity and accuracy.
- **Data aggregation:** Overlay can aggregate attributes or data associated with intersecting features. For example, it can sum the population of census blocks within a specific district.
- **Spatial join:** Overlay operations enable spatial joins, combining attributes from two layers based on their spatial relationships. For example, it can calculate the average income of households within a defined distance from retail stores.
- **Geoprocessing and geocoding:** Overlay is fundamental to geoprocessing tasks, including geocoding addresses, generating service areas, and identifying the nearest features.
- **Land-use planning:** Overlay plays a crucial role in land-use planning by helping to identify areas with land-use conflicts, zoning regulations, or environmental constraints.

Effective implementation of overlay operations requires careful consideration of coordinate systems and data quality, as discrepancies can result in inaccuracies. GIS software offers a comprehensive set of tools and functionalities for conducting overlay analysis, enabling analysts to perform intricate spatial operations across a broad spectrum of applications, including environmental assessment, urban planning, transportation management, and other fields.

6.3.3 Pattern Analysis Aspects

Practical applications of vector data analysis frequently involve pattern analysis, a fundamental aspect of Geographic Information Systems (GIS) aimed at identifying, characterizing, and understanding spatial patterns within vector datasets (Fig. 6.10). Key components of pattern analysis in vector data include:

- **Clustering analysis:** Clustering analysis identifies groups of spatial features with similar characteristics or behaviour, useful in applications such as detecting high-crime areas, analyzing disease clusters, or identifying customer activity hotspots.
- **Dispersion analysis:** Dispersion analysis evaluates the distribution of features across space, helping determine if they are evenly spread, clustered, or dispersed. This is important in fields like urban planning, ecology, and resource management.
- **Spatial autocorrelation:** Spatial autocorrelation quantifies how feature values are related to those of neighbouring features, revealing spatial dependencies. It is crucial in fields such as epidemiology, where disease spread may be influenced by nearby cases.

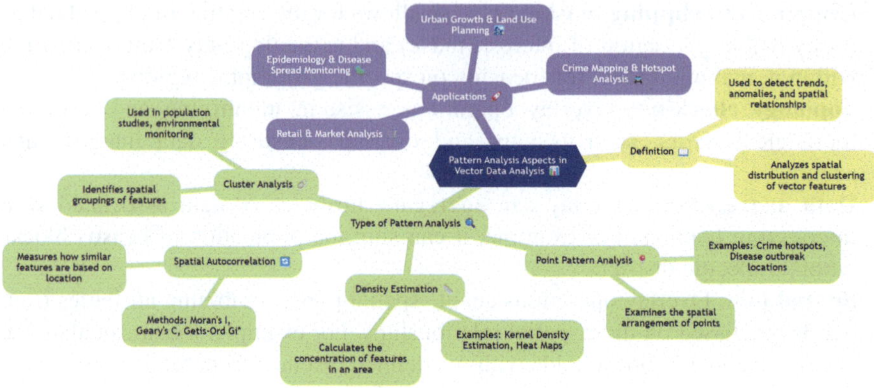

Fig. 6.10 Pattern aspects in vector data analysis (prepared by the authors)

- **Point pattern analysis:** Point pattern analysis examines the distribution of point features, using techniques like nearest neighbours' analysis and Ripley's K-function to determine if patterns are regular, random, or clustered.
- **Spatial trend analysis:** Spatial trend analysis investigates how variables change spatially, helping to identify gradients, trends, and directional patterns, which is beneficial for natural resource management and environmental monitoring.
- **Spatial autoregression models:** These models incorporate spatial relationships into regression analysis to address spatial autocorrelation and heterogeneity in attribute data.
- **Density analysis:** Density analysis calculates the concentration or density of features within defined areas, aiding in assessments such as population density, land-use intensity, and resource utilization.
- **Correlation and covariance analysis:** This analysis examines the relationships between attributes in vector datasets, helping to understand correlations, such as those between income and education levels in a region.
- **Pattern recognition:** Pattern analysis may involve pattern recognition techniques, such as clustering algorithms or machine learning models, to classify and identify spatial data patterns.
- **Data visualization:** Visualization is key to pattern analysis, facilitating quick comprehension of spatial patterns through tools like heatmaps, choropleth maps, and scatter plots.

Effective pattern analysis in vector data necessitates a solid foundation in spatial statistics, proficiency with GIS software, and domain-specific expertise. This approach helps reveal underlying trends, relationships, and anomalies within spatial data, providing valuable insights that support decision-making across a range of disciplines, including urban planning, public health, ecology, and resource management.

6.3.4 Map Manipulation Aspects

Practical applications of vector data analysis frequently include map manipulation, a process that involves creating, editing, and refining maps to effectively communicate spatial information (Fig. 6.11). Key aspects of map manipulation in vector data analysis include:

- **Map design:** The process of map manipulation begins with map design, where decisions are made regarding layout, scale, colour schemes, symbols, and labels. Adhering to effective design principles ensures clarity, readability, and visual appeal of the map.
- **Data classification:** Data classification is a critical aspect of map manipulation, where attribute data is categorized for thematic mapping. The choice of classification methods (e.g., equal intervals, quantiles, natural breaks) is vital in highlighting patterns and variations within the data.
- **Symbolization:** Symbolization involves selecting appropriate symbols, colours, and line styles to represent various features and attributes on the map. Consistency and meaningful representation enhance the interpretability of the map.
- **Labelling:** Effective labelling is essential for map readability. The manipulation process involves determining optimal label placement, preventing overlaps, and using abbreviations or callout boxes where needed to convey additional information.
- **Layer organization:** In GIS software, layer organization helps arrange vector data layers in a specific order to manage visibility and hierarchy, ensuring that critical information remains unobscured by other layers.
- **Map extent and scale:** Manipulating map extent and scale focuses attention on specific geographic areas or features. Appropriate zoom levels help convey varying levels of detail, allowing users to understand both broad and granular perspectives.
- **Cartographic Projections:** Selecting an appropriate cartographic projection is crucial to accurately represent geographic features. Different projections may distort aspects such as shape, area, distance, or direction, so it is important to choose one that minimizes such distortions.

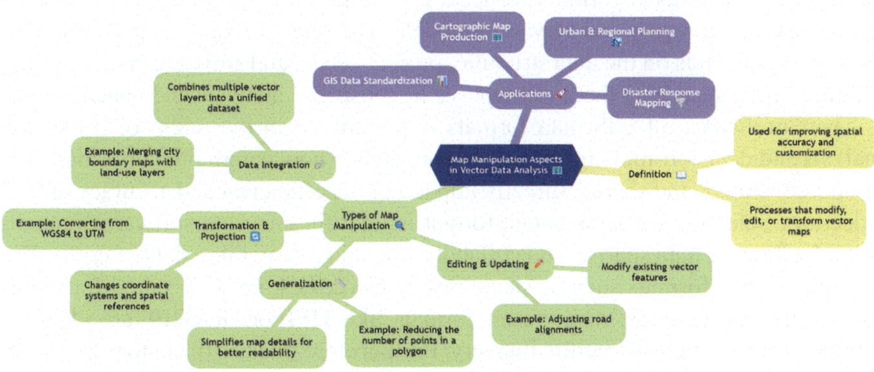

Fig. 6.11 Map manipulation aspects in vector data analysis (prepared by the authors)

- **Map annotations:** Map manipulation may involve adding annotations, legends, scale bars, and notes to provide essential context and explanations for the map reader.
- **Insets and overview maps:** Insets and overview maps are included to offer additional context, helping users better understand specific areas of interest within the larger geographic setting.
- **Map export and sharing:** Once the map is created, it can be exported into various formats (e.g., PDF, image files) for distribution to stakeholders or inclusion in reports and presentations.
- **Interactive maps:** In the digital era, map manipulation extends to creating interactive web maps and applications, enabling users to engage with and explore spatial data online.

Practical map manipulation practices require a blend of cartographic knowledge, proficiency with GIS software, and an understanding of the needs of the target audience. Well-crafted and thoughtfully manipulated maps are crucial tools for effectively conveying spatial information in various fields, including urban planning, environmental assessment, transportation management, and beyond.

6.4 GIS Data Format

The format of Geographic Information System (GIS) data is a critical component of spatial data management and analysis. GIS data is available in a variety of formats, each designed for specific data types and applications. Common GIS data formats include shapefiles, Geodatabases (e.g., Esri File Geodatabase), GeoJSON, KML (Keyhole Markup Language), and raster formats such as TIFF and GeoTIFF. Shapefiles are widely used for vector data storage, facilitating the representation of points, lines, and polygons. Geodatabases provide a more robust and flexible storage solution, supporting various data types, topologies, and relationships. GeoJSON and KML are commonly used for web-based mapping and data sharing, owing to their simplicity and compatibility with platforms like Google Maps. In contrast, raster formats are ideal for representing continuous spatial data, such as satellite imagery and elevation models. The selection of an appropriate GIS data format depends on the data structure, project needs, and software compatibility, with each format serving distinct purposes in geospatial analysis and visualization.

The significance of GIS data formats is paramount in the realm of geospatial analysis and decision-making. These formats dictate how geographic information is structured, stored, and shared, directly impacting the efficiency and accuracy of GIS processes. Selecting the appropriate format is vital for ensuring software compatibility, facilitating data sharing and collaboration, and maintaining data integrity. For example, vector formats such as shapefiles or GeoJSON are well-suited for representing discrete features, while raster formats like TIFF or GeoTIFF are ideal for continuous data, such as satellite imagery. Furthermore, the use of standardized GIS data formats promotes interoperability and allows for the integration of various

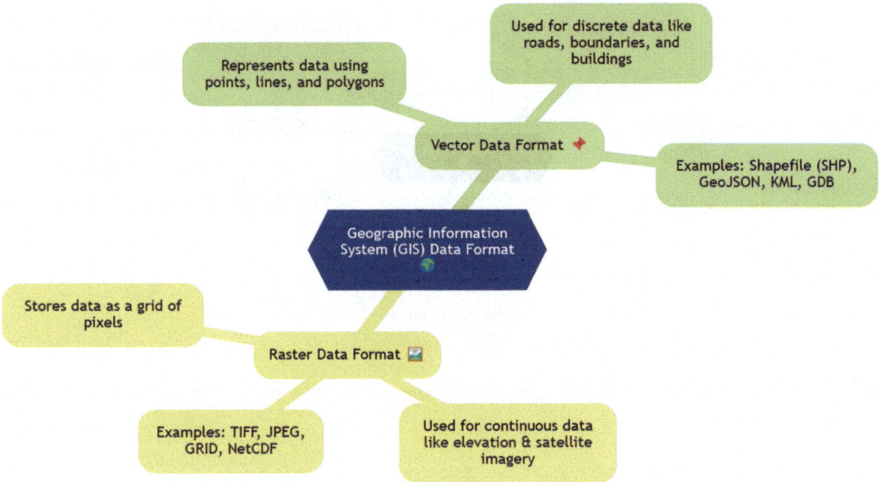

Fig. 6.12 Map manipulation aspects in vector data analysis (prepared by the authors)

datasets, thus enhancing GIS capabilities in fields such as environmental management, urban planning, disaster response, and business analytics (Fig. 6.12).

6.4.1 Raster Data Format

Raster data format is a crucial data structure in Geographic Information Systems (GIS) and remote sensing, representing spatial information as a grid of cells, with each cell containing a value that reflects a specific attribute or characteristic. These formats are extensively used in diverse applications, such as terrain modelling, image analysis, and environmental monitoring (Fig. 6.13). The following discussion provides an in-depth examination of raster data formats:

- **Grid structure:** Raster data is structured as a regular grid, often referred to as a *"raster grid"* or *"pixel grid"*. Each cell in this grid, known as a pixel (short for *"picture element"*), holds a value representing a specific attribute, such as elevation, temperature, land cover type, or colour intensity.
- **Continuous and discrete data:** Raster data can represent both continuous and discrete data. Continuous data, such as elevation or temperature, can take any value within a defined range. In contrast, discrete data, like land cover types or digital images, are represented by distinct categories or classes assigned to each pixel.
- **Common raster formats:** Several standard raster data formats are commonly used in GIS and remote sensing applications (Table 6.1):

    a) JPEG (Joint Photographic Experts Group): JPEG is a frequently used format for compressed raster images, suitable for photographs and visual data. Though it lacks geospatial metadata, it is efficient for storage and sharing purposes.

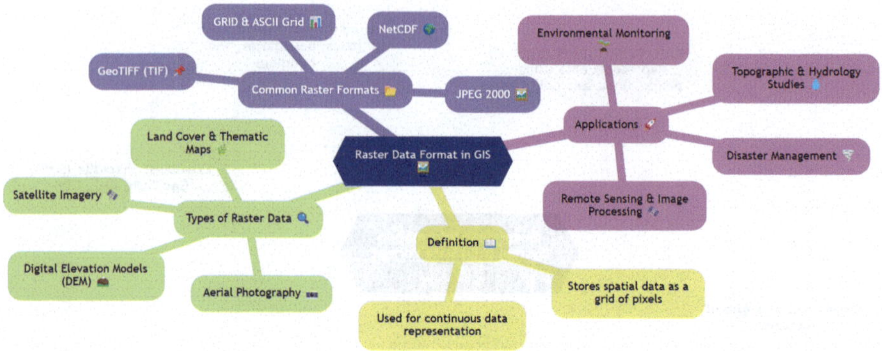

Fig. 6.13 Raster data format in GIS (prepared by the authors)

Table 6.1 Raster data format

Data format	Data format details
ADS	Automated Digitizing System, U.S. Burean of Land Management
ADRG	Arc Digitized Raster Graphics
AMS	Automated Mapping System
ADRI	Arc Digitized Raster Image file
BIL	Band Interleaved by Line
BIP	Band Interleaved by Pixel
BMP	Bitmap File
BSQ	Band Sequential ASCII data description file
CDR	Corel draw format
CADRG	Compressed ADRG
CEOS	Radarsat image file
CIB	Controlled image base file
Dbase	Database file
DEM	Digital Elevation Model
DIB	Device-independent bitmap
DIME	Dual Independent MAP Encoding
DOQ	Digital Orthophoto Quad
DTED	Digital Terrain Elevation Data
ECW	ER mapper compressed image file
EPS	Encapsulated postscript
ERDAS	ERDAS proprietary format
ETAK	Map base file street map (for the United States only)
Geo TIFF	TIFF image with georeferencing
GIF	Graphic Interchange Format
GIRAS	Land use/Land cover data—U.S. geological survey format for land use, land cover
IMG	ERDAS Imagine native format
IMAGINE	ERDAS proprietary format

(continued)

Table 6.1 (continued)

Data format	Data format details
ISIF	Intergraph's standard interchange format
JPEG	Joint Photographic Expert Group
JFIF	JPEG File Interchange Format
LAN	ERDAS Imagine 7.X file
MIDAS	Map information assembly display
MOSS	Moss expert file
MRSID	MRSID compressed file
NTF	National Transfer File—ordnance survey format
NITF	National Imagery Transmission Format
PCX	PC Paintbrush format
PICT	Apple's graphic file
PCT	Macintosh PICT format
PNG	Portable Network Graphics
PSD	Adobe Photoshop image file
RLC	Run-length encoding compressed format
RPF	Raster Product Format
SDTS	Spatial Data Transfer Standard
SLS	Standard linear format
DATA FORMAT	Data Format details
SUN/RAS	Sun rasterfile
TGA	Targa graphics file
TIFF	Tagged Image File Format
WPG	Word Perfect Graphics format

b) TIFF (Tagged Image File Format): TIFF is a versatile and widely supported format for various raster data types, including satellite imagery and scanned maps. It accommodates multiple bands, compression methods, and georeferencing capabilities.
c) GeoTIFF: GeoTIFF is an extension of TIFF that incorporates geospatial metadata, enabling seamless integration with GIS software. It contains information on the coordinate reference system (CRS), pixel size, and georeferencing data.
d) NetCDF (Network Common Data Form): NetCDF is often used for multidimensional datasets, such as climate and atmospheric data. It supports data structures like time series and 3D grids, along with metadata for scientific analysis.
e) HDF (Hierarchical Data Format): HDF is ideal for storing large, complex datasets, including satellite data and remote sensing imagery. It allows for efficient data storage, retrieval, and organization.

- **Multiband raster:** Many raster formats support multiband or multispectral data, where each pixel contains information from multiple bands or spectral channels. This is commonly employed in remote sensing applications, such as satellite and aerial imagery, where different bands capture various aspects of the electromagnetic spectrum.

- **Georeferencing:** Georeferencing is a vital component of raster data, linking each pixel to a specific geographic location on surface. This enables the overlay and integration of raster data with other geographic datasets.
- **Spatial resolution:** Raster data formats specify spatial resolution, which determines the pixel size and level of detail within the dataset. Higher resolution provides more detail but results in larger file sizes.
- **Applications:** Raster data formats are employed in various fields, including:

  a) Natural resource management: Raster data supports the monitoring of forests, agriculture, and water resources for sustainable management practices.
  b) Climate and weather data: Raster data formats store climate variables like temperature, precipitation, and humidity, supporting climate modelling and weather forecasting.
  c) Environmental analysis: Raster data is used to assess habitat suitability, ecological patterns, and conduct environmental impact assessments.
  d) Urban planning: Raster-based land use and land cover data assist in urban planning, transportation network design, and infrastructure development.
  e) Elevation models: Digital elevation models (DEMs) and digital terrain models (DTMs) represent elevation data in raster format, facilitating terrain analysis, hydrology modelling, and viewshed analysis.
  f) Remote sensing: Satellite and aerial imagery in raster format are used for land cover classification, change detection, and environmental monitoring.

- In summary, raster data formats play a pivotal role in the representation and analysis of spatial data within GIS and remote sensing applications. These formats offer a structured approach to organizing and processing diverse geographic information, facilitating a broad spectrum of applications in geospatial science and decision-making. A comprehensive understanding of raster data formats is essential for professionals engaged in spatial data management and geospatial technologies.

6.4.2 *Vector Data Format*

Vector data format is a core data structure in Geographic Information Systems (GIS) and spatial analysis. It represents geographic features and their associated attributes through points, lines, and polygons. This structure is extensively utilized in a variety of applications, including mapping, spatial analysis, urban planning, and natural resource management (Fig. 6.14). The following discussion provides an in-depth examination of vector data formats:

- **Basic elements:** Vector data formats are composed of three primary elements:

  a) Points: Points represent individual geographic locations or discrete features, defined by a single set of coordinates (latitude and longitude).

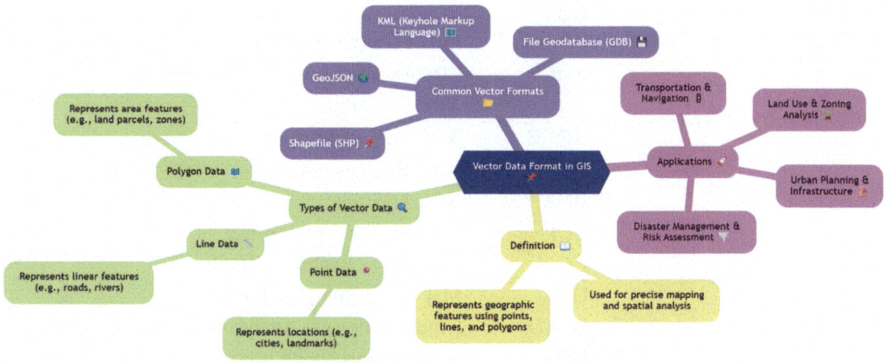

Fig. 6.14 Vector data format in GIS (prepared by the authors)

b) Lines (polylines): Lines depict linear features, such as roads, rivers, and utility networks, composed of a series of connected points.
c) Polygons: Polygons define enclosed areas, such as land parcels, administrative boundaries, or natural features, formed by a series of connected lines that create a closed shape.

- **Attributes:** In vector data formats, each geographic feature is linked to a set of attributes stored in a database table. These attributes provide additional information about the features, such as names, population counts, land-use codes, and more. The combination of geometric shapes and attribute data allows for a detailed and comprehensive representation of geographic information.
- **Common Vector formats:** Several widely used vector data formats are commonly employed in GIS and related fields (Table 6.2):

a) **Georeferencing:** Georeferencing is a crucial aspect of vector data, as each feature in a vector dataset is tied to a specific geographic location on the surface. This is defined by coordinates (latitude and longitude) or a local coordinate system (e.g., easting and northing).
b) **Topology:** Vector data formats often incorporate topological relationships, which define how features interact or connect with each other. These relationships, such as adjacency, containment, and connectivity, are critical for spatial analysis, including network routing and polygon overlay operations.
c) **Geometry types:** Vector data formats support various geometry types, including points, lines, polygons, multipoint, multiline, and multi-polygons, allowing for the representation of complex geographic features and geometries.
d) **Applications:** Vector data formats find diverse applications across multiple fields, including:

   a) Mapping and cartography: Essential for creating detailed maps used in navigation, land surveying, and urban planning.
   b) Spatial analysis: Used in tasks such as proximity analysis, overlay operations, and spatial queries.

Table 6.2 Vector Data Format

Data Format	Data Format Details
AOI	ERDAS imagine area of interest
ARC	ARC export file
ARCE	ARC info export format
ARCG	ARC info generate format
BNA	Atlas GIS file
CDF	Cartographic Data Format
CGM	Computer Graphics Metafile
DAT	Map info native data format
DFAD	Digitized Feature Analysis Data
DGN	Microstation design file
DLG	Digital line graph file
DWG	Autocad drawing file
DXF	Drawing exchange format
EOO	Are info export format
EPS	Encapsulated PostScript file
GDF	Geographic data file
IGDS	Interactive graphic design software export format
IGES	Initial Graphics Exchange Specification
MGE	Intergraph file format
MIF/MID	MapInfo Interchange format
OVR	ERDAS Imagine Annotation Layer
PICT	Apple's graphics file
SHP	ESRI shapefile
SIF	Standard Interchange Format
SDTS	Spatial Data Transfer Standard
TAB	MapInfo format
TIGER	Topologically Integrated Geographic Encoding and Referencing/line census file
VPF	Vector Product Format
VFR	Vector Relational Format
WMF	Windows Metafile Format

c) Environmental management: Supports the monitoring and management of natural resources, habitats, and land use.
d) Infrastructure planning: Aids in the planning and design of infrastructure networks, such as transportation and utilities.
e) Census and demographics: Represents and analyzes demographic data and administrative boundaries.
f) Emergency response: Crucial for disaster management, emergency response, and resource allocation during crises.

Format	Description
Shapefile (.shp)	A widely used format comprising multiple files: .shp (geometry), .shx (index), and .dbf (attributes). Suitable for various types of geographic data
GeoJSON (.geojson)	A lightweight format based on JavaScript Object Notation (JSON), often used in web-based mapping applications
KML (Keyhole Markup Language)	An XML-based format developed by Google, commonly used for custom maps in Google Earth and Google Maps
Geodatabases	Comprehensive data storage solutions, often used with Esri's ArcGIS, supporting various vector data types, relationships, and topological rules

In summary, vector data formats are integral to the representation and analysis of geographic information within GIS and related fields. Their capacity to accurately depict geographic features, coupled with associated attribute data, makes them fundamental to spatial data analysis and decision-making processes.

6.5 Map Data and Input of Map Data

The input of map data is a crucial process in Geographic Information Systems (GIS) and cartography, wherein geographic information is gathered and transformed into digital formats. This process encompasses a variety of methods, including surveying, remote sensing, GPS, and digitization. Surveyors collect ground-truth data using both traditional techniques and advanced surveying equipment, while remote sensing technologies, such as satellites and drones, capture large-scale imagery and geospatial data. GPS technology provides highly accurate location information. Digitization refers to the conversion of paper maps or sketches into digital formats using specialized software. The precision and quality of the input data are fundamental to the reliability of GIS analyses and the creation of accurate maps, making data acquisition a foundational element of geospatial work.

6.5.1 Input of Map Data: Analogue Map

The input of map data from analogue maps is a common procedure in Geographic Information Systems (GIS) and cartography. Analogue maps, such as paper maps or charts, hold valuable geographic information that must be digitized for use in digital systems. The digitization process typically involves scanning or photographing the analogue map to create a digital image. Once the image is acquired, it is georeferenced by aligning control points on the analogue map with their corresponding real-world coordinates, thereby integrating the map with the Earth's surface. Following georeferencing, the digitized map can be converted into either vector or raster format, depending on the nature of the data it contains. This conversion enables the integration of the analogue map's information into GIS databases,

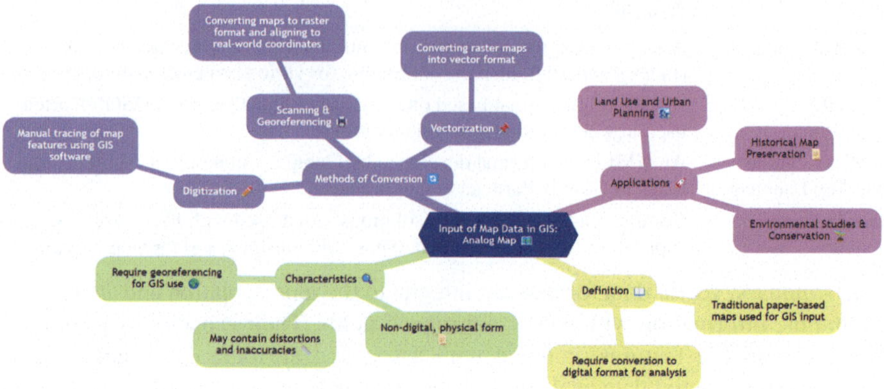

Fig. 6.15 Input of map data: analogue map (prepared by the authors)

supporting spatial analysis, mapping, and decision-making, thereby bridging the gap between traditional paper maps and contemporary digital geospatial technologies (Fig. 6.15).

6.5.2 Input of Map Data: Aerial Photography

The process of inputting map data from aerial photography is a critical component of Geographic Information Systems (GIS) and remote sensing. Aerial photography involves capturing high-resolution images of the surface from elevated platforms, such as airplanes or drones. These images provide essential geospatial information for a wide range of applications. The input process begins with the acquisition of aerial imagery, which is then followed by orthorectification to correct distortions resulting from terrain variations and the camera's perspective. Once orthorectified, the imagery can be utilized to generate detailed and accurate maps, 3D models, and land-use classifications. Aerial photography serves as a fundamental data source for environmental monitoring, urban planning, disaster management, and various other fields, offering a comprehensive and up-to-date representation of the Earth's surface (Fig. 6.16).

6.5.3 Input of Map Data: Satellite Data

Inputting map data from satellite imagery is a key component of Geographic Information Systems (GIS) and remote sensing. Satellites capture high-resolution images of the Earth's surface from space, offering a vast array of geospatial data. This data is essential for a variety of applications, such as environmental monitoring, land-use analysis, disaster management, and urban planning. The input process

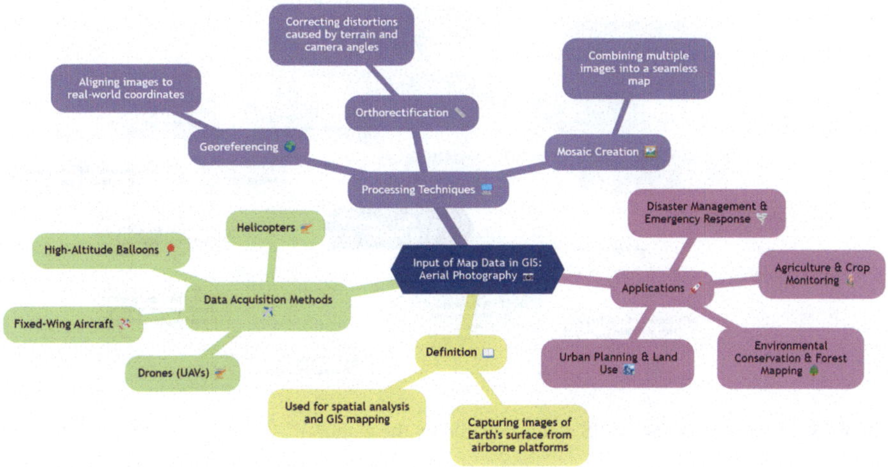

Fig. 6.16 Input of map data: aerial photography (prepared by the authors)

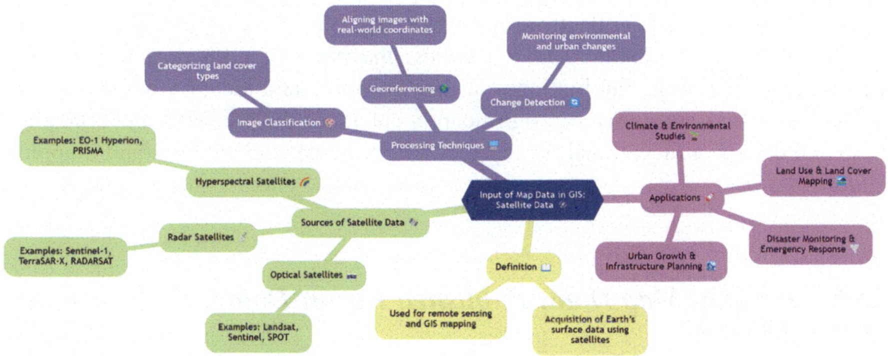

Fig. 6.17 Input of map data: satellite data (prepared by the authors)

begins with the acquisition of satellite imagery, followed by preprocessing procedures like radiometric and geometric corrections. After processing, the satellite data can be employed to create detailed maps, detect temporal changes, and analyze land cover and land-use patterns. Satellite imagery is invaluable for its ability to provide global coverage, regular updates, and a diverse range of spectral and temporal information, establishing it as a crucial resource for geospatial analysis (Fig. 6.17).

6.5.4 *Input of Map Data: GPS Data*

Inputting map data from GPS (Global Positioning System) is a fundamental aspect of modern Geographic Information Systems (GIS) and mapping. GPS receivers determine precise location coordinates by triangulating signals from satellites. This

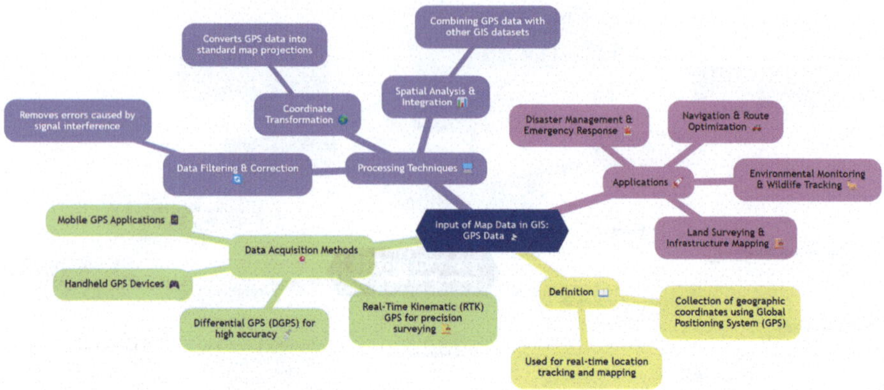

Fig. 6.18 Input of map data: GPS data (prepared by the authors)

location data is then utilized to create new digital maps or update existing ones. GPS data input enables the accurate georeferencing of geographic features and the development of spatial datasets. It plays a critical role in applications such as navigation, asset tracking, field data collection, and surveying. With the widespread availability of smartphones and handheld GPS devices, the input of GPS data has become increasingly accessible, facilitating real-time mapping and location-based services within GIS. This data source has significantly enhanced the precision and efficiency of spatial data accuracy, transforming the field of geospatial analysis and mapping (Fig. 6.18).

6.5.5 *Input of Map Data: Published Report Data*

Incorporating map data from published reports is an important yet often manual task in Geographic Information Systems (GIS). Published reports, such as government publications, scientific studies, and research papers, often contain valuable geographic information in the form of tables, charts, and textual descriptions. To integrate this data into GIS, analysts typically extract the relevant information and convert it into digital formats, such as spreadsheets or geospatial databases. Georeferencing techniques may be applied to associate the extracted data with specific geographic coordinates or regions. This process facilitates the inclusion of critical data from published reports into GIS databases, supporting spatial analysis, mapping, and decision-making. It is particularly valuable in research, environmental assessments, and policy development, where historical or specialized data is necessary for making informed decisions (Fig. 6.19).

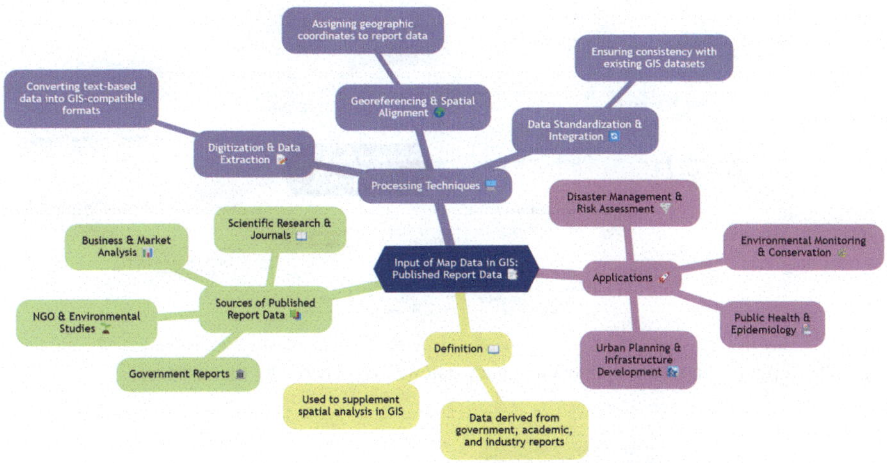

Fig. 6.19 Input of map data: published report data (prepared by the authors)

6.5.6 Input of Map Data: Data from Agency Gateways

Acquiring map data from agency gateways is an efficient and streamlined process within Geographic Information Systems (GIS). Numerous government agencies and organizations offer geospatial data through specialized gateways or web services. These gateways provide convenient access to a variety of spatial datasets, such as maps, aerial imagery, demographic data, and more. Users can directly connect to these sources using GIS software, web mapping applications, or application programming interfaces (APIs) to obtain the most current geospatial data. This approach is time-saving and efficient, as the data is already in digital format and typically includes metadata, making it ready for immediate use in various GIS applications. From urban planning and infrastructure management to environmental monitoring and disaster response, data from agency gateways fosters data sharing, collaboration, and ensures the utilization of reliable and up-to-date geographic information (Fig. 6.20).

6.5.7 Input of Map Data: Participatory GIS

Participatory GIS (PGIS) is a collaborative approach to mapping that actively involves local communities and stakeholders. It empowers individuals and groups to collect, manage, and interpret geographic information, integrating local knowledge and observations into spatial analysis. This approach positions community members as co-producers of spatial data. PGIS methodologies can encompass community mapping activities, participatory data collection through GPS devices or

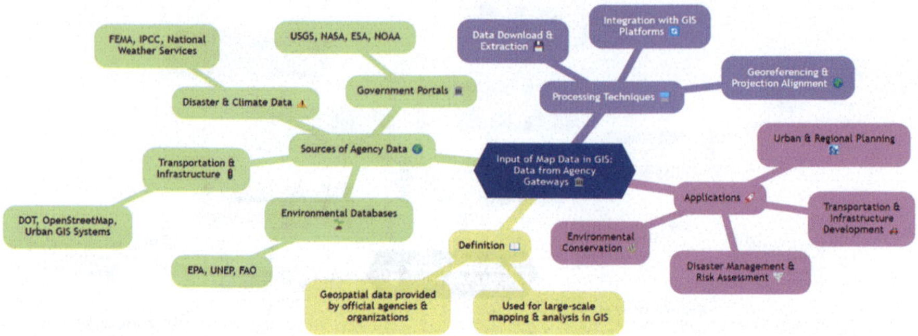

Fig. 6.20 Input of map data: data from agency gateways (prepared by the authors)

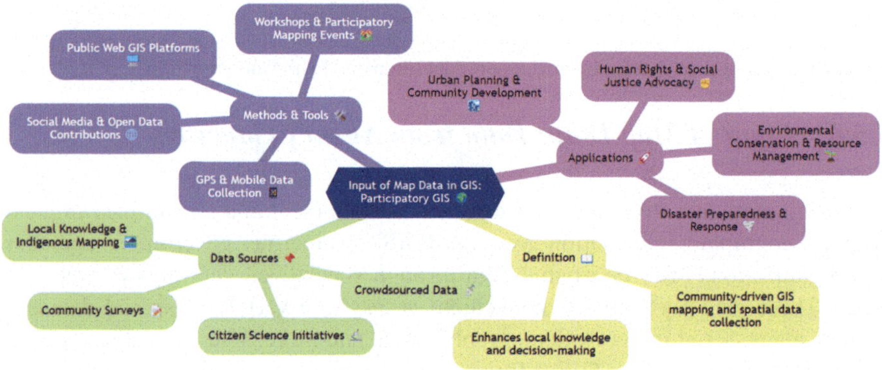

Fig. 6.21 Input of map data: participatory GIS (prepared by the authors)

smartphones, and collaborative spatial analysis, fostering a more comprehensive understanding of geographic phenomena (Fig. 6.21).

6.5.8 *Input of Map Data: Manual Devices, and Others*

Collecting map data using manual devices and alternative methods is a flexible approach in Geographic Information Systems (GIS), particularly when digital sources are unavailable or impractical. This process entails the gathering of geospatial information through traditional techniques, which may include:

- Field surveys: Field surveys involve collecting geospatial data using tools like GPS devices and total stations, with location-specific measurements later entered into GIS software for analysis.

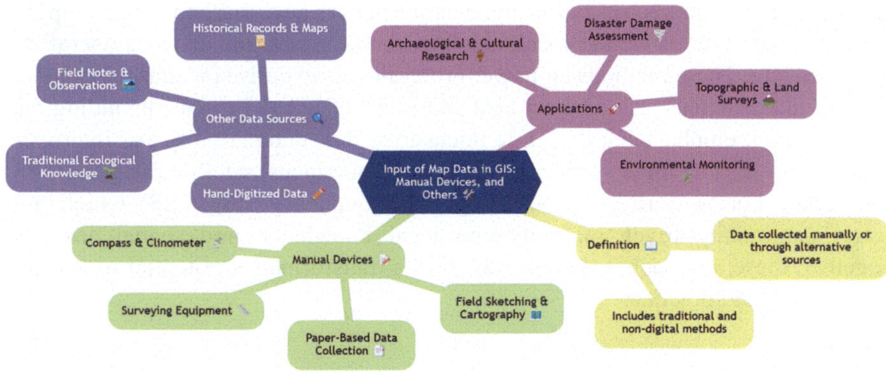

Fig. 6.22 Input of map data: manual devices, and others (prepared by the authors)

- Digitization: The process of converting paper maps or sketches into digital formats, typically using digitizing tablets or scanners, to capture geographic features and their attributes.
- Geocoding: Geocoding assigns geographic coordinates to addresses or place names, converting location-based data into spatial data for mapping.
- GPS-enabled mobile devices: Smartphones and tablets equipped with GPS record location data, photos, and other information for integration into GIS.
- Paper surveys: Traditional paper surveys collect location-based data, which is manually entered into GIS databases for analysis.
- Remote sensing fieldwork: Fieldwork in remote sensing collects ground truth data to validate satellite imagery and monitor environmental conditions, which is then input into GIS systems.

Although manual data input methods are labour-intensive and prone to human error, they remain essential when real-world, on-the-ground data collection is required. These methods complement digital sources by providing accurate, context-specific information crucial for various GIS applications, such as environmental monitoring, land-use planning, asset management, and disaster response (Fig. 6.22).

6.6 Conclusion and Endnotes

In conclusion, this chapter examined key aspects of Geographic Information Systems (GIS), emphasizing their critical role in geospatial analysis and decision-making. GIS technology enables efficient management, analysis, and visualization of spatial data, supporting diverse applications. The discussion highlighted the importance of GIS data analysis in understanding complex geographic phenomena and making informed decisions, focusing on spatial, temporal, and attribute dimensions. Practical techniques for analyzing raster and vector data were explored, including operations such as buffering, overlay, pattern analysis, and map manipulation for vector data, as well as local, neighbourhood, zonal, and distance operations for raster data. These methods empower professionals to derive insights and address real-world challenges. The chapter also addressed GIS data formats, including raster and vector, emphasizing their role in ensuring data compatibility and interoperability. Various data input sources, such as analogue maps, satellite imagery, GPS, and agency reports, were reviewed, alongside Participatory GIS (PGIS), which integrates community-driven mapping and local knowledge into spatial analysis. Together, these approaches demonstrate GIS's transformative potential in geospatial research and practice.

Further Readings

1. Anselin, L., & Getis, A. (1993). Spatial statistical analysis and geographic information systems. In M. M. Fischer & P. Nijkamp (Eds.), *Geographic information systems, spatial modelling and policy evaluation*. Springer. https://doi.org/10.1007/978-3-642-77500-0_3
2. Anselin, L., & Rey, S. J. (Eds.). (2010). *Perspectives on spatial data analysis*. Springer . ISBN: 978-3-642-01975-3. https://doi.org/10.1007/978-3-642-01976-0
3. Bourennane, S., & Kubicek, P. (2022). *Geoinformatics and data analysis: Selected proceedings of ICGDA 2022*. Springer . ISBN: 978-3-031-08016-6. https://doi.org/10.1007/978-3-031-08017-3
4. Goodchild, M. (2008). Data analysis, spatial. In S. Shekhar & H. Xiong (Eds.), *Encyclopedia of GIS*. Springer. https://doi.org/10.1007/978-0-387-35973-1_236
5. Goodchild, M. F., & Haining, R. P. (2004). GIS and spatial data analysis: Converging perspectives. In R. J. G. M. Florax & D. A. Plane (Eds.), *Fifty years of regional science. Advances in spatial science*. Springer. https://doi.org/10.1007/978-3-662-07223-3_16
6. Grekousis, G. (2020). *Spatial analysis methods and practice: Describe – Explore – Explain through GIS*. Cambridge University Press, . ISBN: 9781108614528. https://doi.org/10.1017/9781108614528
7. Konecny, G. (2014). *Geoinformation: Remote sensing, photogrammetry and geographic information systems* (2nd ed.). CRC Press (Taylor and Francis Group). ISBN: 9780429162480. https://doi.org/10.1201/b15765
8. Kresse, W., & Danko, D. (2012). *Springer handbook of geographic information* (Springer handbooks) (2nd ed.). Springer . ISBN: 978-3-030-53124-9. https://doi.org/10.1007/978-3-030-53125-6
9. Reddy, G. P. O. (2018). Spatial data management, analysis, and modeling in GIS: Principles and applications. In G. Reddy & S. Singh (Eds.), *Geospatial technologies in land resources mapping, monitoring and management. Geotechnologies and the environment* (Vol. 21). Springer. https://doi.org/10.1007/978-3-319-78711-4_7
10. Zhu, X. (2024). *Geographical information systems: A practical approach* (2nd ed.). Routledge, (Taylor and Francis Group) ISBN 9781032380469.

Chapter 7
Database Management System: Foundations and Practices

"A well-designed database is like a well-organized library, where information is not just stored but easily accessed and used to shape the future." — Ted Codd.

Abstract A database management system (DBMS) is a software application that facilitates the organized storage, retrieval, and manipulation of data. This chapter explores key DBMS features, database models and languages, architecture, and design. The chapter begins with an introduction to the concept, paradigm shift, purpose, types, database management system, applications, advantages, and disadvantages of database management systems. It defines the function of database management systems (DBMS) in data management, including data integrity, security, and concurrency. Furthermore, this chapter examines certain key DBMS aspects and features, which aid in understanding the concept and application of DBMS. Data definition, database construction, data modelling, data manipulation, data storage and retrieval, concurrency control, data integrity and security, data backup and recovery, data dictionary, database administrators and users, data independence, database structure, and database are keys features of DBMS. The chapter begins by introducing the concept, paradigm shift, purpose, types, and applications of database management systems (DBMS), along with their advantages and disadvantages. It defines the role of DBMS in data management, focusing on aspects such as data integrity, security, and concurrency. Additionally, the chapter explores key DBMS features, including data definition, database construction, data modelling, data manipulation, storage and retrieval, concurrency control, integrity and security, backup and recovery, data dictionaries. It also highlights the significance of data independence and database structure in DBMS. The chapter next goes into database models and languages, laying the groundwork for structuring, organizing, and manipulating data within a database. The chapter goes on to discuss database architecture (two-tier and three-tier) as well as database conceptual, logical, and

K. D. Malakar et al., *Geospatial Technologies in Coastal Ecologies Monitoring and Management*, Advances in Geographic Information Science,
https://doi.org/10.1007/978-3-031-92017-2_7

physical design. The chapter also looks at how DBMS can be integrated with other technologies to help with advanced data analysis and decision-making, such as data warehousing, data mining, and big data analytics.

Keywords DBMS · Database models and languages · Database architecture · ER modelling · Paradigm shift of DBMS

This chapter provides a comprehensive exploration of database management systems (DBMS), highlighting their essential features, models, languages, architecture, and design. It introduces the core concepts of DBMS, including their purpose, types, applications, and advantages. The chapter emphasizes key functions such as data integrity, security, and concurrency control. Additionally, it delves into the various DBMS components, including data definition, manipulation, storage, and retrieval. The chapter also covers database models, architecture, and design phases, while addressing the integration of DBMS with advanced technologies like data warehousing, mining, and big data analytics.

Key Points of the Chapter

- Introduction to DBMS concepts, types, and applications, including their advantages and disadvantages.
- Role of DBMS in data management, emphasizing data integrity, security, and concurrency.
- Key DBMS features: data definition, construction, modelling, manipulation, and storage/retrieval.
- Overview of DBMS architecture (two-tier and three-tier) and database design phases (conceptual, logical, physical).
- Integration of DBMS with advanced data analysis and decision-making.

7.1 Introduction to Database Management System (DBMS)

7.1.1 Concept of DBMS

A database management system (DBMS) is a fundamental software application designed to efficiently organize, store, retrieve, and manipulate data within an organization. Acting as a centralized data repository, it enables multiple users to access and manage information concurrently. The primary objectives of a DBMS are to maintain data integrity, security, and consistency while reducing redundancy. This is achieved through data models such as the relational model, which represents data in tables with defined relationships. DBMS provides a structured framework for storing data, facilitating streamlined querying and analysis. It includes essential features like data backup, recovery, and access control, ensuring reliability and

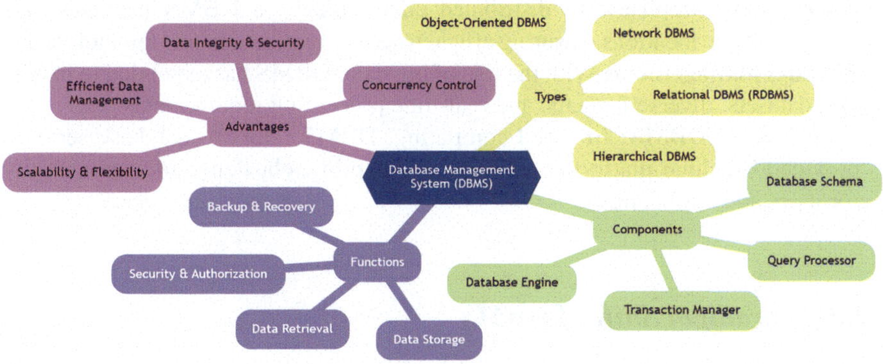

Fig. 7.1 Concept of DBMS (prepared by the authors)

security. DBMS solutions range from lightweight systems like SQLite to enterprise-grade platforms such as Oracle and SQL Server, addressing diverse data management requirements across industries. In the modern data-driven landscape, DBMS plays a pivotal role in enabling organizations to effectively harness and utilize their data for informed decision-making and operational efficiency (Fig. 7.1). A database management system (DBMS) is a sophisticated and multifaceted concept within the field of computer science and data management. The various dimensions of DBMS can be explored from multiple perspectives, as outlined below:

- **Traditional perspective:** A database management system (DBMS) is a software application that facilitates interaction between users, applications, and databases. It provides tools for creating, modifying, organizing, and retrieving data while ensuring data integrity, security, and efficiency.
- **Technical perspective:** A DBMS acts as a software intermediary between physical data storage and applications. It manages storage, retrieval, indexing, and query optimization, ensuring seamless data access and manipulation.
- **User-focused definition:** For users, a DBMS simplifies data management by abstracting complexities, enabling easy definition, storage, and retrieval of data.
- **Administrative perspective:** From an administrative angle, a DBMS provides tools for database administrators to manage databases effectively, offering features such as access control, backup and recovery, performance optimization, and monitoring.
- **Enterprise level:** Within organizations, a DBMS serves as the backbone of data management, offering a centralized repository for data sharing, consistency, and security across departments.
- **Relational databases:** In relational systems, a DBMS facilitates table definition, relationship modelling, and query execution via structured query language (SQL), enforcing constraints and supporting transactions.
- **NoSQL databases:** Modern DBMS also encompasses systems for handling unstructured or semi-structured data, providing flexibility and scalability for contemporary needs.

- **Distributed databases:** In distributed environments, a DBMS manages data across multiple locations, ensuring efficient access, replication, and fault tolerance.
- **Big data management:** With the rise of big data, DBMS has expanded to handle vast datasets, diverse formats, and real-time processing requirements.
- **Cloud-based DBMS:** In cloud computing, DBMS refers to database services provided by cloud platforms, delivering scalability, elasticity, and accessibility for users and applications.

7.1.2 Paradigm Shift of DBMS

The paradigm shift in database management systems (DBMS) signifies a significant evolution in the storage, processing, and management of data. This transformation is propelled by advancements in technology, the growing complexity of data, and changing organizational and business demands. The following are the various dimensions of this paradigm shift:

- **From relational to NoSQL:** The dominance of relational databases has given way to NoSQL systems, offering flexible data models for unstructured, semi-structured, and dynamic data, catering to modern scalability needs.
- **Scalability and distribution:** Big data and cloud computing have driven the rise of scalable and distributed DBMS solutions, enabling efficient management of massive datasets across servers and locations.
- **Schema evolution:** Unlike static schemas, dynamic and schema-less approaches in NoSQL databases allow seamless adaptation to evolving data requirements.
- **Polyglot persistence:** Organizations increasingly adopt multiple database technologies, optimizing data management by using specialized databases for different workloads.
- **Real-time processing:** In-memory databases and stream processing tools meet demands for low-latency access and instant analytics in real-time applications.
- **Open-source innovation:** Open-source DBMS fosters innovation and collaboration, expanding the range of database technologies and tools.
- **Security and compliance:** Enhanced security features and compliance measures address growing concerns over data privacy and breaches.
- **AI and machine learning:** DBMS integrate AI and machine learning for predictive analytics, automation, and intelligent data management.
- **Containerization:** Technologies like Docker and Kubernetes streamline the deployment, scaling, and maintenance of DBMS.
- **User-centric design:** Modern DBMS emphasize accessibility, featuring intuitive interfaces and self-service tools for broader user engagement.

This paradigm shift in database management systems (DBMS) underscores the evolving dynamics of data management, shaped by the growing demands for flexibility, scalability, real-time processing, and adaptability to rapidly changing technological and business environments. It has revolutionized how organizations manage and utilize data, empowering them to unlock the full potential of their information assets.

7.1.3 Purpose of DBMS

The purpose of a database management system (DBMS) is both multifaceted and critical for ensuring efficient and effective data management within organizations (Fig. 7.2). The key objectives of a DBMS include:

- **Data organization and storage:** A DBMS ensures structured storage of large datasets by organizing data into tables, rows, and columns, promoting consistency and reducing redundancy. This structure facilitates efficient data retrieval and manipulation.
- **Data retrieval:** A core function of DBMS is enabling quick and efficient access to specific data through queries, eliminating the need to navigate complex structures.

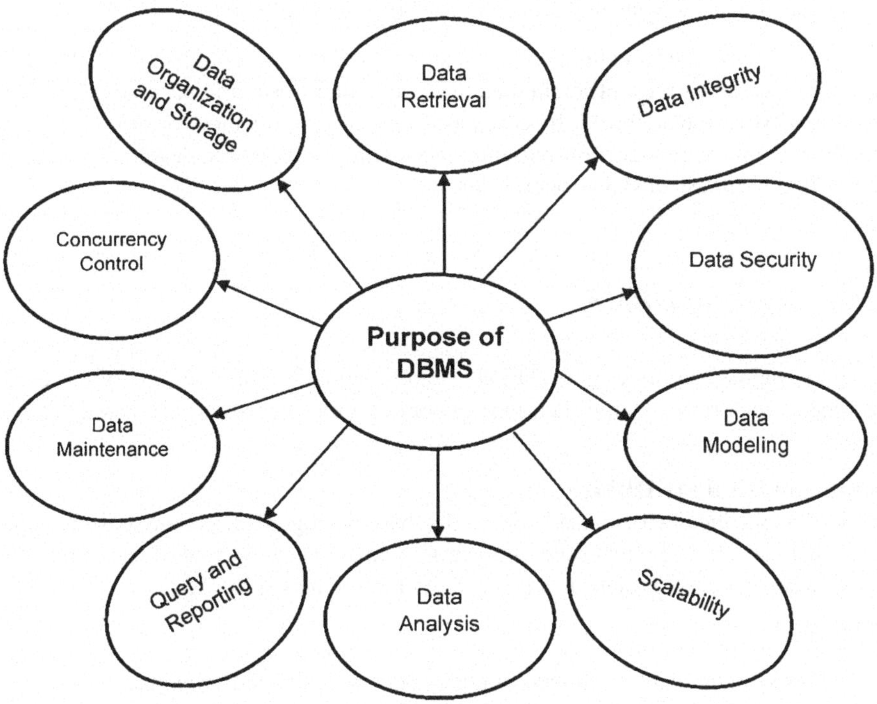

Fig. 7.2 Purpose of DBMS (prepared by the authors)

- **Data integrity:** DBMS enforces rules and constraints to maintain data accuracy and prevent inconsistencies, ensuring the reliability of stored information.
- **Data security:** With robust access control mechanisms, a DBMS restricts unauthorized access by allowing administrators to define roles and permissions, protecting sensitive data.
- **Concurrency control:** DBMS manages simultaneous access by multiple users or applications, ensuring data consistency and resolving potential conflicts.
- **Data maintenance:** Tools for backup and recovery safeguard data availability and ensure restoration in case of failures or corruption.
- **Query and reporting:** DBMS provides query languages (e.g., SQL) to extract and analyze data, enabling users to generate reports for decision-making.
- **Data Modelling:** It supports defining data structures and relationships, simplifying the representation of complex real-world scenarios.
- **Scalability:** Modern DBMS systems can scale to accommodate growing data demands using additional resources or distributed architectures.
- **Application integration:** DBMS enables seamless interaction between databases and software applications, streamlining data manipulation and usage.
- **Data consistency:** By adhering to ACID properties (atomicity, consistency, isolation, durability), DBMS ensures consistent data during transactions or unexpected interruptions.
- **Data analysis:** Integration with analytics tools allows DBMS to support advanced data analysis and reporting for business intelligence.

In conclusion, the fundamental purpose of a DBMS is to establish a reliable and structured framework for efficient data management, prioritizing accuracy, security, accessibility, and availability. It serves as a cornerstone for modern organizations, enabling data-driven decision-making, facilitating application development, and supporting seamless business operations.

7.1.4 Types of DBMS

Database management systems (DBMS) can be classified into various types, each designed to address specific data management requirements and use cases. These include:

Relational DBMS (RDBMS)
RDBMS is the most widely used type of database management system, based on the relational model. It organizes data into structured tables composed of rows and columns, with predefined relationships between these tables (Fig. 7.3).

Features

- Data is systematically arranged in tables with well-defined schemas.
- Ensures data integrity through constraints such as primary and foreign keys.

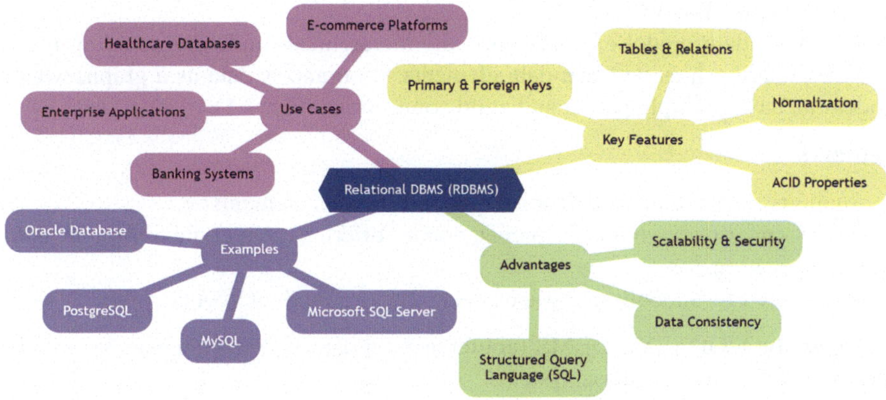

Fig. 7.3 Types of DBMS: relational (prepared by the authors)

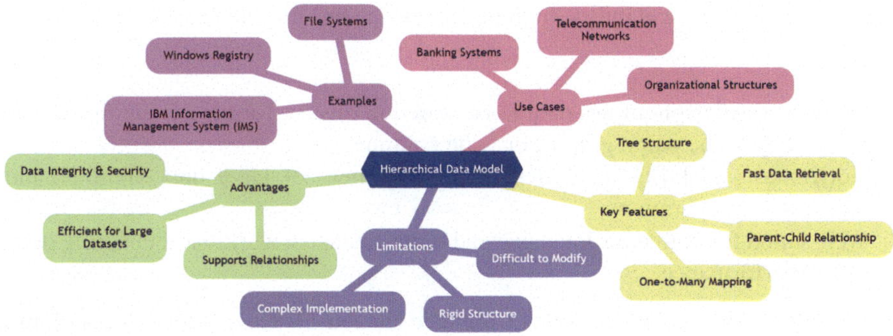

Fig. 7.4 Types of DBMS: hierarchical data model (prepared by the authors)

- Supports robust querying and data manipulation using SQL (structured query language).
- Adheres to ACID properties (atomicity, consistency, isolation, durability) to guarantee data consistency and reliability.

Example: MySQL, PostgreSQL, Oracle Database, Microsoft SQL Server.

Hierarchical Data Model

The hierarchical data model organizes data in a tree-like structure, where each data element has a parent–child relationship, forming levels of hierarchy (Fig. 7.4).

Features

- Data is arranged into hierarchical levels, with parent and child elements.
- Well-suited for structured data with one-to-many relationships.
- Commonly used in early database systems and some niche applications.

Example: IMS (information management system), primarily used in legacy mainframe systems.

Network Data Model

The network data model expands upon the hierarchical model by enabling more complex relationships between data elements. It organizes data as a graph, where nodes can have multiple parent and child connections (Fig. 7.5).

Features

- Supports many-to-many relationships between data elements.
- Nodes can link to multiple parent nodes, offering greater flexibility in data representation.
- Well-suited for modelling intricate data structures and interconnections.

Example: CODASYL DBMS (conference on data systems languages), widely used in some legacy database systems.

NoSQL DBMS

NoSQL databases are tailored for managing unstructured, semi-structured, or rapidly evolving data. They provide flexible data models, making them ideal for applications that demand scalability, high performance, and fast data access (Fig. 7.6).

Features

- Schema-less design allows data to be stored in diverse formats, including documents, key-value pairs, graphs, or wide columns.
- Horizontal scalability facilitates the management of large data volumes and high-velocity operations.
- Well-suited for both read-intensive and write-intensive workloads, ensuring optimized performance.

Example: MongoDB (document-oriented database), Cassandra (wide-column store), Redis (key-value store), and Neo4j (graph database).

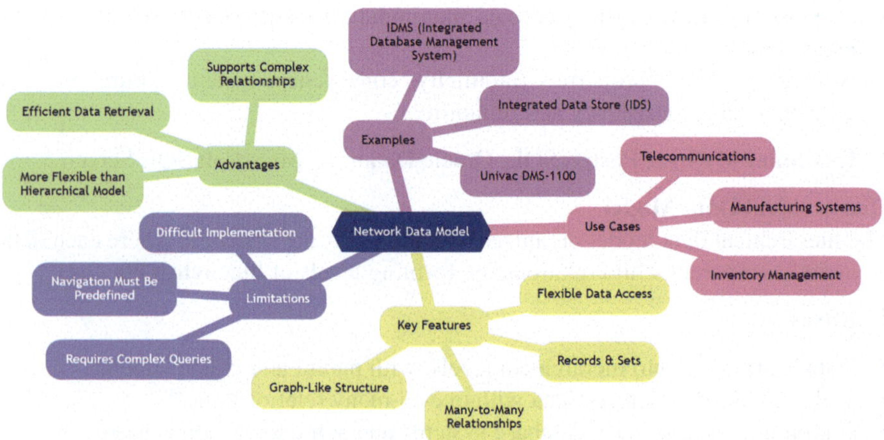

Fig. 7.5 Types of DBMS: network data model (prepared by the authors)

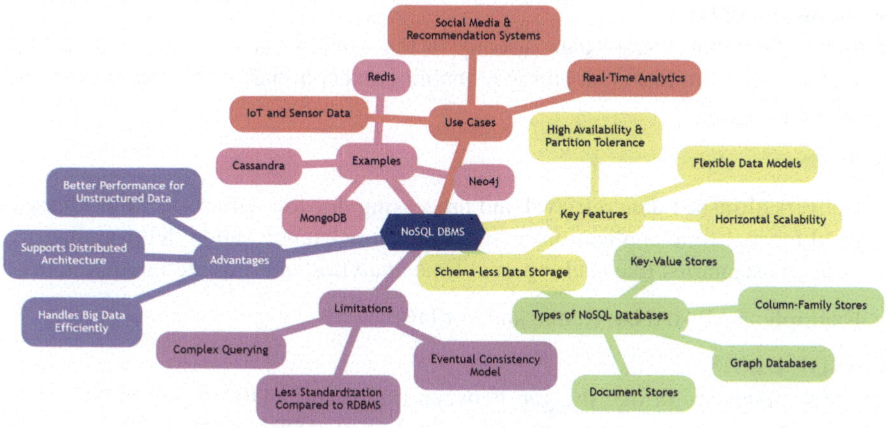

Fig. 7.6 Types of DBMS: NoSQL (prepared by the authors)

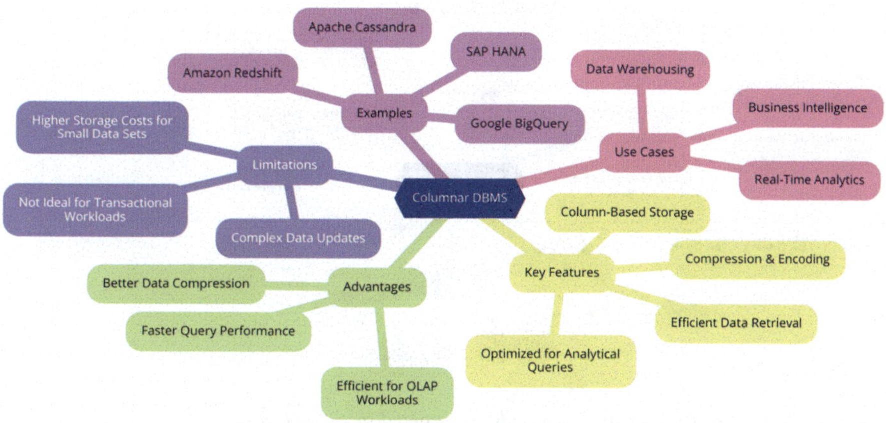

Fig. 7.7 Types of DBMS: columnar (prepared by the authors)

Columnar DBMS

Columnar databases organize data by columns rather than rows, optimizing performance for analytical queries and data warehousing tasks (Fig. 7.7).

Features

- Designed for read-heavy operations and analytical workloads, offering high query performance.
- Employ efficient compression techniques to minimize storage requirements.
- Ideal for business intelligence, data analytics, and processing large datasets.

Example: Amazon Redshift, Google BigQuery, Apache Cassandra (can serve as both a NoSQL and columnar database).

In-Memory DBMS

In-memory databases store data directly in the computer's main memory (RAM) instead of on traditional disk storage, enabling exceptionally fast data access and processing (Fig. 7.8).

Features

- Delivers ultra-fast data retrieval and processing due to memory-resident storage.
- Ideal for real-time applications that demand low latency and high throughput.
- Supports scenarios like caching, real-time analytics, and transactional systems.

Example: SAP HANA, Redis, and VoltDB.

NewSQL DBMS

NewSQL databases bridge the gap between traditional RDBMS and NoSQL systems by offering the consistency and reliability of RDBMS while providing the scalability and performance typical of NoSQL databases (Fig. 7.9).

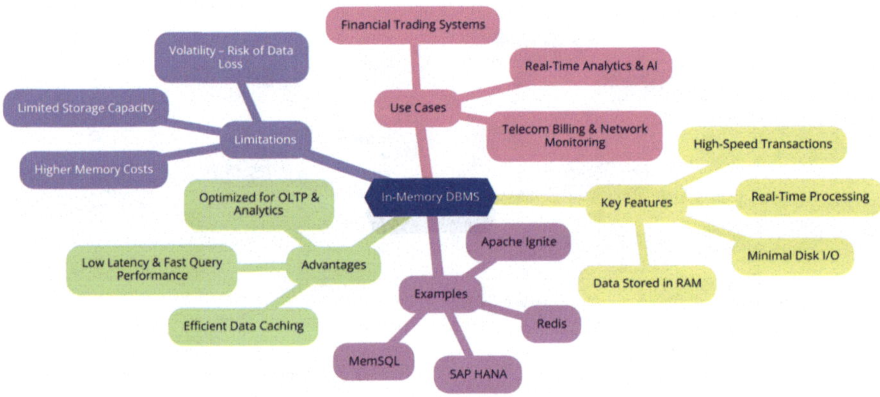

Fig. 7.8 Types of DBMS: in-memory DBMS (prepared by the authors)

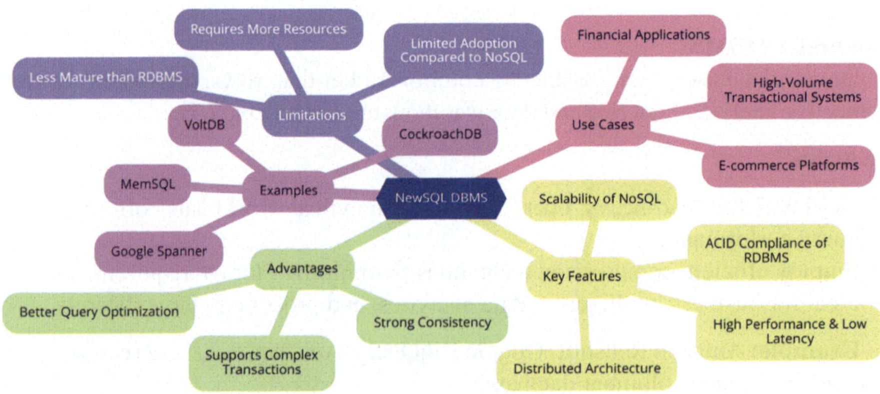

Fig. 7.9 Types of DBMS: NewSQL DBMS (prepared by the authors)

Features

- ACID compliance: Ensures data consistency and reliability for critical applications.
- Horizontal scalability: Supports distributed architectures to handle large-scale data and traffic.
- High performance: Suitable for online transaction processing (OLTP) and other demanding workloads.

Example: Google Spanner, CockroachDB, and NuoDB.

Graph DBMS

Graph databases are specialized database systems designed to manage and analyze highly interconnected data. They are particularly effective for use cases involving complex relationships and network structures (Fig. 7.10).

Features

- Data representation: Stores data as nodes (entities) and edges (relationships), facilitating efficient relationship traversal.
- Optimized for relationships: Ideal for modelling and querying interconnected data such as social networks, recommendation engines, and network graphs.
- High performance: Excels at handling queries that involve deep or complex relationships.

Example: Neo4j, Amazon Neptune, and OrientDB.

Time-Series DBMS

Time-series databases are optimized for storing and retrieving data that is time-stamped, making them ideal for applications that track and analyze data over time, such as monitoring, IoT, and financial analysis (Fig. 7.11).

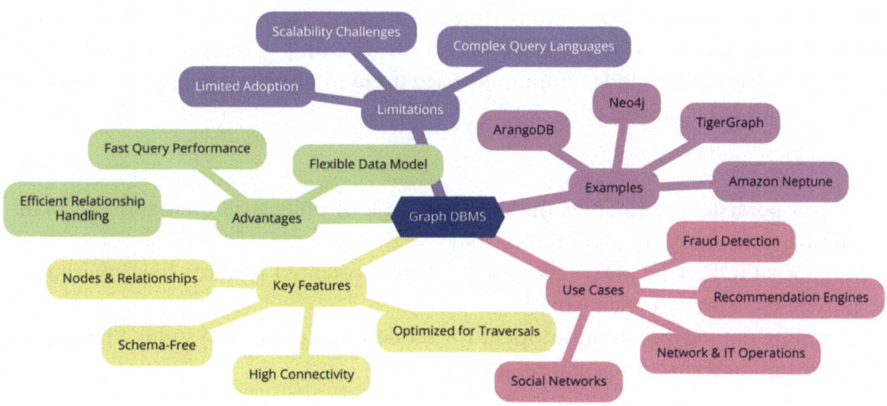

Fig. 7.10 Types of DBMS: graph DBMS (prepared by the authors)

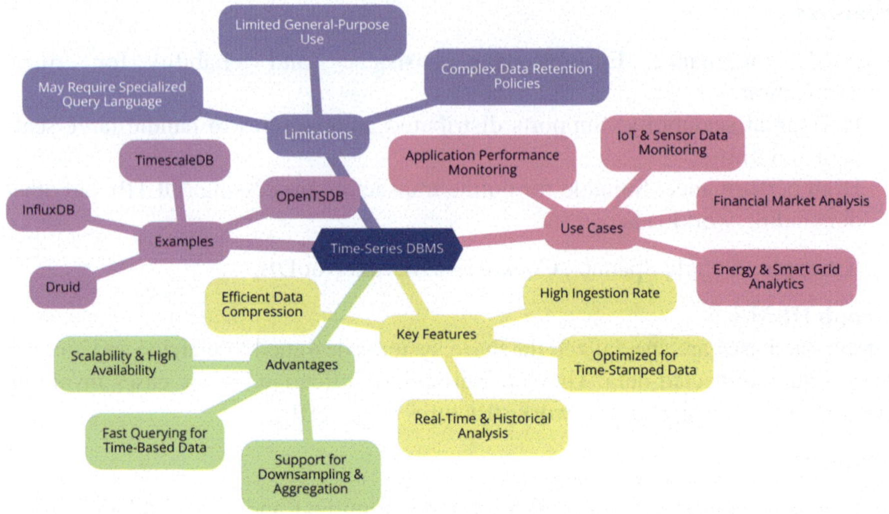

Fig. 7.11 Types of DBMS: time-series DBMS (prepared by the authors)

Features

- Time-optimized storage: Designed to handle high volumes of timestamped data with efficient storage and retrieval mechanisms.
- Real-time analytics: Well-suited for applications that require real-time data analysis over continuous time intervals.
- Scalability: Capable of managing large datasets generated from time-series data streams.

Example: InfluxDB, Prometheus, and TimescaleDB.

Object-Oriented DBMS (OODBMS)

OODBMS stores data in the form of objects, aligning with the principles of object-oriented programming (OOP). This allows for a seamless representation of complex, real-world entities and is commonly used in environments where object-oriented programming is employed (Fig. 7.12).

Features

- Object persistence: Enables the direct storage of objects in the database, maintaining their structure and relationships.
- Supports OOP principles: Incorporates object-oriented features like inheritance, polymorphism, and encapsulation.
- Complex data representation: Ideal for modelling complex systems that require advanced data structures and relationships.

Example: db4o, and Versant Object Database.

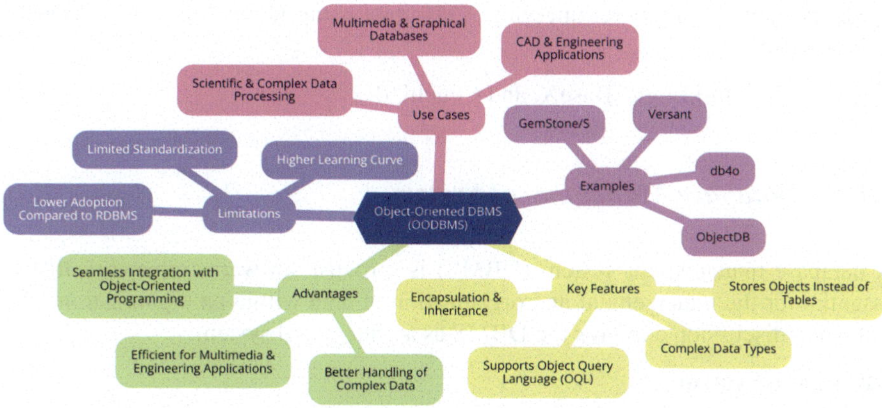

Fig. 7.12 Types of DBMS: object-oriented DBMS (prepared by the authors)

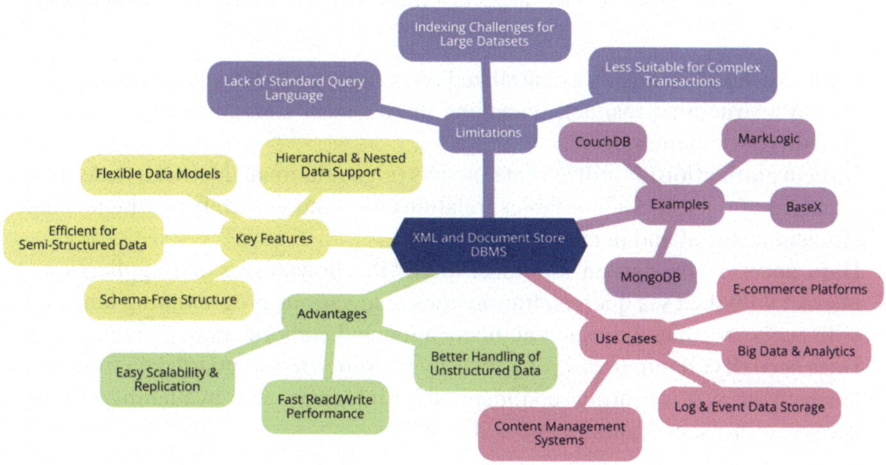

Fig. 7.13 Types of DBMS: XML and document store DBMS (prepared by the authors)

XML and Document Store DBMS

XML databases are specialized for storing, retrieving, and querying XML data and document-based information. They are optimized to handle the unique structure of XML documents, making them ideal for applications that deal with hierarchical and flexible data formats (Fig. 7.13).

Features

- XML data structure handling: Efficiently stores and processes XML data, maintaining the hierarchical relationships within documents.
- Document-oriented: Well-suited for content management systems and applications requiring the storage of document-based information.

- Query support: Supports querying XML data using technologies like XQuery and XPath.

  Example: MarkLogic, BaseX, and CouchDB.

7.1.5 Database Management System

A database management system (DBMS) is a robust software framework that is essential for the management and organization of data within an organization. The following discussion explores the DBMS as an integrated system:

DBMS as a System

A DBMS is a sophisticated and essential component of an organization's information technology infrastructure. It operates as a structured and organized software system, specifically designed to efficiently store, retrieve, update, and manage data (Fig. 7.14). The following key aspects emphasize its nature as a comprehensive system:

- **Data storage:** Serving as a centralized repository, a DBMS organizes and stores data in a structured manner, managing the physical storage on devices such as hard drives or memory.
- **Data organization:** It utilizes data models (e.g., relational, NoSQL) to define the structure of data, including tables, relationships, and constraints, which enables efficient retrieval and manipulation.
- **Data access:** Through an interface, a DBMS allows users and applications to interact with data via queries, transactions, and various programming languages, with access regulated through authentication and authorization.
- **Data security:** It employs strong security measures to safeguard data from unauthorized access, ensuring confidentiality, integrity, and availability through access controls, encryption, and auditing.

Fig. 7.14 DBMS as a system (prepared by the authors)

- **Data integrity:** A DBMS enforces integrity constraints, such as primary and foreign keys, and data validation rules, ensuring the quality and consistency of the stored data.
- **Concurrency control:** In multi-user environments, it manages concurrent access to data, using techniques like locking and isolation to avoid conflicts and ensure data consistency.
- **Data backup and recovery:** It includes features for data backup, replication, and recovery, safeguarding against data loss and system failures, ensuring data resilience.
- **Optimization and performance:** The DBMS optimizes query execution through indexing, caching, and query optimization, ensuring fast and efficient data retrieval.
- **Scalability:** Many DBMS platforms offer scalability options, allowing them to handle growing data volumes and increased workloads, achievable through vertical or horizontal scaling.
- **Monitoring and maintenance:** DBMS provides tools for monitoring performance, identifying bottlenecks, and automating maintenance tasks such as indexing and data compaction.
- **Compatibility and integration:** It supports integration with various programming languages, tools, and applications, making it versatile and adaptable across different software ecosystems.
- **User-friendly interfaces:** Modern DBMS systems offer intuitive interfaces for administrators, developers, and end-users, simplifying database management tasks and enhancing usability.

In summary, a database management system (DBMS) is an integrated and essential component of an organization's data management framework. It ensures the reliability, security, and accessibility of data, providing tools and features that streamline data handling. By supporting data integrity and efficient management, a DBMS plays a crucial role in facilitating informed decision-making and advancing organizational goals.

7.1.6 Applications of DBMS

Database management systems (DBMS) are extensively utilized across diverse industries and sectors, owing to their capacity to efficiently handle, organize, and secure data. The following highlights various applications of DBMS (Fig. 7.15):

Business and Finance

(a) Customer relationship management (CRM): DBMS stores customer data, including interactions and transaction histories, enabling businesses to efficiently manage and enhance customer relationships.

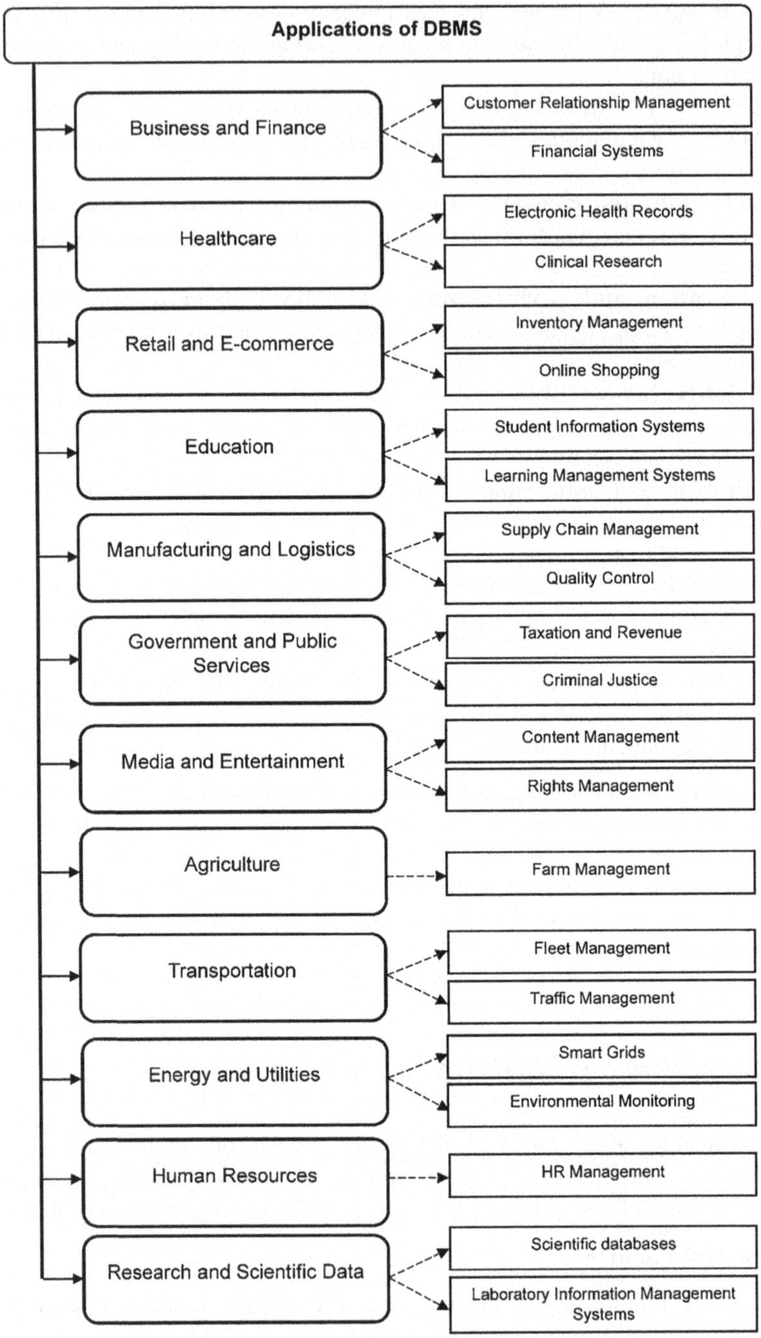

Fig. 7.15 Applications of DBMS (prepared by the authors)

(b) Financial systems: DBMS handles financial data such as transactions, accounts, and assets, ensuring precise and secure record-keeping for financial operations.

Healthcare

(a) Electronic health records (EHR): DBMS maintains patient information, medical histories, and treatment records, facilitating improved care coordination and easy access to healthcare data.
(b) Clinical research: DBMS supports the storage and analysis of data from clinical trials and medical research, advancing drug development and disease understanding.

Retail and E-commerce

(a) Inventory management: DBMS tracks product inventory levels, sales, and stock replenishment, optimizing supply chain and inventory management.
(b) Online shopping: E-commerce platforms utilize DBMS to store product catalogues, customer profiles, and transaction histories, ensuring seamless shopping experiences.

Education

(a) Student information systems (SIS): DBMS organizes student records, enrolment information, grades, and academic data to streamline academic management for educational institutions.
(b) Learning management systems (LMS): DBMS stores course content, user profiles, and learning progress, supporting online education platforms.

Manufacturing and Logistics

(a) Supply chain management: DBMS helps track raw materials, production stages, and distribution channels, improving efficiency throughout the supply chain.
(b) Quality control: DBMS monitors product quality and ensures compliance with industry standards in manufacturing settings.

Government and Public Services

(a) Taxation and revenue: DBMS stores tax records, citizen data, and financial transactions, supporting tax collection and regulatory compliance.
(b) Criminal justice: DBMS manages criminal records, case files, and law enforcement data to streamline operations within the justice system.

Media and Entertainment

(a) Content management: DBMS organizes multimedia content, metadata, and user preferences for streaming services, digital libraries, and content distribution platforms.
(b) Rights management: DBMS tracks copyrights, licenses, and royalties for media and entertainment sectors, ensuring proper rights management.

Agriculture

(a) Farm management: DBMS records crop data, weather patterns, and equipment maintenance schedules, supporting precision agriculture and maximizing crop yields.

Transportation

(a) Fleet management: DBMS monitors vehicle maintenance schedules, routes, and fuel consumption, optimizing logistics and transportation operations.
(b) Traffic management: DBMS stores traffic-related data, supporting intelligent systems that optimize traffic flow and reduce congestion.

Energy and Utilities

(a) Smart grids: DBMS handles data from smart meters to optimize energy distribution, consumption tracking, and outage management.
(b) Environmental monitoring: DBMS stores sensor data for tracking environmental parameters, ensuring compliance with regulations and sustainability efforts.

Human Resources

(a) HR management: DBMS stores employee records, payroll data, and performance evaluations, streamlining HR operations and talent management processes.

Research and Scientific Data

(a) Scientific databases: DBMS facilitates data storage and retrieval in scientific research, particularly in fields like genomics, astronomy, and geology.
(b) Laboratory information management systems (LIMS): DBMS manages sample data, test results, and research workflows, supporting laboratory research and development.

7.1.7 Advantage and Disadvantage of DBMS

Advantages of DBMS

- **Data integrity:** DBMS enforces data integrity constraints, including primary keys, foreign keys, and validation rules, to maintain the accuracy and consistency of the data stored.
- **Data security:** Through access control and authentication mechanisms, DBMS ensures the protection of data from unauthorized access, safeguarding sensitive and confidential information.
- **Data centralization:** DBMS centralizes data storage, reducing redundancy and providing a unified source of truth for the organization, promoting consistency and efficiency.

- **Data accessibility:** DBMS provides a user-friendly interface for data retrieval and manipulation, making it easier for users and applications to access and work with data, thereby simplifying data management processes.
- **Concurrent access:** DBMS manages concurrent data access by multiple users, preventing data conflicts and ensuring consistency during simultaneous transactions.
- **Data backup and recovery:** DBMS offers backup, replication, and recovery tools, ensuring data resilience and minimizing downtime or data loss in case of system failures.
- **Scalability:** Many DBMS platforms support both horizontal and vertical scaling, enabling organizations to manage increasing data volumes and workload demands as they grow.
- **Data Modelling:** DBMS supports various data models, such as relational and NoSQL, providing flexibility for organizations to select the most appropriate model for their specific data management needs.
- **Performance optimization:** DBMS enhances query execution performance through techniques such as indexing, caching, and query optimization, leading to faster data retrieval and processing.
- **Reduced maintenance:** By automating routine maintenance tasks, such as indexing and data compaction, DBMS helps reduce the administrative burden and streamlines ongoing database management.

Disadvantages of DBMS

- **Complexity:** Managing and utilizing a DBMS requires specialized knowledge and expertise, which can lead to increased personnel costs and a steeper learning curve.
- **Cost:** The implementation and maintenance of a DBMS can incur substantial costs, including software licensing, hardware infrastructure, and ongoing maintenance fees.
- **Performance overhead:** The processes involved in query execution, indexing, and transaction management can introduce performance overheads, potentially affecting the efficiency of real-time applications.
- **Vendor lock-in:** Adopting a specific DBMS vendor may result in vendor lock-in, making it difficult and costly to switch to an alternative DBMS solution due to dependencies and proprietary technologies.
- **Resource consumption:** DBMS systems can be resource-intensive, consuming significant CPU, memory, and storage, which may impact overall system performance and scalability.
- **Data security risks:** Inadequate configuration or mismanagement of DBMS systems can expose vulnerabilities, leading to security breaches and the potential exposure of sensitive data.
- **Complex backup and recovery:** Although DBMS provides backup and recovery functionalities, establishing a reliable and comprehensive strategy for backup and recovery can be intricate and require careful planning.

- **Data migration challenges:** Migrating data between different DBMS systems or versions can be complex, necessitating meticulous planning, testing, and validation to ensure data integrity.
- **Scalability limits:** Certain DBMS systems may have inherent limitations when it comes to scaling, and achieving higher scalability could necessitate significant architectural changes or reconfiguration.

7.2 Key Features of DBMS

7.2.1 Data Definition

Data definition in a DBMS involves the process of specifying and managing the structure and attributes of data within a database. This includes the creation, modification, and deletion of database components such as tables, schemas, constraints, indexes, and views (Fig. 7.16). The following outlines the key aspects of data definition within a DBMS:

- **Database schema:** The database schema serves as a comprehensive blueprint that defines the overall structure of the database. It specifies the organization of tables, relationships, constraints, and data types to structure and store the data efficiently.
- **Table definition:** Tables are the core elements of a database. In the process of data definition, the attributes (columns) of each table, along with their data types, constraints (such as primary and foreign keys), and default values, are clearly specified.
- **Constraints:** Constraints are rules that enforce data integrity and consistency within the database. Common types include primary keys (for uniquely identifying records), foreign keys (for establishing relationships between tables), unique constraints (to ensure data uniqueness), and check constraints (to validate data according to predefined conditions).

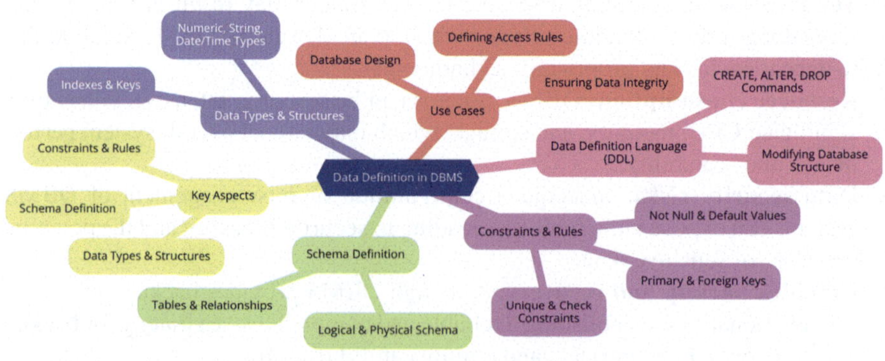

Fig. 7.16 Data definition in DBMS (prepared by the authors)

- **Indexes:** Indexes are structures designed to improve the speed of data retrieval by optimizing search and sorting operations. Data definition involves identifying columns to be indexed to enhance query performance.
- **Views:** Views are virtual tables that offer customized perspectives of the database data. Defining views within data definition helps simplify data access for specific users or applications.
- **Data types:** Data definition involves specifying the appropriate data types (e.g., integer, string, date) for each column, ensuring proper data storage formats and preventing mismatches between expected and actual data.
- **Schema evolution:** Data definition allows for schema evolution, which involves adapting the database structure over time. This can include modifying tables, adding or removing columns, or adjusting constraints to meet evolving data requirements.
- **Data dictionary:** A data dictionary (or data catalogue) is an integral part of data definition, storing metadata about database objects, such as table and column descriptions, constraints, and indexes. It serves as a reference for both users and database administrators.
- **Data validation:** Data definition includes the establishment of validation rules to ensure that the data entered into the database complies with quality standards and business rules, maintaining data accuracy and consistency.
- **Database migration:** During the process of migrating data from one DBMS to another or across different versions, data definition plays a key role by ensuring proper mapping and transformation of database objects for a seamless transition.

Effective data definition is crucial to the success of a database system, as it ensures that data is logically organized, efficiently stored, and maintains its accuracy and consistency. A well-structured database schema, along with the appropriate configuration of associated objects, forms the foundation for data storage, retrieval, and manipulation within the DBMS. This process supports the diverse needs of applications and users across a wide range of domains and industries.

7.2.2 Database Building

Database building, also referred to as database design or creation, is a critical phase in the development of a database management system (DBMS). This process involves the careful planning, design, and implementation of a database, with the goal of efficiently storing, managing, and retrieving data (Fig. 7.17). The following outlines the key aspects of data building within a DBMS:

- **Requirements analysis:** The initial phase of database building involves gathering and analyzing the system's requirements and user needs. This stage focuses on understanding the types of data to be stored, the roles of the users, and the

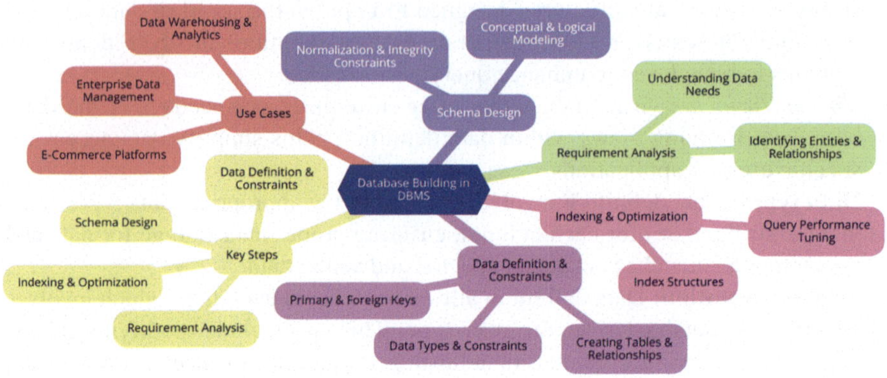

Fig. 7.17 Database building in DBMS (prepared by the authors)

intended purpose of the database. The goal is to identify data storage requirements and how the data will be utilized.

- **Conceptual design:** In this phase, a high-level conceptual model is created for the database. Typically, this involves designing an entity-relationship diagram (ERD) to represent entities, their attributes, and the relationships between them. The focus here is on defining the data structure without addressing technical details.
- **Normalization:** Normalization plays a vital role in optimizing the database schema to minimize redundancy and improve data integrity. It involves breaking down tables and establishing relationships to ensure efficient data organization, free from data anomalies.
- **Logical design:** During the logical design phase, the ERD is converted into a logical schema. This involves defining tables, columns, data types, constraints, and keys. Database schema specification often employs languages such as SQL.
- **Physical design:** In the physical design phase, the logical schema is transformed into physical storage structures. Decisions regarding indexing, storage optimization, and data partitioning are made to ensure efficient data retrieval and storage.
- **Implementation:** This phase involves the actual creation of the database using DBMS-specific commands. Tables are created, constraints are applied, and indexes are defined. Data is loaded into the database through data manipulation tools or scripts.
- **Testing and quality assurance:** Following implementation, comprehensive testing is conducted to verify the database's functionality. This includes checking data integrity, performance, and compliance with business rules.
- **Documentation:** Thorough documentation is critical to understanding the database structure, constraints, and relationships. This includes data dictionaries, schema diagrams, and usage guidelines.
- **Security and access control:** Access control mechanisms are established to ensure that only authorized users can interact with the database. Security measures such as encryption and authentication are implemented to protect sensitive data.

- **Backup and recovery planning:** A robust backup and recovery strategy is developed to safeguard data from loss or corruption. Regular backups and recovery tests are integral to this phase.
- **User training:** Both users and administrators require training to effectively interact with the database. Training programmes should cover data entry, query building, and routine database maintenance tasks.
- **Maintenance and optimization:** Database building is an ongoing process. Continuous monitoring, maintenance, and performance optimization activities are essential to maintaining the database's efficiency and responsiveness over time.

Therefore, effective database building is essential for developing a reliable and efficient data management system. It establishes the foundation for data-driven applications and plays a pivotal role in supporting decision-making processes within organizations. Thorough planning and meticulous attention to detail during the database development phase contribute to the creation of a robust and scalable database capable of adapting to the evolving needs of both users and the organization.

7.2.3 *Data Modelling*

Data modelling is a core process within database management systems (DBMS) that entails constructing an abstract representation of data and its interrelationships within an organization. It functions as a blueprint for the design and organization of the database (Fig. 7.18). The following provides an overview of major aspects of data modelling in DBMS:

- **Purpose of data modelling:** Data modelling seeks to represent the core characteristics of data and its interactions to meet an organization's information requirements. It provides a conceptual and visual framework for understanding how data is structured, stored, and accessed within a database system.

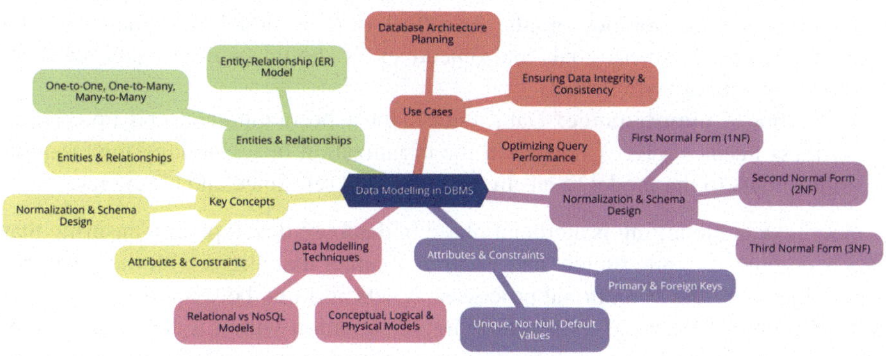

Fig. 7.18 Data modelling in DBMS (prepared by the authors)

- **Entity-relationship diagram (ERD):** The entity-relationship diagram (ERD) is a widely used tool in data modelling. It visually represents entities (objects or concepts) and their relationships using symbols, such as rectangles for entities, diamonds for relationships, and lines for connections. Entities are defined by attributes that describe their characteristics.
- **Types of data models**
  a) Conceptual data model: This high-level model focuses on identifying entities, their attributes, and the relationships among them. It is abstract and technology-agnostic.
  b) Logical data model: This model refines the conceptual model by specifying the structure of tables, columns, keys, and constraints, while still remaining independent of specific technologies.
  c) Physical data model: The physical model defines how data is stored within a database system, including storage mechanisms, indexes, and access paths.
- **Normalization:** Normalization plays a critical role in data modelling, particularly during the logical design phase. It involves organizing data into smaller, related tables to minimize redundancy and enhance data integrity.
- **Data Modelling Tools:** Specialized software and tools, such as ERD drawing tools, are frequently employed to create and manage data models. These tools facilitate diagram creation, attribute definition, and the generation of SQL scripts for database object creation.
- **Notation standards:** Data modelling typically follows established notational standards, such as Crow's foot notation or Chen's notation for ERDs. These standards ensure clarity and consistency in the representation of entities, attributes, and relationships.
- **Data model documentation:** Comprehensive documentation of the data model is essential. This includes detailed descriptions of entities, attributes, relationships, constraints, and associated business rules.
- **Iterative process:** Data modelling is often an iterative process, requiring multiple rounds of refinement and review. It typically involves collaboration with stakeholders to ensure that the model accurately reflects business requirements.
- **Alignment with business goals:** An effective data model is aligned with the organization's business goals and objectives. It should facilitate data retrieval, reporting, and analytics needs.
- **Data model maintenance:** Data models must be maintained and updated as business needs evolve. Changes in the organization or technology may require adjustments to the model to ensure its continued relevance and efficiency.

So, the data modelling is a crucial phase in database development, ensuring that data is organized in an efficient and accurate manner that aligns with the organization's objectives and operational processes. It acts as a link between the conceptual understanding of data and its practical implementation within a database system, offering a structured framework for creating well-organized and effective databases within a DBMS.

7.2.4 Data Manipulation

Data manipulation in database management systems (DBMS) involves the processes of retrieving, inserting, updating, and deleting data within a database. This function is essential for interacting with the data stored in the database, enabling users and applications to effectively manage and utilize the information (Fig. 7.19). The following highlights major aspects of data manipulation in DBMS:

- **Data retrieval (SELECT):** Data retrieval involves querying the database to obtain specific information. Users can use SQL (structured query language) or other query languages to craft queries, which can range from simple requests, such as retrieving all records from a table, to more complex queries that involve filtering, sorting, and aggregating data.
- **Data insertion (INSERT):** Data insertion refers to adding new records or rows to a database table. Users specify the target table, the values for each column, and, optionally, the specific columns where data should be inserted. This operation is commonly used to introduce new data, such as customer records or product details.
- **Data update (UPDATE):** Data updating allows users to modify existing records within a database. The process involves specifying the table, the columns to be updated, and the new values. Users may also define conditions to target specific records for modification. Update operations are essential for maintaining data accuracy and ensuring that information stays current.
- **Data deletion (DELETE):** Data deletion entails removing records from a database table. Users define the target table and set conditions to identify the records to be deleted. Caution is necessary when performing delete operations to prevent unintentional data loss.
- **Transaction management:** DBMS systems typically support transactions, which group multiple data manipulation actions into a single atomic unit. The ACID (atomicity, consistency, isolation, durability) properties ensure that trans-

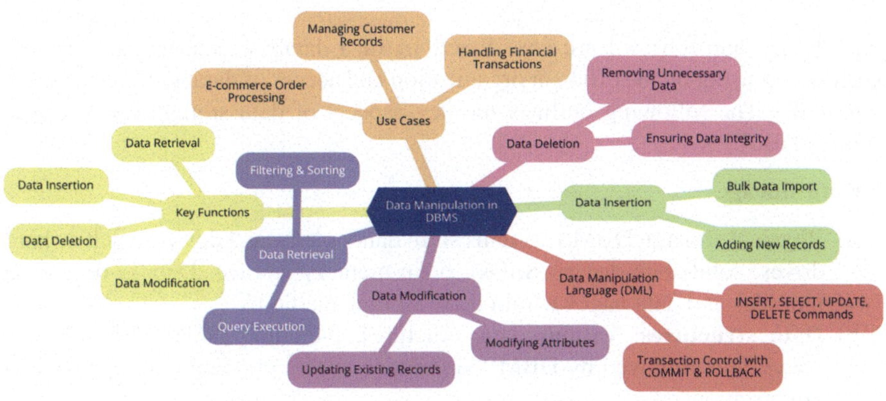

Fig. 7.19 Data manipulation in DBMS (prepared by the authors)

actions are executed reliably, maintaining data integrity and consistency throughout the process.
- **Data validation and integrity constraints:** DBMS systems enforce data validation rules and integrity constraints during data manipulation. Constraints such as primary keys, foreign keys, unique constraints, and check constraints help preserve the accuracy and consistency of data.
- **Query optimization:** To enhance performance, DBMS systems often include query optimization techniques that analyze query execution plans. The goal is to select the most efficient approach for retrieving data, improving the overall speed and responsiveness of data retrieval operations.
- **Access control:** DBMS systems allow administrators to define user roles and permissions to regulate who can perform data manipulation tasks. Access control mechanisms are critical for ensuring data security and privacy, preventing unauthorized modifications.
- **Logging and auditing:** DBMS systems log data manipulation actions, allowing administrators to track changes and maintain an audit trail. Auditing is useful for compliance, troubleshooting, and security monitoring.
- **Error handling:** DBMS systems provide error handling mechanisms to address issues arising during data manipulation, such as constraint violations or transaction failures. Error messages and rollback mechanisms help restore data consistency and prevent disruptions.

Therefore, data manipulation is central to the interaction between users, applications, and databases, enabling organizations to efficiently manage, update, and retrieve data while ensuring data integrity and security. Whether retrieving sales data for reporting purposes, updating customer information, or removing outdated records, data manipulation is essential for the effective utilization of a database management system (DBMS).

7.2.5 Data Storage and Retrieval

Data storage and retrieval are core functions of a database management system (DBMS), focusing on the efficient organization and access of data within a database (Fig. 7.20). The following outlines the key aspects of data storage and retrieval within a DBMS:

- **Data storage**

  a) **Physical storage** DBMS systems store data on physical devices such as hard drives, solid-state drives (SSDs), or in-memory storage. Data is organized into files and data pages within these storage mediums.
  b) **Data structures**: Various data structures, including tables, indexes, and views, are employed by DBMS systems to ensure efficient data storage and organization.

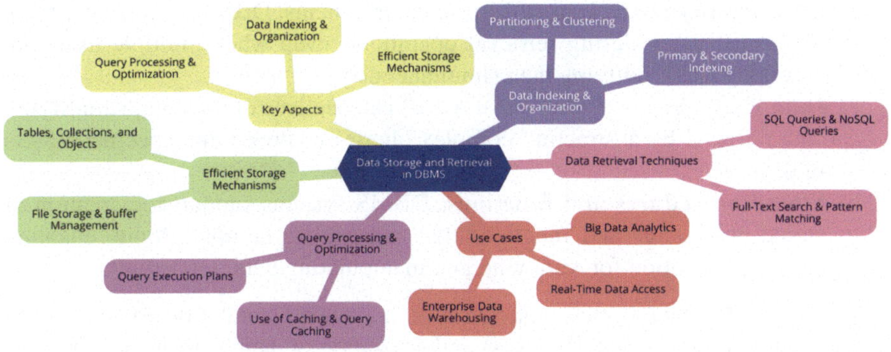

Fig. 7.20 Data storage and retrieval in DBMS (prepared by the authors)

c) **Tables**: Tables are the primary structures for data storage, consisting of rows (records) and columns (attributes), where data is stored in an organized manner.
d) **Indexes**: Indexes are specialized data structures that facilitate faster access to data based on specific columns or attributes. They enhance query performance by reducing the need for full table scans.
e) **Data compression**: To optimize storage capacity, DBMS systems often implement data compression techniques that reduce the physical size of data without compromising its integrity.
f) **Data partitioning**: Large databases may utilize data partitioning, which divides data into smaller, more manageable partitions or tablespaces. This approach helps in efficient storage management and enhances performance.
g) **Storage allocation**: DBMS systems dynamically manage the allocation and deallocation of storage space to accommodate the growth of data.

- **Data retrieval**

a) **Query language**: Users and applications interact with DBMS systems using query languages such as SQL (structured query language) to retrieve data.
b) **SELECT statements**: SELECT statements are used to fetch data from one or more tables. Users can specify the desired columns, conditions, sorting, and grouping criteria within their queries.
c) **Index usage**: Indexes play a crucial role in accelerating data retrieval. When queries involve indexed columns, DBMS can quickly locate relevant data without scanning the entire table.
d) **Query optimization**: DBMS systems employ query optimization techniques to determine the most efficient query execution plan, selecting the best approach for data retrieval.
e) **Caching**: Frequently accessed data may be cached in memory by DBMS systems to reduce the need for repeated disk reads, resulting in faster data retrieval.

f) **Concurrency control**: In multi-user environments, DBMS systems maintain data consistency during retrieval operations, even when multiple users are accessing and modifying data simultaneously.
g) **Data views**: Views provide customized perspectives of the data, simplifying data retrieval by abstracting complex queries or presenting specific subsets of data.
h) **Stored procedures and functions**: DBMS systems support the creation of stored procedures and functions that encapsulate complex logic, enabling efficient execution for retrieving and manipulating data.

Therefore, the data storage and retrieval are fundamental functions within a DBMS, facilitating the efficient access and utilization of data by users and applications. The capability to store data in an organized manner and retrieve it swiftly is crucial for enabling data-driven decision-making, reporting, analytics, and a wide range of business processes within organizations.

7.2.6 Concurrency Control

Concurrency control is a fundamental feature of database management systems (DBMS) that preserves data consistency and integrity in multi-user or multi-application environments, where multiple users or processes may simultaneously access and modify the same data. It plays a crucial role in preventing data conflicts, ensuring that concurrent transactions do not compromise the accuracy and reliability of the database (Fig. 7.21). The following highlights major aspects of concurrency control in DBMS:

- **Concurrency challenges:** In multi-user DBMS environments, multiple transactions may simultaneously attempt to read and modify the same data, leading to

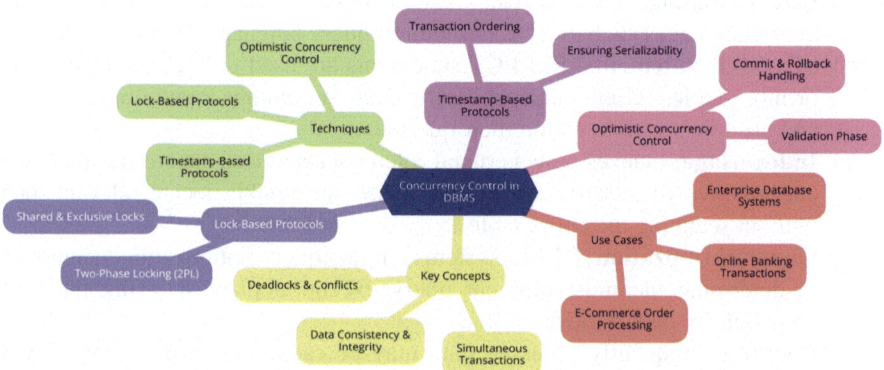

Fig. 7.21 Concurrency control in DBMS (prepared by the authors)

potential issues such as lost updates, uncommitted data, and inconsistent results if not properly managed.

- **ACID properties:** Concurrency control is essential for maintaining the ACID properties (atomicity, consistency, isolation, durability) of transactions:

  a) **Atomicity**: Transactions are executed as indivisible units of work.
  b) **Consistency**: Transactions transform the database from one consistent state to another.
  c) **Isolation**: Transactions operate independently, appearing to execute in isolation from one another.
  d) **Durability**: Once committed, changes made by transactions are permanent and persist even in the event of system failures.

- **Concurrency control mechanisms**

  a) **Locking**: One of the most common methods of concurrency control, where transactions request and release locks on data items to ensure exclusive access. Locks can be shared (for reading) or exclusive (for writing).
  b) **Timestamping**: Each transaction is assigned a timestamp, and the DBMS uses these timestamps to determine the order of execution, thereby avoiding conflicts.
  c) **Two-phase locking (2PL)**: A protocol where transactions must acquire all necessary locks before releasing any, which helps prevent deadlock situations.
  d) **Optimistic concurrency control**: Transactions proceed without locks initially, and conflicts are detected and resolved at the end of the transaction, before it is committed.
  e) **Multi-version concurrency control (MVCC)**: Allows concurrent transactions to access different versions of the same data, thus reducing lock contention.

- **Deadlock prevention and detection**

  a) DBMS systems incorporate deadlock prevention and detection mechanisms to address situations where transactions are indefinitely waiting for resources held by other transactions.
  b) **Prevention** methods include using lock timeouts and establishing resource allocation hierarchies.
  c) **Detection** involves monitoring for circular wait conditions and taking corrective actions when identified.
  d) **Concurrency and performance trade-offs:** Concurrency control mechanisms introduce performance overhead due to the need for locking, timestamping, and conflict resolution. DBMS administrators and developers must strike a balance between ensuring data consistency and maintaining system performance, depending on the specific application's needs.
  e) **Scalability and distributed environments:** Scalability in distributed DBMS systems often presents additional challenges, as concurrency control becomes more complex due to distributed data and multiple sites. Techniques such as

distributed locking and distributed transactions are employed to address these challenges.

In conclusion, concurrency control is a critical component of DBMS, ensuring data integrity and consistency even in environments with multiple users and transactions. By managing access to shared data, these mechanisms prevent conflicts, allowing transactions to execute reliably and predictably within the system.

7.2.7 Data Integrity and Security

Data integrity and security are essential aspects of database system design and management. They ensure that data remains accurate, consistent, and safeguarded against unauthorized access or malicious actions (Fig. 7.22). The following outlines the key aspects of data integrity and security within a DBMS:

Data Integrity

Concept: Data integrity refers to the accuracy, consistency, and reliability of the data stored within a database. It ensures that data remains error-free and retains its quality throughout its life cycle.

Key Aspects

- Entity integrity: Every record in a table must possess a unique identifier, typically established through primary keys, to avoid duplication or incomplete data.
- Referential integrity: Relationships between tables are preserved through foreign keys, ensuring the consistency of related data.
- Domain integrity: Data values must conform to predefined data types and constraints to prevent the entry of incorrect or invalid data.

Fig. 7.22 Data integrity and security in DBMS (prepared by the authors)

- User-defined integrity rules: DBMS systems allow the definition of custom business rules and constraints to uphold data integrity in accordance with specific organizational requirements.

Enforcement

- Data integrity is maintained through the use of constraints, triggers, and validation rules established within the DBMS.
- Constraints such as primary keys, unique constraints, and check constraints are employed to prevent the storage of data that violates integrity rules.

Data Security

- **Concept:** Data security within a DBMS involves safeguarding data from unauthorized access, tampering, disclosure, or destruction, thereby ensuring the confidentiality, integrity, and availability of the data.

  Key Aspects

(a) **Authentication:** Users and entities accessing the database must be authenticated to verify their identity, ensuring that only authorized individuals or systems are granted access.
(b) **Authorization:** Post-authentication, authorization mechanisms regulate the actions that users and entities can perform within the database, specifying their access privileges.
(c) **Encryption:** Sensitive data should be encrypted both during storage and transmission to protect it from unauthorized access or interception.
(d) **Auditing and monitoring:** Many DBMS systems incorporate auditing and monitoring features that track and log user activities, facilitating security analysis and ensuring compliance with security protocols.
(e) **Backup and recovery:** Regular data backups, alongside effective recovery plans, mitigate data loss resulting from system failures, disasters, or cyberattacks.
(f) **Access control lists (ACLs):** ACLs define permissions at the individual object level, offering fine-grained control over who can access specific data or database objects.

  - **Compliance and regulations:** Various industries and jurisdictions impose regulations (e.g., GDPR, HIPAA) that set standards for data security and privacy. DBMS systems assist organizations in meeting these regulations by providing essential security features such as encryption, access controls, and auditing.
  - **Security best practices:** Database administrators should adhere to security best practices, which include regularly updating the DBMS, applying security patches, and enforcing robust password policies. Conducting regular security assessments and penetration testing is essential for identifying vulnerabilities and strengthening the security of the database system.

Therefore, the data integrity and security are critical for preserving the reliability and trustworthiness of a database system. They safeguard sensitive information, prevent data corruption, and ensure that data remains accurate and accessible when required. The implementation of robust security protocols and the enforcement of data integrity constraints are continuous efforts that demand meticulous planning and ongoing attention to address emerging threats and comply with evolving regulatory requirements.

7.2.8 Data Backup and Recovery

Data backup and recovery are essential elements of a database management system (DBMS) that protect data from loss, corruption, or unforeseen disasters. These processes ensure that, in the event of unexpected incidents, data can be restored to a consistent and operational state (Fig. 7.23). The following outlines the key aspects of data backup and recovery within a DBMS:

Data Backup

- **Purpose:** Data backup involves creating duplicate copies of a database or its components (e.g., tables, indexes) to safeguard against potential data loss due to hardware malfunctions, software errors, corruption, or human error.
- **Types of backups**
  (a) Full backup: A comprehensive copy of the entire database, providing the most complete recovery point but requiring substantial time and resources.
  (b) Differential backup: Copies data altered since the last full backup, reducing the time and storage needed in comparison to a full backup.
  (c) Incremental backup: Copies data changed since the last backup, regardless of whether it was a full or incremental backup, offering efficient storage use but potentially requiring a more complex recovery process.

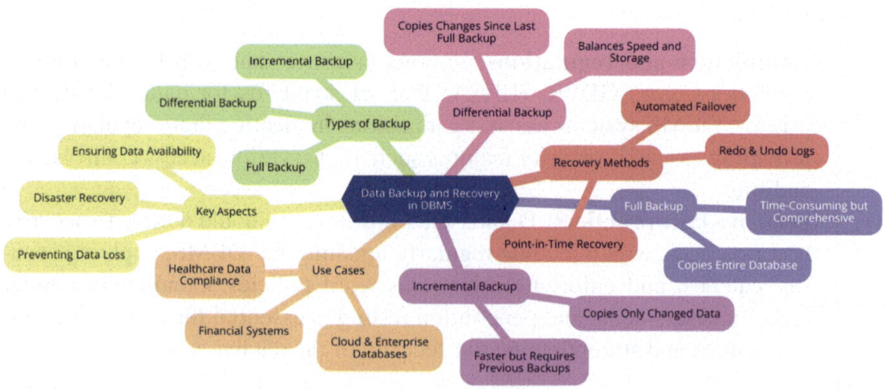

Fig. 7.23 Data backup and recovery in DBMS (prepared by the authors)

(d) Continuous data protection (CDP): Offers real-time, continuous backups, capturing every change in the data, thus minimizing data loss but demanding significant resources.

- **Backup storage:** Backup copies are stored on separate devices or locations, such as external hard drives, tape drives, cloud storage, or off-site data centers. Off-site storage ensures data security even in cases of physical disasters like fires or floods.
- **Backup scheduling:** Regular backup schedules are set to ensure consistent data protection at predefined intervals (e.g., daily, weekly, monthly). Automated backup tools or scripts are often employed for reliability and consistency.
- **Data recovery: Purpose:** Data recovery is the process of restoring data from backup copies to its original or a consistent state following a failure or data loss event.

(a) Recovery methods: Point-in-time recovery: Restores data to a specific point in time, preserving consistency. This is particularly crucial for ensuring transactional integrity.
(b) Rollback and rollforward: Transactions are either rolled back or rolled forward to restore data consistency, often utilizing transaction logs.
(c) Disaster recovery: In the event of a catastrophic failure, disaster recovery plans facilitate the restoration of data and operations from off-site backups.

- **Backup verification:** Regular testing of backups ensures their integrity and recoverability. Test restorations help identify potential issues before a disaster strikes.
- **Data retention policies:** Organizations establish data retention policies to determine how long backup copies are stored. These policies are often influenced by compliance requirements and business needs.
- **Recovery time objective (RTO) and recovery point objective (RPO)**

(a) RTO defines the acceptable time frame for restoring operations after a failure.
(b) RPO specifies the maximum allowable data loss. These metrics help shape recovery strategies.

- So, the data backup and recovery are essential components of disaster recovery planning and business continuity strategies. They ensure that critical organizational data can be swiftly and reliably restored following data loss or system failures, thereby minimizing downtime and mitigating data-related risks.

7.2.9 Data Dictionary

A data dictionary, also known as a metadata repository, is a fundamental component of a database management system (DBMS). It serves as a centralized repository for storing critical information regarding the structure, definitions, and attributes of data within a database (Fig. 7.24). The following outlines the key aspects of data dictionary within a DBMS:

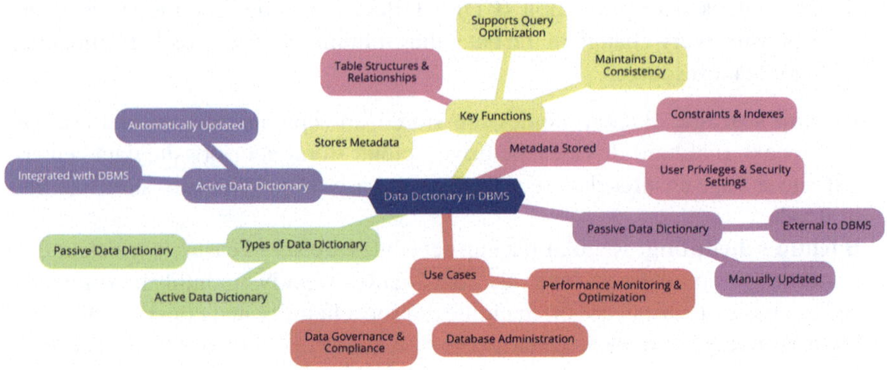

Fig. 7.24 Data dictionary in DBMS (prepared by the authors)

Significance

(a) **Metadata repository:** The Data Dictionary functions as a centralized metadata repository, storing crucial information about the database's schema, data elements, relationships, constraints, and other essential details. Metadata, which refers to data about data, provides a thorough understanding of the database's structure and content.

(b) **Data consistency:** By serving as a reference point for database administrators, developers, and users, the data dictionary ensures data consistency and integrity, which is vital for accurate data management and reporting.

(c) **Data transparency:** The data dictionary promotes data transparency by offering clear and standardized definitions for data elements, eliminating ambiguity and ensuring a unified understanding of data's meaning and usage.

(d) **Data quality:** Through the specification of data validation rules and constraints, the data dictionary supports data quality assurance by enforcing standards and preventing the entry of incorrect or inconsistent data.

Key Aspects

(a) **Data definitions:** The Data Dictionary contains comprehensive definitions of data elements, including names, data types, lengths, and descriptions. It provides a unified understanding of data across the organization.

(b) **Schema information:** It stores essential details about database objects such as tables, views, indexes, and constraints, including their structures, column attributes, and relationships.

(c) **Constraints and rules:** The data dictionary documents information about integrity constraints, business rules, validation rules, and data-related policies, ensuring that data adheres to organizational standards.

(d) **Data usage:** It may include details regarding how data is utilized within applications, reports, and business processes, which helps track data dependencies and understand its context.

(e) **Data ownership:** The Data Dictionary can define data ownership and stewardship responsibilities, clarifying who is responsible for maintaining and managing specific data elements.
(f) **Versioning:** Changes made to the database schema or data definitions are recorded and versioned in the data dictionary, facilitating the tracking of schema evolution and data changes over time.
(g) **Access control:** Access to the Data Dictionary is typically restricted to authorized personnel, such as database administrators, data architects, and data stewards, ensuring data security and integrity.

In summary, a data dictionary is integral to ensuring data consistency, accuracy, and transparency within a database system. It serves as a vital resource for database administrators, developers, and users, guaranteeing that data is clearly defined, thoroughly documented, and aligned with established standards and constraints.

7.2.10 Database Administrators and Users

Database administrators (DBAs) and users play fundamental yet distinct roles within a database management system (DBMS). They work in tandem to manage, maintain, and effectively utilize the database (Fig. 7.25). Following is a brief overview of their roles:

Database Administrators (DBAs)

- **Responsibilities:** DBAs oversee the overall management and administration of the database system. Their responsibilities include tasks such as database installation, configuration, maintenance, and performance tuning. Key duties also involve ensuring data security, implementing access controls, and planning disaster recovery strategies.

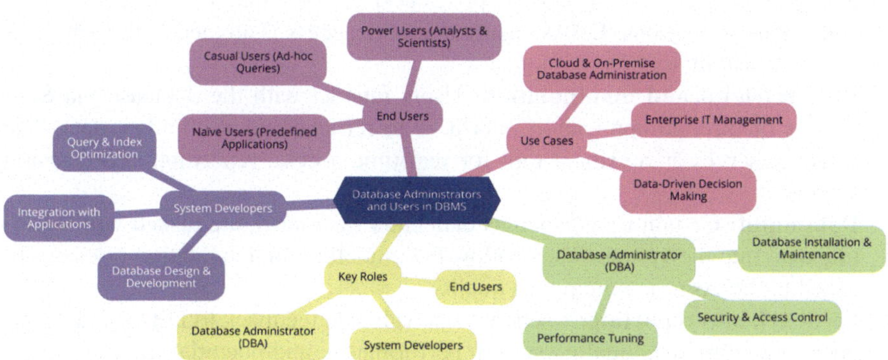

Fig. 7.25 Database administrators and users in DBMS (prepared by the authors)

- DBAs monitor the database's performance, resolve issues, and optimize query execution. Additionally, they are tasked with designing and enforcing data models, schemas, and integrity constraints. In some cases, DBAs may also handle backups, restorations, and data migrations, and may even be involved in database design and development.
- **Access control:** With elevated privileges, DBAs are responsible for managing user access to the database. They grant and revoke permissions based on security policies, ensuring users, applications, and roles have appropriate access levels.
- **Data backup and recovery:** DBAs ensure regular data backups and implement recovery procedures to safeguard against data loss or system failures.
- **Database performance:** They monitor the database for performance issues, identify bottlenecks, and optimize query execution plans. Additionally, DBAs manage hardware and software resources to maintain optimal performance.
- **Security and compliance:** DBAs implement security measures to prevent unauthorized data access or breaches. They collaborate with compliance teams to ensure that data management practices adhere to industry regulations.

Database Users

- **Roles:** Database users encompass a variety of roles, including end-users, application developers, data analysts, and business executives.

  (a) End users interact with the database through applications or query tools to retrieve, modify, or input data.
  (b) Developers build applications that interface with the database, dictating how data is presented and manipulated.
  (c) Data analysts leverage the database for extracting insights and generating reports.
  (d) Business executives use the data to inform decision-making and strategic planning.

- **Access and permissions:** Users are granted varying levels of access to the database, which are determined by the DBAs. These permissions may range from read-only access to full read-write privileges, and they may be restricted to specific database sections. DBAs ensure that user access is aligned with their roles and responsibilities.
- **Data retrieval and manipulation:** Users interact with the database via SQL queries or through application interfaces to retrieve and manipulate data. The database serves as a critical tool for real-time access, reporting, and decision support.
- **Data input:** Certain users, such as data entry personnel, are tasked with inputting data into the system. They follow the guidelines and validation rules established by DBAs to maintain data quality.
- **Feedback and reporting:** Users provide valuable feedback to DBAs and developers regarding data quality, system performance, and potential improvements. They also rely on the database to generate reports and insights that support their day-to-day activities.

In summary, database administrators (DBAs) are responsible for the technical management, maintenance, and security of the DBMS, ensuring its efficient operation. In contrast, database users interact with the database to perform tasks critical to their respective roles. These two roles collaborate closely to ensure that the DBMS meets the organization's data needs while maintaining high standards of data integrity, security, and performance.

7.2.11 Data Independence

Data independence is a core principle in database management systems (DBMS), referring to the ability to modify the schema at one level without affecting the schema at higher levels. It enables changes to the physical storage or organization of data without impacting how data is logically represented or how applications interact with it. Data independence enhances flexibility, scalability, and ease of database maintenance (Fig. 7.26). It is categorized into two types:

Logical Data Independence

- **Concept:** Logical data independence refers to the capacity to modify the conceptual or logical schema of a database without necessitating alterations to the application programmes or queries that interact with the data.
- **Key aspects**
  (a) Application programmes interface with the database via the logical schema, which dictates the organization and structure of the data.
  (b) Logical data independence guarantees that modifications to the logical schema, such as the addition of new tables, changes to attributes, or alterations in relationships, do not require updates to existing application code.

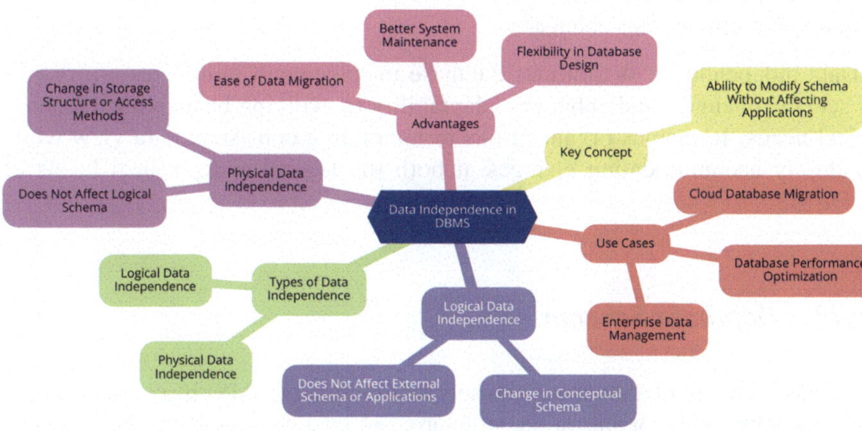

Fig. 7.26 Data independence in DBMS (prepared by the authors)

- **Benefits**
  (a) It simplifies database maintenance, allowing schema changes to accommodate evolving business requirements without disrupting dependent applications.
  (b) Developers are able to focus on enhancing application functionality without the concern of underlying database structural changes.

Physical Data Independence

- **Concept:** Physical data independence is the capability to modify the physical storage and access methods of data without affecting the logical schema or application programmes.
- **Key aspects**
  (a) The physical schema of a database defines how data is stored, including details such as file organization, indexing methods, and data compression techniques.
  (b) Physical data independence ensures that changes to the physical schema, such as transitioning to a different storage system or optimizing storage configurations, do not impact the logical schema or applications.
- **Benefits**
  (a) It empowers database administrators to adapt to evolving hardware and storage technologies without modifying the logical structure of the database.
  (b) Performance optimizations, such as indexing or table partitioning, can be implemented without affecting applications.
- **Examples:** If an organization chooses to partition a large table into smaller segments to improve performance, this can be done at the physical level without altering the logical view of the data or impacting the applications. And when a new column is introduced to an existing table to capture additional data, logical data independence ensures that application programmes do not require modification to incorporate this change.

Data independence streamlines database management, minimizes errors during schema changes, and enhances adaptability to evolving business and technological needs. It enables organizations to maintain a consistent data view while seamlessly accommodating changes at both the logical and physical levels of the DBMS.

7.2.12 Database Schema

A database schema in a database management system (DBMS) is a logical framework that outlines the organization, structure, and relationships of the data within a database. It plays a crucial role in data organization and management (Fig. 7.27). The following highlights its significance and key components:

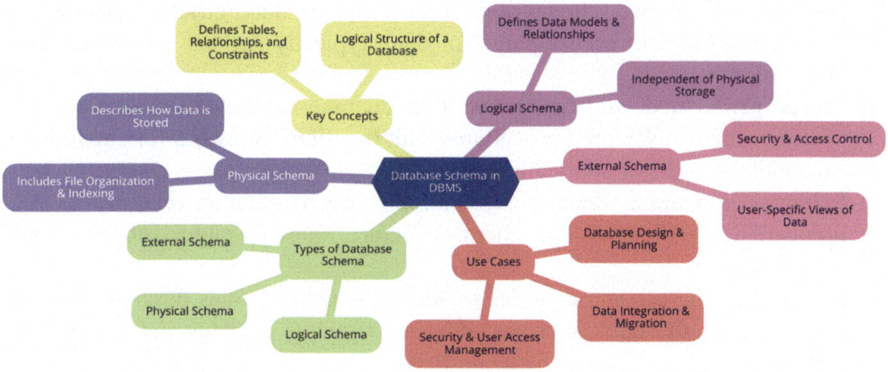

Fig. 7.27 Database schema in DBMS (prepared by the authors)

Significance

(a) Data organization: The database schema provides a structured framework for organizing and representing data. It defines how data is stored, the types of data allowed, and the relationships between data elements.
(b) Data integrity: Schema constraints and rules enforce data integrity by ensuring that the data entered into the database adheres to predefined standards, thus preventing errors and inconsistencies.
(c) Data abstraction: The schema abstracts the complexities of the underlying database, offering a simplified, standardized view of the data for users and applications.
(d) Security: Access control and permissions are often defined at the schema level, allowing administrators to regulate user access to specific sections of the database.

Key Aspects

(a) **Tables:** A schema consists of tables that represent entities or objects, organized into rows (records) and columns (attributes).
(b) **Attributes:** Columns in tables define the properties of data, each with a name, data type (e.g., integer, string), and associated constraints (e.g., primary key, foreign key).
(c) **Relationships:** The schema defines the relationships between tables, indicating how data is interconnected, such as one-to-one, one-to-many, or many-to-many.
(d) **Constraints:** Constraints ensure data integrity by enforcing rules, such as primary keys (unique identifiers), foreign keys (linking tables), check constraints (data validation), and unique constraints (ensuring uniqueness).
(e) **Views:** Views are virtual tables that present a subset of data from one or more tables, enabling users to access specific data without direct interaction with the underlying tables.

(f) **Indexes:** Indexes are schema objects that improve data retrieval performance by creating efficient access paths for data, typically defined on one or more table columns.
(g) **Stored procedures and functions:** The schema may include pre-defined sets of SQL statements, stored procedures, and functions, allowing for streamlined execution by users or applications.
(h) **Schema diagrams:** Visual representations, such as entity-relationship diagrams (ERDs), illustrate the schema structure, showing tables, attributes, and relationships.
(i) **Metadata:** The schema contains metadata, which provides descriptive information about the data, such as table descriptions, attribute definitions, and constraints.
(j) **Evolution:** The schema can evolve over time to meet changing business requirements. Schema modifications are managed to maintain data integrity and consistency.

In summary, a database schema in a DBMS establishes the logical framework and organization of data, ensuring consistent and meaningful data storage. It serves as a vital reference for database administrators and application developers, supporting efficient data management, maintaining data integrity, and enabling data abstraction within the database system.

7.2.13 Database Keys

Database keys are essential elements in a database management system (DBMS) that serve to uniquely identify records and establish relationships between data within the database. They are pivotal in maintaining data integrity, enforcing relational constraints, and ensuring efficient data retrieval (Fig. 7.28). The following highlights its significance and components:

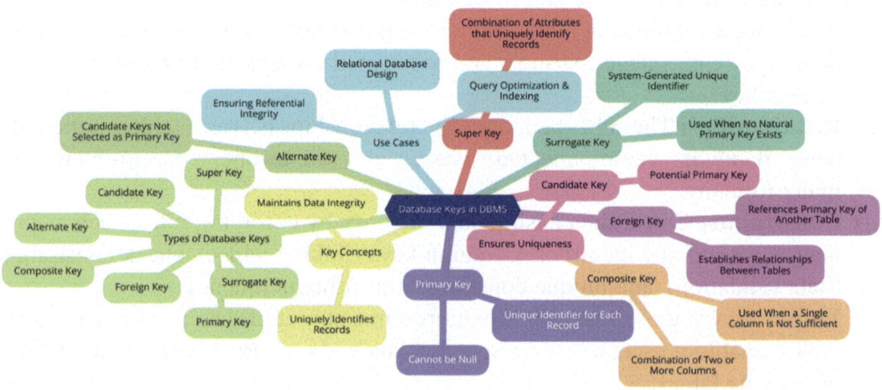

Fig. 7.28 Database keys in DBMS (prepared by the authors)

Significance

(a) Data identification: Database keys serve to uniquely identify individual records within a table, acting as a reference to distinguish one record from another.
(b) Data integrity: By ensuring that values are unique and non-null, keys enforce data integrity, preventing duplicate or inconsistent entries and ensuring that the database maintains accurate and reliable information.
(c) Data relationships: Keys facilitate relationships between tables, maintaining referential integrity by linking related data across tables.
(d) Query performance: As indexing mechanisms, keys enhance query performance by providing efficient data retrieval paths and reducing the need for full table scans.

Types of Database Keys

(a) Primary key (PK): A primary key uniquely identifies each record in a table, ensuring uniqueness and non-null values. It can also be used as a reference in other tables via foreign keys. Each table can have only one primary key.
(b) Foreign key (FK): A foreign key links a field or set of fields in one table to the primary key of another table, establishing relationships between tables and maintaining referential integrity by preventing orphaned records.
(c) Unique key (unique constraint): A unique key enforces uniqueness in one or more columns, allowing for one null value. Unlike the primary key, it does not serve as the primary identifier but can be used to uniquely identify records.
(d) Composite key: A composite key involves two or more columns used together as a unique identifier for records. This is used when no single column can uniquely identify a record but the combination of columns does. It is common in many-to-many relationship junction tables.
(e) Super key: A super key consists of one or more columns capable of uniquely identifying records in a table. It may include primary keys, candidate keys, and other columns, but is not necessarily minimal.
(f) Candidate key: A candidate key is a minimal super key, meaning it is the smallest set of columns that can uniquely identify a record. Multiple candidate keys may exist, from which one is chosen as the primary key.
(g) Alternate key: An alternate key is a candidate key that is not selected as the primary key. While it does not serve as the primary identifier, it can still be used to uniquely identify records.

In summary, database keys in a database management system (DBMS) are critical for ensuring data integrity, defining relationships between tables, and enhancing data retrieval efficiency. They offer a structured approach to identifying and distinguishing records within a database, ensuring the accuracy, consistency, and accessibility of data. The selection of appropriate keys for a database design is contingent upon the specific needs and attributes of the data model.

7.3 Database Models and Languages

Database models, including relational, hierarchical, and network models, define the structure and organization of data within a database system. The relational model, which utilizes tables and SQL (structured query language), is widely favoured for its simplicity and flexibility, allowing data to be organized in rows and columns. In contrast, the hierarchical and network models are more complex, with the hierarchical model being ideal for representing data in tree-like structures, and the network model for depicting interconnected data. These models are often paired with query languages like SQL, which enable users to interact with the database for data retrieval, updates, and manipulation. The selection of a database model and query language depends on the specific needs of the application, with each model offering distinct advantages in terms of data organization and access (Fig. 7.29).

7.3.1 Database Models

7.3.1.1 Database Models: Relational

The relational database model is one of the most widely utilized and established models in database management systems (DBMS). It organizes data using tables, which consist of rows (records) and columns (attributes), forming a structured system for data storage and management (Fig. 7.30). The following discussion outlines the relational database model:

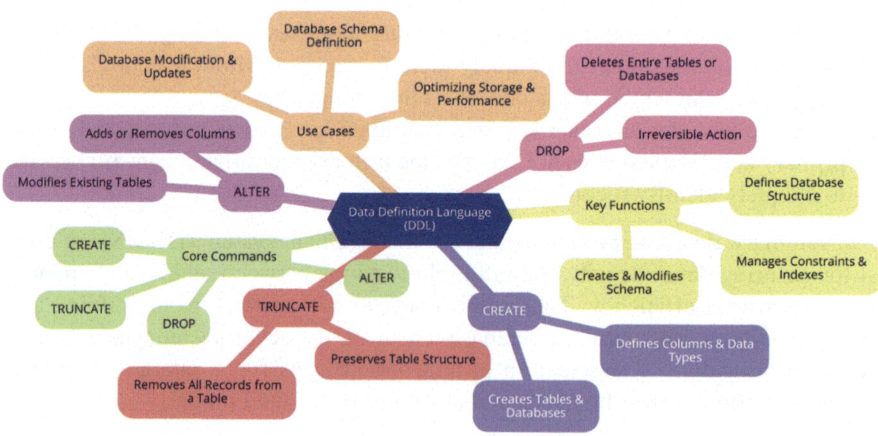

Fig. 7.29 Database models and languages in DBMS (prepared by the authors)

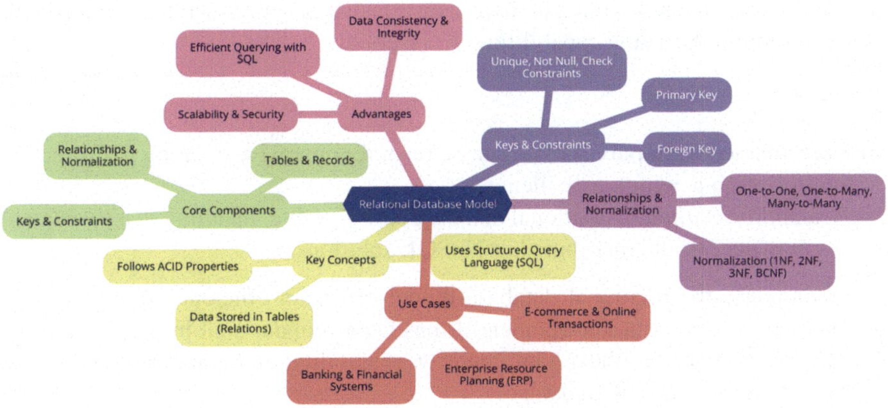

Fig. 7.30 Relational database models in DBMS (prepared by the authors)

Key Characteristics

(a) Tables: Data is organized into tables, or relations, with each table representing a distinct entity or concept. Rows represent individual records, while columns define the attributes or properties of those records.
(b) Structured data: The relational model enforces a structured approach to data storage, ensuring consistency and integrity. Each column is assigned a specific data type, and constraints are applied to ensure valid data.
(c) Primary key: Each table typically includes a primary key, which serves as a unique identifier for each record, ensuring that each row can be uniquely identified.
(d) Relationships: Tables are related through keys. Foreign keys in one table reference the primary key of another, establishing connections and enabling data linkage across tables.
(e) Normalization: The relational model supports normalization techniques that reduce data redundancy and enhance integrity. By decomposing tables into smaller, related tables, normalization minimizes data duplication.

Advantages

(a) Data integrity: The relational model's constraints and primary keys guarantee data accuracy and consistency across the system.
(b) Flexibility: Tables can be modified or extended to meet evolving data requirements, making the model adaptable to changing needs.
(c) SQL: The structured query language (SQL) offers a powerful, standardized method for querying, retrieving, and manipulating data within relational databases.
(d) Wide adoption: Relational databases are supported by numerous DBMS platforms, and the model has a large, active community of users and developers.
(e) Use cases: Relational databases are ideal for a variety of applications, such as business systems, e-commerce platforms, and content management systems.

They excel in environments where data consistency, transaction management, and complex querying capabilities are essential.

Challenges

(a) Performance: Complex queries can be resource-intensive, making performance optimization a common challenge.
(b) Scalability: Although relational databases are scalable, handling massive datasets may necessitate distributed or NoSQL solutions.

In summary, the relational database model offers an efficient, structured, and flexible approach to data management. It has been foundational in the IT industry for decades, providing robust data integrity, scalability, and a standardized query language for interacting with data.

7.3.1.2 Database Models: Entity-Relationship

The entity-relationship (ER) model is a conceptual framework used in database management systems (DBMS) to represent the relationships between entities, their attributes, and the rules governing those relationships (Fig. 7.31). The following discussion outlines the entity-relationship database model.

Key Concepts

(a) Entities: Entities represent objects, concepts, or things with distinct attributes. In a database, they typically correspond to real-world objects like customers, products, or employees, or abstract concepts like invoices or orders.
(b) Attributes: Attributes describe the properties or characteristics of entities. For example, the "*Customer*" entity may include attributes such as "*Name*", "*Address*", and "*Phone Number*".

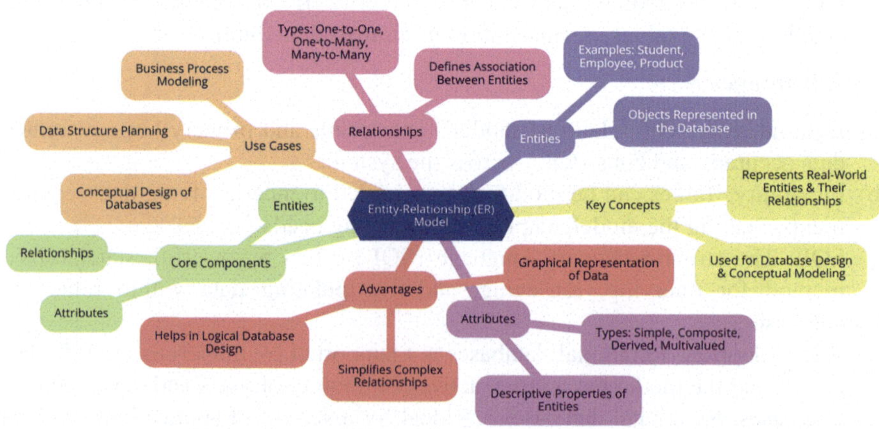

Fig. 7.31 Entity-relationship database models in DBMS (prepared by the authors)

(c) Relationships: Relationships define how entities are connected or interact with one another. For instance, a "Customer" entity may have a relationship with an "*Order*" entity, indicating that customers place orders.
(d) Cardinality: Cardinality refers to the number of instances of one entity that can be associated with instances of another. Cardinality types include one-to-one, one-to-many, and many-to-many.
(e) Weak entities: Weak entities rely on a related strong entity for identification, meaning they cannot be uniquely identified without the reference to another entity.

Advantages

(a) Conceptual clarity: The ER model provides an intuitive, visual representation of data and relationships, simplifying the communication of database requirements.
(b) Data modelling: It is widely used in data modelling, aiding in the design and planning of relational databases.
(c) Normalization: The ER model supports the normalization process, reducing data redundancy and enhancing data integrity.
(d) Use cases: The ER model is particularly useful during the initial stages of database design, helping to define the structure and relationships within the data. It is widely employed in various domains, including finance, healthcare, and e-commerce, to model complex systems and business processes.

Challenges

(a) As a conceptual model, the ER diagram needs to be mapped into a physical database schema.
(b) Complex relationships and intricate data requirements may demand careful and precise modelling to ensure accuracy.

In summary, the entity-relationship model is an essential tool in database design, providing a clear and structured approach to capturing the relationships and attributes of entities within a system. It plays a crucial role in ensuring effective database planning and design.

7.3.1.3 Database Models: Object-Based

The object-based database model is a data model used in database management systems (DBMS) to represent data as objects, which combine both data (attributes) and behaviour (methods or functions). This model is particularly effective for handling complex data structures and relationships, offering a more natural representation compared to traditional relational databases (Fig. 7.32). The following discussion outlines the object-based database model:

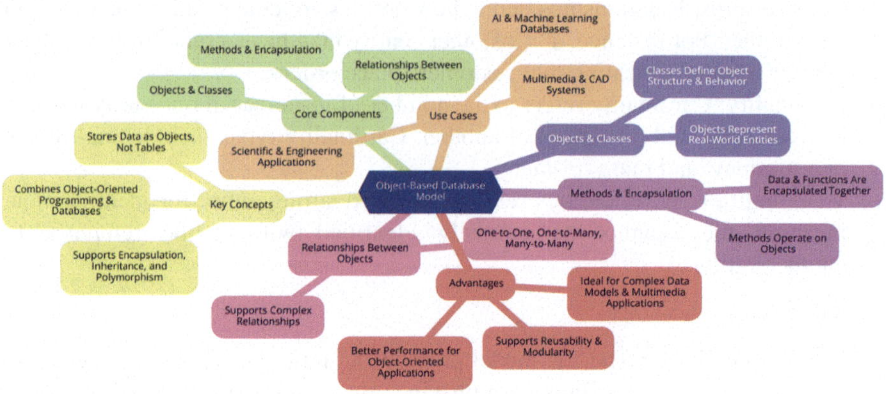

Fig. 7.32 Object-based database models in DBMS (prepared by the authors)

Key Concepts

(a) Objects: In this model, data entities are represented as objects, similar to objects in object-oriented programming. Each object encapsulates both its attributes and its associated methods or behaviours.
(b) Classes: Objects are organized into classes, which define the structure and behaviour of objects within that class. A class serves as a template for creating objects.
(c) Inheritance: Object-based databases support inheritance, allowing classes to inherit attributes and methods from parent classes. This facilitates code reuse and helps model hierarchical relationships.
(d) Complex data types: These databases support complex data types, such as arrays, lists, and user-defined types, making them well-suited for modelling intricate data structures.
(e) Associations: Relationships between objects are represented as associations, which can be one-to-one, one-to-many, or many-to-many, similar to relationships in the entity-relationship model.

Advantages

(a) Complex data modelling: Object-based databases are ideal for modelling data with complex structures, rich relationships, and behaviours, making them suitable for sophisticated applications.
(b) Encapsulation: The integration of data and behaviours within objects allows for better encapsulation, reflecting real-world entities more accurately.
(c) Object-oriented programming integration: These databases align seamlessly with object-oriented programming languages, facilitating application development that interacts directly with the database.
(d) Use cases: Object-based databases are commonly used in fields that require modelling complex data, such as scientific research, engineering, and multimedia applications.

Challenges

(a) Integration: Migrating from relational databases to object-based databases can be challenging due to compatibility issues and differences in data modelling paradigms.
(b) Performance: While object-based databases excel in managing complex data, they may face performance issues when dealing with simpler, tabular data structures.

In summary, the object-based database model is well-suited for managing complex data with rich relationships and behaviours. By leveraging the principles of object-oriented programming, it provides a flexible and intuitive approach to data representation in a database system.

7.3.1.4 Database Models: Semantic-Structure

The semantic-structure database model is a specialized data model designed to capture both the structural and semantic (meaning-related) relationships between data elements. It is primarily used in knowledge representation systems, semantic web applications, and information retrieval (Fig. 7.33). The following discussion outlines the semantic-structure database model:

Key Concepts

(a) Semantic associations: Unlike traditional models that focus on data relationships through keys, the semantic-structure model emphasizes semantic associations between data elements. These associations are often represented through ontologies or controlled vocabularies.

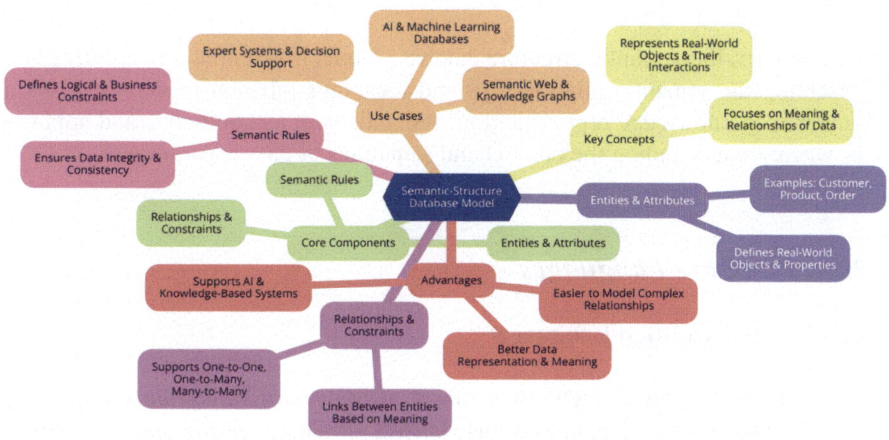

Fig. 7.33 Semantic-structure database models in DBMS (prepared by the authors)

(b) Ontologies: Ontologies are formal representations of knowledge or concepts within a specific domain. They define the entities, attributes, relationships, and constraints governing the data within that domain.
(c) RDF (resource description framework): Semantic-Structure databases often utilize RDF as a data format for representing semantic associations. RDF provides a flexible and extensible way to describe resources and their relationships on the web.
(d) Triples: Data in semantic-structure databases is often expressed as triples, subject-predicate-object statements. These triples capture semantic relationships, where the subject is the resource, the predicate describes the relationship, and the object represents the related resource.
(e) Inferencing: Inferencing or reasoning is a key feature of this model, allowing new information or relationships to be derived from existing semantic associations and ontological rules.

Advantages

(a) Semantic understanding: The semantic-structure model enables systems to understand the meaning of data, facilitating knowledge representation, natural language processing, and semantic search.
(b) Interoperability: It supports data integration and interoperability across diverse systems and domains by providing a standardized approach to expressing semantics.
(c) Use cases: Semantic-structure databases are widely used in the semantic web, playing a crucial role in annotating and linking web resources to improve search and discovery.

Challenges

(a) Complexity: Developing and maintaining ontologies can be complex and requires domain-specific expertise.
(b) Scalability: Managing large-scale ontologies and semantic data can be resource-intensive.

In summary, the semantic-structure database model is essential for capturing and representing the semantic relationships and meanings inherent in data. It plays a critical role in knowledge representation, semantic web technologies, and applications where understanding the context and semantics of data is paramount.

7.3.2 *Database Languages*

7.3.2.1 Data Definition Language

Data definition language (DDL) is a subset of SQL (structured query language) used in database management systems (DBMS) to define and manage the structure and organization of databases. DDL is primarily concerned with schema-related operations, enabling users to create, modify, and delete database objects (Fig. 7.34). The following highlights the significance and key aspects of DDL.

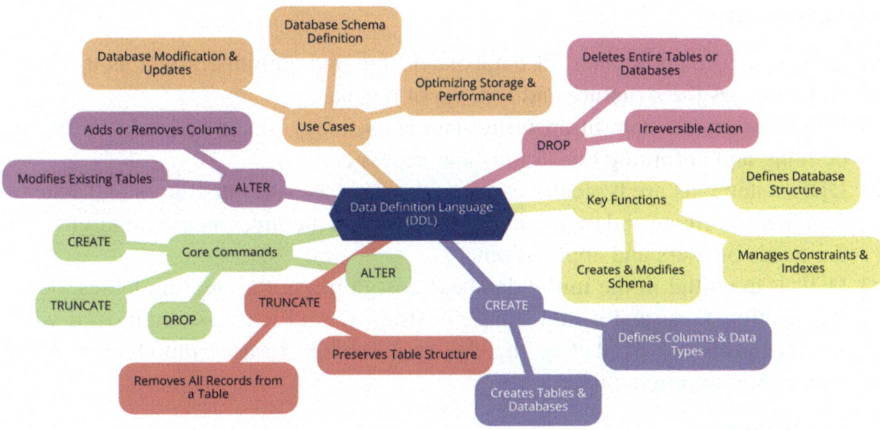

Fig. 7.34 Database languages: data definition language (prepared by the authors)

Key Aspects

(a) Table creation: DDL facilitates the creation of tables by specifying their names, columns, data types, constraints (such as primary key and foreign key), and other attributes. The CREATE TABLE statement is commonly used for this purpose.

(b) Schema modification: DDL allows users to alter existing tables, enabling them to add, modify, or drop columns, constraints, or indexes. The ALTER TABLE statement is employed for such operations.

(c) Database creation and deletion: Through DDL, users can create and delete entire databases using the CREATE DATABASE and DROP DATABASE statements, which define the overarching database structure.

(d) Constraint definition: DDL provides commands to define various constraints, such as PRIMARY KEY, FOREIGN KEY, UNIQUE, and CHECK, ensuring data integrity and enforcing rules on the data stored within tables.

(e) Index creation: Indexes are created using DDL commands like CREATE INDEX to enhance query performance by providing efficient access paths to data within table columns.

(f) View definition: DDL supports the creation of views, which are virtual tables that represent subsets of data from one or more tables. The CREATE VIEW statement is used to define these structures.

(g) User management: Some DBMS systems also include DDL statements for managing users, roles, and permissions, allowing administrators to control access to the database.

Significance

(a) DDL is a core component of database design and administration, as it defines and manages the structure and schema of databases.
(b) It plays a crucial role in ensuring data consistency, integrity, and security by defining and enforcing constraints and access controls.
(c) DDL statements are typically executed by database administrators and schema designers to create and maintain the database's structure, ensuring it aligns with the needs of users and applications.
(d) DDL is essential in the initial database design phase for creating tables, defining relationships, and establishing constraints. And it is also employed when modifications are needed to the database schema to accommodate evolving application requirements.

Challenges

(a) Schema modifications require careful planning, as changes to the structure can affect existing data and applications.
(b) DDL operations generally require administrative privileges and must be executed with caution to prevent unintentional data loss or corruption.

In summary, data definition language (DDL) is an integral part of SQL, enabling users to define and manage the structure of databases. It ensures data integrity, enforces constraints, and facilitates schema design and modification, making it essential for effective database management and administration.

7.3.2.2 Data Manipulation Language

Data manipulation language (DML) is a subset of SQL (structured query language) used in database management systems (DBMS) to retrieve, insert, update, and delete data in a database. It focuses on the manipulation of data stored within database tables (Fig. 7.35). The subsequent discussion provides an overview of the DML:

Key Aspects

(a) Data retrieval: DML enables users to extract data from one or more database tables using the SELECT statement. It allows customization of query results by specifying columns, conditions, and sorting criteria.
(b) Data insertion: The INSERT statement is used to add new records to database tables by providing values for the specified columns.
(c) Data modification: Existing data in a table can be updated using the UPDATE statement, allowing users to specify rows and define the required changes.
(d) Data deletion: The DELETE statement removes data from a table, enabling the deletion of specific rows based on defined criteria.
(e) Transactions: DML operations are often performed within transactions to ensure that multiple statements execute as a single atomic unit, maintaining data consistency and integrity.

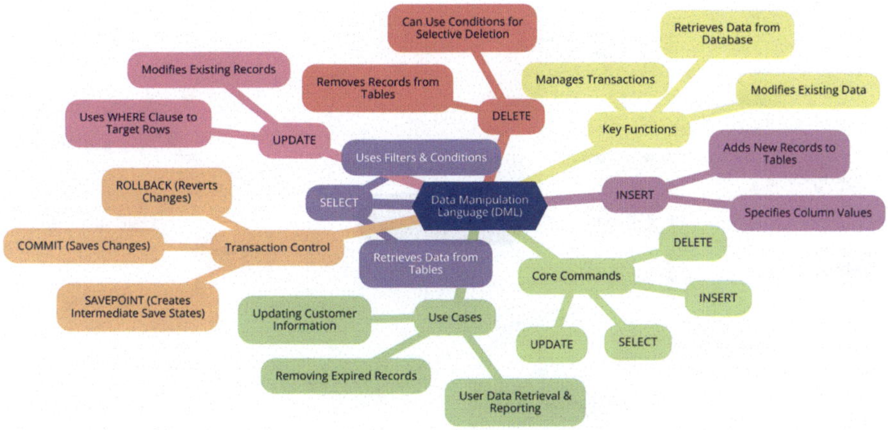

Fig. 7.35 Database languages: data manipulation language (prepared by the authors)

Significance

(a) DML serves as a crucial tool for database interaction, facilitating data retrieval, manipulation, and management to support various applications and business processes.
(b) It offers a standardized and structured approach to handling data, ensuring accessibility for users and applications while maintaining data integrity.
(c) DML queries play a central role in software applications, reporting tools, and business intelligence systems, enabling efficient data extraction and presentation.

Challenges

(a) Data manipulation requires careful execution to avoid unintended data corruption or loss.
(b) Performance optimization is often necessary to ensure efficient execution of DML statements, particularly for large datasets or complex queries.

In summary, data manipulation language (DML) is an essential component of SQL, enabling the retrieval, insertion, updating, and deletion of data in databases. Its structured approach to data manipulation supports a wide range of database-driven applications and ensures efficient data management.

7.3.2.3 Data Control Language

Data control language (DCL) is a subset of SQL (structured query language) used in database management systems (DBMS) to regulate and manage access to data stored in a database. It comprises commands that grant or revoke permissions and privileges for users and roles (Fig. 7.36). The subsequent discussion provides an overview of the DCL:

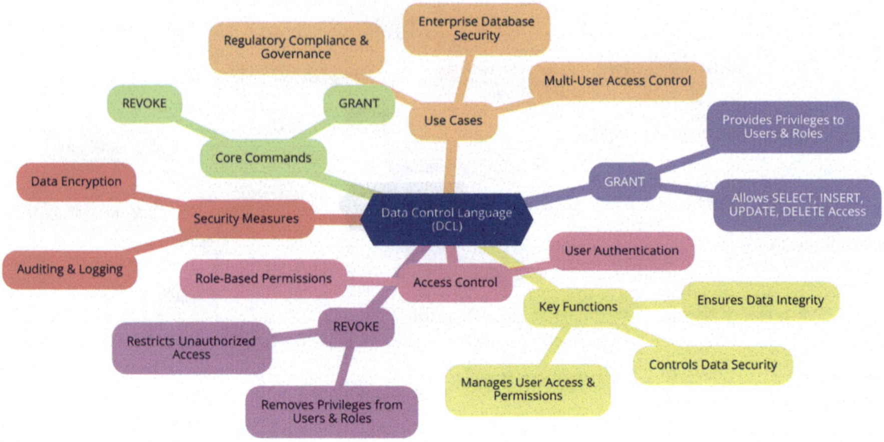

Fig. 7.36 Database languages: data control language (prepared by the authors)

Key Aspects

(a) Permissions and privileges: DCL commands enable database administrators to define and manage user or role-specific permissions. These permissions dictate what actions (e.g., SELECT, INSERT, UPDATE, DELETE) can be performed on specific database objects such as tables or views.
(b) GRANT command: The GRANT statement allows administrators to assign specific privileges to users or roles. It specifies the permitted actions and the database objects to which the permissions apply.
(c) REVOKE command: The REVOKE statement removes previously granted permissions, enabling administrators to restrict or withdraw access to certain database objects.
(d) Roles: DCL often utilizes roles, which are predefined sets of privileges. Roles simplify permission management by grouping privileges and assigning these groups to users, making access control more efficient.

Significance

(a) Data security: DCL plays a pivotal role in enforcing database security by ensuring that only authorized users can access or modify data.
(b) Data integrity and confidentiality: It allows administrators to protect sensitive information and maintain database integrity by controlling who can view, edit, or delete data.
(c) Regulatory compliance: DCL helps organizations adhere to security and privacy regulations by managing access to critical or sensitive data.

Challenges

(a) Permission planning: Striking a balance between accessibility and security requires careful planning of permissions.

(b) Complex systems: Managing intricate permission structures in large or complex database systems can be difficult and time-consuming.

Therefore, the data control language (DCL) is a crucial component of SQL used to manage and control access to a database. By defining and enforcing permissions and privileges, it ensures data security, integrity, and confidentiality. DCL commands are indispensable tools for database administrators, supporting effective implementation of access policies and ensuring compliance with data security standards.

7.3.2.4 Transactional Control Language

Transactional control language (TCL) is a subset of SQL (structured query language) used in database management systems (DBMS) to manage and control transactions within a database. A transaction consists of one or more SQL statements executed as a single unit of work. TCL commands ensure data consistency and integrity during these transactions (Fig. 7.37). The subsequent discussion provides an overview of the TCL:

Key Aspects

(a) Transaction management: TCL commands are primarily designed to handle transactions, ensuring that a sequence of SQL operations is treated as a single, atomic unit.
(b) COMMIT: The COMMIT command finalizes a transaction by saving all changes permanently to the database. Once committed, the transaction is complete, and the changes are permanent.
(c) ROLLBACK: The ROLLBACK command reverses changes made during a transaction, restoring the database to its previous state. It is used to cancel uncommitted transactions.
(d) SAVEPOINT: The SAVEPOINT command allows the creation of specific points within a transaction. These points enable partial rollbacks without undoing the entire transaction.

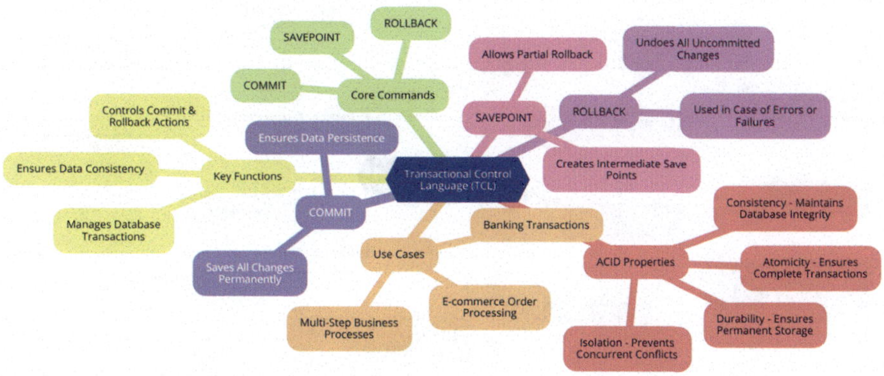

Fig. 7.37 Database languages: transactional control language (prepared by the authors)

Significance

(a) Data consistency: TCL ensures that transactions are executed reliably, maintaining data integrity by committing all changes or rolling them back entirely if errors occur.
(b) Error recovery: It enables recovery from errors or failures by rolling back to a stable state, preventing data corruption.
(c) Concurrency control: TCL facilitates multiple users accessing and manipulating the database simultaneously while maintaining transaction isolation to prevent conflicts.

Challenges

(a) Transaction management: Effective use of TCL requires careful planning to determine when changes should be committed or rolled back.
(b) Complex transactions: Transactions involving multiple steps or interdependencies require robust error-handling and recovery mechanisms.

So, the transactional Control Language (TCL) is an essential component of SQL used to manage and control transactions in a database. By ensuring that data changes are consistently applied or reverted, TCL supports data reliability, error recovery, and concurrency control, making it indispensable for robust database management and application development.

7.4 Database Architecture

Database architecture refers to the structural framework that outlines the methods for storing, accessing, and managing data within a database management system (DBMS). It serves as a cornerstone in maintaining data integrity, optimizing performance, and enabling scalability in database systems (Fig. 7.38). A typical database architecture comprises several essential components:

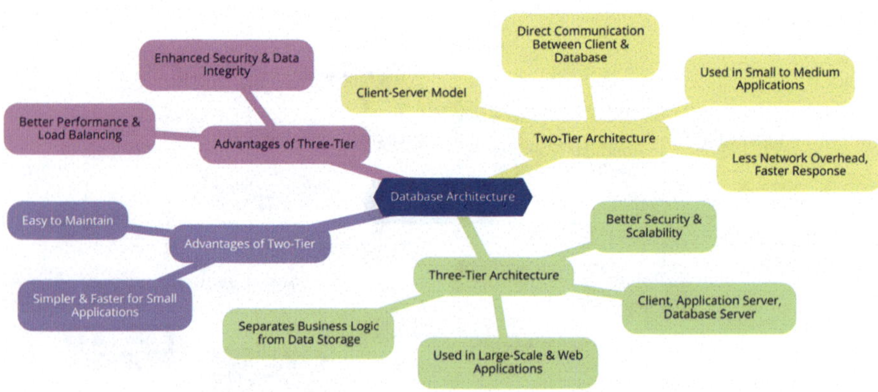

Fig. 7.38 Database architecture in DBMS (prepared by the authors)

- **Database engine:** The database engine, or DBMS, lies at the core of any database architecture. It is responsible for executing SQL queries, managing data storage, enforcing security protocols, and handling transactions efficiently.
- **Storage management:** This component focuses on organizing and storing data on physical storage devices, such as hard drives or solid-state drives. It encompasses elements like data files, indexing structures, and caching mechanisms to enhance data retrieval performance.
- **Query processing:** The architecture integrates a query processor, which interprets and optimizes SQL queries submitted by users or applications. It generates execution plans to efficiently retrieve or modify data.
- **Concurrency control:** In multi-user environments, concurrency control mechanisms are critical for managing simultaneous database access. These mechanisms prevent data inconsistencies arising from concurrent modifications by multiple users.
- **Transaction management:** Transaction management ensures that a sequence of database operations is treated as a single, atomic unit. It guarantees data consistency by enabling transactions to be either committed or rolled back entirely in case of errors.
- **Security:** Security features within the database architecture include authentication, authorization, and encryption mechanisms. These protect sensitive data from unauthorized access, ensuring privacy and compliance with regulatory standards.
- **Backup and recovery:** Backup and recovery mechanisms are essential for ensuring data reliability. These components facilitate the creation of data and log backups, enabling database restoration in the event of system failures.
- **Data models:** The architecture is influenced by the choice of data model—relational, object-based, or others—which determines the structure and algorithms used for data storage and retrieval.
- **Middleware:** Middleware components facilitate seamless communication between applications and the database. They often include features such as connection pooling and abstractions for data access.
- **Scalability:** Architectural design significantly impacts the database system's scalability. Features like replication, clustering, and partitioning are incorporated to accommodate growing workloads and ensure optimal performance.

In summary, database architecture forms the foundation of a DBMS, comprising components and design choices that ensure efficient, secure, and reliable data management. Selecting an appropriate architecture is essential to address an organization's specific data management requirements.

7.4.1 Database Architecture: Two-Tier

The two-tier database architecture, often referred to as the client-server model, is among the earliest and simplest frameworks utilized in database management systems (DBMS). This architecture divides the DBMS into two primary components: the client-side application and the database server (Fig. 7.39). The following outlines its key characteristics and significance.

Key Characteristics

(a) Client-side application: The client-side, often referred to as the front-end or user interface, contains the application logic that facilitates user interaction. It manages functions such as data input, processing, and presentation.
(b) Database server: The database server, typically hosted on a separate machine, is responsible for managing data storage and retrieval. It executes SQL queries, controls data access, and maintains data integrity.
(c) Direct communication: In this architecture, the client communicates directly with the database server through a database-specific protocol. SQL queries are transmitted from the client to the server, which processes them and returns the results.
(d) Minimal middleware: The two-tier architecture involves minimal use of middleware. Most application logic and data processing are performed on the client side, with limited server-side processing.

Significance

(a) Simplicity: The two-tier architecture is simple and easy to implement, making it an ideal choice for small to medium-sized applications with limited data processing demands.

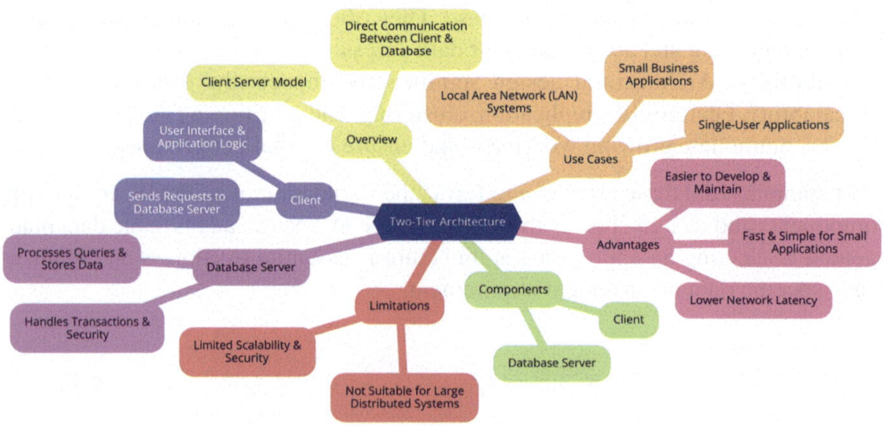

Fig. 7.39 Database architecture: two-tier (prepared by the authors)

(b) Performance: This architecture delivers reliable performance for straightforward applications, particularly when the database server operates under a light workload.

Challenges

(a) Limited scalability: The two-tier architecture faces scalability challenges as applications expand, making it difficult to support a large user base or handle complex data processing efficiently.
(b) Security: In this architecture, security and access control are primarily managed on the client side, which complicates the implementation of robust security measures compared to multi-tier architectures.

In summary, the two-tier database architecture is a simple model where the client-side application interacts directly with the database server. Although suitable for specific applications, it may lack the scalability and security features necessary for larger or more complex systems, which typically demand more advanced multi-tier architectures.

7.4.2 Database Architecture: Three-Tier

The three-tier database architecture is a widely utilized model in modern database management systems (DBMS), offering a structured approach to application design and implementation. This architecture divides the system into three core layers: the presentation layer, the application (or logic) layer, and the data layer (Fig. 7.40). The following outlines its key characteristics and significance:

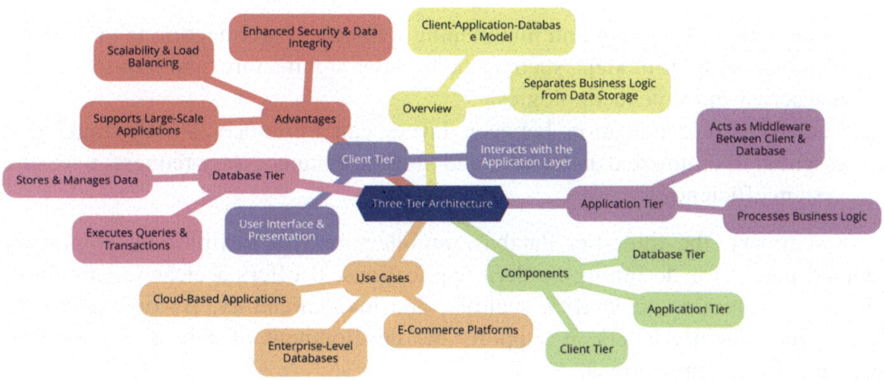

Fig. 7.40 Database architecture: three-tier (prepared by the authors)

Key Characteristics

(a) Presentation layer (client): The presentation layer is the topmost layer of the architecture, responsible for the user interface and user interaction. It encompasses elements such as web browsers, mobile apps, or desktop applications that users directly engage with.
(b) Application (logic) layer (middle tier): The middle tier contains the application logic and processing. It manages tasks such as data validation, business logic execution, and facilitates communication between the presentation layer and the data layer. Common technologies in this layer include web servers and application servers.
(c) Data layer (server): The data layer, positioned at the bottom, hosts the database server. It is responsible for managing data storage, retrieval, and manipulation. The data layer executes SQL queries, ensures data integrity, and oversees transaction management.

Significance

(a) Modularity: The three-tier architecture encourages a modular design, facilitating independent maintenance and scaling of each layer. This separation of concerns improves code reusability and maintainability.
(b) Scalability: The architecture supports scalable deployments, allowing each layer to be scaled as necessary. For example, the application layer can be horizontally scaled by adding more application servers to accommodate higher traffic.
(c) Security: Security can be implemented at each layer, strengthening overall system protection. Access control and authentication mechanisms can be centralized within the application layer.

Challenges

(a) Complexity: Designing and maintaining a three-tier architecture is more complex than simpler models, such as the two-tier architecture, and requires expertise across multiple technologies.
(b) Latency: Communication between layers can introduce latency, making it essential to optimize data transfer and minimize unnecessary requests to ensure system efficiency.

In summary, the three-tier database architecture is a flexible and commonly adopted model for designing modern applications. It offers a clear separation of concerns, promoting scalability, security, and maintainability. By organizing the system into distinct layers, it supports the development of robust and efficient database-driven applications.

7.5 Database Design: Conceptual, Logical, Physical Design

Database design is a crucial process in database management systems (DBMS), focusing on creating an organized plan to manage data efficiently. The process is generally divided into three primary phases, such as conceptual design, logical design, and physical design (Table 7.1 and Fig. 7.41).

Table 7.1 Database design

	Conceptual design	Logical design	Physical design
Objective	The primary objective of conceptual design is to develop a high-level, abstract representation of the data model, focusing on the structure and relationships of the data without addressing implementation specifics	The logical design phase refines the conceptual model into a more detailed and structured representation, making it suitable for implementation within a specific DBMS	The physical design phase converts the logical model into the implementation specific to a database system, taking into account performance, storage, and access requirements
Key activities	Identifying entities: Recognize the key objects or entities within the system Defining relationships: Establish the relationships between entities Attributes: Define the attributes (properties) associated with each entity Output: Entity-relationship diagrams (ERDs) or other conceptual models that visually represent the overall data structure and relationships	Normalization: Structure data to reduce redundancy and enhance data integrity. Data types and constraints: Specify data types for attributes and establish constraints, such as primary keys, foreign keys, and unique constraints Indexes: Identify fields to be indexed in order to optimize query performance Output: Entity-relationship diagrams (ERDs) with detailed attributes, tables, relationships, and a data dictionary	Storage structures: Determine storage mechanisms, file organization, and tablespaces Data partitioning: Decide on data partitioning methods to enhance performance Indexing strategy: Select an appropriate indexing strategy and create physical indexes Query optimization: Plan for efficient query execution to optimize performance Output: The physical database schema, including tables, indexes, views, and constraints, tailored to a specific DBMS
Significance	Conceptual design emphasizes understanding the data requirements from a business viewpoint, ensuring that the database model aligns with the organization's objectives and needs	Logical design focuses on data modelling by defining the structure and relationships in a DBMS-agnostic manner, aiming to ensure data integrity and minimize redundancy	Physical design focuses on the technical aspects of database implementation, ensuring the system operates efficiently and meets performance requirements

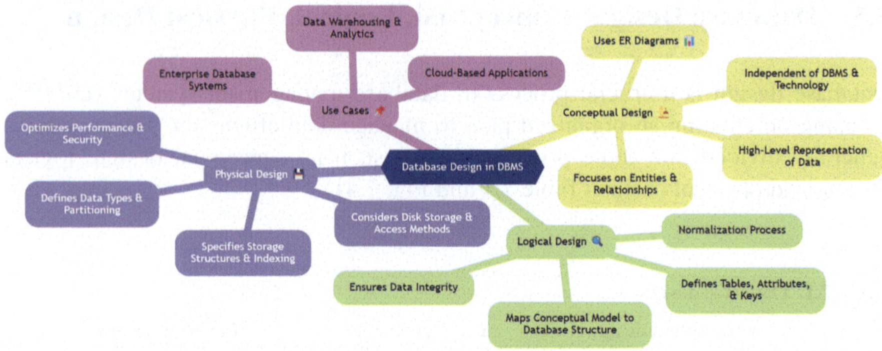

Fig. 7.41 Database design in DBMS (prepared by the authors)

Significance

(a) The three phases of database design, such as conceptual, logical, and physical, ensure that the database system is effectively structured, efficient, and aligned with the organization's business requirements.
(b) These phases promote data consistency, integrity, and performance, contributing to efficient data management and the development of reliable applications.
(c) An iterative approach may be adopted, allowing for continuous refinement and enhancement of the design as the understanding of the system and its needs evolves.

In summary, the database design process involves a gradual refinement of data representation, transitioning from high-level concepts to a detailed, efficient, and optimized database schema. Each phase is essential in ensuring the database system effectively fulfils its intended purpose.

7.5.1 Entity-Relationship (ER) Modelling

Entity-relationship (ER) modelling is a widely utilized and effective technique in database management systems (DBMS) for designing database schemas and representing the structure and relationships of data entities. It offers a visual and conceptual approach to capturing the data requirements of an organization or application (Fig. 7.42). The following discussion explores ER modelling and its significance:

Concepts

(a) Entities: Entities represent real-world objects, concepts, or things that possess distinct attributes. In a database, entities are typically depicted as tables, with each row in the table representing an instance of the entity.
(b) Attributes: Attributes are the properties or characteristics associated with entities, detailing the data that can be linked to an entity. For instance, a "Customer"

Fig. 7.42 Database design: entity-relationship (ER) modelling (prepared by the authors)

entity might include attributes such as “Name”, “Address”, and “Phone Number”.

(c) Relationships: Relationships illustrate the associations between entities, describing how entities are connected or interact. For example, a “Customer” entity can have a relationship with an “Order” entity, signifying that customers place orders.

(d) Cardinality: Cardinality defines the number of instances of one entity that can be associated with a specific number of instances of another entity. Cardinality types include one-to-one, one-to-many, and many-to-many relationships.

Process

(a) Identification of entities and attributes: The initial step in ER modelling involves identifying the key entities and their attributes within the relevant domain. This is typically achieved through consultations with stakeholders and subject matter experts.

(b) Defining relationships: After identifying entities and attributes, the relationships between them are defined. These relationships clarify how entities are connected and how they interact with one another.

(c) Diagram creation: ER diagrams are created to visually represent entities, attributes, and relationships. Standard notations, such as crow’s foot notation, are employed to illustrate these components.

(d) Normalization: ER modelling often incorporates normalization, a process that reduces data redundancy and enhances data integrity by organizing data into suitable tables and relationships.

Significance

(a) Clarity: ER modelling offers a clear and visual representation of data and relationships, facilitating understanding and communication of database requirements among stakeholders.

(b) Data modelling: It is an essential step in the design and planning of relational databases, ensuring that the database schema accurately reflects real-world data and relationships.
(c) Normalization: ER modelling serves as the foundation for normalization, a process that optimizes database structure for efficient data storage and retrieval.

In summary, entity-relationship modelling is a core technique in database design that facilitates the creation of a clear and accurate representation of data and its relationships. It acts as a bridge between business requirements and database implementation, ensuring that databases are properly structured and aligned with the needs of applications and users.

7.5.2 Relational Modelling

Relational modelling is a widely adopted approach in database management systems (DBMS) for designing database schemas, rooted in the principles of the relational model introduced by Edgar F. Codd in 1970. This methodology structures data into tables (relations) comprising rows and columns, where rows represent entities and columns define their attributes (Fig. 7.43). The following discussion explores relational modelling and its significance:

Concepts

(a) Tables (relations): In relational modelling, tables serve as the foundational components. Each table represents an entity, with rows corresponding to individual instances of that entity and columns defining its attributes.
(b) Attributes: Attributes are represented as columns in a table, defining the properties or characteristics of the entity. Each attribute is assigned a name and a data type (e.g., text, number, date) that specifies the type of data it can hold.

Fig. 7.43 Database design: relational modelling (prepared by the authors)

(c) Primary keys: A primary key is a unique identifier assigned to each row in a table, ensuring that every instance of an entity is distinct and can be uniquely identified.
(d) Foreign keys: Foreign keys establish relationships between tables by referencing the primary key of another table, connecting related data and defining associations between entities.

Process

(a) Entity identification: The initial step in relational modelling involves identifying entities and their corresponding attributes within the domain. This is typically achieved through consultations with stakeholders and subject matter experts.
(b) Table design: Identified entities are translated into tables within the database schema, where attributes are represented as columns in these tables.
(c) Normalization: The database schema is normalized to organize data efficiently, reduce redundancy, and enhance data integrity.
(d) Defining relationships: Relationships among entities are established using foreign keys, which link tables and define how data in different tables is interconnected.
(e) Data integrity constraints: Constraints such as primary keys, foreign keys, unique constraints, and check constraints are implemented to enforce data integrity and adhere to business rules.

Significance

(a) Data organization: Relational modelling offers a systematic approach to organizing data into tables while defining relationships, ensuring accuracy and consistency.
(b) Query efficiency: Relational databases are optimized for efficient data retrieval and support complex querying and reporting needs.
(c) Flexibility: The relational model's adaptability to diverse data types and structures makes it applicable across various industries and use cases.

In summary, relational modelling is a core methodology in database design, organizing data into structured tables and establishing relationships between them. This approach offers a transparent and efficient framework for managing and retrieving data, making it one of the most prevalent and effective techniques in database management.

7.6 Conclusion and Endnotes

This chapter explored various facets of database management systems (DBMS) and database design, both of which are essential to effective data management. These concepts are pivotal in modern applications, organizations, and information systems. The following key conclusion emerges from the preceding discussion:

- DBMS types and paradigms: The chapter examined various DBMS types, including relational, document-oriented, key-value, and graph databases, each tailored to specific use cases. It also highlighted the progression from traditional file-based systems to structured DBMS paradigms, emphasizing advantages such as data centralization and independence.
- Purpose and types of DBMS: The primary objective of a DBMS, which is efficient data management, retrieval, and manipulation, was emphasized. Different categories of DBMS, including single-user, multi-user, centralized, distributed, and cloud-based systems, were discussed, addressing scalability, accessibility, and reliability requirements.
- Database models and languages: The discussion covered relational, entity-relationship, object-based, and semantic-structure models, along with essential database languages such as DDL, DML, DCL, and TCL.
- Database architecture: Two-tier and three-tier architectures were explored for their role in ensuring scalability, security, and maintainability in modern applications.
- Database design: The chapter described the stages of database design, including conceptual, logical, and physical, emphasizing their importance in developing efficient and business-aligned schemas.
- Entity-relationship and relational modelling: ER modelling was discussed as a tool for visualizing data entities, attributes, and relationships, while relational modelling was presented as a foundation for structuring data into tables and ensuring accuracy and query efficiency.

In conclusion, a well-structured DBMS and a robust database design are vital for efficient data management, integrity, and meeting the demands of modern applications. The selection of DBMS type, data model, and architecture must align with domain-specific needs and scalability requirements, underpinning the success of businesses and applications in the digital era.

Endnotes

The preceding discussions offer a thorough overview of DBMS and database design principles, which are fundamental to developing robust, efficient, and scalable database systems. Further exploration and practical experience are crucial for mastering the complexities of database management and design, as they play a key role in modern information technology and data-driven decision-making processes.

Further Readings

1. Adam, N. R., & Gangopadhyay, A. (1997). *Database issues in geographic information systems*. Springer . ISBN: 978-0-7923-9924-7. https://doi.org/10.1007/978-1-4615-6109-5
2. Bamford, C., & Curran, P. (1991). Database management systems. In *Data structures, files and databases* (Macmillan computer science series). Palgrave. https://doi.org/10.1007/978-1-349-12615-6_9

3. Chopra, R. (2016). *Database management systems (DBMS): A practical approach*. S Chand Publishing ISBN: 978-9385676345.
4. Elmasri, R. (2018). Data Management Fundamentals: Database Management System. In L. Liu & M. T. Özsu (Eds.), *Encyclopedia of database systems*. Springer. https://doi.org/10.1007/978-1-4614-8265-9_80674
5. Gambosi, G., Scholl, M., & Six, H. (Eds.). (2011). *Geographic database management systems: Workshop proceedings Capri, Italy, May 1991*. Springer ISBN: 978-3-642-77607-6.
6. Geisler, S., & Quix, C. (2020). Database management systems (DBMS). In L. Schintler & C. McNeely (Eds.), *Encyclopedia of big data*. Springer. https://doi.org/10.1007/978-3-319-32001-4_538-1
7. Helland, P. (2009). Database management system. In L. Liu & M. T. Özsu (Eds.), *Encyclopedia of database systems*. Springer. https://doi.org/10.1007/978-0-387-39940-9_662
8. Lemahieu, W., Broucke, S. V., & Baesens, B. (2018). *Principles of database management: The practical guide to storing, managing and analyzing big and small data*. Cambridge University Press, . ISBN: 9781107186125. https://doi.org/10.1017/9781316888773
9. Patni, J. C., Sharma, H. K., Tomar, R., & Katal, A. (2022). *Database management system: An evolutionary approach*. Chapman and Hall/CRC, . ISBN: 9780429282843. https://doi.org/10.1201/9780429282843
10. Taipalus, T. (2024). Database management system performance comparisons: A systematic literature review. *Journal of Systems and Software, 208*, 111872. https://doi.org/10.1016/j.jss.2023.111872
11. Zlatanova, S., & Stoter, J. (2006). The role of DBMS in the new generation GIS architecture. In S. Rana & J. Sharma (Eds.), *Frontiers of geographic information technology*. Springer. https://doi.org/10.1007/3-540-31305-2_8

Chapter 8
Digital Imaging: Processing and Analysis

"Digital technology is a bridge to a new way of seeing the world, turning data into clarity, and complexity into simplicity." — Jonas Bødtker.

Abstract The chapter "Digital Image Processing and Analysis" provides an in-depth discussion of the principles, techniques, and applications of digital image processing and analysis. It looks into the acquisition, manipulation, and extraction of critical information from digital pictures. The chapter begins by introducing the concept, characteristics, and format of digital photos, as well as the fundamental image processing processes. The chapter also looks at how digital image properties like resolution, bit depth, and colour spaces affect image processing and analysis. Following that, the chapter delves into digital image processing techniques. Image enhancement techniques such as spatial domain and frequency domain approaches for boosting image quality, contrast, and sharpness are covered. The chapter discusses picture restoration techniques for reducing image noise, blur, and other flaws. Image compression techniques for lowering digital image storage and transmission requirements, as well as image processing and classification, are also covered. The chapter also goes over digital image processing and interpretation. It investigates ways to feature extraction for recognizing and quantifying specific visual features, such as edge detection, texture analysis, and shape analysis. This chapter also discusses pattern recognition and machine learning approaches used in image categorization and object detection applications. The chapter explains how digital image analysis can be used to extract relevant information for decision-making, research, and problem-solving.

Keywords Digital image processing · Image transformation · Image analysis · Image enhancement · Modern techniques of image classification

K. D. Malakar et al., *Geospatial Technologies in Coastal Ecologies Monitoring and Management*, Advances in Geographic Information Science,
https://doi.org/10.1007/978-3-031-92017-2_8

This chapter offers a comprehensive exploration of the concepts, techniques, and applications essential to understanding and utilizing digital image data. It covers the acquisition, manipulation, and extraction of meaningful information from digital images, with a focus on key elements such as resolution, bit depth, and colour spaces. The chapter delves into a range of processing techniques and analysis steps, including enhancement, restoration, compression, and classification. It also examines feature extraction and machine learning methods for pattern recognition, providing a solid foundation for image analysis applications in various research and practical fields.

Key Points of the Chapter

- Introduces the concept, characteristics, and formats of digital images, highlighting processing and analysis steps.
- Discusses the impact of image properties like resolution, bit depth, and colour spaces on analysis.
- Explores image enhancement techniques for improving quality, contrast, and sharpness in both spatial and frequency domains.
- Covers image restoration and compression methods to address noise, blur, and reduce storage requirements.
- Discusses feature extraction and machine learning approaches for pattern recognition, categorization, and object detection.

8.1 Introduction to Digital Image

8.1.1 Concept of Digital Image

The concept of a digital image (Fig. 8.1) pertains to the representation of visual information in a digital format, typically structured as a grid of pixels. In contrast to traditional photographs or paintings, digital images consist of discrete units called pixels, each encoding specific information about colour and brightness. These images are electronically stored and can be manipulated, transmitted, and displayed through devices such as computers, smartphones, and digital cameras. Digital images encompass various forms, including photographs, graphics, and scanned representations of physical objects. They have become pervasive in contemporary society, serving diverse functions ranging from artistic creation and entertainment to scientific research and data visualization. The digital format facilitates seamless editing, enhancement, and sharing, rendering digital images a cornerstone of modern communication and creative expression. A comprehensive understanding of digital images is essential in disciplines such as photography, graphic design, computer science, and multimedia communication.

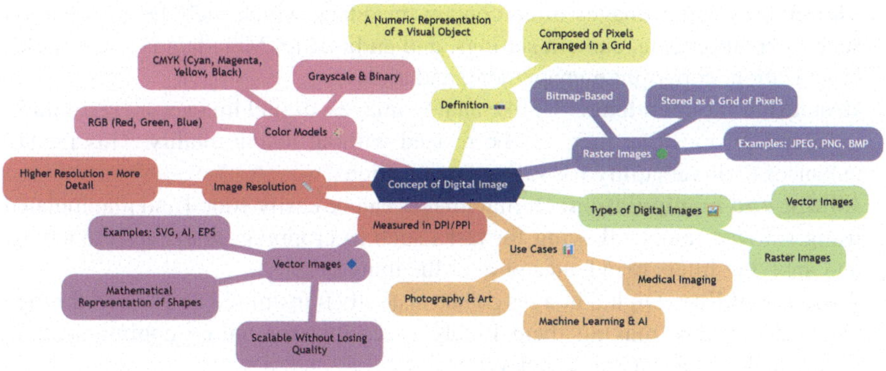

Fig. 8.1 Concept of digital image (prepared by the authors)

8.1.2 Properties of Digital Image

Digital images possess several fundamental properties. A thorough understanding of these properties is crucial for effectively working with digital images, whether in photography, graphic design, or any discipline that depends on visual content, including:

- **Resolution:** Resolution defines the level of detail in a digital image and is typically measured in pixels per inch (PPI) or dots per inch (DPI). Images with higher resolution contain more pixels, resulting in sharper and more detailed visuals.
- **Colour depth:** Colour depth also known as bit depth, colour depth determines the number of colours a digital image can represent. Common depths include 8-bit (256 colours), 24-bit (16.7 million colours), and 32-bit (which includes transparency). A higher colour depth enables more precise and vibrant colour reproduction.
- **File format:** Digital images are stored in various file formats such as JPEG, PNG, GIF, and TIFF, each with distinct features like compression levels, transparency support, and suitability for specific purposes.
- **Size:** The size of a digital image refers to its file size, which is influenced by resolution and compression. Smaller file sizes are ideal for web usage, while larger files are often required for high-quality printing.
- **Compression:** Compression reduces the file size of digital images and can be either lossy (resulting in some quality loss) or lossless (retaining all original quality). The choice of compression method depends on the image's intended application.
- **Aspect ratio:** The aspect ratio describes the proportional relationship between an image's width and height. Different ratios, such as 4:3 for monitors and 16:9 for widescreen displays, are suited for specific uses.

- **Metadata:** Digital images often contain metadata, which includes information such as creation date, camera settings, and authorship. Metadata is essential for organization, copyright management, and archiving.
- **Resolution independence:** Unlike analog images, digital images are resolution-independent, meaning they can be resized without losing quality. This feature enhances their versatility for various applications.
- **Editing and manipulation:** Digital images can be easily edited and manipulated using software tools, allowing for tasks such as cropping, resizing, retouching, and applying filters to enhance or alter the image.
- **Transmissibility:** Digital images can be instantly transmitted and shared through electronic means, making them highly accessible for online communication, social media, and remote collaboration.

8.1.3 Format of Digital Image

Digital images exist in various formats, and selecting the appropriate format depends on factors such as the image's content, intended purpose, file size limitations, and compatibility with software and devices. A thorough understanding of these formats enables users to choose the most suitable option for their specific requirements. Some of the common formats include:

- **JPEG (Joint Photographic Experts Group):** A widely used format for photographs and continuous-tone images, JPEG employs lossy compression to balance file size and image quality. It is ideal for web use and digital photography.
- **PNG (Portable Network Graphics):** A lossless format supporting transparency, PNG is suitable for sharp-edged images like logos and web graphics, offering high-quality compression.
- **GIF (Graphics Interchange Format):** GIF known for supporting animations, GIF uses lossless compression and a limited colour palette (up to 256 colours), making it ideal for simple animations and small graphics online.
- **TIFF (Tagged Image File Format):** A high-quality, versatile format used in professional photography and design, TIFF supports lossless compression, layers, and metadata, making it suitable for archival purposes.
- **BMP (Bitmap):** A basic, uncompressed format native to Windows, BMP delivers high-quality images but large file sizes, often used in professional applications for precise storage.
- **PSD (Adobe Photoshop Document):** Adobe Photoshop's native format, PSD retains layers and editing elements, making it ideal for non-destructive editing in graphic design.
- **SVG (Scalable Vector Graphics):** A vector format described via XML, SVG is perfect for scalable designs like logos and icons without quality loss.
- **HEIF (High-Efficiency Image File Format):** A modern format offering efficient compression and advanced features like HDR and animation, HEIF is increasingly popular for mobile photography.

- **WebP:** Developed by Google, WebP supports both lossy and lossless compression, delivering smaller file sizes with good quality, making it ideal for web use.

8.2 Fundamental Steps in Image Processing

Image processing is a vital field focused on manipulating and analyzing digital images to extract information or enhance quality. The process begins with image acquisition, where images are captured using devices like cameras or scanners. This is followed by pre-processing, which corrects imperfections and enhances quality. Image enhancement improves visual appeal through techniques like sharpening and contrast adjustment, while image restoration aims to recover degraded visuals. Image segmentation divides the image into meaningful regions, and feature extraction identifies key characteristics within these regions. Image recognition and classification, often using machine learning, categorize objects or patterns. Post-processing refines results or integrates multiple images. Finally, analysis and interpretation present findings for understanding. These steps are foundational to diverse applications such as remote sensing, medicine, and computer vision, profoundly influencing how digital images are utilized in the modern world.

8.2.1 Image Acquisition

Image acquisition is the foundational step in image processing, involving the capture and conversion of real-world scenes or objects into digital images using devices such as digital cameras, scanners, or sensors (Fig. 8.2). The quality of the acquired image plays a pivotal role, as it directly impacts all subsequent processing stages. Key considerations in image acquisition include the selection of appropriate hardware, camera settings (e.g., exposure, focus), and environmental conditions such as lighting and camera angles. The unique characteristics of imaging devices,

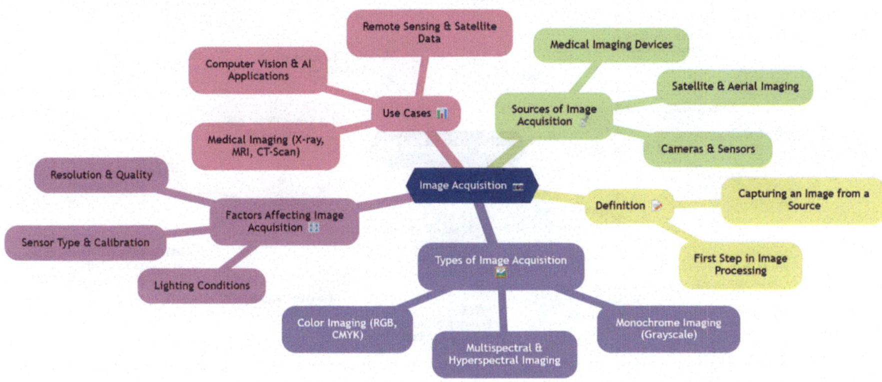

Fig. 8.2 Fundamental steps in image processing: image acquisition (prepared by the authors)

including resolution, colour depth, and light sensitivity, also significantly influence the output. Challenges such as noise, distortion, and artifacts may arise during acquisition, often requiring correction in later stages. In fields like medical imaging, remote sensing, and digital photography, careful planning and control during image acquisition are essential to ensure high-quality input data for accurate analysis, interpretation, and manipulation.

8.2.2 Image Enhancement

Image enhancement is a critical step in image processing aimed at improving the visual quality of digital images. This process is especially important when images are affected by poor lighting, low contrast, or other imperfections that reduce their clarity and interpretability. The primary goal is to emphasize key details and make images more visually appealing while avoiding unrealistic modifications (Fig. 8.3). Several techniques are employed in image enhancement, including:

- Contrast adjustment: Enhances the distinction between light and dark areas in an image, making features more prominent and easier to identify.
- Brightness correction: Modifies the overall brightness levels to ensure the image is well-balanced, avoiding excessive darkness or overexposure.
- Histogram equalization: Redistributes pixel intensities across the image, improving detail visibility in both bright and dark areas.
- Noise reduction: Minimizes unwanted distortions, such as graininess or speckles, to enhance overall image clarity.
- Sharpening: Increases edge contrast, making boundaries and fine details more defined and visually striking.

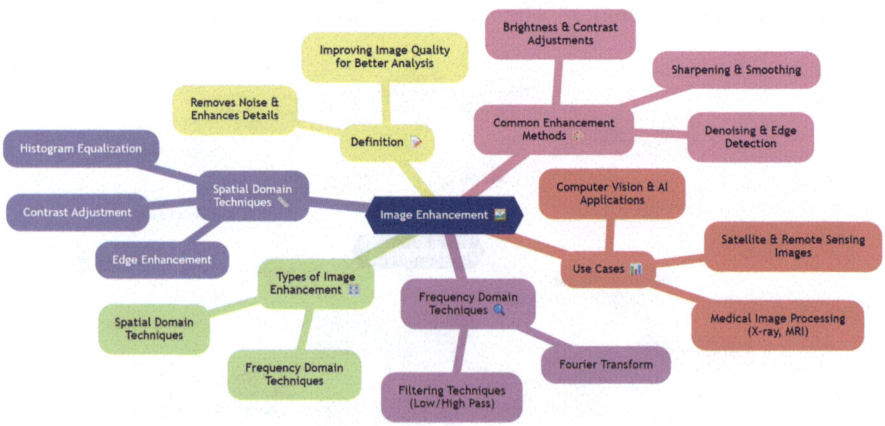

Fig. 8.3 Fundamental steps in image processing: image enhancement (prepared by the authors)

- Colour enhancement: Improves colour balance, vibrancy, and saturation, creating a more visually appealing image.
- Spatial filtering: Applies convolution techniques to emphasize or suppress specific features or patterns within the image.

Image enhancement techniques are commonly applied in fields such as medical imaging, satellite imagery analysis, and photography to improve the clarity and interpretability of images. However, it is crucial to maintain a balance during enhancement to prevent overemphasizing details or introducing artifacts that could compromise the image's integrity or mislead viewers. The selection of enhancement methods depends on the specific characteristics of the image and the objectives of the processing task.

8.2.3 Image Compression

Image compression is an essential process in image processing designed to reduce the storage space and transmission bandwidth required for digital images, while maintaining an acceptable level of image quality. This is especially important in situations with limited storage capacity or network bandwidth, such as storing large image collections on devices or transmitting images over the Internet (Fig. 8.4). There are two main types of image compression:

- **Lossless compression:** This method reduces image file size without compromising image quality by eliminating redundancy in the image data, while preserving all original information. It is typically used for images where quality retention is critical, such as in medical imaging or technical drawings.
- **Lossy compression:** In contrast, lossy compression achieves greater file size reduction but at the expense of some image quality. This is done by discarding details that are less perceptible to the human eye. Lossy compression techniques, such as JPEG, are widely used for photographs and images on the web.

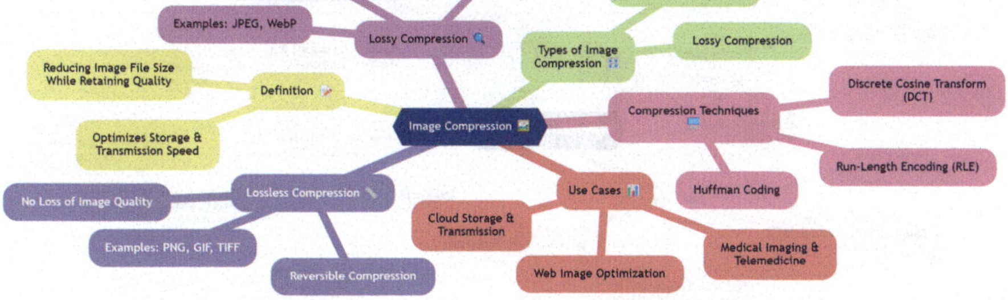

Fig. 8.4 Fundamental steps in image processing: image compression (prepared by the authors)

The choice between lossless and lossy compression depends on the image's intended use. For example, lossy compression is suitable for web images where some quality loss is acceptable, but not for medical images where detail is essential. Compression techniques use various algorithms, such as discrete cosine transform (DCT) in JPEG and run-length encoding (RLE) in lossless compression. The balance between compression ratio and image quality is crucial in selecting the appropriate method. Overall, image compression is vital for efficient storage and transmission, with applications ranging from web design and multimedia to medical imaging and remote sensing.

8.2.4 Image Classification

Image Classification is a fundamental step in image processing, focused on categorizing or labelling images according to their content or features. This process, which falls under computer vision and machine learning, involves training algorithms to recognize patterns and assign predefined labels or categories to images. Image classification has a wide range of applications, including object recognition, medical diagnostics, autonomous vehicles, and content tagging on social media (Fig. 8.5). The key elements of image classification are as follows:

- **Training data:** The effectiveness of image classification depends on a comprehensive dataset containing labelled examples of various classes. These images are used to train machine learning models, enabling them to learn the patterns and features associated with each class.
- **Feature extraction:** Relevant features, such as colour histograms, texture attributes, and edge information, are extracted from images prior to classification. This process helps the model identify distinguishing characteristics for each class.
- **Model selection:** A range of machine learning algorithms can be utilized for image classification, including deep neural networks (e.g., convolutional neural

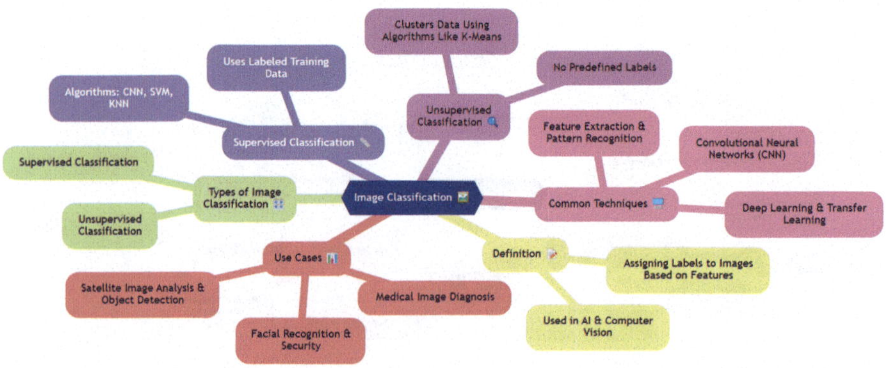

Fig. 8.5 Fundamental steps in image processing: image classification (prepared by the authors)

networks or CNNs), support vector machines (SVM), and decision trees. The choice of algorithm depends on the complexity of the task and the nature of the available data.

- **Training:** During training, the model adjusts its internal parameters to differentiate between classes based on the features extracted from the training data. The goal is to optimize the model's accuracy.
- **Testing and validation:** After training, the model's performance is evaluated using a separate testing dataset. Cross-validation techniques are often applied to ensure the model's ability to generalize to new, unseen data.
- **Classification:** Once trained and validated, the model is used for classification. It assigns the most probable class label to new, unlabelled images based on the patterns it has learned.
- **Post-processing:** Depending on the application, post-processing techniques such as thresholding, filtering, or combining multiple classifiers may be employed to enhance the classification results.

Image classification has undergone significant advancements with the introduction of deep learning techniques, particularly convolutional neural networks (CNNs), which are highly effective at extracting hierarchical features from images. These models have demonstrated exceptional accuracy in tackling complex classification tasks, establishing image classification as a fundamental component of many contemporary AI applications.

8.2.5 Image Processing

Image processing is a core discipline in computer science and digital technology, focusing on the manipulation and analysis of digital images to extract meaningful information or improve their visual quality (Fig. 8.6). The fundamental steps in image processing are as follows:

- **Image acquisition:** The first step involves capturing a real-world scene using digital devices like cameras or sensors, with image quality influencing the entire processing workflow.
- **Pre-processing:** This phase improves the acquired image by performing operations such as noise reduction, contrast adjustment, and resizing to prepare it for further processing.
- **Image enhancement:** Techniques such as adjusting brightness, contrast, colour balance, and sharpening are applied to improve the visual quality of the image.
- **Image restoration:** Employed to recover degraded images due to blurriness or noise, techniques like deblurring and denoising are used to restore image clarity.
- **Image segmentation:** This step divides the image into distinct regions or objects, essential for object recognition and analysis, using methods like thresholding, edge detection, and clustering.

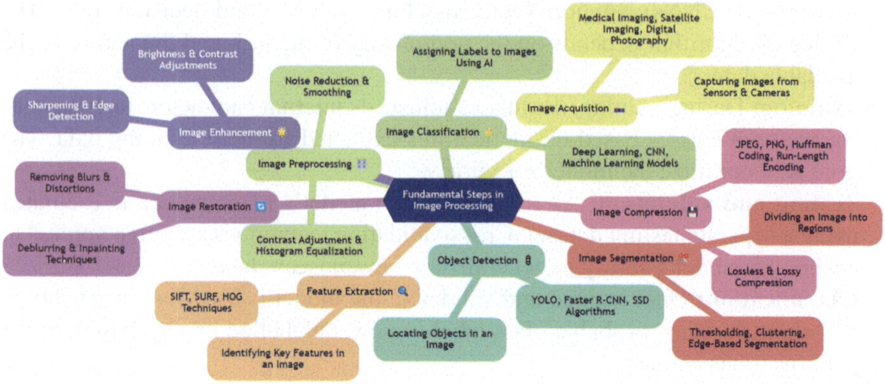

Fig. 8.6 Fundamental steps in image processing: image processing (prepared by the authors)

- **Feature extraction:** Relevant features, such as shape, texture, or colour, are extracted from segmented regions to describe the image content.
- **Image recognition and analysis:** After feature extraction, algorithms are used to classify objects, detect patterns, or perform specific tasks through machine learning or computer vision.
- **Post-processing:** Additional operations, such as filtering or morphological operations, refine the results or combine outputs from different processing stages.
- **Analysis and interpretation:** The final processed image is analyzed and interpreted to derive insights or support decision-making within the context of the application.
- **Visualization and presentation:** The processed results are presented in a comprehensible format, using techniques like overlays, graphs, or reports for end users or researchers.

These fundamental steps in image processing provide a systematic framework for handling digital images across various fields, such as computer vision, medical imaging, remote sensing, and digital photography. The selection of techniques and algorithms at each stage is determined by the specific objectives of the task and the characteristics of the input data.

8.2.6 Image Transformation

Image transformation is a critical step in image processing that involves modifying the spatial arrangement of pixels to achieve specific objectives. These transformations can improve image quality, emphasize particular features, or prepare images for subsequent analysis (Fig. 8.7). The fundamental steps in image transformation are as follows:

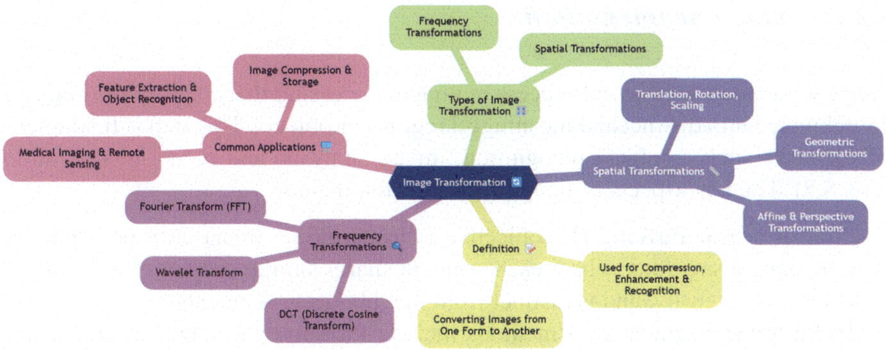

Fig. 8.7 Fundamental steps in image processing: image transformation (prepared by the authors)

- **Spatial vs. frequency domain:** Image transformations can be performed in the spatial domain (direct manipulation of pixel values) or the frequency domain (manipulation of frequency components). Spatial domain transformations include operations like rotation, scaling, and geometric changes, while frequency domain transformations, such as Fourier analysis, are typically used for filtering tasks.
- **Geometric transformations:** These modify the spatial relationships between pixels, including translation, rotation, scaling, and shearing. They are essential for tasks like image registration, where accurate alignment of images is required.
- **Histogram equalization:** This technique enhances image contrast by redistributing pixel intensity values across the full range, improving details in both dark and light areas.
- **Image rotation:** Rotation adjusts the image orientation or aligns it with a specific axis by altering the spatial arrangement of pixels based on a set angle.
- **Image scaling:** Scaling resizes the image, either enlarging it for detailed viewing or reducing its size for storage or display.
- **Affine transformations:** These transformations combine translation, rotation, scaling, and shearing, commonly used for image manipulation tasks like registration and distortion correction.
- **Frequency domain transformations:** These transformations involve converting an image to its frequency components, often used for filtering or deconvolution to enhance or suppress specific features.
- **Image warping:** Warping applies non-linear transformations to deform the image, useful in image morphing, rectification, or correcting lens distortions.

Image transformations are critical techniques in image processing, enabling the modification and adaptation of images to achieve specific goals. The selection of a transformation method is determined by the intended outcome and the nature of the processing task, whether it involves enhancing image quality, extracting relevant information, or preparing images for subsequent analysis.

8.2.7 *Image Segmentation*

Image segmentation is a vital process in image processing that entails partitioning a digital image into distinct and meaningful regions or objects. This step is fundamental for tasks such as object recognition, image analysis, and scene interpretation (Fig. 8.8). The key aspects of image segmentation include:

- **Goal of segmentation:** The objective is to divide an image into perceptually homogeneous regions, where each segment shares similar properties like colour, texture, or intensity, and is distinct from neighbouring segments.
- **Techniques:** Segmentation methods include thresholding, edge-based, region-based, and clustering approaches, each suited to different image types and tasks.
- **Thresholding:** A simple method that divides pixels based on intensity values relative to a set threshold, ideal for binary segmentation.
- **Edge-based methods:** These methods detect boundaries within an image, valuable for object recognition and tracking.
- **Region-based methods:** Pixels are grouped based on similarity in properties like colour or texture, using techniques like region growing or split-and-merge.
- **Clustering techniques:** Methods like k-means clustering group pixels by feature similarity, useful for handling complex segmentation tasks.
- **Multi-level segmentation:** This hierarchical approach segments images into sub-regions, aiding in the analysis of complex scenes.
- **Challenges:** Factors such as noise, occlusion, and varying lighting complicate segmentation, requiring careful method selection and parameter tuning.
- **Applications:** Segmentation is widely used in fields like medical imaging, autonomous driving, satellite imagery, and industrial automation.
- **Post-processing:** Refining segmentation results may involve removing small regions, merging adjacent segments, or filling gaps.

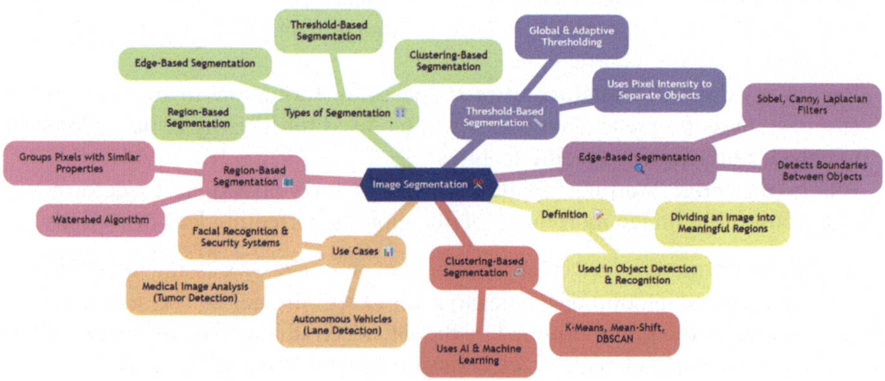

Fig. 8.8 Fundamental steps in image processing: image segmentation (prepared by the authors)

Image segmentation is a foundational component of computer vision and image analysis, essential for subsequent processing and interpretation of visual data. The selection of an appropriate segmentation method is determined by the specific characteristics of the image and the goals of the analysis or application.

8.2.8 Image Representation and Description

Image representation and description is a critical step in image processing, where raw pixel data is converted into a format suitable for computer analysis, interpretation, and understanding. This step is vital for tasks such as object recognition, feature extraction, and content-based image retrieval (Fig. 8.9). The key aspects of image representation and description include:

- **Pixel representation:** Images in digital processing are represented as grids of pixels, each containing information on colour or intensity. Pixels form the foundational data for further representation and description.
- **Feature extraction:** This process identifies and extracts key information, such as colour histograms, texture patterns, edge data, and shape descriptors, to characterize image content.
- **Colour representation:** Colour information can be represented through models like RGB, CMYK, or HSV, depending on the specific analysis or processing needs.
- **Texture description:** Texture, describing surface patterns, is quantified using measures like the grey-level co-occurrence matrix (GLCM) or wavelet transforms, distinguishing smooth from textured regions.
- **Shape analysis:** Shape features describe geometric properties of objects, such as area, perimeter, aspect ratio, and moments.

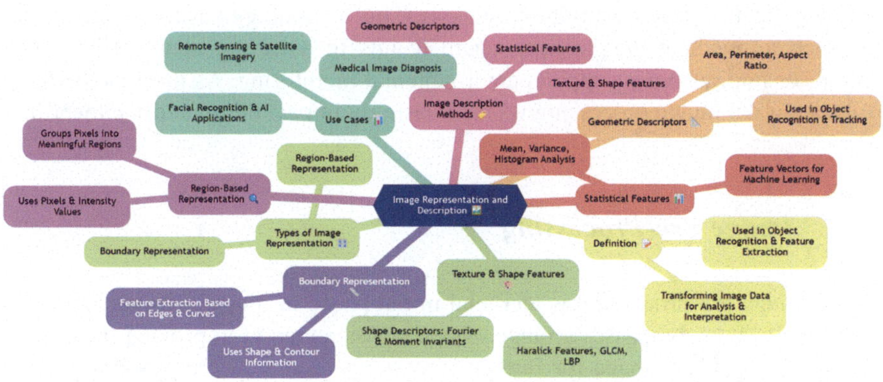

Fig. 8.9 Fundamental steps in image processing: image representation and description (prepared by the authors)

- **Histograms:** Histograms summarize pixel intensity or colour distribution and are used for tasks like image enhancement and classification.
- **Local descriptors:** Local descriptors, such as SIFT and SURF, focus on features from specific regions or key points, often applied in object recognition and matching.
- **Vectorization:** Image data is often converted into numerical vectors for processing by machine learning models, with each vector element representing a specific feature.
- **Dimensionality reduction:** High-dimensional feature vectors can be simplified using methods like principal component analysis (PCA), reducing complexity while preserving essential information.
- **Applications:** These techniques are fundamental in image classification, content-based retrieval, segmentation, and object recognition, enabling computers to interpret image content.

The selection of suitable image representation and description methods is contingent upon the specific objectives of the image analysis task and the inherent characteristics of the data. Efficient feature extraction and representation are critical for precise image analysis and pattern recognition across diverse domains, including computer vision and medical imaging.

8.3 Digital Image Processing and Analysis

Digital image processing and analysis involves the manipulation, enhancement, and interpretation of digital images through algorithms and techniques. It is critical in various fields, including medicine, remote sensing, computer vision, and multimedia. The process typically includes steps such as acquisition, pre-processing, enhancement, segmentation, feature extraction, classification, and post-processing, which facilitate the extraction of valuable information like object recognition, pattern identification, and quantitative analysis. This technology has transformed sectors such as medical diagnostics, where it aids in detecting abnormalities, and remote sensing, where it supports environmental monitoring. As machine learning evolves, digital image processing will increasingly impact our data-driven world.

8.3.1 Image Pre-Processing

Digital image pre-processing is a crucial initial phase in image analysis and computer vision. It encompasses a range of operations aimed at improving the quality and suitability of an image for further processing. Common pre-processing tasks include noise reduction, contrast adjustment, resizing, and colour correction, all of which enhance image quality and facilitate more accurate feature detection, pattern recognition, and object identification. Effective pre-processing is vital for optimizing the performance of subsequent tasks such as object recognition, segmentation, and classification, making it an essential component of any successful image processing pipeline.

8.3.1.1 Feature Extraction

Feature extraction is a vital step in digital image pre-processing, focused on identifying and isolating meaningful information or features from an image (Fig. 8.10). These features, such as edges, textures, shapes, or other distinctive characteristics, provide a concise representation of the image content and are essential for advanced tasks like object recognition and classification. Techniques for feature extraction vary based on the nature of the image and analytical goals. Common methods include edge detection algorithms (e.g., Sobel or Canny), texture analysis (e.g., co-occurrence matrices), and shape descriptors (e.g., Hu moments). By reducing an image's complexity to a set of relevant features, feature extraction enables subsequent algorithms to efficiently process and interpret the data. These features are foundational inputs for applications such as machine learning-based classification and pattern recognition.

8.3.1.2 Radiometric Correction

Radiometric correction is an essential component of digital image pre-processing, particularly in remote sensing and satellite imagery analysis. It aims to rectify inconsistencies in radiometric values (pixel intensities) that arise from factors such as sensor characteristics, atmospheric interference, and variations in illumination (Fig. 8.11). The following are the key aspects of radiometric correction:

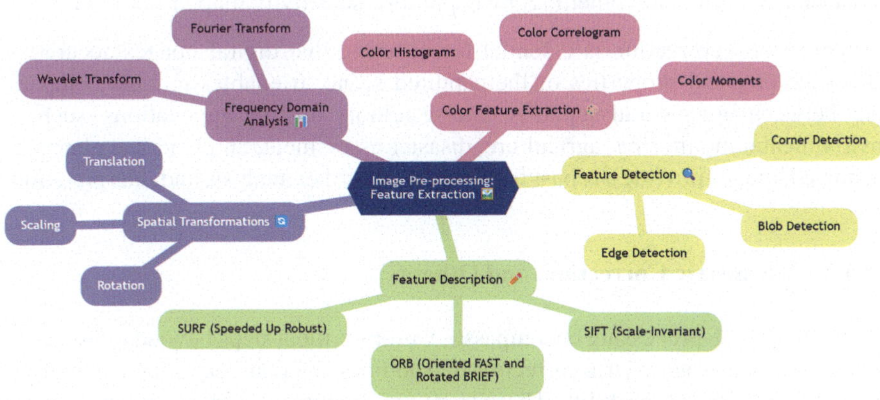

Fig. 8.10 Image pre-processing: feature extraction (prepared by the authors)

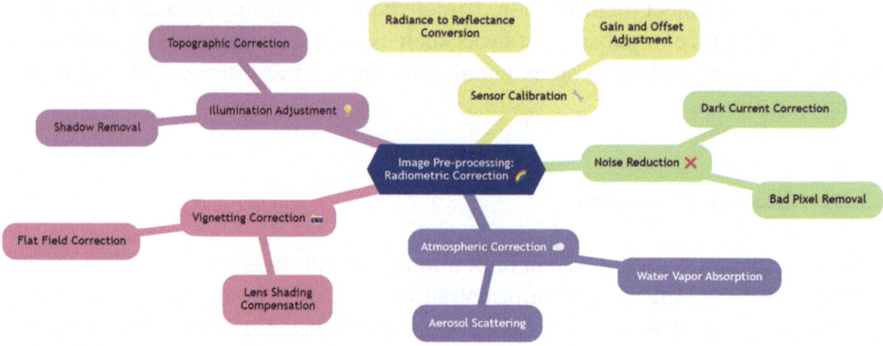

Fig. 8.11 Image pre-processing: radiometric correction (prepared by the authors)

- **Sensor calibration:** Converts raw sensor data into meaningful values (e.g., radiance or reflectance) by correcting for sensor-specific issues like dark current, gain, and non-linearity.
- **Atmospheric correction:** Mitigates effects of atmospheric absorption, scattering, and reflection to retrieve accurate surface reflectance, critical for environmental monitoring.
- **Illumination correction:** Normalizes radiometric differences caused by varying lighting conditions using methods like histogram equalization or shading correction to enhance feature clarity.
- **Noise reduction:** Improves image quality by removing noise and artifacts through filtering or statistical techniques, enhancing the signal-to-noise ratio.
- **Dynamic range adjustment:** Expands pixel value ranges to enhance contrast and detail visibility, particularly when values are compressed.
- **Normalization:** Ensures radiometric consistency across images from different sources or times, enabling reliable comparisons and time-series analysis.
- **Flat field correction:** Compensates for sensor artifacts, such as uneven pixel sensitivity or optical vignetting, by applying a sensitivity map or flat field.

Radiometric correction is essential for ensuring that digital images accurately reflect the physical properties of the captured scene. It enables reliable comparisons between images and supports diverse remote sensing applications, such as environmental monitoring, agriculture, disaster assessment, and land-use planning. Corrected images provide a robust foundation for further analysis and interpretation.

8.3.1.3 Geometric Correction, and Others

Digital image pre-processing encompasses various critical steps beyond radiometric correction, such as geometric correction and additional techniques aimed at improving image quality and usability. These steps are fundamental for ensuring accurate analysis and interpretation in image processing tasks. These include:

- **Geometric correction:** Geometric correction also known as orthorectification or image registration, this process corrects geometric distortions caused by sensor orientation, terrain relief, or Earth's curvature. It transforms images to a standardized coordinate system, enabling precise spatial analysis and GIS applications.
- **Resampling:** During geometric correction, pixel values are interpolated using methods like nearest-neighbour, bilinear, or cubic convolution. The chosen method balances computational efficiency with image accuracy.
- **Mosaicking:** Combines multiple image tiles into a seamless, unified image, essential for large-area analysis in remote sensing by eliminating boundary artifacts.
- **Pan-sharpening:** Enhances multispectral images by fusing them with higher-resolution panchromatic images, improving spatial and spectral detail for applications like land cover classification.
- **Image enhancement:** Techniques such as contrast stretching, histogram equalization, and sharpening improve visual quality and emphasize critical features, tailored to specific analysis goals.
- **Image compression:** Reduces file sizes for efficient storage and transmission. Lossless compression (e.g., PNG) retains all information, while lossy compression (e.g., JPEG) sacrifices detail for smaller sizes.
- **Noise filtering:** Applies spatial or frequency domain filters to remove unwanted noise and artifacts, ensuring cleaner images.
- **Colour balancing:** Adjusts colour distributions to correct for casts or uneven lighting, ensuring accurate and visually appealing representation.
- **Format conversion:** Converts images into formats like TIFF, JPEG, or PNG, depending on application needs.
- **Quality control:** Ensures processed images meet accuracy standards by validating geometric precision, radiometric consistency, and artifact absence.

Digital image pre-processing is an essential step in optimizing images for diverse applications, such as environmental monitoring, agriculture, urban planning, and disaster management. These processes ensure that images are accurate, visually enhanced, and well-suited for detailed analysis and interpretation.

8.3.2 Image Enhancement

Image enhancement is a fundamental process in digital image processing designed to improve the visual quality and interpretability of images. It encompasses techniques such as brightness and contrast adjustment, colour balancing, and noise reduction to highlight finer details and enhance overall image aesthetics. Widely applied across fields such as medical imaging for accurate diagnosis, satellite imagery for clearer Earth observation, and photography for visually compelling results, image enhancement optimizes image appearance, ensuring that critical features are more distinct and accessible for analysis and interpretation.

8.3.2.1 Image Magnification and Reduction

Image magnification and reduction are essential processes in digital image enhancement that involve resizing an image to enlarge (magnification) or shrink (reduction) its dimensions (Fig. 8.12). These operations serve various purposes and are critical for a wide range of applications:

- **Image magnification (upscaling):** Magnification enlarges an image, enhancing its detail and suitability for zooming into specific regions. Techniques like bilinear and bicubic interpolation are commonly used to estimate pixel values, ensuring smooth transitions and maintaining image quality. This process is essential for applications like digital zoom in photography, where clarity is needed for close-up views of distant subjects.
- **Image reduction (downscaling):** Reduction shrinks an image, often to save storage space, improve processing speed, or create thumbnails. Down sampling methods, such as averaging or selecting representative pixels, are used, which may result in a loss of fine details but are suitable for scenarios prioritizing efficiency over resolution.

The decision to magnify or reduce an image depends on the specific needs of the image processing task and the trade-off between size and detail retention. Both processes are integral to image enhancement, enabling diverse applications across fields such as multimedia, web design, medical imaging, and remote sensing.

Fig. 8.12 Image enhancement: image magnification and reduction (prepared by the authors)

8.3.2.2 Colour Composite

Image colour compositing is a technique that combines multiple single-channel (grayscale) or multi-channel (multispectral) images into a single colour image to create visually informative and aesthetically appealing representations. It is widely applied in remote sensing, medical imaging, and scientific visualization (Fig. 8.13). Key aspects of the process include:

- **Multi-channel Data:** Sensors capture multiple channels of data across different parts of the electromagnetic spectrum, such as RGB for visible light or infrared, red, green, and blue for multispectral imagery.
- **Colour Mapping:** Each channel is assigned a specific colour, often using the RGB model (e.g., red for the first channel, green for the second, and blue for the third), with variations based on application needs.
- **Combination:** The channels are merged into a single colour image, allowing viewers to perceive and interpret complex data visually through a unified representation.

Examples of colour composites include false-colour images in remote sensing, where infrared and near-infrared bands are mapped to RGB channels to highlight vegetation health or geological features. In medical imaging, colour composites enhance the visibility of specific tissues or structures. This technique effectively conveys information and facilitates the analysis of complex datasets, enabling valuable insights.

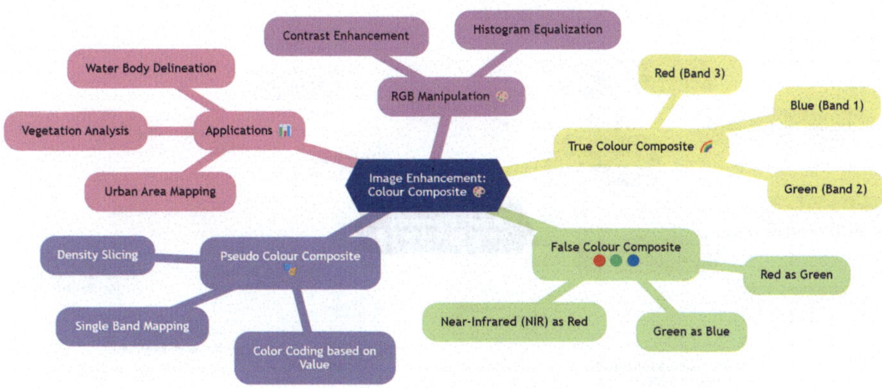

Fig. 8.13 Image enhancement: colour composite (prepared by the authors)

8.3.2.3 Contrast Enhancement

Digital Image Contrast Enhancement is a vital technique aimed at improving the visibility of details and structures by amplifying intensity or colour differences between image regions. It is particularly effective for low-contrast images where distinguishing objects or features is difficult (Fig. 8.14). Following are the common contrast enhancement methods include:

- **Histogram equalization:** This technique redistributes pixel intensities to enhance global contrast, often used for improving overall image visibility.
- **Contrast stretching:** Expands the intensity range of an image to enhance distinctions between features.
- **Adaptive contrast enhancement:** Adjusts contrast locally based on image regions, suitable for images with varying illumination.
- **CLAHE (contrast-limited adaptive histogram equalization):** An advanced method that prevents over-amplification of noise by limiting contrast enhancement in localized regions.
- **Gamma correction:** Adjusts brightness by applying a power-law transformation, particularly useful for correcting non-linear intensity variations.

These methods enhance image interpretability across fields like medical imaging, satellite imagery, and photography. Therefore, the image contrast enhancement is essential in medical imaging for identifying abnormalities, in photography for improving aesthetics, and in computer vision for enhancing object detection and recognition. It ensures vital details are preserved in low-contrast images, making them more visually informative and accessible.

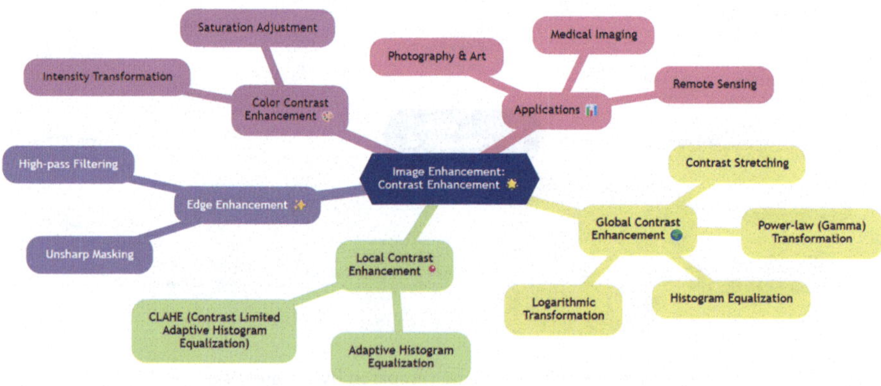

Fig. 8.14 Image enhancement: contrast enhancement (prepared by the authors)

8.3.2.4 Transect Extraction

Image transects extraction is a specialized image enhancement technique commonly applied in remote sensing and geospatial analysis. It involves generating linear cross sections or transects within an image to extract and analyze data along a predefined path (Fig. 8.15). Following key aspects of this technique include:

- **Purpose:** Focuses on studying features like roads, rivers, or vegetation boundaries by extracting data along linear paths.
- **Method:** Transects are defined by specifying start and end points or drawing a line, extracting information such as pixel values, spectral signatures, or statistical data along the path.
- **Software tools:** Remote sensing and GIS tools enable manual or automated transect creation, supporting diverse analysis needs.
- **Data analysis:** Extracted transect data can provide insights like average reflectance of land cover classes or vegetation health variations along the path.
- **Visualization:** Transects are visualized alongside the image to correlate extracted data with spatial patterns.
- **Applications:** Used in environmental monitoring, land cover classification, and geological studies, such as measuring river width or analyzing vegetation health.

In conclusion, image transect extraction is a crucial technique in remote sensing and geospatial analysis, enabling the targeted examination of linear features within images. This method offers valuable insights for diverse applications in fields such as agriculture, environmental science, and urban planning.

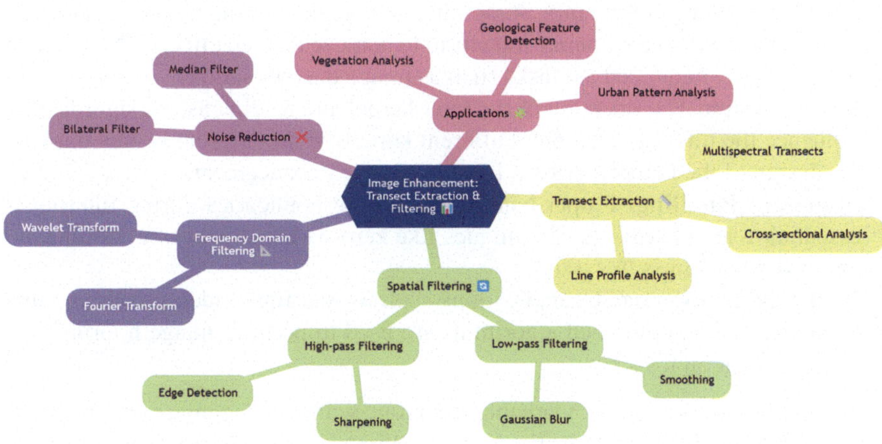

Fig. 8.15 Image enhancement: transect extraction and filtering (prepared by the authors)

8.3.2.5 Filtering

Image filtering is a core technique in digital image enhancement, where a filter or convolution kernel is applied to an image to alter its pixel values. This process is essential for enhancing specific features or reducing noise, thereby improving image quality and facilitating better information extraction (Fig. 8.15). The key aspects of image filtering are as follows:

- **Purpose:** Image filtering serves multiple functions, such as reducing noise, enhancing edges, sharpening, blurring, and extracting features. The selection of a specific filter is determined by the intended enhancement objective.
- **Convolution:** Image filtering is executed via a mathematical operation known as convolution. In this process, a filter (or kernel) is applied to each pixel in the image, generating a weighted sum of pixel values within a specified neighborhood surrounding each pixel.
- **Noise reduction:** Filters such as Gaussian and median filters are frequently employed for noise reduction. These filters smooth pixel values, mitigating random noise variations, and resulting in a cleaner image.
- **Edge enhancement:** Filters like Laplacian or gradient filters are applied to enhance edges and accentuate sudden intensity changes within the image. This is crucial for tasks such as feature extraction and object detection.
- **Sharpening:** Sharpening filters, including Sobel and Prewitt filters, highlight fine details and edges by accentuating high-frequency components in the image.
- **Blurring:** Filters like Gaussian and average filters blur the image, reducing sharpness and detail by smoothing pixel values. These filters are often used for image deblurring or generating artistic effects.
- **Frequency domain filtering:** Some filters are applied in the frequency domain, modifying the image's Fourier transform to achieve desired effects. This method is particularly beneficial for tasks such as image deconvolution.
- **Kernel design:** The selection of the filter kernel and its dimensions significantly influence the filtering outcome. Different kernels yield different results, making the design of the kernel a critical factor in image enhancement.
- **Boundary handling:** Proper handling of image boundaries during filtering is essential to avoid artifacts. Techniques like zero-padding or mirror-padding are employed to address this issue.
- **Real-time processing:** In applications such as real-time video processing and computer vision, optimized algorithms are used to perform image filtering efficiently and at high speeds.

Image filtering is a highly adaptable technique with a wide range of applications in image processing, computer vision, and multimedia. It is essential for enhancing image quality, extracting pertinent information, and preparing images for further analysis or visualization.

8.3.3 *Image Transformation*

Image transformation is a core process in digital image processing, involving the modification of pixel positions or values to achieve particular goals. These transformations encompass geometric alterations, such as rotation, scaling, and shearing, as well as frequency domain operations, including Fourier transforms. Image transformation is crucial for tasks such as image registration, correction of geometric distortions, and the generation of artistic effects. It plays a key role in tailoring images to meet specific needs, whether aligning images for medical analysis or applying creative filters in digital photography, thereby making it an indispensable tool in image processing and computer vision (Fig. 8.16).

8.3.3.1 Image Arithmetic Operations

Image arithmetic operations involve applying mathematical functions to the pixel values of two or more images to generate a new image. These operations are fundamental in digital image processing and serve a variety of purposes:

- **Addition and subtraction:** The addition or subtraction of images is useful for tasks such as image fusion, where data from multiple sources or sensors are integrated to form a composite image. For instance, adding a thermal infrared image to a visible light image can provide valuable insights in remote sensing applications.
- **Multiplication and division:** Multiplying or dividing images allows for adjustments in intensity or the enhancement of specific features. These operations are commonly employed for tasks like contrast adjustment or applying masks to isolate regions of interest.
- **Bitwise operations:** Bitwise operations, such as AND, OR, XOR, and NOT are applied to binary or monochromatic images. These operations are essential for image masking, pattern recognition, and image segmentation.

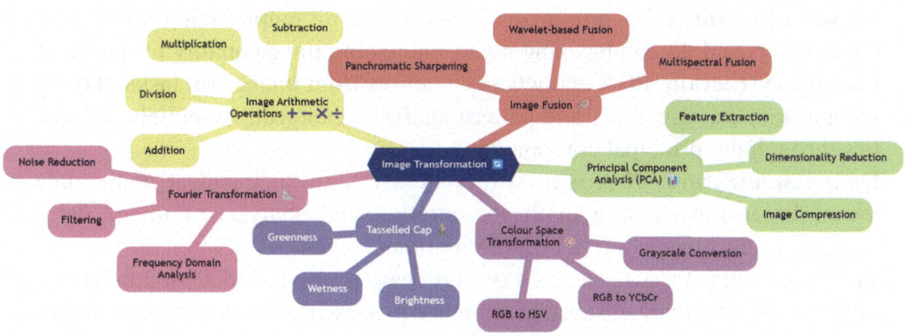

Fig. 8.16 Image transformation (prepared by the authors)

- **Blending:** Blending involves merging two or more images with varying weights or transparency levels to create a seamless transition between them. This technique is frequently used in image editing and compositing to produce special effects.
- **Arithmetic mean and weighted mean:** Computing the mean pixel value within a specific region helps reduce noise or generate smoother images. Weighted means assign varying levels of importance to individual pixels, enabling more controlled smoothing.
- **Image registration:** In image registration, arithmetic operations are employed to align or overlay images for subsequent analysis or comparison.

Image arithmetic operations are versatile tools with broad applications across fields such as remote sensing, computer vision, medical imaging, and multimedia. These operations enable the creation of composite images, improve image quality, and support a range of image manipulation tasks to extract meaningful information from visual data.

8.3.3.2 Principal Component

Principal component analysis (PCA) is a robust mathematical method widely used in digital image transformation and analysis. PCA is primarily employed to reduce the dimensionality of image data while retaining its key information. The following are the key aspects of PCA in image processing:

- **Dimensionality reduction**: PCA reduces the high-dimensionality of image data by identifying principal components, which capture the majority of variance, enabling data reduction with minimal loss of information.
- **Orthogonal transformation**: PCA transforms correlated pixel values into linearly uncorrelated principal components, ranked by their variance-explaining capacity.
- **Image compression**: By retaining only a subset of principal components, PCA facilitates image compression, reducing data size while preserving image quality, especially in bandwidth-limited applications.
- **Noise reduction**: PCA minimizes image noise by emphasizing signal-related components and de-emphasizing noise, enhancing image quality for analysis.
- **Feature extraction**: PCA extracts key features from images for tasks like object recognition, classification, and pattern analysis, capturing essential characteristics through the principal components.
- **Data visualization**: PCA is used to visualize high-dimensional image data in lower dimensions, aiding in the exploration and analysis of image datasets through scatterplots and other representations.
- **Pre-processing**: PCA serves as a pre-processing step to reduce redundancy and improve the efficiency of subsequent image processing or machine learning tasks.

Therefore, PCA is a versatile tool in digital image transformation, enabling dimensionality reduction, noise reduction, data visualization, and feature extraction. Its applications extend across fields such as remote sensing, medical imaging, facial recognition, and data compression, assisting analysts and researchers in extracting valuable insights from complex image data.

8.3.3.3 Tasselled Cap

The tasselled cap (TCA) transformation is a widely used technique in remote sensing and digital image processing. It aims to extract meaningful information from multispectral satellite or aerial imagery by reducing data dimensionality while preserving key features. The following are the key aspects of the tasselled cap transformation:

- **Concept:** The TCA transformation summarizes the spectral reflectance of each pixel using three orthogonal components: brightness, greenness, and wetness. These components are linear combinations of the original spectral bands.
- **Component interpretation:** Each TCA component represents distinct reflectance characteristics of the scene: brightness indicates overall illumination, greenness reflects vegetation, and wetness represents moisture content.
- **Data fusion:** TCA enables the integration of multispectral data from different satellite sensors, allowing for consistent comparison and analysis across various sources.
- **Normalization:** TCA inherently normalizes data, reducing sensitivity to variations in illumination and sensor calibration, ensuring consistency across images and time periods.
- **Mathematical formulation:** The TCA transformation involves matrix operations, where original spectral bands are multiplied by a transformation matrix derived through statistical analysis of the image data.
- **Component interpretability:** TCA components are interpretable, linking them to real-world land cover and environmental features, such as using the wetness component to identify water bodies or soil moisture levels.
- **Applications:** TCA-transformed images are widely used for land cover classification, environmental monitoring, agricultural assessments, and change detection, simplifying analysis by reducing dimensionality and emphasizing relevant features.

The tasselled cap transformation is an essential tool in remote sensing and geospatial analysis, enabling the extraction of significant information from multispectral imagery. It streamlines data interpretation and supports a wide range of applications, including land-use mapping, forestry assessments, environmental monitoring, and disaster management.

8.3.3.4 Colour Space

Colour spaces, also referred to as colour models or colour representations, are mathematical frameworks that define how colours are represented and organized in digital images. They offer a structured approach to representing and manipulating colour information. The following are key aspects of colour spaces in digital image transformation:

- **RGB colour space:** The RGB colour space is the most widely used model in digital imaging, representing colours through combinations of red, green, and blue channels. Each channel specifies the intensity of its respective colour, and any colour can be created by varying the values of these channels. RGB is commonly employed in computer displays, digital cameras, and image editing software.
- **CMYK colour space:** The CMYK colour space, comprising cyan, magenta, yellow, and key (black)—is primarily used in colour printing and reproduction. As a subtractive model, CMYK generates colours by subtracting varying amounts of ink or pigment from a white background.
- **HSV and HSL colour spaces:** The HSV (hue, saturation, value) and HSL (hue, saturation, lightness) colour spaces are cylindrical models representing colour based on perceptual attributes. Hue indicates the colour type (e.g., red or blue), saturation controls the intensity, and value (or lightness) refers to brightness. These models are widely used in image editing for colour adjustment and enhancement.
- **YUV and YCbCr colour spaces:** YUV and YCbCr are colour spaces used in video compression and transmission. Y (luma) represents brightness, while U and V (chroma) convey colour information. Separating luminance and chrominance improves compression and efficiency in video data transmission.
- **Lab colour space:** The lab colour space is designed to be perceptually uniform, where equal distances in the Lab space correspond to approximately equal perceptual differences in colour. It is widely used in colourimetry, colour correction, and colour management.
- **Grayscale:** Although not a colour space in the traditional sense, grayscale represents images in varying shades of grey, from black to white. It is commonly used in applications where colour information is unnecessary, such as in medical imaging and document processing.
- **Colour transformations:** Converting between colour spaces is a common image processing task to optimize specific operations. For example, converting RGB to grayscale simplifies image analysis by reducing colour complexity.

Colour spaces are fundamental in digital image transformation, colour correction, image analysis, and numerous applications where precise representation and manipulation of colour information are critical. The selection of a colour space is determined by the specific requirements of the task and the intended output medium.

8.3.3.5 Fourier Transformation

Fourier transformation is a fundamental mathematical technique in digital image processing, used to analyze and convert spatial information into frequency domain data. The following are key points regarding Fourier transformation in the context of digital images:

- **Basis of frequency analysis:** Fourier transformation decomposes an image into its constituent frequencies, representing complex images as sums of sinusoidal components with varying frequencies and amplitudes.
- **2D Fourier transform:** In digital image processing, the 2D Fourier transform generates a complex-valued representation, where each pixel corresponds to a specific frequency component. The magnitude represents amplitude, while the phase encodes spatial information.
- **Frequency components:** Low-frequency components capture coarse features, while high-frequency components represent fine details and edges in the image.
- **Inverse Fourier transform:** The inverse Fourier transform reconstructs the original image from its frequency representation, useful in image synthesis and reconstruction.
- **Fast Fourier transform (FFT):** The FFT is an efficient algorithm that speeds up the computation of the 2D Fourier Transform, making it suitable for real-time and large-scale image processing.
- **Frequency domain filtering:** Fourier transformation enables frequency domain filtering, where specific frequency components are modified to enhance or suppress image features for enhancement and restoration.
- **Fourier descriptors:** Fourier descriptors capture shape information by extracting Fourier series coefficients, commonly used in object recognition and shape analysis.
- **Applications:** Fourier transformation is used in tasks like image filtering, restoration, compression, and pattern recognition. It enhances or removes specific frequency components, such as in Fourier domain filtering.

Fourier transformation is a vital technique in digital image processing that connects spatial and frequency domain analysis. It enhances the understanding of image content and supports various image manipulation and analysis tasks by revealing the underlying spatial frequencies within the image data.

8.3.3.6 Image Fusion

Image fusion is a process in digital image transformation that integrates multiple images or data sources into a single composite, offering a more detailed and informative representation of a scene. It is widely used in applications such as remote sensing, medical imaging, and computer vision. The following are key points about image fusion:

- **Purpose:** Image fusion enhances scene visibility by combining complementary data from multiple sources or sensors, improving image quality, reducing noise, and providing more detailed information.
- **Multisensory fusion:** In remote sensing, image fusion merges data from different sensors (e.g., optical, thermal, radar), offering both visual and thermal insights for applications like disaster monitoring and land cover classification.
- **Multimodal fusion:** In medical imaging, fusion combines data from different modalities (e.g., MRI, CT, PET) to provide a comprehensive view of a patient's anatomy, aiding diagnosis and treatment planning.
- **Multitemporal fusion:** Fusion can also merge images captured at different times, helping to monitor changes or detect trends, particularly in environmental monitoring and change detection.
- **Fusion techniques:** Image fusion techniques include pixel-level fusion (directly combining pixel values), feature-level fusion (merging extracted features), and decision-level fusion (combining results from independent algorithms).
- **Applications:** Image fusion is used in object detection, target recognition, and surveillance, enhancing the accuracy of automated systems by providing richer data.
- **Challenges:** Key challenges include aligning input images, managing differences in resolution and scale, and selecting the appropriate fusion method for each task.

Image fusion is essential for improving the interpretability and utility of images across various domains. By combining information from diverse sources or time points, it enables analysts and researchers to extract meaningful insights and make well-informed decisions based on a more comprehensive and informative image representation.

8.3.4 Image Classification

Image classification is a key task in digital image processing and computer vision, where predefined categories or labels are assigned to pixels or regions based on visual characteristics. Using machine learning algorithms trained on labelled data, it automatically identifies and classifies objects or features in new images. Image classification has diverse applications, including land cover mapping from satellite imagery, disease detection in medical images, and object recognition in autonomous driving. It significantly automates tasks that would otherwise be time-consuming for manual analysis.

8.3.4.1 Clustering

Clustering is a widely used technique in image classification, particularly in unsupervised learning, where labelled data is not available. It involves grouping pixels or image regions with similar visual characteristics into clusters. Key points about clustering in image classification include:

- **Unsupervised learning:** Clustering does not rely on labelled data; it identifies natural groupings based on the inherent similarities within the data.
- **Feature-based clustering:** Features such as colour, texture, or other visual properties are used to define similarity, often measured through metrics like Euclidean distance or cosine similarity.
- **K-means clustering:** A widely used algorithm that divides data into K clusters based on feature similarity, minimizing intra-cluster variance.
- **Hierarchical clustering:** Builds a tree-like structure of clusters, providing both detailed and coarse segmentation for multi-scale analysis.
- **Challenges:** Key challenges include selecting the right algorithm, determining the optimal number of clusters, and defining effective feature representations.
- **Hybrid approaches:** Some methods combine clustering with supervised learning, where clustering identifies regions of interest, and a supervised classifier is applied to labelled data within those regions.
- **Applications:** Clustering is used in tasks like image segmentation, object detection, and content-based retrieval to identify regions of interest for further analysis.

So, clustering is an essential technique in image classification, particularly for large datasets or when the number of classes is unknown or extensive. It helps reveal inherent groupings within images, providing a basis for further classification or labelling tasks in image analysis.

8.3.4.2 Classification Tools

Image classification involves the use of various tools and software to develop, train, and evaluate models. These tools are critical for processing and analyzing image data. Following are the commonly used classification tools include:

- **Machine learning libraries**: Libraries such as Scikit-Learn, TensorFlow, and PyTorch offer a wide range of machine learning algorithms and deep learning frameworks, enabling researchers and developers to construct and train classification models effectively.
- **Pre-trained models**: Pre-trained models, available through platforms like TensorFlow Hub and PyTorch's model zoo, provide a strong starting point for image classification tasks. These models, trained on large datasets, can be fine-tuned for specific use cases.
- **Deep learning frameworks**: Frameworks like TensorFlow and PyTorch provide extensive support for designing, training, and optimizing neural networks for

image classification. These tools include functions for building network architectures, data processing, and training workflows.

- **Image annotation tools**: Tools such as LabelImg, VGG Image Annotator (VIA), and Labelbox facilitate the annotation of images with labels and bounding boxes, which are essential for training and evaluating classification models.
- **Data augmentation libraries**: Libraries like Augmentor and Albumentations enable the application of data augmentation techniques to training datasets, enhancing model robustness and generalization by simulating diverse scenarios.
- **Model evaluation tools**: Tools like Scikit-Learn offer functions for evaluating model performance using metrics like accuracy, precision, recall, F1-score, and ROC curves. Confusion matrix visualization tools are also useful for performance assessment.
- **GPU acceleration**: Frameworks such as TensorFlow and PyTorch support GPU acceleration, significantly speeding up the training process of deep learning models, which is vital for large-scale image classification tasks.
- **Hyperparameter tuning platforms**: Platforms like Google Cloud AI Platform, AWS SageMaker, and Microsoft Azure Machine Learning assist in hyperparameter tuning, automating the search for optimal model configurations and scaling up workflows.
- **Deployment tools**: Deployment tools like TensorFlow Serving, TensorFlow Lite, and ONNX Runtime allow trained models to be deployed across diverse platforms, from web applications to edge devices.
- **Model Interpretation Tools**: Tools like SHAP and Lime provide interpretable insights into the decision-making process of classification models, helping to explain predictions and ensure model transparency.

These classification tools are essential for developing, training, evaluating, and deploying image classification models across various applications, from computer vision research to commercial products and services. They streamline the development process, ensuring efficient analysis of visual data and enhancing the effectiveness of image classification tasks.

8.3.4.3 Modern Techniques, and Statistics

Modern image classification techniques utilize advanced machine learning and deep learning algorithms to achieve high accuracy in detecting objects, patterns, and features in images. These methods are often integrated with statistical approaches to enhance performance and interpret results. The following are key aspects:

- **Deep learning**: Convolutional neural networks (CNNs) have transformed image classification by automatically learning hierarchical features, excelling in complex recognition tasks. Architectures like ResNet, Inception, and VGG have significantly advanced classification accuracy.
- **Transfer learning**: Transfer learning involves fine-tuning pre-trained models, such as those trained on large datasets like ImageNet, for specific tasks. This

approach reduces training time and data requirements while achieving strong performance.

- **Ensemble methods**: Techniques like Random Forests and Gradient Boosting combine multiple classifiers to enhance accuracy, reduce overfitting, and increase model robustness.
- **Statistical metrics**: Metrics such as precision, recall, F1-score, and ROC curves evaluate classification model performance, helping to quantify trade-offs between true positives, false positives, and false negatives.
- **Confusion matrix**: A confusion matrix visually summarizes classification model performance, showing the correct and incorrect predictions for each class.
- **Cross-validation**: Methods like k-fold cross-validation assess model generalization by partitioning data into training and testing subsets, estimating performance on unseen data.
- **Overfitting mitigation**: Techniques like regularization, dropout, and data augmentation prevent overfitting, which occurs when a model becomes too complex and fits noise in training data.
- **Class imbalance handling**: To address class imbalance, methods such as oversampling, under sampling, or weighted loss functions are used in image classification tasks.
- **Interpretability**: Efforts to enhance model interpretability include techniques like attention mechanisms and gradient-based saliency maps, which help visualize and understand neural network decision-making processes.

Modern image classification techniques are central to advancements in computer vision, driving applications in autonomous vehicles, medical diagnosis, facial recognition, and beyond. By combining sophisticated machine learning algorithms with robust statistical methods, these approaches offer powerful tools for deriving meaningful insights from visual data.

8.4 Conclusion and Endnotes

In conclusion, the exploration of digital image processing and analysis has encompassed a broad range of topics and techniques essential for extracting meaningful information from images. From foundational processes such as image acquisition, enhancement, compression, and classification, to advanced methods like image transformation, radiometric correction, and image fusion, we have examined the diverse approaches in image analysis. Modern techniques, including deep learning and ensemble methods, along with the critical role of statistical evaluation in model performance, have also been emphasized. These discussions highlight that digital image processing is a multidisciplinary field with wide-ranging applications across remote sensing, medical imaging, computer vision, and beyond. It enables the extraction of insights, pattern detection, and informed decision-making in various sectors.

Endnotes:

- Digital images are composed of pixels, which are fundamental to all image processing operations.
- Image properties such as resolution, bit depth, and colour representation determine the quality and characteristics of the data.
- Different image formats (JPEG, PNG, TIFF) serve various storage and transmission needs, each with unique advantages.
- The image processing workflow, from acquisition to classification, provides a structured approach for image analysis.
- Image enhancement techniques, like contrast adjustment and radiometric correction, improve quality for further analysis.
- Image compression reduces file size while preserving quality, essential for efficient storage and transmission.
- Image classification methods, including clustering and deep learning, categorize objects and features automatically.
- Statistical metrics and tools assess the effectiveness of image classification models.
- Image fusion integrates data from multiple sources or sensors, providing a more comprehensive representation.
- Classification tools, supported by GPUs and cloud platforms, facilitate the development, training, and deployment of image models.

In an era where visual data is becoming ever more integral, the knowledge and techniques outlined here form a fundamental basis for researchers, engineers, and practitioners in various fields that rely on images as a key source of information.

Further Readings

1. Blackledge, J. M. (2006). *Digital image processing: Mathematical and computational methods*. Woodhead Publishing ISBN: 978-1-898563-49-5.
2. Dwivedi, R. S. (2017). Digital image processing. In *Remote sensing of soils*. Springer. https://doi.org/10.1007/978-3-662-53740-4_3
3. Gomarasca, M. A. (2009). Digital image processing. In *Basics of geomatics*. Springer. https://doi.org/10.1007/978-1-4020-9014-1_8
4. Kemp, Z. (1999). A framework for integrating GIS and digital images. In P. Agouris & A. Stefanidis (Eds.), *Integrated spatial databases. ISD 1999* (Lecture notes in computer science) (Vol. 1737). Springer. https://doi.org/10.1007/3-540-46621-5_10
5. Khorram, S., Nelson, S., Cakir, H., & Van Der Wiele, C. (2013). Digital image processing: Post-processing and data integration. In J. N. Pelton, S. Madry, & S. Camacho-Lara (Eds.), *Handbook of satellite applications*. Springer. https://doi.org/10.1007/978-1-4419-7671-0_92
6. Reddy, G. P. O. (2018). Digital image processing: Principles and applications. In G. Reddy & S. Singh (Eds.), *Geospatial technologies in land resources mapping, monitoring and management* (Geotechnologies and the environment, 21). Springer. https://doi.org/10.1007/978-3-319-78711-4_6
7. Richards, J. A. (2022). *Remote sensing digital image analysis*. Springer, . ISBN: 978-3-030-82326-9. https://doi.org/10.1007/978-3-030-82327-6

8. Tan, H. Y., Li, L., Wu, G. L., et al. (2019). Digital imaging detection and image analysis of internal structural defects in GIS. *Optoelectronics, Instrumentation and Data Processing, 55*, 592–596. https://doi.org/10.3103/S8756699019060098
9. Trussell, H. J., & Vrhel, M. J. (2008). *Fundamentals of digital imaging*. Cambridge University Press ISBN: 9780521868532.
10. Umbaugh, S. E. (2010). *Digital image processing and analysis: Human and computer vision applications with CVIPtools* (2nd ed.). CRC Press, (Taylor and Francis Group) ISBN: 9780429196324.

Chapter 9
Participatory GIS: A Solution to Traditional GIS Challenges

> *"GIS is not complete until it becomes 'people's GIS,' where citizens actively contribute to, engage with, and benefit from the spatial information ecosystem." — Dara O'Rourke.*

Abstract This chapter digs into the concept, technique, and benefits of Participatory Geographic Information Systems (PGIS) as a solution to traditional GIS challenges. PGIS actively engages local communities and stakeholders in data collection, analysis, and decision-making, aiming to address power dynamics, promote inclusion, and integrate local knowledge into GIS. This chapter emphasizes the concepts of PGIS, PPGIS, VGI, CIGIS, and Web GIS. Outline the methodologies, uses, limitations, and future possibilities of PGIS as well. It demonstrates the significance of community participation, knowledge co-production, and capacity building in PGIS activities. The chapter discusses participatory mapping techniques such as sketch mapping, aerial image interpretation, and mobile mapping, which allow communities to spatially represent their knowledge and perspectives. Furthermore, the chapter explores the advantages of PGIS in solving a variety of difficulties. It goes over how PGIS can help with community empowerment, social inclusion, and local decision-making. The chapter emphasizes the importance of indigenous knowledge systems and local knowledge in improving the accuracy and usefulness of GIS data. In addition, the chapter also explores the integration of participatory methodologies with technologies such as mobile applications and social media to enhance data collection and community engagement. It examines PGIS applications across various fields, including urban planning, natural resource management, disaster management, and community development, highlighting its role in evidence-based decision-making, conflict resolution, and sustainable development. The chapter concludes by discussing the challenges and future prospects of PGIS, highlighting the importance of capacity building, institutional support, and integrating traditional and participatory approaches. It underscores PGIS's potential in tackling issues like social justice, and community-driven sustainable development.

K. D. Malakar et al., *Geospatial Technologies in Coastal Ecologies Monitoring and Management*, Advances in Geographic Information Science,
https://doi.org/10.1007/978-3-031-92017-2_9

Keywords Participatory GIS · Traditional GIS challenges · Public participation geographic information system · Web GIS · Future of PGIS

This chapter delves into the transformative potential of Participatory Geographic Information Systems (PGIS) as a response to traditional GIS limitations. By engaging local communities in spatial data collection, analysis, and decision-making, PGIS fosters inclusivity, empowerment, and knowledge co-production. Key methodologies, including participatory mapping and mobile applications, are explored alongside applications in various fields. The chapter highlights the critical role of indigenous and local knowledge in enhancing GIS accuracy and relevance. Finally, it addresses challenges and future opportunities, emphasizing PGIS's promise in advancing social justice and community-led development.

Key Points of the Chapter

- Understanding the concepts of traditional GIS challenges and PGIS.
- Explores the conceptual foundation of participatory GIS as solutions to traditional GIS challenges.
- Highlights the participatory techniques and methodologies for spatially representing community knowledge.
- Thinking about geospatial citizenship, voice mapping, and participatory decision-making.

9.1 Introduction to Participatory GIS (PGIS)

Participatory Geographic Information Systems (PGIS) represent a transformative evolution of traditional Geographic Information Systems (GIS), addressing several challenges inherent in the field. PGIS involves the active engagement of local communities and stakeholders in data collection, mapping, and decision-making, democratizing GIS and broadening its application. A key challenge in traditional GIS is ensuring data accuracy and currency, which PGIS overcomes by incorporating local knowledge to provide contextually relevant and up-to-date spatial information. Moreover, PGIS promotes accessibility and inclusivity by empowering marginalized communities to participate in the mapping process, ensuring their perspectives are considered in decision-making. It also fosters a sense of ownership and pride among participants, enhancing the sustainability of GIS projects and strengthening community engagement. Ultimately, PGIS addresses traditional GIS limitations while promoting a more collaborative, inclusive, and locally grounded approach to spatial information management.

9.1.1 Concept of PGIS

Participatory Geographic Information Systems (PGIS) is an inclusive approach to geographic information management that involves local communities, stakeholders, and experts in spatial data collection, mapping, and analysis. PGIS values local knowledge, empowering communities to actively participate in decisions regarding their geographic environment. It combines traditional and participatory mapping methods, integrates local and scientific knowledge, and addresses local priorities. PGIS is applied in areas such as natural resource management, land-use planning, disaster risk reduction, and social development, promoting informed decision-making and sustainable development by bridging the gap between GIS and community-based knowledge (Fig. 9.1). Key applications of PGIS include:

- **Natural resource management:** PGIS engages local communities in managing resources like water bodies, forests, and agricultural lands, combining local knowledge with scientific data for sustainable decision-making.
- **Land-use planning:** PGIS supports community-driven land-use planning by helping residents identify zones for development or conservation, ensuring fair land allocation.
- **Environmental conservation:** PGIS aids environmental conservation by involving local communities in wildlife monitoring and habitat mapping to protect vital areas.
- **Disaster risk reduction:** PGIS enables communities to map hazard-prone areas, evacuation routes, and resources, aiding in disaster preparedness and reducing risk.
- **Healthcare:** PGIS helps identify health risks, map disease distribution, and plan healthcare facilities, contributing to effective public health initiatives.
- **Indigenous knowledge preservation:** PGIS preserves indigenous knowledge related to land, biodiversity, and cultural heritage, protecting community rights and traditions.
- **Education:** PGIS serves as an educational tool, enabling students and communities to learn about their surroundings through participatory mapping activities.

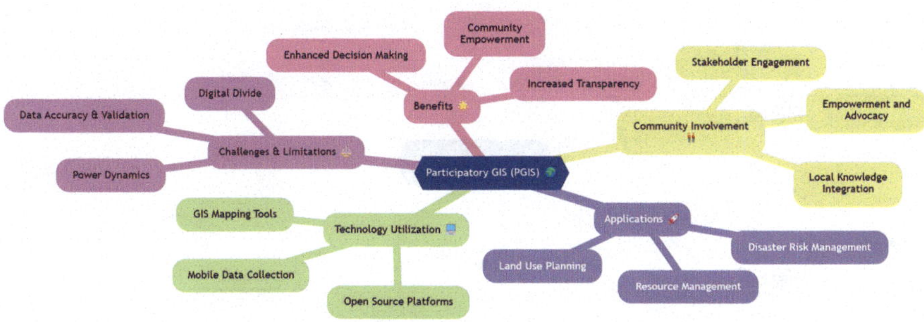

Fig. 9.1 Concept of PGIS (prepared by the authors)

- **Agriculture and food security:** PGIS supports agricultural communities in optimizing crop planning, soil management, and resource allocation to improve food security.
- **Community development:** PGIS fosters community engagement in development projects, identifying locations for schools, clinics, and infrastructure.
- **Social justice:** PGIS advocates for social justice by helping marginalized communities secure land rights and resources, contributing to land reform processes.

Therefore, PGIS is a multifaceted approach that enables communities and stakeholders to engage actively in decision-making processes, fostering more sustainable and inclusive development across diverse sectors. By integrating traditional and local knowledge with contemporary technology, PGIS addresses complex challenges and promotes collaborative solutions.

9.1.2 Concept of Participatory Public Geographic Information Systems (PPGIS)

Participatory Public Geographic Information Systems (PPGIS) is an expansion of Participatory Geographic Information Systems (PGIS) that focuses on engaging the broader public and citizens in geospatial decision-making. PPGIS harnesses crowdsourcing and citizen science to gather, share, and apply geographic data for various objectives (Fig. 9.2). Key elements of PPGIS include:

- **Citizen participation:** PPGIS promotes active involvement of citizens, community groups, and the general public in the generation and management of geographic data, encompassing tasks such as data collection, validation, and sharing local knowledge.
- **Technology and crowdsourcing:** PPGIS utilizes modern technologies like smartphones and online mapping platforms to facilitate public contributions of geospatial data. Crowdsourcing platforms, such as OpenStreetMap, exemplify this method.

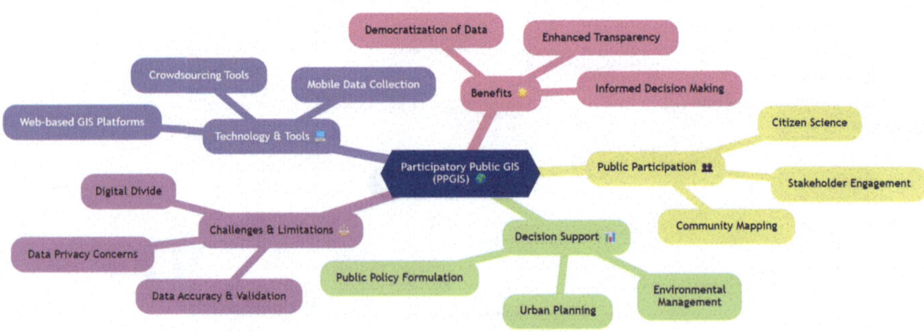

Fig. 9.2 Concept of PPGIS (prepared by the authors)

- **Data quality and verification:** Ensuring data accuracy remains a challenge in PPGIS, with verification techniques like peer review or expert validation used to maintain data reliability.
- **Transparency and accountability:** PPGIS enhances transparency and accountability by engaging citizens and making geospatial data and decision-making processes publicly accessible.
- **Applications:** PPGIS is applied in diverse fields, including disaster management, urban planning, environmental monitoring, and community development. Citizens can map issues such as potholes, pollution, or deforestation, providing essential data for decision-making.
- **Challenges:** Key challenges in PPGIS include concerns around data privacy, maintaining data quality and consistency, and managing issues related to data ownership and intellectual property.

PPGIS signifies the democratization of geographic information, enabling a wider range of individuals to engage in spatial data collection and analysis. It empowers citizens to actively participate in decisions that impact their communities, promoting a more inclusive and informed approach to decision-making. Therefore, the PPGIS has diverse applications across multiple fields, such as environmental management, urban planning, public health, and social justice. The following provides a detailed discussion of its applications:

- **Natural resource management:** Communities and indigenous groups contribute local knowledge to sustainable management of resources like forests and watersheds, addressing issues such as deforestation and water quality.
- **Environmental conservation and biodiversity mapping:** PPGIS enables communities to map biodiversity hotspots and monitor illegal activities, contributing to effective conservation and ecosystem protection efforts.
- **Urban planning and land-use management:** PPGIS engages local communities in urban planning, helping to identify spaces for parks, schools, and infrastructure while supporting participatory zoning and land-use decisions.
- **Disaster risk reduction and management:** PPGIS aids in mapping vulnerable areas, creating hazard maps, and enhancing disaster preparedness through community-driven data collection and response planning.
- **Public health planning:** PPGIS helps identify health disparities, track disease spread, and plan healthcare access by involving communities in mapping health-related issues and services.
- **Community development and social justice:** PPGIS supports community development by identifying infrastructure needs and empowering marginalized groups in decision-making, particularly on issues like land tenure and education access.
- **Tourism planning:** PPGIS facilitates sustainable tourism development by involving local communities in planning, ensuring that tourism benefits the local population.

- **Transportation planning:** By gathering community input on transportation needs, PPGIS helps improve infrastructure and services to address mobility challenges.
- **Cultural heritage preservation:** PPGIS enables communities to actively protect and manage cultural heritage sites, ensuring their preservation for future generations.

9.1.3 Concept of Volunteered Geographic Information (VGI)

Volunteered geographic information (VGI) refers to the collaborative and participatory creation of geographic data by individuals or groups, typically not professional geographers or cartographers. Leveraging crowdsourcing and technology, VGI enables the collection, creation, and sharing of geospatial information (Fig. 9.3). Following is a basic exploration of VGI within the context of Geographic Information Systems (GIS):

Elements of VGI

- **User-generated data:** VGI is based on contributions from the public, typically via digital platforms and technologies. Individuals can contribute a variety of geospatial data, including GPS tracks, location-tagged photos, and textual descriptions.
- **Crowdsourcing:** VGI exemplifies crowdsourcing, where a broad range of individuals collectively contribute to the creation and updating of geographic datasets. This collaborative model facilitates the rapid accumulation of diverse geographic information.
- **Open and accessible:** VGI datasets are generally open and accessible, shared through online platforms for free use by researchers, policymakers, and the public.

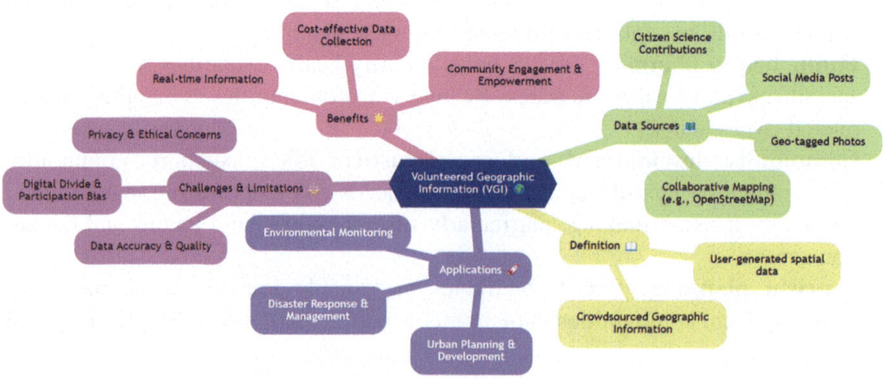

Fig. 9.3 Concept of VGI (prepared by the authors)

Examples of VGI

- **OpenStreetMap (OSM):** OSM is a prominent example of VGI, functioning as a collaborative mapping platform where users contribute to the creation of comprehensive and current maps. Contributors add data on roads, buildings, points of interest, and more.
- **Social media geotagging:** Platforms such as Twitter, Instagram, and Facebook allow users to attach location data to their posts, generating valuable geographic information about locations, events, and activities.
- **Wikipedia geotagging:** Many Wikipedia entries include geospatial coordinates, allowing users to access articles related to specific geographic areas. This feature of geotagging is a form of VGI.
- **Citizen science projects:** Numerous citizen science initiatives engage volunteers in gathering environmental data, such as wildlife sightings, weather data, and pollution levels, often utilizing mobile apps and GPS devices.

Benefits of VGI

- **Rapid data updates:** VGI enables the swift and frequent updating of geographic data, ensuring maps and datasets remain current.
- **Local knowledge:** VGI captures local insights and context, which may be absent in traditional data sources, improving the richness and accuracy of geospatial information.
- **Cost-effective:** VGI offers a cost-efficient method for gathering extensive geographic data, in contrast to traditional data acquisition techniques.
- **Community engagement:** Involving the public in data collection nurtures a sense of ownership and engagement in geographic information and decision-making.
- **Diverse data types:** VGI encompasses various data formats, such as location-based photos and textual descriptions, making it versatile for multiple applications.

9.1.4 Concept of Web GIS

Web GIS is a specialized branch of Geographic Information Systems (GIS) that utilizes Internet technology to provide geospatial data, tools, and services to users via web browsers. By integrating GIS technology with web-based functionalities, Web GIS enhances the accessibility, interactivity, and collaboration of geographic information (Fig. 9.4). Following is a basic exploration of Web GIS:

Components of Web GIS

- **Web-based mapping:** Web GIS platforms enable users to interact with dynamic maps via web browsers, displaying various spatial data such as maps, satellite imagery, and aerial photos.

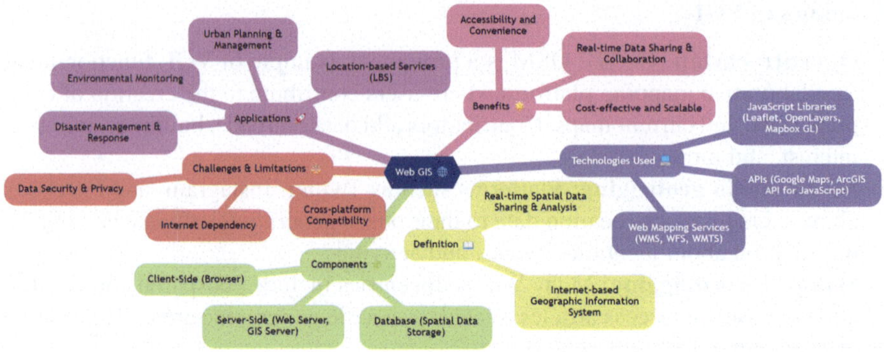

Fig. 9.4 Concept of Web GIS (prepared by the authors)

- **Data integration:** Web GIS integrates diverse data sources, including geographic, demographic, and environmental information, into a unified platform. This integration allows users to overlay and analyze multiple data layers for informed decision-making.
- **Web services:** Web GIS utilizes services like web map services (WMS), web feature services (WFS), and web processing services (WPS) to facilitate the sharing and retrieval of geospatial data and to perform spatial analysis tasks online.
- **User interaction:** Web GIS applications enable users to engage in activities such as querying, buffering, spatial analysis, and visualization. Additionally, users can contribute data, create custom maps, and share their work with others.
- **Collaboration:** Web GIS fosters collaboration by enabling multiple users to simultaneously access and work with the same geographic data, promoting data sharing and cooperative decision-making across teams and organizations.

Benefits of Web GIS

- **Accessibility**: Web GIS broadens access to geographic data, making it available to both experts and non-experts via the Internet, thereby promoting data democratization.
- **Scalability**: Web GIS is capable of handling large, complex datasets, making it adaptable for a wide range of industries and applications.
- **Real-time data**: Web GIS supports the integration of real-time data, providing users with the most current and relevant information.
- **Collaboration**: It enables collaboration across geographically dispersed teams and stakeholders, fostering cooperative decision-making.
- **Cost-efficiency**: By eliminating the need for extensive software installations and updates, Web GIS reduces IT infrastructure costs.
- **Customization**: Web GIS allows users to tailor maps and applications to meet their specific needs and preferences, enhancing usability.

9.2 Applications, Limitations, and Future of PGIS

9.2.1 Applications of PGIS

Participatory Geographic Information Systems (PGIS) integrates GIS technology with participatory methods to engage local stakeholders and communities in spatial information and resource management decisions. PGIS is widely applied to empower communities, improve data collection, and foster collaborative decision-making (Fig. 9.5). Key applications of PGIS include:

- **Natural resource management:** PGIS facilitates sustainable natural resource management by involving local communities in mapping and monitoring activities, addressing issues such as deforestation, wildlife conservation, and resource allocation.
- **Environmental conservation and biodiversity mapping:** Communities contribute to mapping biodiversity hotspots and ecosystems, informing conservation efforts and combating illegal activities like poaching and logging.
- **Land-use planning and land tenure:** PGIS aids in land-use planning by involving communities in delineating boundaries, documenting land tenure, and making decisions about land use, thereby reducing conflicts and improving management.

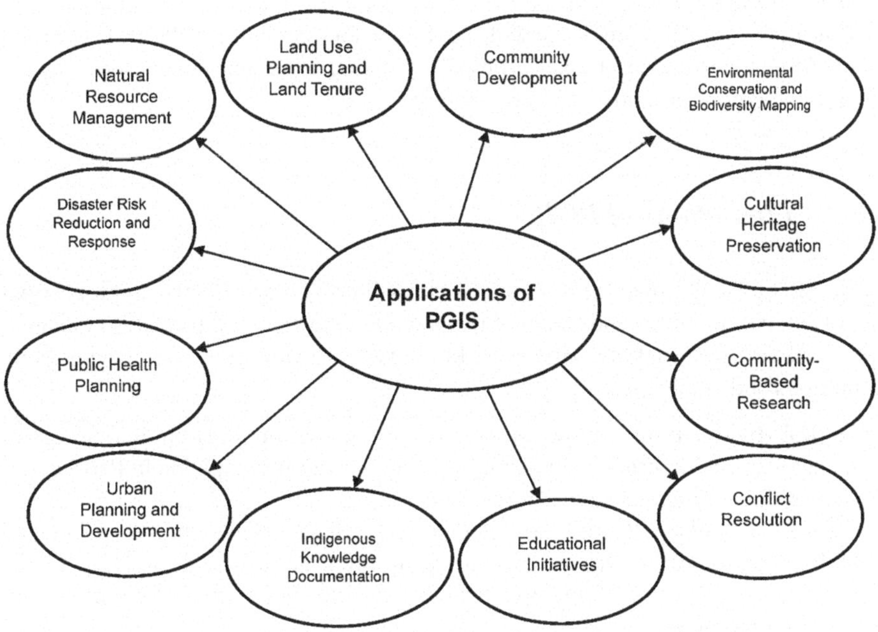

Fig. 9.5 Application of participatory GIS (prepared by the authors)

- **Disaster risk reduction and response:** In disaster-prone areas, PGIS supports mapping vulnerable zones, hazards, and evacuation routes, allowing communities to participate in disaster preparedness and response, enhancing resilience.
- **Community development:** PGIS helps communities identify infrastructure needs, prioritize projects, and monitor progress, addressing issues such as sanitation, housing, and access to services.
- **Public health planning:** PGIS maps health disparities, locates healthcare facilities, and aids in planning interventions based on community health needs, improving resource allocation and disease prevention.
- **Urban planning and development:** PGIS empowers residents to participate in urban planning by identifying optimal locations for public facilities, green spaces, and housing projects.
- **Cultural heritage preservation:** PGIS helps preserve cultural heritage by involving local communities in documenting and managing heritage sites, ensuring respect for indigenous knowledge and practices.
- **Community-based research:** PGIS supports community-based research, enabling residents to collect and analyze spatial data for various social and environmental studies.
- **Indigenous knowledge documentation:** PGIS assists indigenous communities in documenting and preserving traditional knowledge related to land, resources, and cultural practices, supporting land rights assertion.
- **Educational initiatives:** PGIS is integrated into educational programmes to teach geographic concepts, community engagement, and the application of technology to real-world problems.
- **Conflict resolution:** In areas with land or resource conflicts, PGIS provides a platform for transparent and participatory dialogue among stakeholders, facilitating negotiations and conflict resolution.

9.2.2 Limitations of PGIS

Participatory Geographic Information Systems (PGIS) is an effective approach that empowers communities and stakeholders to engage in spatial information management and decision-making. However, like any methodology, PGIS faces certain limitations and challenges:

- **Digital divide:** In many low-income or remote communities, limited access to technology and Internet connectivity may hinder full participation in PGIS activities, reducing the inclusivity of the process.
- **Technical skills:** Effective use of GIS technology often requires specialized skills. Communities may need extensive training and capacity-building to use tools and software efficiently, which can be both time-consuming and resource-intensive.

- **Data quality and accuracy:** PGIS depends on data contributed by non-experts, which may vary in quality and accuracy. Ensuring the reliability of community-generated data is challenging, as errors can impact decision-making.
- **Data privacy and security:** Sharing geospatial data, particularly in sensitive contexts, can raise concerns over privacy and security. Protecting individuals' personal information and mitigating security risks is crucial.
- **Power imbalance:** Power imbalances within communities or between communities and external organizations can hinder meaningful participation. Dominant voices may overshadow marginalized groups, limiting equitable involvement.
- **Cultural and linguistic barriers:** Language and cultural differences can impede effective communication and collaboration in PGIS initiatives. Ensuring comfortable and meaningful engagement for all participants is essential.
- **Data ownership and control:** Ownership and control over the data generated through PGIS can become contentious. Clarifying data ownership and usage rights is vital to avoid disputes.
- **Sustainability:** PGIS initiatives often rely on external funding and support, which may not be sustainable in the long term. Ensuring the continuity of PGIS activities after external support ends remains a challenge.
- **Data integration:** Integrating community-generated data with official or authoritative data can be challenging due to variations in formats, standards, and protocols.
- **Scale and complexity:** PGIS can be resource-intensive, particularly in large-scale projects or complex issues that require extensive data collection and analysis, making scaling up challenging.
- **Resistance to change:** Communities and organizations may resist adopting new technologies or participatory methods. Overcoming this resistance and fostering trust in PGIS processes can take time.
- **Conflict and disagreements:** PGIS can reveal existing conflicts or disagreements within communities or between stakeholders. Managing and resolving these conflicts during the participatory process is a delicate task.

9.2.3 Future of PGIS

The future of Participatory Geographic Information Systems (PGIS) is marked by several promising developments and trends that are expected to influence its progression. Following is a brief overview of the potential directions for the future of PGIS:

- **Advanced technologies:** PGIS will benefit from mobile and cloud computing, enhancing data collection and sharing. Wearable devices and augmented reality may also support spatial data interaction.
- **Internet of Things (IoT):** Integrating IoT devices will enable real-time data collection for environmental, infrastructure, and public service monitoring.

- **Artificial intelligence (AI) and machine learning:** AI will enhance data analysis through automated image recognition, spatial pattern detection, and predictive modelling.
- **Data privacy and security:** Growing concerns will drive PGIS to adopt robust data protection measures and clear guidelines for handling sensitive information.
- **Blockchain technology:** Blockchain could ensure data integrity and transparency, particularly around ownership and usage rights.
- **Decentralized systems:** Decentralized data storage and peer-to-peer networks may reduce reliance on centralized platforms, giving communities more control.
- **Open data and interoperability:** Increased data sharing and standardized interoperability will allow easier integration of community data with official datasets.
- **Community empowerment:** PGIS will continue to empower communities by enabling participation in decision-making processes and recognizing local knowledge.
- **Sustainability and resilience:** PGIS will support sustainability efforts, especially in climate adaptation, disaster management, and resource management.
- **Ethical considerations:** Ethical guidelines will ensure PGIS respects cultural norms, indigenous knowledge, and the rights of marginalized communities.
- **Global collaboration:** PGIS will foster cross-border collaboration, allowing communities and organizations to share experiences and best practices.
- **Education and capacity building:** There will be a greater focus on educating communities and building their capacity to engage with PGIS through workshops and training programmes.

The future of PGIS is promising and dynamic, propelled by technological advancements, shifting societal needs, and an increasing acknowledgment of the importance of community engagement and local knowledge in tackling complex spatial challenges. As these trends evolve, PGIS is poised to become an even more essential tool for informed decision-making and sustainable development.

9.3 Methodologies of PGIS

Participatory Geographic Information Systems (PGIS) encompasses a range of methodologies and approaches aimed at engaging stakeholders and communities in the collection, management, and utilization of geographic information. These methods are designed to empower participants, integrate local knowledge, and foster collaboration in decision-making processes (Fig. 9.6). The following discussion explores the key methodologies used in PGIS:

- **Community mapping:** This core PGIS methodology involves engaging community members in map creation. Using tools such as paper maps, GPS devices, or digital mapping software, participants document local geography, landmarks, resources, and issues based on their knowledge.

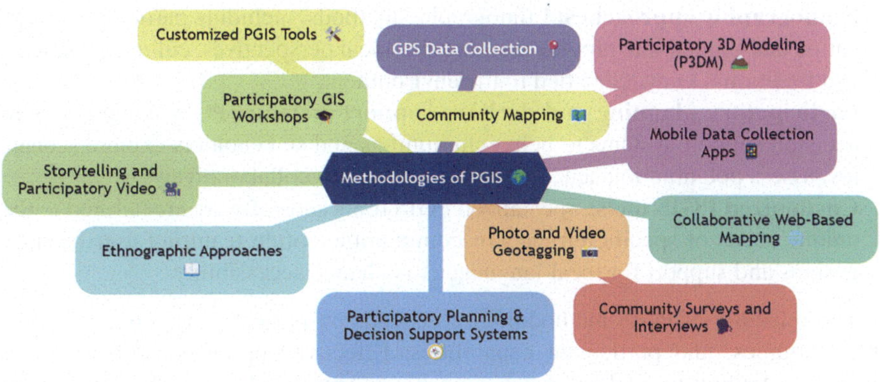

Fig. 9.6 Methodologies of participatory GIS (prepared by the authors)

- **Participatory GIS workshops:** Structured workshops facilitate discussions among community members, guided by facilitators, to share spatial knowledge, identify challenges, and collaboratively create maps or geographic representations. These sessions encourage dialogue and exchange of ideas.
- **GPS data collection:** Communities use GPS technology to record precise geographic coordinates of features or areas of interest. The collected data is then integrated into GIS databases for further analysis and decision-making.
- **Mobile data collection apps:** Mobile applications equipped with GPS capabilities allow community members to gather field data using smartphones or tablets. Data can be instantly transmitted to central databases for visualization and analysis.
- **Participatory 3D modelling (P3DM):** This method combines physical and digital techniques to create 3D models of landscapes, enabling communities to visualize and interact with their environment. It fosters a deeper understanding of spatial relationships and environmental dynamics.
- **Community surveys and interviews:** Surveys and interviews gather both qualitative and quantitative data from community members about land use, resource management, or spatial challenges. The data is georeferenced for mapping and analysis.
- **Photo and video geotagging:** Participants capture photographs or videos embedded with geographic coordinates to document local conditions, environmental changes, or concerns. These multimedia data enrich maps and facilitate visual storytelling.
- **Storytelling and participatory video:** Communities share their experiences and narratives about their environment through participatory video or storytelling exercises. This qualitative data complements the spatial information and offers context to maps.
- **Collaborative web-based mapping:** Platforms like OpenStreetMap and Ushahidi enable communities to collaboratively map their areas online. Users can update and refine maps while engaging with a wider audience.

- **Ethnographic approaches:** Ethnographic methods, including participant observation and immersion, provide insights into local perspectives, cultural practices, and traditional knowledge tied to the environment.
- **Participatory planning and decision support systems:** PGIS integrates with decision support systems to help communities and stakeholders explore scenarios, assess potential impacts, and prioritize actions collaboratively.
- **Customized PGIS tools:** Specialized PGIS tools and software are tailored to the unique needs of specific projects or communities, often featuring user-friendly designs and support for local languages to enhance accessibility.

The selection of PGIS methodologies depends on project goals, context, available resources, and participant capabilities. Effective approaches should foster inclusivity, build trust, and generate actionable geographic data to support informed decision-making and empower communities.

9.4 Traditional GIS Challenges and PGIS

Traditional Geographic Information Systems (GIS) and Participatory Geographic Information Systems (PGIS) each present unique challenges, shaped by their differing approaches to spatial data and information management. The following discussion highlights and compares these challenges:

Traditional GIS Challenges

- **Expertise and technical complexity:** Traditional GIS demands specialized knowledge and technical proficiency to operate effectively, often requiring skilled professionals for data management and analysis.
- **Cost:** The acquisition and maintenance of GIS software, hardware, and geospatial data can be prohibitively expensive, particularly for small organizations or communities with limited financial resources.
- **Data accessibility:** Access to high-quality or authoritative geographic data is often restricted, especially in developing regions, with licensing agreements further limiting availability.
- **Data quality:** Maintaining accuracy and consistency in geospatial data is challenging, as errors or outdated information can lead to flawed analyses and poor decision-making.
- **Data ownership and control:** Traditional GIS systems often centralize data ownership within government agencies or private organizations, raising concerns over equitable access and control.
- **Data privacy:** Sensitive geographic data present privacy challenges, necessitating robust measures to protect individual and community information during collection and analysis.
- **Community engagement:** Traditional GIS is sometimes criticized for its exclusivity, as it may not sufficiently incorporate local knowledge or actively engage communities in the decision-making process.

PGIS Challenges

- **Digital divide:** The reliance of PGIS on technology and Internet connectivity poses challenges for marginalized or remote communities that lack access to these resources, limiting participation.
- **Technical skills:** PGIS tools often require training and capacity-building to enable effective community involvement. Without addressing technical barriers, inclusivity and participation may be constrained.
- **Data quality and validation:** Community-generated data may vary in accuracy and reliability, necessitating rigorous validation and verification to ensure its usefulness for decision-making.
- **Power imbalance:** Unequal power dynamics within communities or between external organizations and local stakeholders can undermine meaningful participation, with dominant voices potentially sidelining marginalized groups.
- **Data ownership and control:** The complexities of determining ownership and usage rights for data generated through PGIS can lead to disputes and tensions over control and access.
- **Cultural and linguistic barriers:** Differences in language and cultural practices can hinder effective communication and collaboration, requiring cultural sensitivity and tailored approaches to foster inclusivity.
- **Conflict resolution:** PGIS processes can highlight pre-existing conflicts or disagreements among stakeholders, making conflict management and resolution integral to successful project outcomes.

Therefore, despite their respective challenges, traditional GIS and PGIS offer distinct strengths that are often complementary. Traditional GIS excels in advanced data analysis and management, while PGIS prioritizes community engagement and the integration of local knowledge. Leveraging the strengths of both approaches can result in more inclusive and holistic spatial information management and decision-making processes.

9.4.1 Participatory GIS: A Solution to Traditional GIS Challenges

Traditional Geographic Information Systems (GIS) have been indispensable for managing and analyzing spatial data. However, they face challenges related to accessibility, community engagement, and data quality. Participatory Geographic Information Systems (PGIS) provide an alternative approach, addressing many of these limitations by prioritizing collaboration, inclusivity, and the integration of local knowledge. The following discussion explores in detail how PGIS overcomes the challenges associated with traditional GIS:

a) **Accessibility**

- **Traditional GIS challenges:** GIS tools are often costly and require specialized skills, limiting access, particularly in developing regions. This creates a digital divide, restricting its use to a select few.
- **PGIS solutions:** PGIS uses accessible technology, such as user-friendly mobile apps and open-source platforms, reducing cost and skill barriers. This democratizes GIS tools, enabling wider community participation in spatial data management.

b) **Community engagement**

- **Traditional GIS challenges:** Traditional GIS often follows a top-down approach, sidelining local knowledge and perspectives, resulting in incomplete or biased data.
- **PGIS solutions:** PGIS emphasizes collaboration and empowers communities as active participants in data collection and decision-making. It integrates local knowledge, ensuring that data reflects community priorities and fosters meaningful dialogue among stakeholders.

c) **Data quality**

- **Traditional GIS challenges:** Maintaining data accuracy and reliability is a persistent issue, with errors or outdated information impacting decision-making.
- **PGIS solutions:** PGIS incorporates validation processes by cross-referencing community-generated data with authoritative sources. Local knowledge enhances data accuracy, and frequent updates ensure relevance and timeliness.

d) **Data ownership and control**

- **Traditional GIS challenges:** Data ownership is often centralized with government agencies or private entities, leaving communities disenfranchised.
- **PGIS solutions:** PGIS prioritizes community control over data collection, use, and sharing. It fosters transparency through clear agreements and open data principles, empowering communities to safeguard their data rights.

e) **Cultural and linguistic barriers**

- **Traditional GIS challenges:** Traditional GIS often overlooks cultural and linguistic diversity, leading to communication challenges.
- **PGIS solutions:** PGIS adopts culturally sensitive and linguistically inclusive approaches, using local languages and culturally appropriate methods. Cross-cultural facilitation further ensures meaningful community participation.

f) **Conflict resolution**

- **Traditional GIS challenges:** GIS may intensify conflicts by highlighting competing interests within or between communities.

- **PGIS solutions:** PGIS promotes open dialogue and consensus-building, involving all stakeholders in participatory processes. By addressing differences constructively, it facilitates conflict resolution and shared decision-making.

So, Participatory Geographic Information Systems (PGIS) mark a departure from the expert-driven model of traditional GIS, emphasizing accessibility, inclusivity, and integration of local knowledge. By empowering communities to participate in decision-making, PGIS enhances collaboration, data accuracy, and equity in spatial information management. With advancing technologies and methodologies, PGIS is set to play an increasingly vital role in tackling spatial challenges and advancing sustainable development.

9.5 Geospatial Citizenship, Voice Mapping and PGIS: Research Scope

Geospatial citizenship highlights the active participation of individuals and communities, especially marginalized groups, in creating and utilizing spatial information to influence decision-making. It promotes their right to shape geographic data impacting their lives, aligning closely with Participatory Geographic Information Systems (PGIS), which empowers communities to contribute directly to spatial data collection and management. On the other hand, the voice mapping, a PGIS approaches, enables communities to document their knowledge, experiences, and concerns through geographic representations. Using tools like photos, videos, and geotagged narratives, it amplifies under-represented voices in spatial decision-making processes.

The research scope of geospatial citizenship, voice mapping, and PGIS is broad, involving studies on the ethical implications of data ownership, the digital divide, power dynamics, and the integration of local knowledge in formal decision-making. Researchers will also explore how these approaches can enhance community resilience, address environmental and social justice issues, and improve governance through participatory processes. Moreover, advancements in mobile technology, data privacy concerns, and the potential for global collaborations provide a dynamic landscape for future research in these fields.

9.6 Conclusion and Endnotes

In conclusion, Geographic Information Systems (GIS) are essential tools for managing and analyzing spatial data across disciplines such as urban planning and environmental management. However, traditional GIS approaches face challenges, including limited accessibility, insufficient community engagement, data quality concerns, and issues related to data ownership and control. Participatory Geographic Information Systems (PGIS) have emerged as a transformative approach, addressing these limitations by prioritizing inclusivity, community empowerment, and the integration of local knowledge. PGIS enhances accessibility through user-friendly tools, fosters meaningful community engagement, improves data quality via validation and verification processes, ensures community control over data, overcomes cultural and linguistic barriers, and incorporates mechanisms for conflict resolution. By involving communities in spatial data collection, analysis, and application, PGIS supports more informed and equitable decision-making processes. Looking ahead, PGIS is poised to evolve further with advancements in technologies such as mobile and cloud computing, the Internet of Things (IoT), artificial intelligence (AI), and augmented reality (AR). These innovations will enhance the accessibility, functionality, and impact of PGIS, making it an increasingly vital tool for addressing complex spatial challenges and driving sustainable development.

Further Readings

1. Bullen, J., & Miles, A. (2024). Exploring local perspectives on flood risk: A participatory GIS approach for bridging the gap between modelled and perceived flood risk zones. *Applied Geography, 163*, 103176. https://doi.org/10.1016/j.apgeog.2023.103176
2. Burnett, C. M. (2023). *Evaluating participatory mapping software*. Springer . ISBN: 978-3-031-19593-8. https://doi.org/10.1007/978-3-031-19594-5
3. Craig, W. J., Harris, T. M., & Weiner, D. (2002). *Community participation and geographical information systems*. CRC Press, (Taylor and Francis Group) ISBN: 9780367578626.
4. de Carvalho, C. M., Luiz Giatti, L., Fagerholm, N., Bedran-Martins, A. M., & Kytta, M. (2021). Participatory Geographic Information Systems (PGIS) to assess water, energy and food availability in a vulnerable community in Guarulhos (Brazil). *International Journal of Urban Sustainable Development, 13*(3), 516–529. https://doi.org/10.1080/19463138.2021.2019041
5. Dunn, C. E. (2007). Participatory GIS—A people's GIS? *Progress in Human Geography, 31*(5), 616–637. https://doi.org/10.1177/0309132507081493
6. Fagerholm, N., Raymond, C. M., Olafsson, A. S., Brown, G., Rinne, T., Hasanzadeh, K., et al. (2021). A methodological framework for analysis of participatory mapping data in research, planning, and management. *International Journal of Geographical Information Science, 35*(9), 1848–1875. https://doi.org/10.1080/13658816.2020.1869747
7. Jankowski, P. (2009). Towards participatory geographic information systems for community-based environmental decision making. *Journal of Environmental Management, 90*(6), 1966–1971. https://doi.org/10.1016/j.jenvman.2007.08.028
8. Malakar, K. D., & Roy, S. (2024). Mapping geospatial citizenship: The power of participatory GIS. *Springer Cham*. ISBN: 978-3-031-63106-1. https://doi.org/10.1007/978-3-031-63107-8
9. Ndzabandzaba, C. (2020). Participatory geographic information system (PGIS): A discourse toward a solution to traditional GIS challenges. In S. Brunn & R. Kehrein (Eds.), *Handbook of the changing world language map*. Springer. https://doi.org/10.1007/978-3-030-02438-3_122
10. Norris, T. B. (2017). Public participation GIS, participatory GIS, and participatory mapping. *Oxford Bibliographies*. https://doi.org/10.1093/obo/9780199874002-0160

Part III
Fundamentals of Geospatial Technologies: Concepts and Techniques of GNSS and GPS

Chapter 10
Global Navigation Satellite Systems: An Introductory Guide

"The future of navigation is not in knowing where we are, but in knowing where we can go." — David C. Gee.

Abstract This chapter explores the evolution, functionality, and significance of Global Navigation Satellite Systems (GNSS) as a foundational pillar of modern technology. Moving beyond the foundational GPS, GNSS integrates multiple satellite constellations such as GLONASS, Galileo, and BeiDou, offering enhanced accuracy, redundancy, and global interoperability. It explores the components of GNSS, including satellite constellations, ground control systems, and user devices, highlighting their coordinated roles in delivering reliable positioning, navigation, and timing (PNT) services. The chapter analyses the advantages of GNSS over standalone GPS, focusing on improved signal resilience, precision, and multi-platform compatibility. It discusses GNSS applications across diverse sectors such as transportation, disaster management, agriculture, and environmental monitoring, emphasizing its contribution to technological innovation and societal advancement. Future developments in GNSS are also addressed, including the expansion of satellite constellations, integration with emerging technologies such as AI, IoT, and 5G, and advancements aimed at enhancing signal reliability and sustainability. The chapter concludes by underscoring GNSS's vital role in fostering global connectivity, supporting sustainable development, and addressing global challenges such as urbanization and climate change. It emphasizes the importance of continuous innovation and international collaboration to fully realize GNSS's potential in shaping a more connected, resilient, and efficient future. This chapter provides a comprehensive understanding of GNSS's transformative impact, offering insights into its historical evolution, current applications, and future prospects.

Keywords Global navigation satellite systems · Global positioning system · Satellite constellation · Functionality and interoperability of GNSS · GLONASS

K. D. Malakar et al., *Geospatial Technologies in Coastal Ecologies Monitoring and Management*, Advances in Geographic Information Science,
https://doi.org/10.1007/978-3-031-92017-2_10

Global Navigation Satellite Systems (GNSS) have become indispensable in modern technology, shaping global connectivity and innovation. This chapter delves into the evolution of GNSS beyond GPS, encompassing systems like GLONASS, Galileo, and BeiDou, and highlights their enhanced accuracy, interoperability, and redundancy. It explores the components and applications of GNSS across diverse sectors, emphasizing its transformative role in societal advancement. The discussion extends to future developments, including satellite expansions and integration with AI, IoT, and 5G, underscoring GNSS's potential in addressing global challenges and fostering sustainable development in an increasingly connected world.

Key Points of the Chapter

- Transition from GPS to multi-constellation systems.
- Integration of satellite constellations and user devices.
- Enhanced precision and multi-platform compatibility.
- Applications in diverse sectors like agriculture and disaster management.
- Future integration with AI, IoT, and 5G technologies.

10.1 Introduction to Global Navigation Satellite Systems (GNSS)

10.1.1 GNSS as an Umbrella Term

The term Global Navigation Satellite Systems (GNSS) refers to a comprehensive category of satellite-based systems engineered to deliver geospatial positioning, navigation, and timing data worldwide. This encompasses both fully operational global systems and regional augmentation systems, which support a diverse array of civilian and military applications. GNSS plays a pivotal role in various sectors, including transportation, agriculture, defence, disaster management, and geospatial sciences, highlighting its extensive relevance and integral role in contemporary technological advancements.

Core Global GNSS Systems

At its essence, GNSS comprises a network of satellite constellations managed by various nations or alliances, which collectively provide autonomous global positioning capabilities:

- **GPS (Global Positioning System):** Originating from the United States, GPS was the first fully operational GNSS, consisting of a constellation of Earth-orbiting satellites that offer location, velocity, and time data.
- **GLONASS (Global Navigation Satellite System):** Russia's counterpart to GPS, GLONASS provides similar functionalities and is commonly used alongside GPS to enhance positional accuracy.

- **Galileo:** Developed by the European Union, Galileo aims to establish an independent global navigation system with superior precision, particularly for civilian applications.
- **BeiDou Navigation Satellite System (BDS):** Managed by China, BeiDou has quickly become a major player in GNSS, offering both regional and global coverage.

Regional Augmentation Systems Within GNSS

GNSS also encompasses regional satellite systems that offer complementary or supplementary services to global constellations. Although these systems do not provide independent global coverage, they significantly enhance positioning accuracy and reliability in specific regions. Notable examples include:

- **QZSS (Quasi-Zenith Satellite System):** A Japanese system designed to augment GNSS signals in the Asia-Pacific region, particularly in urban environments where signal obstructions are common.
- **NavIC (Navigation with Indian Constellation):** India's regional navigation system, which delivers precise positioning across the Indian subcontinent and its surrounding areas.
- **SBAS (Satellite-Based Augmentation Systems):** Systems such as WAAS (U.S.), EGNOS (Europe), and GAGAN (India) enhance global GNSS performance by correcting errors and improving overall accuracy.

GNSS as a Framework

GNSS functions as an integrated framework that combines various systems and technological layers. It consists of the following components:

- **Space Segment:** The satellite constellations located in medium Earth orbit that broadcast positioning signals.
- **Control Segment:** Ground-based control stations responsible for monitoring satellite performance, updating satellite positions, and ensuring the reliability of signals.
- **User Segment:** GNSS receivers embedded in devices ranging from smartphones and vehicles to sophisticated surveying instruments and autonomous systems.

Applications of GNSS

The comprehensive nature of GNSS renders it essential across a wide range of domains:

- **Transportation:** GNSS facilitates navigation and route optimization across road, rail, aviation, and maritime sectors. Autonomous vehicles, in particular, depend on multi-GNSS data for accurate location tracking and path planning.
- **Agriculture:** Precision farming techniques leverage GNSS for tasks such as soil mapping, automated machinery guidance, and yield monitoring, thereby enhancing productivity.
- **Disaster Management:** GNSS plays a critical role in disaster response and recovery by enabling real-time tracking, mapping of affected areas, and supporting search-and-rescue operations.

- **Surveying and Mapping:** Geospatial professionals utilize GNSS to obtain high-precision location data for map creation and land surveying.
- **Telecommunications and Finance:** GNSS provides precise timing, which is crucial for synchronizing communication networks and ensuring the accuracy of financial transactions.

Challenges and Advancements Under the GNSS Umbrella
Although GNSS systems are highly effective, they are not without their challenges:

- **Signal Interference and Multipath:** Physical obstructions such as buildings, trees, and other structures can interfere with signal quality, leading to reduced accuracy.
- **Vulnerability to Jamming and Spoofing:** GNSS signals, being relatively weak, are susceptible to intentional disruption through jamming or spoofing techniques.
- **Dependency on Infrastructure:** GNSS systems depend on the reliable operation of satellites and ground stations, which are expensive and complex to maintain.

To mitigate these challenges, continuous advancements are improving the capabilities of GNSS systems:

- **Multi-GNSS Receivers:** The integration of signals from multiple systems enhances both accuracy and reliability.
- **Augmented GNSS Systems:** Satellite-Based Augmentation Systems (SBAS) and Ground-Based Augmentation Systems (GBAS) provide real-time error corrections to improve performance.
- **Integration with Emerging Technologies:** GNSS is progressively being combined with artificial intelligence, the Internet of Things (IoT), and 5G networks, facilitating the development of innovative applications.

As a comprehensive term, GNSS encapsulates a dynamic and evolving framework that integrates both global and regional navigation systems to fulfill a wide array of geospatial requirements. Its capacity to combine various technologies and methodologies guarantees its continued prominence in positioning, navigation, and timing solutions. By overcoming existing challenges and embracing emerging advancements, GNSS maintains a pivotal role in fostering innovation and promoting sustainable development worldwide.

10.1.2 GPS as a GNSS

The Global Positioning System (GPS) is one of the most widely recognised and extensively utilised Global Navigation Satellite Systems (GNSS). Developed and maintained by the United States Department of Defence, GPS is a satellite-based navigation system that delivers accurate positioning, navigation, and timing (PNT) information on a global scale. Initially designed for military purposes, GPS has evolved into a fundamental component of modern technology, supporting a wide range of civilian and commercial applications (Fig. 10.1).

Features of GPS

- **Satellite Constellation:** GPS operates with a constellation of at least 24 satellites orbiting at an altitude of approximately 20,200 km. These satellites are arranged in six orbital planes to ensure comprehensive global coverage.
- **Positioning Capability:** GPS delivers real-time location data by triangulating signals received from multiple satellites. By utilizing signals from a minimum of four satellites, a GPS receiver can accurately determine the user's position (latitude, longitude, altitude) and time.
- **Control Segment:** A global network of ground control stations oversees the operation of GPS satellites. These stations monitor the health of the satellites, update their orbital parameters, and ensure precise timing through the use of atomic clocks.
- **Civilian Accessibility:** Since the early 2000s, GPS signals have been made freely accessible for civilian use, leading to widespread adoption in fields such as navigation, geospatial applications, and timing services.
- **Accuracy and Reliability:** GPS provides positioning accuracy within a few meters for civilian applications. Additionally, augmented services like the U.S. Wide Area Augmentation System (WAAS) further enhance accuracy, achieving sub-meter precision.

GPS in the Context of GNSS

As a GNSS, GPS is one of several systems that offer global navigation services. Although GPS operates independently, it is often used in conjunction with other GNSS constellations, such as GLONASS, Galileo, and BeiDou. The integration of multiple GNSS systems improves signal availability, accuracy, and reliability, particularly in challenging environments like urban canyons or dense forests.

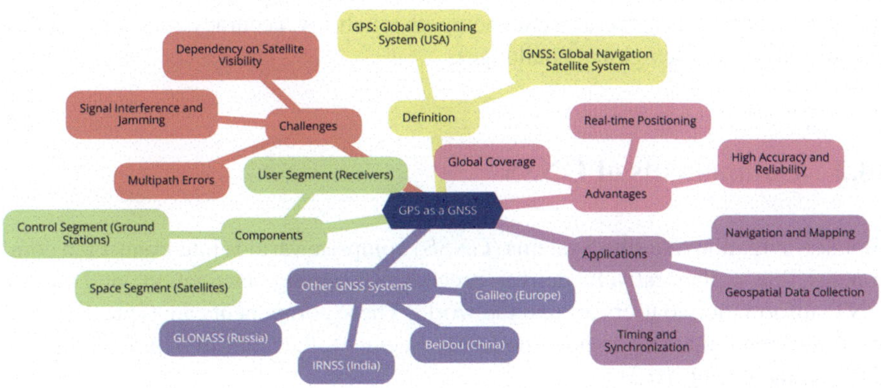

Fig. 10.1 GPS as a GNSS (Prepared by the authors)

Applications of GPS

- **Transportation:** GPS is widely employed in road navigation, aviation, maritime, and rail systems for tasks such as route planning, vehicle tracking, and traffic management.
- **Surveying and Mapping:** Geospatial professionals depend on GPS for precise land surveys, creating digital maps, and updating geospatial databases.
- **Agriculture:** Precision farming utilizes GPS for tasks like field mapping, soil sampling, and guiding automated machinery, enhancing efficiency and crop yields.
- **Disaster Management:** GPS supports search-and-rescue operations, mapping disaster-stricken areas, and coordinating emergency responses.
- **Telecommunications and Timing:** GPS provides accurate timing for synchronizing telecommunications networks, power grids, and financial systems.

Advantages of GPS

- Global Coverage: Offers continuous, reliable coverage across the globe.
- Free Accessibility: Available for civilian use without subscription or licensing fees.
- High Accuracy: Delivers dependable positioning and timing data.

Challenges of GPS

- Signal Vulnerability: GPS signals are weak and can be obstructed or degraded by buildings, trees, or adverse atmospheric conditions.
- Jamming and Spoofing: GPS is susceptible to interference and intentional disruption through jamming or spoofing.
- Dependency: Excessive reliance on GPS may create vulnerabilities in critical systems if disruptions occur.

GPS exemplifies the foundational role of GNSS technologies in contemporary society. Its capacity to deliver precise, global positioning and timing data has revolutionized industries and daily activities. As an integral component of the broader GNSS framework, GPS continues to evolve by integrating with other systems and emerging technologies, thereby enhancing its reliability, accuracy, and applicability in addressing complex global challenges.

10.2 Components of GNSS

Global Navigation Satellite Systems (GNSS) comprise various interconnected components that collaboratively deliver precise positioning, navigation, and timing (PNT) information to users around the world. These components are typically organized into three main segments: the Space Segment, the Control Segment, and the User Segment (Fig. 10.2).

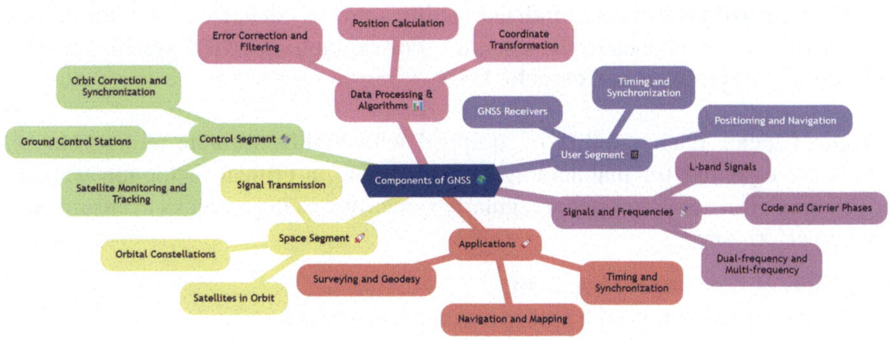

Fig. 10.2 Components of GNSS (Prepared by the authors)

a) **Space Segment**

The space segment encompasses the satellites that orbit the Earth and transmit signals to ground-based receivers.

Satellite Constellation

- GNSS systems typically rely on a constellation of satellites to ensure global coverage. For example:
- GPS: At least 24 satellites in medium Earth orbit (MEO).
- GLONASS: 24 operational satellites in MEO.
- Galileo: 24 operational satellites, along with spare satellites.
- BeiDou: Over 30 satellites in geostationary, inclined geosynchronous, and MEO orbits.

Satellite Orbits

Satellites are positioned in various orbits depending on the GNSS. Most GNSS satellites are situated in medium Earth orbit (MEO), at altitudes ranging from 20,000 to 24,000 km, ensuring extensive coverage of Earth's surface.

Signal Transmission

Satellites transmit signals that contain information about their position, time, and status. These signals are used by receivers to calculate distances and determine the user's location through trilateration.

b) **Control Segment**

The control segment is responsible for managing the operation of the satellite constellation and ensuring the accuracy of the signals transmitted by the satellites.

Ground Control Stations GNSS systems utilize a network of ground stations that monitor and manage the satellites. These stations track satellite orbits, assess signal health, and update satellite position and timing data.

Master Control Centers Centralized facilities process data from monitoring stations and issue commands to the satellites. For example, the GPS system's master control center is located in Colorado, USA.

Atomic Clocks Each satellite is equipped with highly accurate atomic clocks, which are essential for maintaining precise timing, a critical factor for accurate positioning. The control segment regularly synchronizes these clocks to ensure consistent performance.

Functions of the Control Segment

- Monitoring satellite health and operational performance
- Updating satellite ephemeris (orbital data)
- Ensuring signal accuracy and integrity

c) **User Segment**

The user segment encompasses GNSS receivers and the end users who rely on the system for a variety of applications.

GNSS Receivers

These are devices equipped with antennas and processors that receive and decode satellite signals to determine position, velocity, and time. GNSS receivers are found in a wide range of devices, from handheld GPS units to integrated systems in smartphones, vehicles, and aircraft.

User Applications

- GNSS is extensively applied across various sectors, including:
- Transportation: Navigation systems for aviation, maritime, and road vehicles.
- Surveying and Mapping: Geospatial data collection and land surveying activities.
- Agriculture: Precision farming techniques for optimized resource management.
- Disaster Management: Emergency response coordination and disaster area mapping.
- Telecommunications: Network and system synchronization for reliable communication.

The components of GNSS, space, control, and user segments, function in unison to provide accurate and reliable positioning and navigation services. Each segment is essential, from the transmission of satellite signals to ground-based control and user-level applications, collectively ensuring that GNSS continues to serve as a fundamental element of modern technological infrastructure.

10.3 Functionality and Interoperability

Global Navigation Satellite Systems (GNSS) deliver precise positioning, navigation, and timing (PNT) information by utilizing satellite constellations, ground control networks, and user devices. The effectiveness of GNSS stems from its capacity to support a wide range of applications across sectors such as transportation, agriculture, disaster management, and military operations. Additionally, interoperability facilitates the seamless integration of multiple GNSS constellations, enhancing performance and ensuring global coverage (Fig. 10.3).

Functionality of GNSS

Positioning

- GNSS receivers determine the user's position by processing signals from at least four satellites.
- Applications: Location-based services (LBS), surveying, and geospatial mapping.

Navigation

- GNSS provides real-time guidance for navigation between locations.
- Applications: Automotive navigation systems, maritime route planning, and aviation flight paths.

Timing Synchronization

- GNSS offers highly accurate timing signals, sourced from atomic clocks on satellites.
- Applications: Synchronization of telecommunication networks, power grids, and financial transactions.

Monitoring and Tracking

- GNSS enables real-time tracking of assets, vehicles, and individuals.
- Applications: Fleet management, supply chain logistics, and emergency response systems.

Environmental and Scientific Applications

- GNSS is utilized in atmospheric studies, tectonic plate movement analysis, and climate monitoring.
- Applications: Earthquake prediction, meteorology, and space exploration.

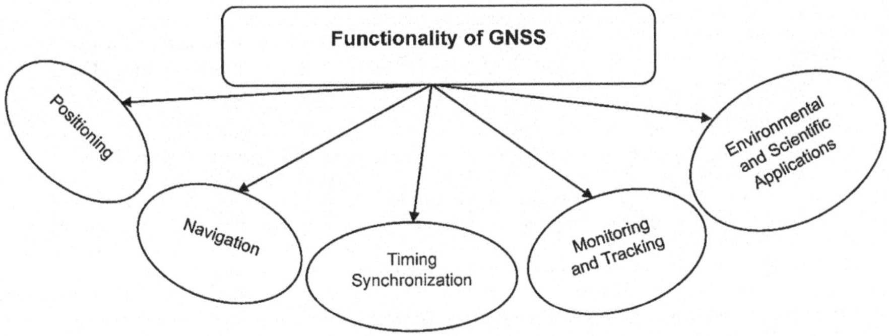

Fig. 10.3 Functionality of GNSS (Prepared by the authors)

Interoperability of GNSS

Interoperability denotes the capacity of multiple GNSS systems to function cohesively, thereby enhancing their collective performance. Key GNSS constellations, such as GPS (USA), GLONASS (Russia), Galileo (EU), and BeiDou (China), are designed to be interoperable, optimizing their availability, accuracy, and reliability (Fig. 10.4).

Signal Compatibility

- GNSS systems transmit signals within overlapping frequency bands to enable receivers to process signals from multiple constellations.
- Example: The L1/E1 band used by GPS and Galileo.

Global Coverage

- The integration of multiple GNSS constellations enhances the number of visible satellites, improving positioning accuracy in challenging environments such as urban canyons or dense forests.

Improved Accuracy

- By combining data from several systems, users can minimize errors caused by atmospheric interference, satellite geometry, or clock inaccuracies.

Redundancy and Resilience

- Interoperability ensures continued service in the event of outages or disruptions in one GNSS system.
- Example: If GPS signals are unavailable, receivers can switch to GLONASS, Galileo, or BeiDou signals.

Augmentation Systems Integration

- Satellite-based augmentation systems (e.g., WAAS, EGNOS) and ground-based systems (e.g., DGNSS) enhance interoperability by providing corrections applicable to multiple GNSS systems.

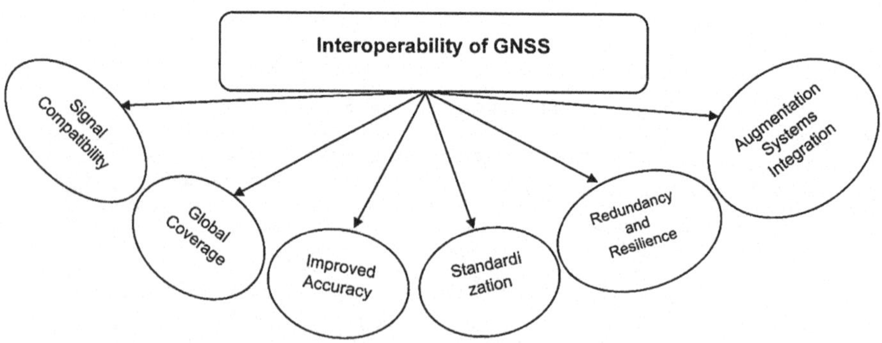

Fig. 10.4 Interoperability of GNSS (Prepared by the authors)

Standardization

- Collaboration among international GNSS providers through organizations like the International Committee on GNSS (ICG) ensures the harmonization of standards and protocols.

The functionality of GNSS is rooted in its capacity to deliver precise positioning, navigation, and timing services, with interoperability further enhancing these capabilities through the integration of multiple constellations. Together, they create a robust system that underpins a wide range of applications worldwide, providing accurate, reliable, and resilient solutions to complex spatial challenges. As GNSS technologies advance, interoperability will remain fundamental to expanding their potential and ensuring global accessibility.

10.4 Advantages of GNSS over GPS

Although GPS (Global Positioning System) is the most widely recognized satellite navigation system, GNSS (Global Navigation Satellite Systems) includes several satellite constellations, such as GPS, GLONASS, Galileo, and BeiDou. GNSS provides notable advantages over relying exclusively on GPS, including broader coverage, enhanced accuracy, and improved reliability. The following are the key benefits of GNSS compared to GPS:

Enhanced Global Coverage

- GPS may encounter difficulties providing accurate positioning in specific regions, such as high-latitude areas or urban canyons.
- By integrating multiple systems, GNSS offers a greater number of satellites in view, ensuring comprehensive global coverage, even in challenging environments like dense forests, mountainous regions, or polar zones.

Increased Satellite Availability

- GPS relies on a constellation of 31 satellites, which may not always yield optimal satellite geometry, particularly in obstructed areas.
- GNSS, by leveraging multiple constellations, significantly increases satellite availability, leading to better satellite geometry, thereby enhancing accuracy and reliability.

Higher Accuracy

- GPS accuracy is constrained by factors like signal multipath, satellite geometry, and atmospheric interference.
- GNSS reduces positioning errors by combining data from multiple constellations. For instance, dual-frequency or multi-constellation GNSS receivers can minimize ionospheric errors, offering centimetre-level accuracy in certain cases.

Improved Reliability and Redundancy

- Exclusive reliance on GPS exposes systems to risks from failures, signal jamming, or satellite outages.
- GNSS ensures redundancy by enabling receivers to switch between systems (e.g., GPS, GLONASS, Galileo, or BeiDou), improving reliability, particularly in mission-critical applications such as aviation, defence, and disaster management.

Resilience in Challenging Environments

- GPS performance may degrade in environments with limited satellite visibility, such as urban canyons, tunnels, or dense forests.
- GNSS mitigates this limitation by providing access to a larger number of satellites, ensuring uninterrupted positioning and navigation even in such obstructed environments.

Better Performance for High-Dynamic Applications

- GNSS offers superior tracking for high-speed or high-altitude applications (e.g., aviation, space exploration, and military operations) due to its multi-constellation framework.
- This advantage is particularly crucial for applications requiring real-time, highly accurate positioning.

Integration with Augmentation Systems

- GPS is compatible with augmentation systems like WAAS (Wide Area Augmentation System) or EGNOS (European Geostationary Navigation Overlay Service).
- GNSS, as a multi-system approach, benefits from a broader range of augmentation systems, enhancing positioning accuracy and reliability for a variety of global applications.

Adaptability to Regional Needs

- While GPS is a global system, it does not address specific regional needs.
- GNSS systems, such as Galileo (Europe), BeiDou (China), and QZSS (Japan), are tailored to regional applications, offering specialized services like high-precision signals and disaster management capabilities for their respective regions.

Greater Application Versatility

- GPS is primarily used for general navigation and geospatial data collection.
- GNSS is better suited for precision-demanding applications such as geodetic surveying, precision agriculture, autonomous vehicles, and environmental monitoring, due to its multi-constellation integration and enhanced accuracy.

Future-Proof Technology

- While GPS continues to advance, it is limited by the upgrades within its own system.
- GNSS, being a combination of multiple systems, benefits from continuous technological advancements across various constellations, ensuring adaptability to evolving needs and applications.

The key advantage of GNSS over GPS is its integration of multiple satellite systems, offering enhanced accuracy, coverage, reliability, and resilience. As demand for precise positioning increases in sectors like autonomous vehicles, aviation, and disaster management, GNSS's multi-constellation approach provides a more versatile and future-proof solution than GPS alone.

10.5 Applications Leveraging GNSS

Global Navigation Satellite Systems (GNSS) play a crucial role across various industries by providing accurate positioning, navigation, and timing services. Prominent applications include (Fig. 10.5):

- **Transportation and Navigation:** GNSS is utilized in aviation, maritime, and road navigation for route optimization, real-time tracking, and the operation of autonomous vehicles.
- **Agriculture:** GNSS enables precision farming through applications such as automated machinery guidance, crop monitoring, and yield mapping, thus boosting productivity.
- **Disaster Management:** GNSS supports early warning systems, coordinates emergency response efforts, and aids in post-disaster damage assessment.
- **Surveying and Mapping:** GNSS provides high-precision geospatial data for infrastructure development, urban planning, and environmental monitoring.
- **Telecommunications and Timing:** GNSS ensures accurate time synchronization across networks, financial transactions, and power grids.

GNSS's adaptability fosters innovation in various sectors, enhancing operational efficiency and contributing to sustainable development goals.

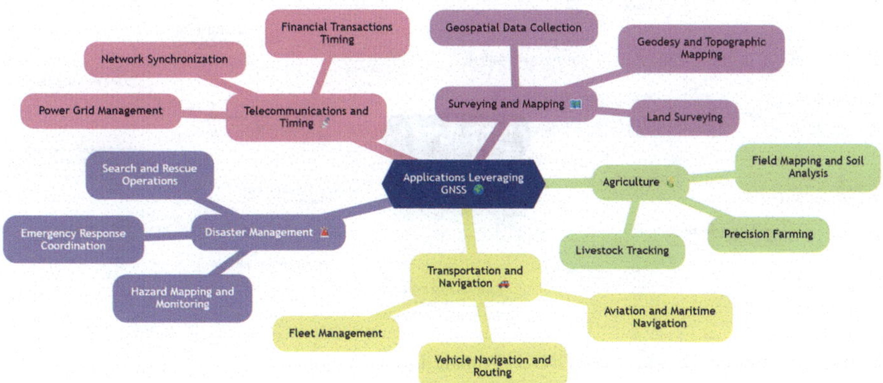

Fig. 10.5 Applications leveraging GNSS (Prepared by the authors)

10.6 Future Developments

Global Navigation Satellite Systems (GNSS) are progressively advancing to address the increasing demand for more precise positioning, navigation, and timing (PNT) services. The primary focus of upcoming developments includes (Fig. 10.6):

- **Expanded Satellite Constellations:** Growth of existing systems (e.g., GPS, GLONASS, Galileo, BeiDou) and establishment of regional systems to enhance global coverage, redundancy, and accuracy.
- **Improved Accuracy and Precision:** Integration of next-generation atomic clocks, multi-frequency signals, and advanced error correction algorithms to achieve centimetres-level accuracy, particularly for critical applications like autonomous vehicles.
- **Enhanced Interoperability and Integration:** Strengthened interoperability among GNSS systems to ensure seamless global coverage and improved performance, facilitating reliable multi-GNSS solutions.
- **Resilience and Robustness:** Development of anti-jamming, anti-spoofing, and signal authentication technologies to address vulnerabilities and secure GNSS operations.
- **Integration with Emerging Technologies:** Collaboration with Artificial Intelligence (AI), the Internet of Things (IoT), and 5G networks to enable real-time, intelligent positioning solutions across diverse industries.
- **Space Applications:** Expanded use of GNSS in deep-space missions, satellite navigation beyond Earth's orbit, and precise positioning for space exploration and research.
- **Sustainability and Energy Efficiency:** Design of energy-efficient satellites and ground stations to reduce the environmental impact of GNSS infrastructure.

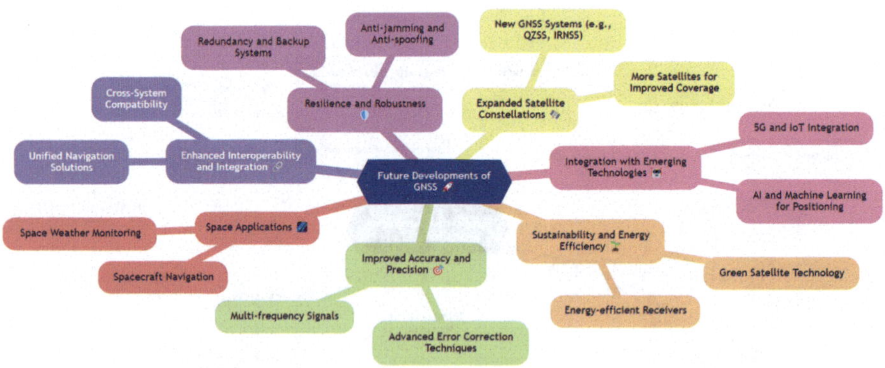

Fig. 10.6 Future developments of GNSS (Prepared by the authors)

10.7 Conclusion and Endnotes

Global Navigation Satellite Systems (GNSS) have become a critical technological foundation for modern society, providing precise positioning, navigation, and timing (PNT) services across a wide range of sectors. From the pioneering GPS to multi-constellation systems like GLONASS, Galileo, and BeiDou, GNSS has evolved into a comprehensive and dependable framework that underpins global connectivity. Its core components, satellite constellations, ground control networks, and user equipment, function cohesively to deliver exceptional accuracy and reliability. The integration of multiple constellations has elevated GNSS beyond standalone GPS, offering enhanced accuracy, redundancy, and global interoperability. This has expanded its applications in transportation, agriculture, defence, disaster management, and beyond. Looking ahead, GNSS advancements, such as the expansion of satellite constellations, improved signal resilience, and integration with technologies like AI, IoT, and 5G, are set to further enhance its capabilities. These innovations will also address global challenges, including urbanization, climate change, and sustainable development. As GNSS continues to evolve, it remains at the forefront of technological progress, enabling a more connected, resilient, and efficient future. Its ongoing development ensures its role as an indispensable tool for advancing industries, tackling global issues, and improving the quality of life worldwide.

Endnotes

- **GNSS as an Integrated System:** GNSS comprises multiple global and regional satellite navigation systems, offering enhanced redundancy and precision beyond the limitations of standalone GPS.
- **Technological Integration:** The convergence of GNSS with advanced technologies like AI and IoT is revolutionizing industries by enabling intelligent, real-time solutions.
- **Broad Applications and Societal Benefits:** GNSS applications extend well beyond navigation, playing a vital role in agriculture, disaster response, smart city development, and environmental monitoring.
- **Global Significance:** By delivering precise and secure positioning worldwide, GNSS fosters international collaboration and drives progress toward sustainable development goals.
- **Advancing Future Capabilities:** The future of GNSS is shaped by expanding satellite constellations, improved signal resilience, and a focus on sustainability to meet emerging global demands.

Further Readings

1. Banerjee, P., & Matsakis, D. (2023). Global navigation satellite systems (GNSS). In *An introduction to modern timekeeping and time transfer* (Springer series in measurement science and technology). Springer. https://doi.org/10.1007/978-3-031-30780-5_8
2. Bhatta, B. (2021). *Global navigation satellite systems: New technologies and applications* (2nd ed.). CRC Press, (Taylor and Francis Group) ISBN: 9780367709723.
3. Bonnor, N. (2011). A brief history of global navigation satellite systems. *The Journal of Navigation, 65*(1), 1–14. https://doi.org/10.1017/S0373463311000506
4. Enríquez-Caldera, R. (2016). Global navigation satellite systems: Orbital parameters, time and space reference systems, and signal structures. In J. Pelton, S. Madry, & S. Camacho-Lara (Eds.), *Handbook of satellite applications*. Springer. https://doi.org/10.1007/978-1-4614-6423-5_93-3
5. Grapenthin, R. (2024). The global navigation satellite system (GNSS): positioning, velocities, and reflections. In E. Chaussard, C. Jones, J. A. Chen, & A. Donnellan (Eds.), *Remote sensing for characterization of geohazards and natural resources* (Springer remote sensing/ photogrammetry). Springer. https://doi.org/10.1007/978-3-031-59306-2_2
6. Hofmann-Wellenhof, B., Lichtenegger, H., & Wasle, E. (2008). *GNSS—Global navigation satellite systems: GPS, GLONASS, Galileo, and more*. Springer Vienna. ISBN 978-3-211-73012-6. https://doi.org/10.1007/978-3-211-73017-1
7. Madry, S. (2024). Global navigation satellite systems and their applications. Springer Cham, (2nd ed., p. 978-3-031-74487-7). ISBN. https://doi.org/10.1007/978-3-031-74488-4
8. Sholarin, E. A., & Awange, J. L. (2015). Global navigation satellite system (GNSS). In *Environmental project management. Environmental science and engineering*. Springer. https://doi.org/10.1007/978-3-319-27651-9_9
9. Teunissen, P. J. G., & Montenbruck, O. (Eds.). (2017). *Springer handbook of global navigation satellite systems*. Springer. https://doi.org/10.1007/978-3-319-42928-1
10. Walker, J., & Awange, J. (2020). Global navigation satellite system. In *Surveying for civil and mine engineers*. Springer. https://doi.org/10.1007/978-3-030-45803-4_14

Chapter 11
GPS 101: Introduction to Global Positioning System

"The value of location is not just in the place, but in knowing exactly where you stand." — Michael Goodchild.

Abstract This chapter explored the fundamentals of Global Positioning System (GPS) technology, including its historical development, the evolution of navigation techniques, its operational mechanisms, GPS chronology, satellite systems, and functional segments. Over time, GPS has become an indispensable tool for navigation, surveying, tracking, and numerous other applications. By delving into the principles and concepts underlying this pervasive technology, readers gain valuable insights into its foundational ideas and transformative impact. The chapter examined the operation and components of the GPS system, comprising the space, control, and user segments. The space segment includes a constellation of Earth-orbiting satellites, while the control segment consists of ground-based stations that monitor and manage satellite orbits. The user segment involves GPS receivers used by individuals or devices to process satellite signals for positioning and navigation. The GPS system operates on the principle of trilateration, where a receiver determines its precise location by measuring distances to multiple satellites. By receiving signals from at least four satellites and calculating the time taken for these signals to travel, the receiver determines distances using the speed of light. Accurate timekeeping is essential for GPS functionality, as satellite transmissions provide precise timing data to synchronize with receivers. The chapter also addressed factors affecting GPS accuracy, including signal blockages, atmospheric interference, and satellite geometry. The chapter concluded by examining the diverse applications of GPS technology. From personal navigation and vehicle tracking to aviation, agriculture, and emergency services, GPS has transformed how people navigate and interact with the world. It has become an essential part of daily life, facilitating efficient transportation, outdoor activities, geolocation services, and precise timekeeping for financial and telecommunications systems.

K. D. Malakar et al., *Geospatial Technologies in Coastal Ecologies Monitoring and Management*, Advances in Geographic Information Science,
https://doi.org/10.1007/978-3-031-92017-2_11

Keywords Global Positioning System · GPS segments · Error of GPS · Navigation and Satellite · Differential GPS (DGPS)

This chapter provides an in-depth exploration of the Global Positioning System (GPS) technology, tracing its historical development and the evolution of navigation techniques. It examines GPS's operational principles, including trilateration, satellite systems, and the interdependent space, control, and user segments. By highlighting the accuracy and timekeeping mechanisms vital to its functionality, the chapter addresses factors influencing performance, such as atmospheric interference and signal blockages. Concluding with diverse applications across navigation, agriculture, aviation, and telecommunications, the chapter underscores GPS's transformative impact as an indispensable tool in modern life.

Key Points of the Chapter

- Concept and operational framework of GPS technology.
- Various dimensions of errors and accuracy in GPS technology.
- Understanding the applications and scope of GPS in modern life.

11.1 Introduction to Global Positioning System (GPS)

The Global Positioning System (GPS) is a groundbreaking technology that has become an indispensable part of modern life. Fundamentally, GPS is a satellite-based navigation system designed to provide precise location information anywhere on Earth. The system operates by triangulating signals from a constellation of 24–32 satellites orbiting the planet. These satellites transmit data containing their position and the exact time the signal was sent. A GPS receiver on the ground determines its location by analysing signals from at least four satellites, accounting for the time each signal takes to reach the receiver. This process enables the calculation of latitude, longitude, altitude, and speed, offering users highly accurate positioning and navigation capabilities.

The applications of GPS are vast and diverse, ranging from assisting drivers on daily commutes to supporting military operations, conducting geological surveys, and tracking wildlife. A clear understanding of the fundamentals of GPS is essential for maximizing its potential, whether for personal or professional purposes. This discussion will explore the components of GPS, its accuracy, and its pivotal role in contemporary society, highlighting how it has revolutionized navigation and our interaction with the environment.

11.1.1 Concept of GPS

The Global Positioning System (GPS) is a satellite-based navigation technology that enables precise location determination anywhere on Earth. It operates through a network of satellites that transmit signals containing their position and the exact time the signals were sent. GPS receivers calculate their position by analysing signals from at least four satellites, using the time delays to determine latitude, longitude, altitude, and velocity.

This triangulation and precise timing form the core of GPS technology, which has diverse applications, including navigation, land surveying, search and rescue, and military operations. GPS has revolutionized modern life, becoming essential for navigation, logistics, and numerous other fields. Understanding its fundamental principles is key to leveraging its full potential and recognizing its impact on our interconnected world (Fig. 11.1).

11.1.2 History of Navigation

Navigation, the discipline of guiding one's journey across land and sea, encompasses a rich and dynamic history spanning millennia. Early navigation methods were basic, relying on natural cues such as the position of the sun, stars, and coastlines. The following provides an overview of the evolution of navigation:

- **Early Human Navigation:** Early humans, as nomadic travellers, relied on local terrain knowledge and natural cues like the sun and stars. Coastal and river navigation emerged as the first systematic travel methods.
- **Celestial Navigation:** Ancient civilizations, such as the Egyptians, Greeks, and Polynesians, developed techniques using celestial bodies, including the sun, moon, and stars. Instruments like the astrolabe, quadrant, and sextant were later introduced to improve accuracy.

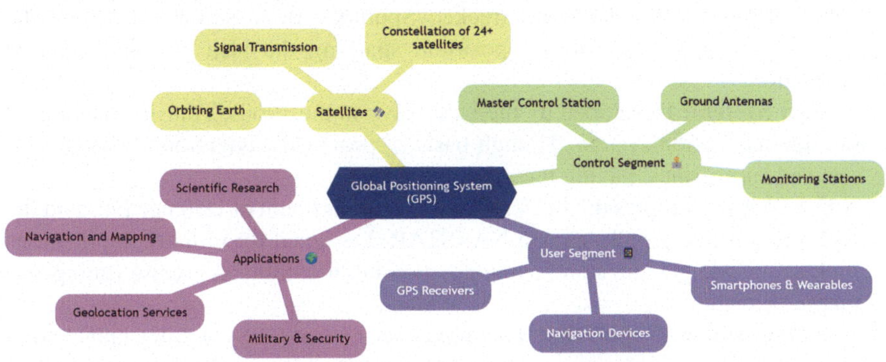

Fig. 11.1 Concept of GPS (Prepared by the authors)

- **Compass Invention:** The Chinese invented the magnetic compass in the twelfth century, revolutionizing navigation by enabling direction-finding even in poor visibility.
- **Age of Exploration:** During the fifteenth and sixteenth centuries, explorers like Columbus and Magellan utilized advanced tools like the compass and astrolabe to discover new lands and navigate the globe.
- **Longitude Problem:** Longitude determination remained a challenge until John Harrison's marine chronometer in the eighteenth century provided a precise method, transforming oceanic navigation.
- **Modern Navigation:** The twentieth century introduced electronic navigation systems, including LORAN and GPS, with GPS providing highly accurate, satellite-based positioning.
- **Aviation Navigation:** The rise of aviation in the twentieth century led to radio-based systems such as VOR (VHF Omnidirectional Range) and ILS (Instrument Landing System) for precise air navigation.
- **Maritime Navigation:** Modern maritime navigation integrates GPS, radar, and electronic charts, significantly enhancing accuracy and safety while facilitating detailed oceanic mapping.

Navigation today continues to evolve through the integration of satellite and digital technologies. Systems such as GPS and other satellite-based platforms have become indispensable for global positioning, making navigation accessible to a wide range of users, including hikers, drivers, and pilots. As technology progresses, navigation remains a cornerstone of our increasingly interconnected world.

11.1.3 History of GPS

The history of the Global Positioning System (GPS) is both intriguing and rooted in developments from the mid-twentieth century.

- **Sputnik's Influence:** The launch of the Soviet satellite Sputnik in 1957 indirectly inspired GPS. Scientists tracked Sputnik's radio signals and used the Doppler effect to calculate its position, sparking the idea of satellite-based navigation.
- **Transit System:** Developed by the U.S. Navy in the 1960s, Transit was the first satellite navigation system. Though basic, it provided navigation for naval vessels and submarines, paving the way for GPS.
- **NAVSTAR Development:** In 1973, the U.S. Department of Defence initiated the GPS program, originally named NAVSTAR. Designed for military applications such as missile guidance and troop positioning, it combined precise timing and ranging capabilities.
- **Full Operational Capability:** The first experimental GPS satellites launched in the late 1970s, with the system achieving Full Operational Capability (FOC) in

1995. A full constellation of 24 satellites ensured global coverage and accuracy for military and civilian users.

- **Selective Availability (SA):** To maintain military superiority, the DoD initially degraded civilian GPS accuracy through Selective Availability. SA was discontinued in 2000, significantly improving civilian applications.
- **Civilian Expansion:** After SA's removal, GPS became integral to daily life, powering personal navigation, smartphones, and logistics tracking systems.
- **Modernization:** Continuous upgrades, including new satellite signals (L2C and L5) and additional satellites, have improved GPS accuracy and reliability.
- **Global Integration:** GPS is now a global standard, while other nations have developed systems like Russia's GLONASS, the EU's Galileo, and China's BeiDou, ensuring worldwide navigation capabilities.

Today, GPS has revolutionized navigation, geolocation, and numerous industries, including transportation, agriculture, and emergency services. It has become an essential tool for both civilian and military purposes, exemplifying the extraordinary advancement of a technology that began with the tracking of a single orbiting satellite.

11.1.4 Timeline and Satellite of GPS

- **1957:** The concept of GPS traces back to the launch of the Soviet satellite Sputnik. Scientists used the Doppler effect to track its position, inspiring the idea of satellite-based navigation.
- **Early 1960s:** The U.S. Navy developed the Transit system, an early satellite navigation system using low Earth orbit satellites for basic navigation.
- **1973:** The U.S. Department of Defence (DoD) initiated the development of GPS, originally named NAVSTAR, focusing on military applications.
- **Late 1970s:** The first experimental GPS satellites were launched.
- **1983:** Following the Korean Air Flight 007 incident, President Ronald Reagan authorized GPS for civilian use to enhance global navigation safety.
- **1993:** The launch of the final Block II satellite completed the 24-satellite constellation, achieving full operational capability (FOC).
- **2000:** The U.S. deactivated Selective Availability (SA), significantly improving GPS accuracy for civilian users.
- **2005:** The first Block IIR-M satellite was launched, marking the start of GPS modernization with enhanced system capabilities.
- **2010:** The launch of the first Block IIF satellite introduced new signals and advanced features.
- **2016:** The first GPS III satellite was launched, offering improved accuracy, anti-jamming capabilities, and advanced technologies.
- **2020s:** Modernization efforts continue with the deployment of additional GPS III satellites and further advancements in system features.

11.1.5 Segments of GPS

The Global Positioning System (GPS) consists of multiple segments that collaboratively ensure precise positioning and navigation services (Fig. 11.2). These segments include:

- **Space Segment:** The Space Segment consists of GPS satellites orbiting Earth, forming a constellation of satellites equipped with atomic clocks. These satellites continuously transmit signals containing precise timing and positioning information, ensuring global coverage.
- **Control Segment:** The Control Segment includes a network of ground-based monitoring and control stations managed by the U.S. Air Force. These stations track the satellites, monitor their signals, and maintain the accuracy of their atomic clocks. The Master Control Station (MCS), located at Schriever Air Force Base in Colorado, plays a key role in managing the satellite constellation.
- **User Segment:** The User Segment encompasses all GPS receivers and devices used by individuals, businesses, the military, and various industries. These devices receive satellite signals and calculate precise positions based on the signals' time of arrival, including applications such as smartphones, vehicle navigation systems, and specialized equipment for surveying, agriculture, and research.
- **Augmentation Segment:** The Augmentation Segment includes systems that enhance GPS accuracy and reliability, such as Differential GPS (DGPS) and Satellite-Based Augmentation Systems (SBAS). DGPS stations provide correction signals to improve accuracy, while SBAS networks offer wide-area corrections, supporting applications like aviation.
- **Policy and Control Segment:** This segment involves government agencies, primarily the U.S. Department of Defence (DoD), responsible for overseeing the GPS system. It establishes policies, security protocols, and operational standards to ensure the system's proper functioning and protect against potential threats.

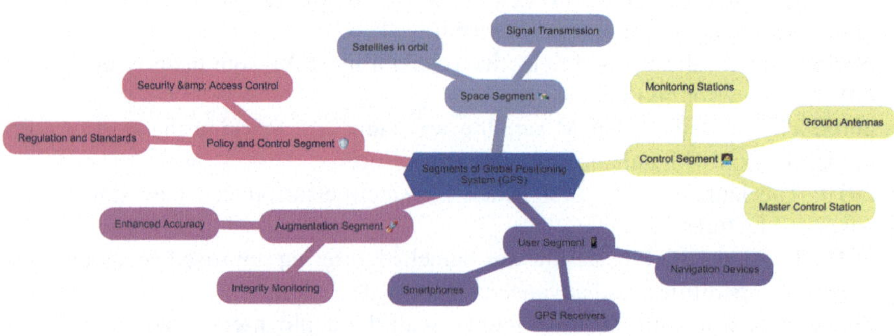

Fig. 11.2 Segments of GPS (Prepared by the authors)

Therefore, these segments make GPS a highly reliable and adaptable system for positioning and navigation, serving a wide range of applications in everyday life and across industries. The seamless coordination among these segments ensures that GPS users globally have access to accurate location data and precise timing information, supporting activities from basic navigation to critical operations.

11.2 Principles of GPS

The Global Positioning System (GPS) is based on several fundamental principles that enable accurate location determination on Earth. The following outlines the key principles of GPS (Fig. 11.3):

- **Trilateration:** The core principle of GPS, trilateration, determines a receiver's position by measuring its distance from multiple satellites. By calculating the time, it takes for signals from at least four satellites to reach the receiver, the device can pinpoint its 3D location (latitude, longitude, and altitude).
- **Satellite Constellation:** GPS relies on a constellation of 24–32 satellites orbiting Earth, distributed across six orbital planes. The receiver must have at least four operational satellites in view to ensure accurate triangulation.
- **Precise Timing:** Accurate timing is essential for GPS. Satellites equipped with atomic clocks broadcast signals that include the transmission time, allowing receivers to calculate the distance from each satellite by comparing signal arrival times.
- **Signals from Space:** GPS satellites transmit signals on frequencies such as L1 (1575.42 MHz) and L2 (1227.60 MHz), containing identification, ephemeris data, and the precise transmission time. These signals enable position calculations and error corrections.

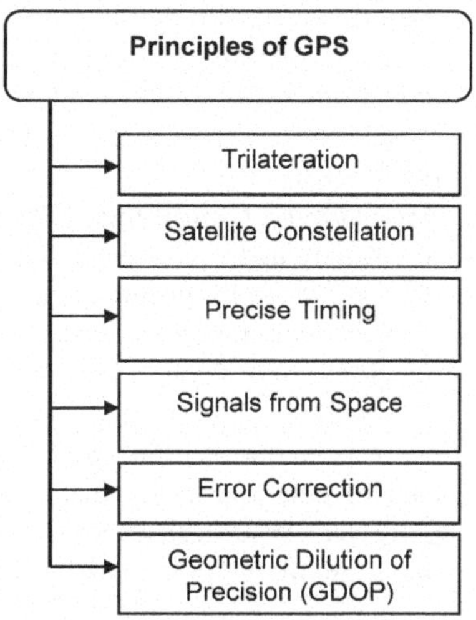

Fig. 11.3 Principles of GPS (Prepared by the authors)

- **Error Correction:** GPS signals can be affected by factors like atmospheric conditions and signal reflections. To improve accuracy, receivers use error correction techniques such as Differential GPS (DGPS) and Satellite-Based Augmentation Systems (SBAS).
- **Geometric Dilution of Precision (GDOP):** GDOP measures the relative positions of satellites and their impact on positioning accuracy. Lower GDOP values indicate a more favourable satellite arrangement, improving GPS precision.

In summary, the integration of these foundational principles has enabled GPS technology to revolutionize navigation, mapping, surveying, and numerous other applications across various industries. Whether used for driving directions, precise surveying, or tracking fitness activities, understanding how GPS operates provides a deeper appreciation for its remarkable ability to accurately determine location on Earth.

11.3 Accuracy and Errors of GPS

The accuracy of the Global Positioning System (GPS) is vital for its diverse applications, including navigation, surveying, and scientific research. However, various factors can impact GPS accuracy, resulting in errors that must be considered when using the technology. A thorough understanding of these factors is essential to optimize GPS performance and reduce potential inaccuracies. The following factors influence the accuracy of GPS:

- **Satellite Geometry and Geometric Dilution of Precision (GDOP):** The positioning of GPS satellites relative to each other significantly impacts the accuracy of location calculations. When satellites are clustered together, the GDOP value increases, resulting in less precise positioning. A more evenly distributed satellite constellation reduces GDOP and enhances accuracy.
- **Signal Multipath:** Multipath errors occur when GPS signals reflect off nearby surfaces, such as buildings or mountains, before reaching the receiver. These reflections cause delays in signal reception, leading to inaccuracies in position calculations. Urban areas with tall structures are particularly prone to multipath errors.
- **Atmospheric Conditions:** GPS signals can be delayed as they travel through the ionosphere and troposphere. The ionosphere, in particular, can cause significant delays, especially during solar storms. These atmospheric influences can introduce errors in range measurements, thereby affecting accuracy.
- **Clock Errors:** Though GPS satellites are equipped with highly accurate atomic clocks, even minor discrepancies in the clocks of the receiver or satellite can lead to errors. While satellite clocks are generally precise, receiver clocks may not be, causing potential timing discrepancies.
- **Selective Availability (SA):** Before 2000, the U.S. Department of Defence intentionally degraded GPS signal accuracy for civilian users through Selective

Availability (SA). While SA was disabled in May 2000, its historical effects on GPS accuracy should still be considered when analysing data from that period.
- **Receiver Quality:** The accuracy of GPS positioning also depends on the quality of the receiver. High-end receivers used for professional surveying or military applications can achieve centimetres-level accuracy, while consumer-grade receivers may have accuracy within several meters.

Errors in the Global Positioning System (GPS) refer to the discrepancies between the position calculated by a GPS receiver and the actual, true location. These inaccuracies can arise from various factors, impacting the precision of GPS-based positioning. The following are some common types of errors associated with GPS:

- **Satellite Clock Errors:** GPS accuracy depends on precise satellite clock synchronization. Minor timing discrepancies in satellite clocks can result in position calculation errors.
- **Orbital Errors:** Inaccurate knowledge of satellite orbital parameters can lead to orbital errors, as GPS satellites move rapidly in their orbits.
- **Atmospheric Delays:** GPS signals experience delays passing through the ionosphere and troposphere due to variations in atmospheric density and composition, causing positional inaccuracies.
- **Multipath Effects:** Signals reflecting off nearby surfaces or objects before reaching the receiver can create multiple paths with varying travel times, reducing positional accuracy.
- **Selective Availability (SA):** Historically, the U.S. Department of Defence intentionally degraded civilian GPS accuracy through SA, causing errors of up to 100 m. SA was discontinued in 2000.
- **Geometric Dilution of Precision (GDOP):** Poor satellite geometry, where satellites are not optimally positioned relative to the receiver, can reduce accuracy.
- **Receiver Errors:** Hardware imperfections or suboptimal processing in GPS receivers, particularly low-quality models, can introduce additional inaccuracies.

Modern GPS receivers and systems use advanced techniques to reduce and correct errors, improving accuracy and reliability. Methods like Differential GPS (DGPS), Satellite-Based Augmentation Systems (SBAS), and Real-Time Kinematic (RTK) GPS enhance position data precision through error corrections. Despite inherent challenges, GPS continues to be an indispensable and highly accurate tool for navigation, mapping, surveying, and numerous other applications across industries.

Error Correction Methods

- **Differential GPS (DGPS):** DGPS enhances GPS accuracy using a network of ground-based reference stations with known positions. These stations determine discrepancies between their actual locations and received GPS signals, transmitting correction data to nearby GPS receivers. This method reduces errors caused by atmospheric conditions, multipath, and clock inaccuracies.

- **Satellite-Based Augmentation Systems (SBAS):** SBAS, such as the Wide Area Augmentation System (WAAS) in the U.S., transmits correction data via geostationary satellites. These systems enhance GPS accuracy and reliability, particularly in aviation, by correcting signal errors and improving positioning precision.
- **Real-Time Kinematic (RTK) and Post-Processing:** RTK GPS delivers centimetres-level accuracy for high-precision tasks like land surveying by using real-time corrections from a local base station. Similarly, post-processing techniques allow users to correct GPS data by comparing it with reference points after data collection, ensuring enhanced positional accuracy.

In summary, the GPS accuracy can be affected by factors like satellite geometry, atmospheric conditions, and signal multipath. Techniques such as DGPS, SBAS, and RTK improve precision, making GPS a reliable tool for diverse applications.

11.4 Application and Scope of GPS

GPS technology has become integral to numerous facets of modern life, transforming industries, enhancing safety, and increasing convenience. The following provides a detailed exploration of its various applications and scopes:

a) **Navigation and Transportation:**

- **Automotive Navigation**: GPS systems in vehicles and smartphones provide turn-by-turn directions, real-time traffic updates, and estimated arrival times.
- **Aviation**: GPS is crucial for aircraft navigation, assisting with take-off, landing, and en-route guidance. It enhances safety and improves airspace management efficiency.
- **Maritime Navigation**: GPS ensures precise navigation for ships, boats, and vessels, promoting safe and efficient travel on seas, oceans, and waterways.
- **Rail Transportation**: Rail systems use GPS for tracking trains, enhancing safety, and optimizing schedules.
- **Public Transportation**: Public transit systems rely on GPS for real-time tracking, route planning, and providing passenger information.

b) **Agriculture:**

- **Precision Agriculture**: GPS is applied in agriculture to optimize planting, irrigation, and fertilization, boosting crop yields, conserving resources, and reducing environmental impact.
- **Precision Livestock Farming**: GPS helps in the precise tracking and management of livestock, improving monitoring of animal health and behaviour.

c) **Outdoor Activities and Recreation:**

- **Hiking and Camping**: GPS devices assist outdoor enthusiasts with route planning, tracking, and ensuring safety in wilderness areas.

- **Geocaching**: GPS coordinates are used by geocachers to locate hidden containers, or *"geocaches,"* in various outdoor locations around the world.
- **Fitness Tracking**: GPS technology in smartphones and wearable devices tracks exercise routes, measures distance, and calculates speed and elevation.

d) **Mapping and Surveying:**

- **Cartography**: GPS aids in the creation of highly accurate maps and geographic data.
- **Land Surveying**: GPS is utilized by surveyors for precise land measurements and determining property boundaries.
- **3D Modeling**: GPS supports the creation of 3D models of terrain and structures, beneficial for engineering and architecture.

e) **Search and Rescue:**

- **Wilderness and Maritime Rescue**: GPS assists search and rescue teams in locating individuals in distress, especially in remote areas or at sea.
- **Military and Defence:**
- **Navigation and Targeting**: The military use GPS for precise positioning, navigation, and timing for troop deployment, missile guidance, and situational awareness.
- **Secure Communication**: GPS plays a role in secure military communication systems, ensuring time synchronization.

f) **Emergency Services:**

- **First Responders**: GPS enables emergency responders to efficiently reach accident scenes and locate distressed individuals.
- **Disaster Relief**: During natural disasters, GPS supports the coordination of relief efforts and mapping affected regions.

g) **Fleet Management:**

- **Commercial Fleets**: Businesses use GPS tracking to monitor and manage vehicle fleets, optimize routes, and improve safety.
- **Logistics**: GPS enhances the logistics and supply chain industry by tracking shipments and optimizing delivery routes.

h) **Time Synchronization:**

- **Telecommunications**: GPS is critical in synchronizing time across telecommunications networks and financial transactions, ensuring accurate data transmission.

i) **Mining and Construction:**

- **Heavy Equipment**: GPS supports the management of mining and construction equipment, ensuring precise operations such as digging, grading, and excavation.

j) **Smartphone Applications:**

- **Location-Based Services**: Numerous smartphone apps rely on GPS for services like finding nearby restaurants, mapping routes, tracking lost devices, and checking in on social media.

k) **Scientific Research and Environmental Monitoring:**

- **Geology**: GPS is essential for monitoring tectonic plate movements and land deformations.
- **Wildlife Tracking**: Researchers employ GPS to track wildlife migrations and study animal behaviour.
- **Climate Change**: GPS is used to measure land movements, sea level rise, and glacial changes, contributing to climate change research.

In summary, the versatility and precision of GPS have established it as an essential tool across a wide range of fields. Ongoing advancements and its integration with emerging technologies guarantee the expansion of its applications, further influencing how we navigate, work, and engage with our environment.

11.5 Advantages and Disadvantages of GPS Application

Advantages of GPS Applications

The application of GPS encompasses a range of advantages and disadvantages, as illustrated in Fig. 11.4, including:

- **Accurate Positioning:** GPS provides precise, real-time location data with meter-level accuracy, essential for navigation, mapping, and surveying.
- **Enhanced Safety:** By delivering accurate navigation, traffic updates, and real-time tracking, GPS enhances safety in transportation, aviation, maritime, and search-and-rescue operations.
- **Efficiency and Productivity:** In agriculture, GPS optimizes practices, reduces resource waste, and boosts yields. In logistics and fleet management, it streamlines route planning, saving time and fuel costs.
- **Environmental Benefits:** GPS-driven precision agriculture reduces environmental impact by minimizing fertilizer and pesticide use. Efficient transportation routing lowers carbon emissions.
- **Recreation and Fitness:** GPS supports outdoor activities like hiking and cycling while tracking fitness metrics, promoting physical health.
- **Scientific Research:** GPS aids in studies of plate tectonics, climate change, and wildlife migration, providing valuable data for environmental monitoring.
- **Emergency Response:** GPS accelerates emergency responses by pinpointing locations of accidents, lost individuals, and disaster areas, improving crisis management.

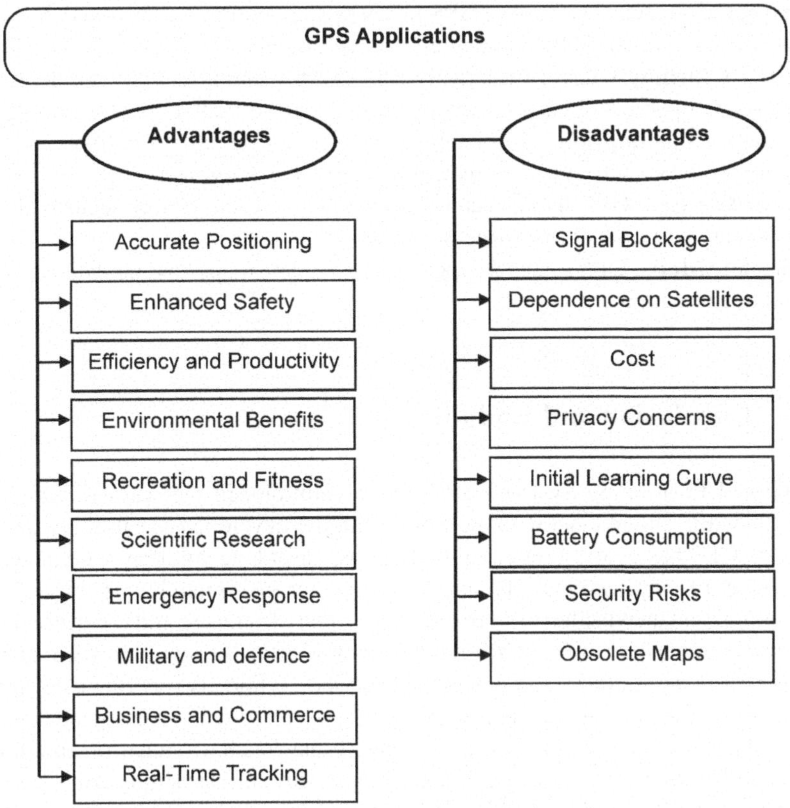

Fig. 11.4 Advantages and Disadvantages of GPS Application (Prepared by the authors)

- **Military and defence:** GPS is critical for military operations, ensuring precise positioning, navigation, and communication for situational awareness and accurate targeting.
- **Business and Commerce:** In fleet-based businesses, GPS optimizes logistics, reduces costs, and enhances customer service with accurate delivery estimates.
- **Real-Time Tracking:** GPS enables real-time tracking of assets such as vehicles, equipment, and packages, improving security and asset management.

Disadvantages of GPS Applications

- **Signal Blockage:** Tall buildings, dense forests, and natural obstacles can block or degrade GPS signals, reducing accuracy or causing signal loss in urban or heavily forested areas.
- **Dependence on Satellites:** GPS's reliance on satellites makes it vulnerable to disruptions from space weather or potential attacks on the satellite network.
- **Cost:** High-quality GPS receivers and systems can be expensive, limiting accessibility for some users or organizations.

- **Privacy Concerns:** The use of GPS raises privacy issues, as location data can be collected and misused by companies or governments.
- **Initial Learning Curve:** Effectively using GPS technology may require learning how to operate GPS devices or apps, which can be challenging for some users.
- **Battery Consumption:** GPS-enabled devices, particularly smartphones, consume significant battery power, requiring frequent recharging during extended use.
- **Security Risks:** GPS signals can be jammed or spoofed, posing security threats in military and critical infrastructure contexts.
- **Obsolete Maps:** GPS systems depend on up-to-date maps, and outdated data can lead to navigation errors.

11.6 Conclusion and Endnotes

The Global Positioning System (GPS) has revolutionized the ways in which we navigate, work, and interact with the world. Initially developed as a military navigation tool, GPS has evolved into a globally indispensable technology with an extensive range of applications. Its numerous advantages include highly accurate positioning, enhanced safety, improved operational efficiency, and contributions to environmental sustainability. GPS has profoundly transformed industries such as transportation, agriculture, and emergency services, while also serving as a cornerstone for scientific research and military operations. Despite its widespread benefits, GPS technology faces significant challenges. Issues such as signal obstruction in urban and forested environments, reliance on satellite infrastructure, concerns over privacy, and security vulnerabilities underscore the need for continuous innovation and robust safeguards to maintain its reliability and ethical use. Nevertheless, GPS remains an invaluable technological asset with immense potential for future advancements. As it integrates with emerging technologies like augmented reality and autonomous systems, GPS is poised to further shape our world, unlocking new opportunities for industries and individuals alike.

Endnotes

- GPS technology is founded on the principles of satellite-based positioning and precise timing, enabling accurate location determination for users.
- It finds extensive applications across various industries, including navigation, agriculture, aviation, and scientific research, enhancing both efficiency and safety.
- The history of GPS is compelling, encompassing its early development stages to the modern constellation of satellites that support diverse functionalities.
- The evolution of navigation techniques, from natural cues to sophisticated systems like GPS, has revolutionized exploration and transportation.
- GPS operates through interconnected segments, including the space, control, user, augmentation, and policy and control components, which collectively ensure precise positioning.

- Its principles involve trilateration, satellite constellations, precise timing, space-based signals, error correction mechanisms, and the concept of geometric dilution of precision (GDOP).
- GPS accuracy can be impacted by various factors, such as satellite clock and orbital errors, atmospheric interference, multipath effects, and other environmental conditions.
- The range of GPS applications is extensive, spanning navigation, agriculture, outdoor recreation, scientific research, and military operations, delivering significant benefits to diverse fields.
- Despite its advantages in accuracy, safety, and efficiency, GPS faces challenges, including signal blockages, privacy concerns, and security vulnerabilities.

Further Readings

1. Boeser, S. M. (2019). Global positioning systems (GPS). In C. W. Finkl & C. Makowski (Eds.), *Encyclopedia of coastal science. Encyclopedia of earth sciences series*. Springer. https://doi.org/10.1007/978-3-319-93806-6_154
2. Capderou, M. (2014). Global positioning systems (GPS). In *Handbook of satellite orbits*. Springer. https://doi.org/10.1007/978-3-319-03416-4_14
3. El-Rabbany, A. (2002). Introduction to GPS: The global positioning system. *Artech House*. ISBN: 9781580531832.
4. Hegarty, C. J. (2017). The global positioning system (GPS). In P. J. Teunissen & O. Montenbruck (Eds.), *Springer handbook of global navigation satellite systems* (Springer handbooks). Springer. https://doi.org/10.1007/978-3-319-42928-1_7
5. Hofmann-Wellenhof, B., Lichtenegger, H., & Collins, J. (2001). *Global positioning system: Theory and practice*. Springer Vienna. ISBN: 978-3-211-83534-0. https://doi.org/10.1007/978-3-7091-6199-9
6. Hofmann-Wellenhof, B., Lichtenegger, H., & Wasle, E. (2008). *GNSS—Global navigation satellite systems: GPS, GLONASS, Galileo, and more*. Springer Vienna. ISBN 978-3-211-73012-6. https://doi.org/10.1007/978-3-211-73017-1
7. Hofmann-Wellenhof, B., Lichtenegger, H., & Collins, J. (2001). Overview of GPS. In *Global positioning system*. Springer. https://doi.org/10.1007/978-3-7091-6199-9_2
8. Loweth, R. P. (1997). GPS. In *Manual of offshore surveying for geoscientists and engineers*. Springer. https://doi.org/10.1007/978-94-011-5826-8_7
9. Mansfeld, W. (2004). Global positioning system (GPS). In *Satellitenortung und Navigation*. Vieweg+Teubner Verlag. https://doi.org/10.1007/978-3-663-11328-7_3
10. Misra, P., & Enge, P. (2010). *Global positioning system: Signals, measurements, and performance* (2nd ed.). Ganga-Jamuna Press ISBN: 978-0970954428.

Chapter 12
GPS for All: Challenges and Future Directions

> *"The important thing is not where you've been, but where you are going and GPS is our compass to the future." — Jack Dangermond.*

Abstract The chapter *"GPS for All: Challenges and Future Directions"* explores the pivotal role of Global Positioning System (GPS) technology in modern society, emphasizing its widespread use in everyday life. From navigation and emergency services to logistics and agriculture, GPS has become integral in enhancing operational efficiency and improving quality of life. However, despite its numerous advantages, GPS has inherent limitations, including signal blockage, accuracy issues, and dependence on satellite visibility, which affect its reliability in certain environments. The chapter highlights the importance of fostering GPS literacy, ensuring users are well-informed about the system's functionalities and limitations to ensure safe and effective utilization. It also explores Differential GPS (DGPS), an advanced form of GPS that offers enhanced positioning accuracy, thereby expanding the system's capabilities. The chapter concludes by discussing the future trajectory of GPS, exploring technological innovations such as the integration with 5G, IoT, and quantum computing, and highlighting new applications in autonomous systems, environmental monitoring, and more. Through this exploration, the chapter provides a comprehensive overview of the challenges and opportunities that lie ahead for GPS technology in shaping the future of navigation and communication.

Keywords Future of GPS technology · Promoting GPS literacy · Technological advancements · Differential GPS · Internet of Things (IoT) and GPS

K. D. Malakar et al., *Geospatial Technologies in Coastal Ecologies Monitoring and Management*, Advances in Geographic Information Science,
https://doi.org/10.1007/978-3-031-92017-2_12

The Global Positioning System (GPS) has profoundly transformed modern life, underpinning countless applications from navigation and logistics to agriculture and emergency services. While GPS has enhanced operational efficiency and quality of life, its limitations, such as signal interference and accuracy issues, necessitate deeper understanding and innovation. This chapter delves into the significance of fostering GPS literacy, examines Differential GPS (DGPS) as a solution for improved accuracy, and envisions the future of GPS with advancements like 5G, IoT, and quantum computing. By addressing challenges and exploring emerging opportunities, the chapter highlights GPS's evolving role in shaping technology and society.

Key Points of the Chapter

- Understanding the applications, limitations, and future directions of GPS.
- Promoting GPS literacy and addressing its challenges.
- Exploring the concept of Differential GPS (DGPS) as an advanced form of GPS.

12.1 GPS in Everyday Life

The Global Positioning System (GPS) has become a fundamental component of contemporary life, providing accurate location and navigation services that impact various facets of daily activities (Fig. 12.1). Originally developed by the U.S. Department of Defence, this satellite-based system is now widely accessible and facilitates a broad range of civilian applications.

(a) **Navigation and Transportation:** GPS is essential for navigation across various modes of transport, including cars, bicycles, and walking. Applications such as Google Maps and Waze offer real-time directions, traffic updates, and estimated arrival times. Public transportation systems also utilize GPS to track and share the locations of buses, trains, and taxis, enhancing scheduling and reducing wait times.

(b) **Travel and Tourism:** Travelers depend on GPS to explore unfamiliar destinations, locate nearby attractions, restaurants, and accommodations, and plan efficient travel routes. GPS-enabled devices and apps have largely replaced traditional paper maps, making travel more convenient and accessible.

(c) **Emergency Services:** GPS is critical in emergency response, allowing first responders to quickly locate incidents. Police, fire departments, and medical services use GPS to track vehicles and optimize response times. Additionally, mobile devices with GPS enable users to share their precise location during emergencies.

(d) **Weather and Disaster Management:** GPS is instrumental in monitoring weather conditions and predicting natural disasters. It aids in tracking the movement of storms, earthquakes, and floods, enabling authorities to issue timely warnings and effectively manage evacuations.

Fig. 12.1 GPS in Everyday Life (Prepared by the authors)

(e) **Fitness and Health:** GPS-enabled wearables, such as smartwatches and fitness trackers, are popular among fitness enthusiasts for tracking activities like running, cycling, or hiking. These devices measure distance, speed, and routes, helping users set and achieve their fitness goals.

(f) **Agriculture:** GPS technology is transforming agriculture through precision farming. It allows for the mapping of fields, monitoring of crops, and guiding of machinery for planting, fertilizing, and harvesting, thereby improving efficiency and reducing waste.

(g) **Logistics and Delivery:** GPS enhances logistics and supply chain management by enabling real-time tracking of shipments. Delivery services use GPS to optimize routes, ensure timely deliveries, and provide customers with live updates on their packages.

(h) **Personal Safety and Security:** GPS tracking devices are widely used for personal safety. Parents use GPS-enabled devices to monitor their children's location, while pet owners track their pets with GPS collars. GPS also plays a role in anti-theft systems for vehicles and personal belongings.

(i) **Entertainment and social media:** GPS powers location-based games like Pokémon GO and geocaching, offering engaging recreational activities. Social media platforms leverage GPS for geotagging photos, sharing location-specific content, and discovering nearby friends or events.

(j) **Scientific Research:** Researchers utilize GPS in various studies, including tracking animal migrations, monitoring geological changes, and conducting environmental surveys.

In summary, GPS has revolutionized the way we live, work, and engage with the world. Its applications are continuously expanding, enhancing efficiency, connectivity, and safety. As technology progresses, GPS is expected to play an increasingly pivotal role in shaping the future.

12.2 GPS and Its Limitations

The Global Positioning System (GPS) has become an essential tool in modern society, with diverse applications ranging from navigation to scientific research. However, despite its many benefits, GPS technology has limitations that can impact its accuracy, reliability, and accessibility. Recognizing these limitations is crucial for users who depend on GPS for various everyday tasks. There are such as;

(a) **Signal Blockage and Obstruction:** GPS signals can be obstructed by physical structures such as tall buildings, dense forests, or mountainous terrain. This is particularly problematic in urban canyons, where skyscrapers block the line of sight to satellites, leading to weak or inaccurate positioning. In such situations, GPS receivers may struggle to obtain a signal or provide incorrect location data.
(b) **Accuracy Issues:** While GPS typically offers accuracy within 5–10 m under ideal conditions, this can be compromised due to various factors. Atmospheric interference, multipath signals (signals reflected from surfaces), and the quality of the GPS receiver can all contribute to positioning errors. For applications requiring higher precision, such as surveying or autonomous vehicles, additional correction systems like Differential GPS are necessary.
(c) **Signal Interference and Jamming:** GPS signals are vulnerable to interference from sources such as radio frequency jamming or electromagnetic interference (EMI). Deliberate jamming, often used in military or sensitive contexts, can disrupt GPS functionality, rendering navigation and tracking impossible. Additionally, solar storms and space weather can affect satellite signals, causing temporary service interruptions.
(d) **Dependence on Satellite Visibility:** GPS requires a clear line of sight to at least four satellites to function effectively. In environments where this visibility is obstructed, such as indoors, underground, or in areas with dense foliage, GPS signals cannot be received, limiting its usability in certain locations like basements, parking garages, or remote regions.
(e) **Environmental Factors:** GPS accuracy can be influenced by environmental conditions like ionospheric delays, which may cause slight errors in positioning. These atmospheric factors are difficult to predict and mitigate, potentially affecting the precision of GPS measurements.

(f) **Power Consumption:** GPS-enabled devices typically require significant power to maintain a continuous connection with satellites, which can quickly deplete battery life, especially in mobile devices like smartphones and wearables. Prolonged GPS use for navigation or tracking often necessitates frequent recharging.
(g) **Limited Availability in Remote Areas:** Although GPS operates globally, its performance can be compromised in remote regions or locations with limited infrastructure. For example, in polar regions or deep oceanic zones, satellite visibility may be reduced, resulting in degraded signal strength or a lack of reliable coverage.
(h) **Vulnerability to Spoofing:** GPS signals are not encrypted, making them susceptible to spoofing, a process where false GPS signals are transmitted to deceive receivers. This vulnerability poses risks in sensitive applications, such as aviation, transportation, or military operations, where compromised GPS data could lead to safety and security concerns.
(i) **Dependency on External Infrastructure:** GPS relies on a constellation of satellites operated by the U.S. government. Any disruption, such as satellite malfunctions or government-imposed restrictions, could lead to a loss of GPS services. Regional GPS outages could severely affect navigation for millions of users.
(j) **Privacy Concerns:** The continuous tracking of users' locations by GPS-enabled devices raises privacy concerns. Unauthorized tracking, location-based advertising, or data misuse can be problematic, especially if location data is collected without explicit user consent.

So, while GPS is a widely used and powerful technology, it has limitations, including signal obstruction, accuracy issues, interference susceptibility, and reliance on external infrastructure. Users must be aware of these constraints and may need to integrate GPS with other technologies (e.g., Wi-Fi positioning, inertial navigation systems) for more reliable and accurate location services. As GPS technology evolves, some limitations may be addressed, but others are inherent to its design and external factors.

12.3 Promoting GPS Literacy

In today's technology-driven society, the Global Positioning System (GPS) has become a crucial tool in everyday life, supporting applications ranging from personal navigation to scientific research. However, despite its extensive use, many individuals lack a comprehensive understanding of how GPS functions and how to utilize it effectively. Enhancing GPS literacy is vital for optimizing the advantages of this technology and ensuring that users can navigate its complexities with confidence and safety (Fig. 12.2).

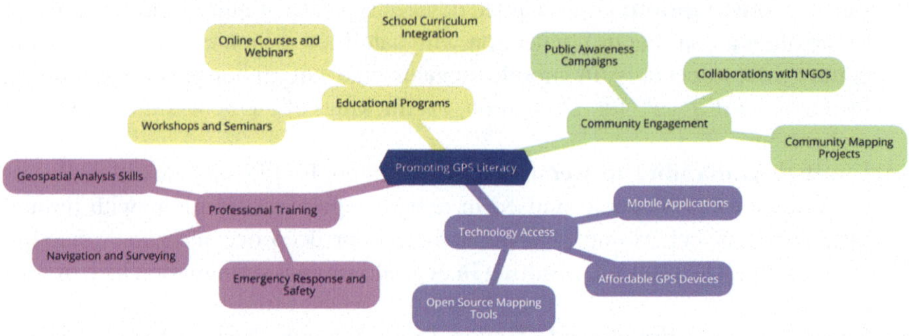

Fig. 12.2 Promoting GPS Literacy (Prepared by the authors)

The Importance of GPS Literacy

(a) **Improved Navigation Skills:** GPS literacy empowers individuals to effectively use navigation tools across various platforms, including mobile devices, vehicles, and outdoor settings. A clear understanding of GPS helps users interpret maps, follow directions, and make informed travel decisions, thereby enhancing efficiency and reducing dependence on traditional maps.
(b) **Safety in Emergency Situations:** In emergencies, knowing how to utilize GPS to share one's precise location can be critical. GPS literacy enables individuals to accurately track their position and communicate effectively with emergency responders, ensuring timely assistance.
(c) **Awareness of Privacy and Security:** With the growing prevalence of GPS tracking in mobile devices, vehicles, and wearables, it is essential to understand how GPS operates to protect privacy. Users should be aware of location tracking settings, potential misuse, and how to manage access to their location data.
(d) **Advancement of Digital Literacy:** GPS is an integral digital tool, and understanding it contributes to overall digital literacy, which is crucial in today's technological landscape. Just as individuals need to navigate the internet, social media, and email, they must also comprehend how to effectively use technologies like GPS in their daily routines.

Strategies for Promoting GPS Literacy

(a) **Educational Programs and Awareness Campaigns:** Educational institutions, community centers, and online platforms can offer workshops, tutorials, and resources to educate people on the workings of GPS and how to effectively use GPS-enabled devices. This could include hands-on training in navigation apps, map interfaces, and interpreting GPS data in practical scenarios.
(b) **Integrating GPS into Academic Curricula:** GPS literacy can be integrated into the curriculum of schools and universities, particularly in subjects such as geography, environmental science, and technology. Students can gain knowledge of satellite systems, map reading, spatial analysis, and geospatial tech-

nologies, empowering them to use GPS in daily life and pursue careers in geospatial sciences.

(c) **Public Awareness Initiatives:** Governments, NGOs, and private organizations can conduct awareness campaigns to highlight the importance of GPS literacy. These campaigns could focus on navigation safety, GPS-enabled safety apps, and the environmental and social implications of location tracking.

(d) **Targeted Education for Specific Groups:** Certain populations, such as older adults, people living in remote areas, or those with disabilities, may benefit from tailored GPS education. For example, seniors might need simplified instructions for using GPS-enabled devices for navigation or emergency alerts, while individuals in rural areas may benefit from learning how GPS can help them access essential services or connect with others.

(e) **Professional Training in Specialized Fields:** Professionals in sectors such as disaster management, agriculture, logistics, and transportation can receive specialized GPS training to optimize the system's capabilities. Knowledge of advanced features like real-time tracking, geofencing, and geospatial analysis can enhance decision-making and improve operational efficiency in these fields.

(f) **Interactive Learning Platforms:** Online platforms, mobile apps, and interactive tools can make GPS learning more engaging. For instance, virtual simulations could allow users to practice using GPS in various environments or situations, helping them build confidence in utilizing the technology.

(g) **Collaborating with Technology Developers:** Collaboration with technology providers of GPS-based applications can ensure that products are user-friendly and come with instructional resources. In-app tutorials and videos on GPS functionality can enhance accessibility for users across different age groups and technical abilities.

(h) **Addressing Ethical and Legal Concerns:** As GPS technology raises questions regarding privacy and data security, promoting GPS literacy should also include educating the public about its ethical and legal aspects. This involves informing individuals on how their location data is used, how to control or share this data, and understanding the potential risks of misuse.

Challenges in Promoting GPS Literacy

(a) **Technological Barriers:** Limited access to devices with GPS capabilities or reliable internet connectivity in certain regions may impede efforts to promote GPS literacy.

(b) **Digital Divide:** Not all individuals possess the same level of familiarity or comfort with digital technologies. For some, GPS literacy may appear challenging or unfamiliar, necessitating targeted efforts to engage these populations effectively.

(c) **Resource Limitations:** In certain contexts, resources and training materials for GPS literacy may be scarce or overly complex, making it difficult for the average user to grasp the concepts without proper guidance.

Promoting GPS literacy is essential for improving individual navigation abilities, ensuring personal safety, and advancing digital inclusion. By integrating GPS education into curricula, conducting public awareness campaigns, and offering targeted training for diverse groups, society can empower individuals to fully leverage this transformative technology. As GPS becomes ever more integral to modern life, widespread literacy in its usage will be vital for navigating the complexities of the digital era.

12.4 Differential GPS (DGPS): Advance Form of GPS

Differential GPS (DGPS) is an advanced variation of the Global Positioning System (GPS) that improves the accuracy of location data provided by GPS receivers. It corrects the inherent errors in standard GPS measurements, which may arise from factors such as atmospheric interference, signal delays, and satellite clock discrepancies. Therefore, the following discussion provides an overview of the conceptual understanding of DGPS, as also depicted in Fig. 12.3.

Key Concepts of DGPS

- **Error Correction:** DGPS functions using a network of fixed ground-based reference stations with accurately surveyed, known locations. These stations receive GPS signals in the same way as standard GPS receivers.
- **Differential Corrections:** By knowing their precise location, reference stations can identify errors in received GPS signals. They then transmit these corrections to nearby GPS receivers, allowing real-time adjustments to position calculations and significantly improving accuracy.

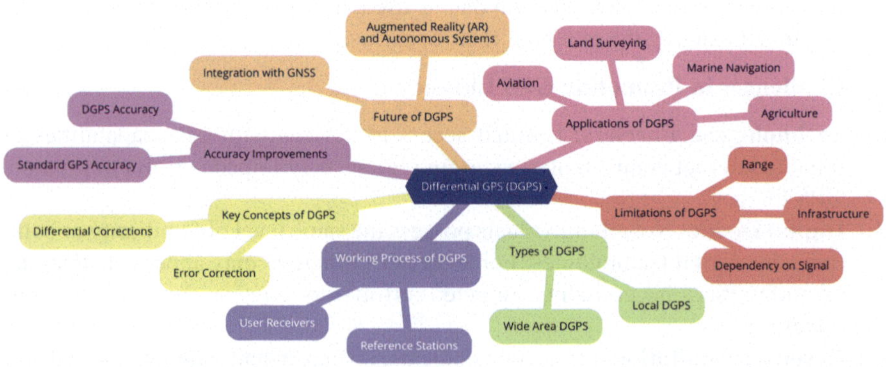

Fig. 12.3 Concept of Differential GPS (DGPS) as an Advance form of GPS (Prepared by the authors)

Types of DGPS

- **Local DGPS:** Typically employed within a confined geographical area, local DGPS relies on corrections transmitted from a single reference station or a limited network of stations. It is commonly utilized in applications such as harbour navigation and land surveying.
- **Wide Area DGPS (WADGPS):** WADGPS operates over extensive regions, often encompassing entire countries or large areas. It leverages a network of reference stations in conjunction with satellite or ground-based broadcasting systems to provide corrections across vast expanses.

Working Process of DGPS

- **Reference Stations:** These stations continuously track GPS signals, calculate the discrepancy between the expected and actual positions (i.e., errors), and broadcast the correction data to nearby users.
- **User Receivers:** DGPS-capable GPS receivers can receive these corrections via radio signals or other communication methods. By applying the correction data, they adjust their readings to provide a more accurate location.

Accuracy Improvements

- **Standard GPS Accuracy:** Conventional GPS systems generally offer positional accuracy within a range of 5–10 m.
- **DGPS Accuracy:** Through the use of differential corrections, DGPS enhances accuracy to approximately 1–3 m, with potential for even greater precision, contingent upon system quality and environmental factors.

Applications of DGPS

- **Marine Navigation:** DGPS is extensively utilized in maritime contexts to ensure safe navigation, particularly in restricted areas such as harbours or channels where high precision is essential.
- **Aviation:** DGPS improves the accuracy of GPS systems in aviation, facilitating procedures like precision approaches and landings.
- **Agriculture:** In precision farming, DGPS provides accurate guidance for machinery such as tractors, enhancing the efficiency of planting, fertilization, and harvesting processes.
- **Land Surveying:** Surveyors depend on DGPS to attain the high levels of accuracy necessary for tasks such as mapping and boundary delineation.

Limitations of DGPS

- **Dependency on Signal:** DGPS requires an unobstructed line of sight to both GPS satellites and the reference station signal. This requirement can be hindered by physical barriers such as buildings, trees, or uneven terrain.
- **Range:** The accuracy of DGPS diminishes as the distance from the reference station increases, with optimal performance typically observed within a 200–300-km radius of the station.

- **Infrastructure:** The establishment and upkeep of DGPS infrastructure can incur significant costs, which may restrict its accessibility in certain regions.

Future of DGPS

- **Integration with GNSS:** With the increasing adoption of Global Navigation Satellite Systems (GNSS) such as Galileo and BeiDou, DGPS is likely to be integrated with these systems, providing enhanced accuracy and reliability.
- **Augmented Reality (AR) and Autonomous Systems:** DGPS is anticipated to be pivotal in the development of emerging technologies that demand precise location data, including autonomous vehicles and augmented reality applications.

Therefore, the DGPS constitutes a significant advancement over traditional GPS, facilitating a broad spectrum of applications where high precision is crucial.

12.5 Future of GPS

The Global Positioning System (GPS) has transformed navigation and location-based services, becoming indispensable in various sectors, ranging from everyday transportation to cutting-edge scientific research. As technological advancements continue and new innovations surface, the future of GPS is poised to bring even more transformative changes, characterized by enhanced capabilities, broader applications, and improved precision. The following discussion explores the potential evolution of GPS in the coming years:

(a) **Enhanced Accuracy and Precision:** Future advancements in GPS are expected to significantly enhance its accuracy. While current systems typically provide accuracy within a range of 5–10 m, future developments may achieve centimetre-level precision. This could be realized through:

   - Next-generation satellite constellations: The deployment of advanced GPS satellites, such as the U.S. GPS III series, promises to enhance accuracy, signal strength, and resistance to interference.
   - Real-time corrections: Advancements in Differential GPS (DGPS) and Real-Time Kinematic (RTK) technologies will enable even more precise positioning, crucial for applications such as autonomous vehicles, drone navigation, and precision agriculture.

(b) **Integration with Other Satellite Systems:** GPS will likely integrate with other global navigation satellite systems (GNSS) such as GLONASS (Russia), Galileo (EU), and BeiDou (China) to provide:

   - Improved coverage and resilience: Combining signals from multiple GNSS systems will enhance service, particularly in areas where GPS alone is limited, such as urban canyons, dense forests, or remote locations.

- Redundancy and reliability: Increased interconnectivity between different GNSS networks will ensure fewer disruptions, even in cases where one system is compromised or underperforming.

(c) **Augmented Reality (AR) and GPS Integration:** The integration of GPS with augmented reality (AR) is set to transform user interaction with their environment. In the future:

- AR navigation: GPS-enabled AR applications will overlay digital directions, landmarks, and real-time information onto the user's view, enhancing navigation in both familiar and unfamiliar areas.
- Interactive experiences: Combining GPS and AR will enrich entertainment and tourism, offering location-based storytelling, interactive city guides, and gamified experiences, making interactions more engaging.

(d) **Autonomous Vehicles and Drones:** GPS will be critical in the navigation and positioning of autonomous vehicles (AVs) and drones:

- Autonomous vehicles: GPS, alongside sensors like LiDAR, cameras, and radar, will guide self-driving cars in navigating urban environments, adhering to traffic regulations, and avoiding obstacles.
- Drones: GPS will play a central role in the precise navigation of drones for commercial purposes, such as deliveries, surveying, and infrastructure monitoring, ensuring accurate flights and safe operations in complex environments.

(e) **Internet of Things (IoT) and GPS:** The widespread adoption of IoT devices will make GPS increasingly ubiquitous, with sensors embedded in everything from wearables to household appliances:

- Smart cities: GPS-enabled IoT devices will optimize traffic management, waste collection, environmental monitoring, and real-time data on public transportation, enhancing urban planning and service efficiency.
- Personal and asset tracking: GPS will continue to be used for tracking personal belongings, such as children or pets, as well as assets in logistics and supply chain management.

(f) **GPS for Precision Agriculture:** In agriculture, GPS will continue to evolve, driving more efficient farming practices:

- Autonomous farming equipment: GPS will guide autonomous tractors, harvesters, and drones to perform tasks such as planting, fertilization, and harvesting with pinpoint accuracy, minimizing waste and maximizing yields.
- Field-level data: GPS will enable the collection of geospatial data that helps farmers analyse soil conditions, monitor crops, and make informed decisions, improving sustainability and reducing environmental impact.

(g) **GPS and 5G Integration:** The rollout of 5G technology will further enhance GPS capabilities, providing more reliable, faster, and accurate location services:

- Real-time location tracking: The combination of GPS and 5G will facilitate ultra-low latency, real-time location tracking, benefiting industries like logistics, transportation, and emergency response.
- Improved positioning in dense environments: GPS, combined with 5G's high-speed connectivity, will improve positioning in areas where traditional GPS signals struggle, such as indoors or in cities with dense infrastructure.

(h) **GPS and Quantum Technologies:** As quantum computing and quantum sensors advance, they may revolutionize GPS systems:

- Quantum-enhanced positioning: Quantum sensors could enable significantly greater precision than current GPS technologies, allowing for more accurate measurements of position, velocity, and time.
- Quantum communication: Quantum encryption may be used to secure GPS signals, preventing interference and spoofing, thereby enhancing the security and reliability of navigation systems.

(i) **GPS and Environmental Monitoring:** In response to climate change and natural disasters, GPS will play a vital role in environmental monitoring:

- Climate monitoring: GPS will be instrumental in tracking the sea-level rise, movement of glaciers, and other environmental changes, providing critical data for climate change studies and disaster mitigation efforts.
- Disaster management: GPS will support faster, more accurate responses during natural disasters, such as earthquakes and floods, by guiding rescue operations and assisting in recovery efforts.

(j) **Improved Privacy and Security Features:** As GPS technology becomes increasingly integrated into daily life, addressing privacy and security concerns will be paramount:

- Stronger encryption: Future GPS systems are expected to feature more robust encryption methods to secure user location data, making it more difficult for malicious actors to intercept or spoof signals.
- User control: GPS systems will likely evolve to give users greater control over when and how their location is shared, enhancing privacy while still providing the benefits of location-based services.

The future of GPS holds promising advancements that will enhance its capabilities, making it increasingly vital in daily life. With centimetre-level accuracy, support for autonomous systems, and contributions to smart cities, GPS will drive innovation across diverse sectors. As it evolves, GPS will become more precise, integrated, and secure, playing a key role in the future of navigation, communication, and technology.

12.6 Conclusion and Endnotes

The Global Positioning System (GPS) has transformed from a military tool into a vital, everyday technology with far-reaching applications, ranging from navigation and transportation to emergency services and scientific research. Its impact is evident across numerous industries, such as agriculture, logistics, healthcare, and entertainment, where it enhances convenience, safety, and efficiency. As technology advances, GPS will continue to evolve, offering greater accuracy, reliability, and integration with emerging technologies like 5G, IoT, and quantum computing. The future of GPS holds promising advancements, including more precise positioning, autonomous systems, augmented reality integration, and improved privacy features. However, challenges such as signal interference, environmental constraints, and security vulnerabilities remain. To maximize GPS's potential, fostering GPS literacy is crucial, ensuring users can effectively and responsibly utilize the system while addressing privacy and security concerns. As GPS technology continues to evolve, its role in shaping the digital landscape will become increasingly significant, contributing to smarter cities, safer navigation, and more efficient operations across various sectors. Emphasizing GPS literacy and addressing its limitations will be essential to fully realizing the benefits of this indispensable technology in the future.

Endnotes

- GPS is widely used in daily life, from navigation to emergency services, demonstrating its versatility in enhancing quality of life and operational efficiency.
- While GPS is reliable, its limitations, such as signal blockage, accuracy issues, interference, and satellite visibility dependence, must be understood to fully grasp its constraints in various environments.
- Educating users on GPS functionality, ethical use, and applications is essential for maximizing benefits and ensuring safe, efficient, and responsible use.
- The future of GPS holds promising advancements, including higher accuracy, integration with other satellite systems, and new applications in fields like autonomous vehicles, agriculture, and environmental monitoring.
- As GPS use increases, there is a growing need for stronger privacy protections and security measures to prevent misuse, interference, and unauthorized tracking, safeguarding user trust.
- The integration of GPS with technologies such as 5G, IoT, AR, and quantum computing will enable more advanced systems, offering new opportunities and improving the overall user experience.

Further Readings

1. Bock, Y., & Leppard, N. (1990). *Global positioning system—An overview: Symposium no. 102 Edinburgh, Scotland, August 7–8, 1989*. Springer, . ISBN: 978-0-387-97266-4. https://doi.org/10.1007/978-1-4615-7111-7
2. Brozena, J. M. (1991). GPS and airborne gravimetry: Recent Progress and future plans. In K. P. Schwarz & G. Lachapelle (Eds.), *Kinematic systems in Geodesy, surveying, and remote sensing. International Association of Geodesy Symposia* (Vol. 107). Springer. https://doi.org/10.1007/978-1-4612-3102-8_44
3. Chen, D., & Cheng, R. (2019). Key scientific problems based on GPS positioning system. In *Intelligent processing algorithms and applications for GPS positioning data of Qinghai-Tibet railway*. Springer. https://doi.org/10.1007/978-3-662-58970-0_4
4. Hofmann-Wellenhof, B., Lichtenegger, H., & Collins, J. (1993). Future of GPS. In *Global positioning system*. Springer. https://doi.org/10.1007/978-3-7091-3293-7_13
5. Ibrahim, A., Zhang, H., Clinch, S., et al. (2021). From GPS to semantic data: How and why—A framework for enriching smartphone trajectories. *Computing, 103*, 2763–2787. https://doi.org/10.1007/s00607-021-00993-z
6. Raza, A., Hameed, S., & Macintyre, T. (2008). Global positioning system—Working and its applications. In K. Elleithy (Ed.), *Innovations and advanced techniques in systems, computing sciences and software engineering*. Springer. https://doi.org/10.1007/978-1-4020-8735-6_84
7. Shekhar, S., Feiner, S., & Aref, W. G. (2015). From GPS and virtual globes to spatial computing—2020. *GeoInformatica, 19*, 799–832. https://doi.org/10.1007/s10707-015-0235-9
8. Smith, B. J., Hart, K. M., Mazzotti, F. J., et al. (2018). Evaluating GPS biologging technology for studying spatial ecology of large constricting snakes. *Animal Biotelemetry, 6*, 1. https://doi.org/10.1186/s40317-018-0145-3
9. Teunissen, P. J. G., & Montenbruck, O. (Eds.). (2017). *Springer handbook of global navigation satellite systems*. Springer . ISBN: 978-3-319-42928-1. https://doi.org/10.1007/978-3-319-42928-1
10. Xu, P. (2002). New challenges in connection with GPS precise positioning. In J. Ádám & K. P. Schwarz (Eds.), *Vistas for geodesy in the new millennium. International Association of Geodesy Symposia* (Vol. 125). Springer. https://doi.org/10.1007/978-3-662-04709-5_59

Part IV
Geospatial Solutions in Coastal Ecological Monitoring and Management: A Theoretical Exploration

Chapter 13
Coastal Socio-ecological Systems: GIScience in Monitoring and Management

> *"Healthy coasts require healthy communities, scientific tools must help connect the dots between the environment and the people who depend on it."* — *Sylvia A. Earle.*

Abstract This chapter explores the application of geospatial technologies in the monitoring and management of coastal socio-ecological systems. Coastal ecosystems, among the Earth's most productive and diverse, are increasingly threatened by human activities and climate change. Geospatial technologies provide a toolkit for understanding and addressing these challenges, enabling the collection, analysis, and visualization of spatial data across vast areas and timeframes. The chapter begins by examining the state of coastal ecosystems and the issues they encounter, such as climate change, erosion, pollution, habitat loss, and other socio-ecological concerns. It then details the geospatial technologies employed in coastal ecology monitoring and management, including geographic information systems (GIS), remote sensing, and global positioning systems (GPS). The focus is on the application of these technologies in micro-level coastal planning, before expanding to cover broader topics such as coastal urbanization, forestry mapping, forest resource management, coastal disaster monitoring (e.g., flooding, storms, sea-level rise, erosion), socio-ecological vulnerability, hydrological management, shoreline change detection, oceanographic data collection, hazard mapping, and risk assessment. Additionally, the chapter addresses the role of geospatial technologies in sustainable coastal development, marine fisheries management, tourism, and land use planning. Therefore, this chapter highlights how geospatial technologies are transforming coastal ecology monitoring and management, offering new tools for improved decision-making and sustainability.

Keywords Coastal ecologies · GIScience · Socio-ecological systems · Coastal micro-level planning · Socio-ecological vulnerability · Coastal monitoring and management · Climate crisis · Application of geospatial technologies

K. D. Malakar et al., *Geospatial Technologies in Coastal Ecologies Monitoring and Management*, Advances in Geographic Information Science,
https://doi.org/10.1007/978-3-031-92017-2_13

This chapter delves into the pivotal role of geospatial technologies in monitoring and managing coastal socio-ecological systems. As coastal ecosystems face growing threats from human activities and climate change, geospatial tools like GIS, remote sensing, and GPS provide essential capabilities for addressing these challenges. The chapter examines the state of coastal ecosystems and highlights key issues such as erosion, pollution, and habitat loss. It further explores the application of these technologies in various aspects of coastal management, including disaster monitoring, vulnerability assessment, sustainable development, and marine resource management, offering innovative solutions for more informed decision-making.

Key Points of the Chapter

- A theoretical understanding of coastal micro-level planning.
- Understanding the application of geospatial technologies in coastal socio-ecological systems.
- Explains how various human and natural environments are related in a socio-ecological context.

13.1 Introduction

Coastal socio-ecological systems represent dynamic interfaces between land and sea, characterized by diverse ecosystems and the sustenance of millions of livelihoods. These regions play a vital role in global biodiversity, climate regulation, and economic development but face escalating pressures from human activities, climate change, and natural hazards. Effective understanding and management of these complex systems necessitate innovative approaches that integrate ecological, social, and technological dimensions. This chapter examines the transformative potential of Geographic Information Science (GIScience) in monitoring and managing coastal systems. By utilizing spatial data, remote sensing technologies, and geospatial analysis, GIScience provides critical tools for assessing environmental changes, resource distribution, and socio-economic impacts in these vulnerable regions. It further explores the role of GIScience in supporting evidence-based decision-making, ranging from disaster risk reduction and habitat restoration to sustainable resource management. The chapter also addresses the challenges and opportunities associated with adopting geospatial technologies for coastal governance, highlighting the importance of interdisciplinary collaboration and community participation. By tackling the intricacies of coastal socio-ecological systems, this chapter presents a comprehensive framework for leveraging GIScience to foster resilience, sustainability, and adaptive management in these vital areas.

13.2 Application of Geospatial Technologies in Coastal Micro-level Planning

Geospatial technologies are integral to micro-level planning in coastal regions, providing precise insights and solutions tailored to the specific challenges these areas face. By leveraging Geographic Information Systems (GIS), Remote Sensing, and Global Positioning Systems (GPS), these tools enable data-driven decision-making to address local needs effectively. GIS facilitates the creation of detailed maps and databases that capture the complexities of coastal environments, including land use, topography, infrastructure, and natural resources. By analysing and overlaying these data layers, planners can identify vulnerable zones, prioritize conservation efforts, and design strategies for sustainable development. Remote sensing, through satellite imagery and aerial surveys, offers real-time information on land cover changes, erosion patterns, and shoreline dynamics, supporting adaptive planning in response to environmental changes. GPS technology further enhances planning by enabling precise, location-based data collection essential for infrastructure development, habitat restoration, and disaster preparedness. By integrating these geospatial tools, micro-level planning for coastal regions ensures that local characteristics and vulnerabilities are addressed, promoting resilience and sustainable development while preserving these critical ecosystems for future generations.

The following discussions provide a theoretical exploration of the application of geospatial technologies in the monitoring and management of coastal socio-ecological systems. There are such as;

13.3 Use of Geospatial Technologies in Coastal Ecologies Monitoring and Management

13.3.1 Assessment of Coastal Urbanization and Urban Sprawl

Assessing coastal urbanization and urban sprawl is crucial for understanding the changing dynamics of coastal areas, where rapid development frequently intersects with vulnerable ecosystems and risks associated with sea-level rise and climate change. Geospatial technologies provide essential tools for evaluating these trends and their associated impacts (Fig. 13.1).

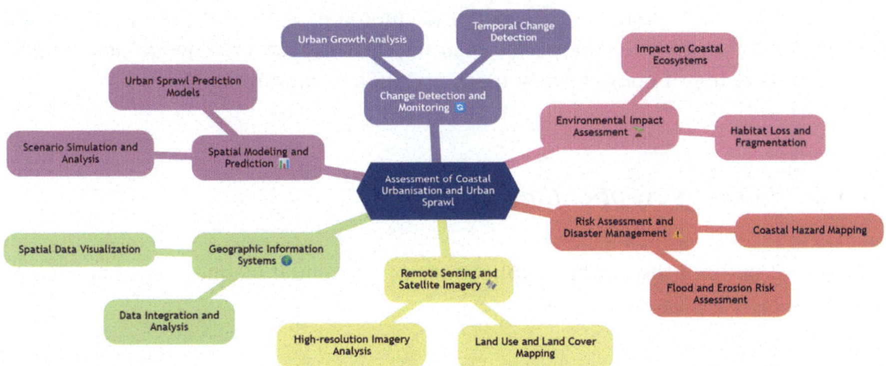

Fig. 13.1 Assessment of Coastal Urbanization and Urban Sprawl using Geospatial Technologies (Prepared by the authors)

(a) **Remote Sensing and Satellite Imagery:** Geospatial technologies utilize satellite-based remote sensing to provide extensive data for analysing urban growth over time. High-resolution satellite imagery distinguishes between natural land cover and built-up areas, offering critical insights into the scale and extent of coastal urbanization while identifying regions vulnerable to urban sprawl.
(b) **Geographic Information Systems (GIS):** GIS integrates and analyses diverse datasets, including land use, infrastructure, population density, and environmental factors. This spatial analysis helps uncover urban expansion patterns, evaluate land-use changes, and detect urban encroachment into ecologically sensitive zones.
(c) **Change Detection and Monitoring:** Geospatial technologies enable change detection by comparing historical and contemporary satellite imagery. This capability allows for quantifying land-cover changes and infrastructure growth over time, supporting the development of effective land-use policies.
(d) **Spatial Modeling and Prediction:** GIS-based spatial models predict future urbanization trends using historical data and socio-economic indicators. These models assist urban planners and policymakers in making evidence-based decisions regarding zoning, infrastructure, and disaster preparedness in coastal urban regions.
(e) **Environmental Impact Assessment:** Geospatial tools evaluate the environmental consequences of coastal urbanization, such as habitat degradation, water pollution, and coastal erosion, aiding sustainable urban planning and conservation initiatives.
(f) **Risk Assessment and Disaster Management:** Coastal urban areas face significant risks from natural disasters like hurricanes and tsunamis. Geospatial technologies identify high-risk zones, support disaster preparedness, and enhance strategies such as evacuation planning and early warning systems.

In summary, geospatial technologies are essential for assessing coastal urbanization and urban sprawl, providing valuable insights into the scale, patterns, and effects of urban expansion. By facilitating sustainable development practices, these tools help maintain the delicate equilibrium between human settlements and coastal ecosystems, especially in the context of global challenges such as climate change. Through data-driven analyses, geospatial technologies empower policymakers and planners to make informed decisions, ensuring the long-term resilience, sustainability, and well-being of coastal urban areas and their communities.

13.3.2 Forestry Application

Figure 13.2 outlines the various applications of geospatial technologies in different aspects of forestry. There are such as:

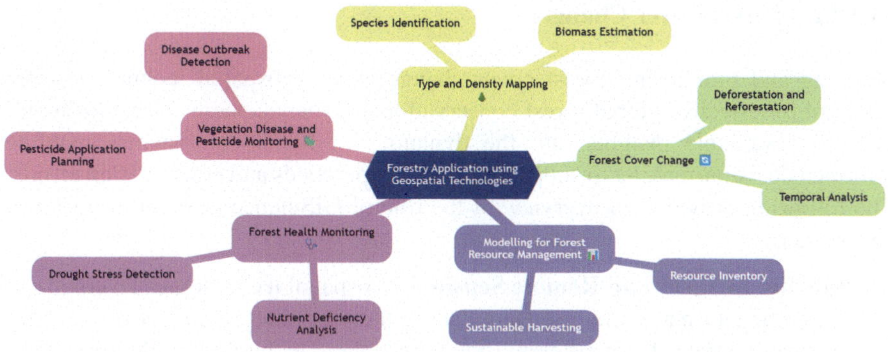

Fig. 13.2 Application of Geospatial Technologies in Forestry management (Prepared by the authors)

13.3.2.1 Type and Density Mapping

Forest type and density mapping, enabled by remote sensing and GIS, is vital for effective forest management, conservation, and land-use planning.

(a) **Remote Sensing:** Satellites with advanced sensors capture data across visible, infrared, and microwave wavelengths to differentiate forest types based on spectral signatures. Distinct reflectance patterns enable the mapping of deciduous, coniferous, and mixed forests.
(b) **Forest Density Estimation:** Technologies like LiDAR provide detailed canopy height and structure data, enabling accurate forest density assessments critical for evaluating forest health and carbon sequestration.
(c) **GIS Integration:** GIS combines remote sensing imagery, topographic data, and field inventories, offering a comprehensive understanding of forest type and density to support forestry and land-use planning.
(d) **Monitoring Changes:** Geospatial tools track forest changes over time, identifying deforestation, degradation, and regrowth to assess conservation efforts and target interventions.
(e) **Conservation and Management:** Accurate maps guide conservation strategies and sustainable management by identifying high-biodiversity areas, reforestation potential, and deforestation risks.

In summary, geospatial technologies serve as indispensable tools for mapping forest types and densities, offering critical insights to support informed decision-making in forest conservation, land-use planning, and sustainable management. These technologies play a vital role in preserving biodiversity, enhancing carbon sequestration, and ensuring the sustainable utilization of forest resources.

13.3.2.2 Forest Cover Change

Mapping and monitoring forest cover change using geospatial technologies are essential for effective global forest conservation and management. These technologies offer valuable insights into the dynamics of forest ecosystems, enabling informed decision-making to support sustainable practices and conservation efforts. The following provides an overview of the role of GIScience in monitoring forest cover change:

(a) **Satellite Imagery and Remote Sensing:** Geospatial technologies leverage satellite imagery and remote sensing data to monitor changes in forest cover. By analysing images from different time points, these technologies identify alterations such as deforestation, reforestation, or degradation.
(b) **Change Detection Algorithms:** Advanced algorithms in geospatial tools automatically detect and quantify changes in forest cover using spectral, temporal, and spatial data from satellite imagery.
(c) **Historical Data Comparison:** Comparing historical satellite data with current images helps assess long-term trends, pinpoint areas of concern, and track deforestation or afforestation.
(d) **Forest Loss and Gain Assessment:** Geospatial tools identify areas of forest loss due to activities like logging, agriculture, or urbanization, as well as regions undergoing regeneration or restoration.
(e) **Monitoring Deforestation:** Essential for conservation efforts, geospatial technologies track deforestation rates and identify hotspots, enabling targeted protection efforts.
(f) **Biodiversity Conservation:** Forest cover change mapping aids in evaluating the impact on biodiversity, helping to identify critical areas for conservation.
(g) **Carbon Sequestration:** Monitoring forest cover changes is key to estimating carbon sequestration levels, which is crucial for climate change mitigation.
(h) **Policy and Decision Support:** Data from geospatial technologies provides policymakers and land managers with evidence for formulating effective forest conservation and management strategies.

In summary, mapping and monitoring forest cover changes through geospatial technologies are essential for forest conservation, climate change mitigation, and biodiversity protection. These technologies provide critical insights that enable stakeholders to make informed decisions and implement proactive strategies for the sustainable management of forest resources.

13.3.2.3 Modelling for Forest Resource Management

Modelling for forest resource management utilizing geospatial technologies is an effective approach that integrates Geographic Information Systems (GIS), remote sensing, and advanced modelling techniques to enhance the sustainable use and conservation of forest resources. The following is an overview of the key components:

- **Spatial Analysis:** Geospatial technologies facilitate the integration of various spatial datasets, such as topography, land cover, and forest inventory data, into GIS platforms, forming the foundation for spatial analysis and modelling.
- **Forest Inventory and Assessment:** Comprehensive Forest inventories are vital for modelling. Geospatial tools enable the collection of essential forest data, including tree species, age, density, and volume, which help assess the current state of forest ecosystems.
- **Growth and Yield Models:** Geospatial technologies support the development of growth and yield models that predict forest development under various management scenarios. These models incorporate factors like tree growth rates, environmental conditions, and forest management practices.
- **Biodiversity Modeling:** Geospatial technologies assist in mapping and predicting biodiversity within forested areas, supporting efforts to conserve species distribution and their habitats.
- **Fire and Pest Risk Assessment:** Geospatial models assess forest fire and pest outbreak risks by analysing factors such as climate, vegetation, and historical fire data, enabling proactive risk management.
- **Carbon Sequestration Modeling:** Geospatial technologies help estimate the carbon sequestration potential of forests, contributing to climate change mitigation strategies.
- **Optimal Harvest Planning:** Forest management models optimize the timing and intensity of timber harvesting, balancing economic returns with sustainable forest practices.
- **Land Use Planning:** Geospatial technologies support land use planning by evaluating factors such as forest cover, soil quality, and environmental regulations, assisting in the integration of conservation and development goals.
- **Decision Support Systems:** Geospatial models function as decision support systems for forest managers and policymakers, enabling them to explore various management scenarios and make well-informed decisions regarding forest resource utilization.

In summary, geospatial technologies enable advanced modelling techniques for forest resource management, helping balance economic and environmental goals while promoting the sustainable use of forest ecosystems.

13.3.2.4 Forest Health Monitoring

Forest health monitoring through geospatial technologies is a critical aspect of sustainable forest management and conservation. Tools such as Geographic Information Systems (GIS), remote sensing, and Global Positioning Systems (GPS) play a pivotal role in evaluating and ensuring the vitality of forest ecosystems. The following is a concise overview of forest health monitoring:

(a) **Remote Sensing for Early Detection:** Remote sensing technologies, such as satellite and aerial imagery, provide a broad perspective on forests. By assessing vegetation health, forest managers can identify early signs of stress, disease, or pest infestations, allowing for timely interventions to mitigate potential damage.

(b) **Deforestation and Land-Use Changes:** Geospatial technologies track changes in forest cover and land use, helping identify threats like illegal logging and encroachment, which impact forest health. Monitoring these changes is crucial for taking timely corrective actions.

(c) **Vegetation Health Assessment:** Remote sensing data, particularly multispectral and hyperspectral imagery, are instrumental in assessing forest health. Indices like the Normalized Difference Vegetation Index (NDVI) offer valuable insights into forest vigor and stress levels, highlighting areas needing attention.

(d) **Pest and Disease Management:** Geospatial technologies assist in identifying pest and disease outbreaks. Mapping the extent of infestations allows for targeted control measures, helping to mitigate the spread of pests and diseases.

(e) **Carbon Sequestration Estimation:** Monitoring forest health is key for estimating carbon sequestration rates. Healthy forests absorb more CO2, contributing to climate change mitigation. Geospatial technologies enable the quantification of carbon stocks and track changes over time.

(f) **Fire Risk Assessment:** Forest fires pose a significant threat to forest health. Geospatial technologies assess fire risk by considering factors such as weather, topography, and fuel load, providing essential data for fire management and prevention strategies.

(g) **Biodiversity Conservation:** Healthy forests support biodiversity. Geospatial technologies monitor habitat quality, track endangered species distributions, and identify areas needing protection or restoration, contributing to conservation efforts.

(h) **Data Integration and Decision Support:** Geographic Information Systems (GIS) integrate various data sources, enabling forest managers to create comprehensive maps and decision support systems that inform effective, data-driven forest management strategies.

In conclusion, forest health monitoring through geospatial technologies is essential for the protection of forests, climate change mitigation, biodiversity conservation, and the preservation of the ecological, economic, and social benefits provided by healthy forests. These technologies facilitate proactive management, ensuring the long-term resilience and vitality of forest ecosystems.

13.3.2.5 Vegetation Disease and Pesticide Monitoring, and Others

Monitoring vegetation diseases and pesticide application using geospatial technologies is a vital component of contemporary agriculture and environmental management. Tools such as Geographic Information Systems (GIS), remote sensing, and GPS facilitate efficient surveillance and targeted responses to disease outbreaks and pesticide use. Following is an overview of their applications:

Vegetation Disease Monitoring

(a) **Remote Sensing for Early Detection:** Satellite and aerial imagery allow for the early detection of subtle vegetation health changes caused by diseases, often before they become visible to the naked eye. By analysing spectral indices like the Normalized Difference Vegetation Index (NDVI), geospatial technologies enable the timely identification of affected areas.
(b) **Spatial Analysis and Mapping:** GIS provides detailed maps illustrating the extent and severity of disease outbreaks. These spatial insights support targeted response strategies and efficient resource allocation for managing vegetation diseases.
(c) **Temporal Analysis:** Geospatial technologies enable the monitoring of disease progression over time. This temporal assessment aids in evaluating the effectiveness of management interventions and forecasting potential future outbreaks.

Pesticide Monitoring

(a) **GPS-Enabled Applications:** GPS technology integrated into pesticide application equipment ensures precise mapping of pesticide use, enabling accurate records of application locations and methods. This enhances regulatory compliance while reducing the risk of overuse and environmental harm.
(b) **Data Integration:** GIS facilitates the integration of pesticide application data with environmental and land-use datasets. This comprehensive analysis provides insights into application patterns and their potential effects on ecosystems and water quality.
(c) **Environmental Impact Assessment:** Geospatial technologies support the evaluation of pesticide use impacts by mapping runoff and drift, which can affect nearby water bodies and non-target species. These insights inform mitigation measures and guide regulatory policies.

Other Applications

(a) **Crop Health Assessment:** Geospatial technologies facilitate the monitoring of crop health by detecting issues such as nutrient deficiencies, water stress, and pest infestations. These insights enable precision agriculture practices, optimizing resource allocation and enhancing crop productivity.
(b) **Invasive Species Monitoring:** GIS and remote sensing play a vital role in tracking the spread of invasive plant species and pests. Early detection and targeted control efforts help mitigate ecosystem disruptions and protect biodiversity.
(c) **Wildfire Risk Assessment:** By analysing vegetation health and fuel load, geospatial technologies contribute to wildfire risk assessment. This information aids in prioritizing fire prevention measures and implementing effective mitigation strategies.

In summary, geospatial technologies are indispensable for monitoring vegetation diseases, pesticide use, and other critical aspects of agriculture and environmental management. These tools support data-driven decision-making, improve resource efficiency, and foster sustainable and environmentally responsible practices in agriculture and natural resource stewardship.

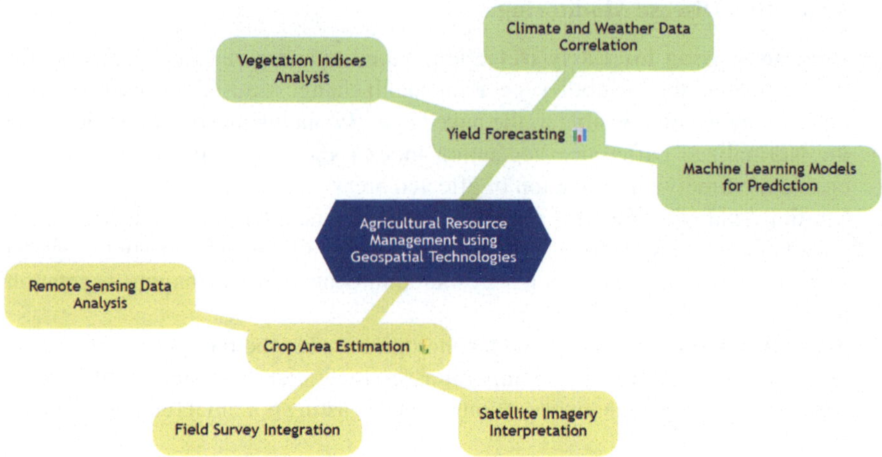

Fig. 13.3 Application of Geospatial Technologies in Agricultural Resource Management (Prepared by the authors)

13.3.3 Agricultural Resource Management

Figure 13.3 illustrates the application of geospatial technologies in agricultural resource management, particularly in crop area estimation and yield forecasting. These applications include:

13.3.3.1 Crop Area Estimation

Crop area estimation through geospatial technologies integrates remote sensing, GIS, and data analytics to deliver accurate and timely information on the extent and distribution of agricultural crops. Below is a concise overview of key components:

(a) **Satellite Imagery and Remote Sensing:** Remote sensing captures detailed surface data, detecting variations in crop characteristics such as colour, texture, and spectral properties.
(b) **Image Classification:** Algorithms classify satellite imagery into land cover types, including crops, forests, and urban areas. Crop types are identified based on distinct spectral signatures.
(c) **Spectral Indices:** Indices like NDVI assess vegetation density and health, enabling differentiation of crop types and growth stages.
(d) **Ground Truthing:** Field verification ensures classification accuracy by comparing remote sensing data with ground observations, calibrating results.
(e) **GIS Integration:** GIS combines remote sensing data, ground observations, and historical records to create precise crop distribution maps.
(f) **Temporal Analysis:** Multiple satellite images track crop development and land-use changes over time.
(g) **Statistical Modeling:** Models extrapolate crop estimates using factors like weather, soil conditions, and historical yields.

(h) **Benefits:**

- Provides accurate and timely data for policymakers and farmers
- Enhances crop yield forecasts and food security assessments
- Supports efficient resource allocation for agricultural planning and disaster management

Therefore, the crop area estimation via geospatial technologies is an indispensable tool for agriculture, supporting sustainable land use, informed decision-making, and improved resilience to challenges in the agricultural sector.

13.3.3.2 Yield Forecasting

Crop yield forecasting is a data-driven method that integrates remote sensing, Geographic Information Systems (GIS), and advanced analytics to predict agricultural harvest outputs. This approach plays a crucial role in food security, agricultural planning, and risk management. The following is a concise overview of the key processes involved:

(a) **Remote Sensing Data**: Geospatial technologies utilize satellite and aerial imagery, equipped with diverse sensors, to capture vital information regarding crops and their surrounding environment. These sensors gather data on factors such as crop health, vegetation density, and land surface temperature.
(b) **Vegetation Indices**: A crucial aspect of crop yield forecasting is the application of vegetation indices, with the Normalized Difference Vegetation Index (NDVI) being the most prevalent. NDVI assesses the greenness and vigor of vegetation, offering valuable insights into crop health and growth dynamics.
(c) **GIS Integration**: Geographic Information Systems (GIS) facilitate the integration of remote sensing data with other pertinent information, including weather patterns, soil characteristics, historical crop yields, and agricultural practices. This integration forms a comprehensive spatial database that supports in-depth analysis.
(d) **Statistical Models**: Various statistical and machine learning models are employed to analyse the integrated geospatial data, identifying relationships between crop performance and environmental factors. These models leverage historical data to train algorithms that predict crop yields based on current environmental conditions.
(e) **Temporal Analysis**: Crop yield forecasting is a continuous process that involves periodic assessments throughout the growing season. Analysts compare present crop conditions with historical data and evaluate the potential impacts of weather and other variables on final yields.
(f) **Spatial Variability**: Geospatial technologies allow for the identification of spatial variability within fields, enhancing the accuracy of crop performance predictions across different sections of the same area. This capability is particularly beneficial for precision agriculture and the efficient allocation of resources.

(g) **Benefits**: The use of geospatial technologies in crop yield forecasting offers numerous advantages, such as:

- Early detection of potential yield-limiting factors like drought, pests, or disease
- Improved decision-making for farmers, enabling adjustments to crop management practices in response to forecasts
- Enhanced food security by assisting governments and organizations in anticipating and planning for potential crop production shortfalls
- Efficient resource allocation, such as water, fertilizers, and pesticides, to optimize crop yields

In summary, crop yield forecasting through the use of geospatial technologies serves as a crucial tool for agricultural stakeholders. It facilitates more informed and timely decision-making, which is vital for optimizing crop productivity, mitigating risks, and safeguarding global food security.

13.3.4 Water Resource-Related Application

Figure 13.4 illustrates the application of geospatial technologies in water resource-related applications, particularly in land surface water and subsurface water. These applications include:

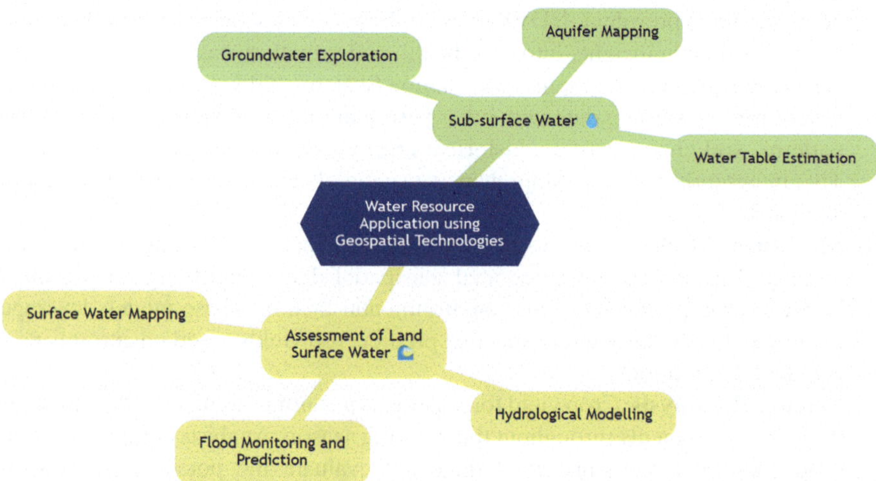

Fig. 13.4 Application of Geospatial Technologies in Water Resource Management (Prepared by the authors)

13.3.4.1 Assessment of Land Surface Water

The evaluation of land surface water through geospatial technologies is an essential aspect of environmental monitoring and water resource management. Technologies such as remote sensing, GIS, and GPS offer robust tools for the precise and efficient assessment of land surface water. The following provides a concise overview of this process:

- **Remote Sensing for Water Detection.**
  (a) **Satellite Imagery:** Geospatial technologies use satellite imagery to identify and map surface water bodies such as lakes, rivers, and wetlands, leveraging the spectral properties of water.
  (b) **Water Indices:** Remote sensing data is processed to generate indices like the Normalized Difference Water Index (NDWI) and Modified NDWI, enhancing the visibility of water features in imagery.
- **Water Extent Mapping:**
  (a) **GIS Integration:** Geographic Information Systems integrate remote sensing data with geographic and hydrological information, enabling detailed mapping of land surface water extent.
  (b) **Change Detection:** Geospatial technologies monitor changes in water extent by comparing historical and current data, revealing trends such as seasonal fluctuations or long-term water level changes.
- **Flood Mapping and Risk Assessment:**
  (a) **Flood Inundation Mapping:** Remote sensing and GIS are instrumental in creating flood inundation maps, aiding flood risk identification during extreme weather events.
  (b) **Early Warning Systems:** Geospatial technologies support real-time monitoring of water levels and rainfall, contributing to early warning systems for flood mitigation.
- **Water Quality Assessment:**
  (a) **Water Quality Monitoring:** Geospatial technologies track water quality by analysing parameters like turbidity and temperature, detecting pollution or algal blooms through changes in water colour and clarity.
  (b) **Eutrophication Detection:** Remote sensing and GIS help identify areas affected by eutrophication, signalling nutrient overload and potential ecosystem degradation.
- **Resource Management:**
  (a) **Water Resource Allocation:** Geospatial technologies provide accurate data on surface water availability, guiding decisions for water allocation in agriculture, industry, and domestic use.

In summary, geospatial technologies are crucial for water resource management, flood monitoring, environmental protection, and disaster risk reduction, offering valuable insights for sustainable water management.

13.3.4.2 Sub-surface Water, and Others

Sub-surface water, commonly known as groundwater, is vital for drinking water supply, agricultural support, and ecosystem maintenance. Geospatial technologies, such as Geographic Information Systems (GIS), remote sensing, and geophysical methods, are essential tools for assessing, managing, and safeguarding sub-surface water resources. The following provides a concise overview of the role of these technologies in sub-surface water management and related applications:

- **Groundwater Assessment and Monitoring:**
  (a) **Geophysical Surveys:** Geospatial technologies use geophysical methods such as resistivity, electromagnetic induction, and seismic surveys to map subsurface geology and identify groundwater sources, aiding in the assessment of both quantity and quality.
  (b) **Aquifer Mapping:** GIS is employed to map aquifers, providing insights into their boundaries, depth, thickness, and fluctuations in the water table.
  (c) **Groundwater Quality Assessment:** Remote sensing and GIS help monitor groundwater quality by analysing land use patterns, pollution sources, and hydrogeological conditions, essential for safeguarding drinking water.
- **Groundwater Modeling and Simulation:**
  (a) **Numerical Models:** Geospatial technologies support numerical models that simulate groundwater flow and transport, helping predict changes in groundwater levels and assess impacts of factors like pumping and recharge.
  (b) **Scenario Analysis:** Groundwater models enable scenario analysis to evaluate the effects of management strategies and land use changes on groundwater resources.
- **Land Subsidence Detection:**
  (a) **InSAR:** Interferometric Synthetic Aperture Radar (InSAR) detects land subsidence caused by over-extraction of groundwater, identifying areas at risk of infrastructure damage and increased flood vulnerability.
- **Sustainable Water Resource Management:**
  (a) **Resource Allocation:** GIS plays a key role in managing groundwater allocation for agriculture, industry, and municipal use, promoting sustainable practices.
  (b) **Drought Monitoring:** Geospatial technologies help monitor drought conditions, assess groundwater levels, and provide early warnings for potential water shortages, supporting mitigation and conservation efforts.

- **Environmental Conservation:**
  (a) **Wetland Protection:** Geospatial tools help identify and protect wetlands dependent on groundwater, evaluating the impacts of groundwater extraction on these ecosystems.
  (b) **Ecosystem Restoration:** GIS aids in planning groundwater-based ecosystem restoration, such as aquifer recharge projects to support natural habitats.

In summary, geospatial technologies are essential for effective groundwater management, enabling the assessment, modelling, and protection of this vital resource while supporting sustainable water use and ecosystem conservation.

13.3.5 Assessment of Coastal Wasteland

Monitoring and modelling coastal wasteland ecosystems using geospatial technologies is essential for comprehending and managing these delicate and frequently degraded environments. Advanced geospatial tools, such as Geographic Information Systems (GIS), remote sensing, and Global Positioning Systems (GPS), provide invaluable resources for assessing, monitoring, and modelling coastal wastelands. The following is a concise overview:

- **Remote Sensing for Coastal Wasteland Monitoring**
  (a) **Satellite Imagery:** Geospatial technologies employ satellite imagery to capture high-resolution data on coastal wasteland areas. This data is instrumental in identifying land cover changes, monitoring erosion, and tracking the expansion of degraded or barren coastal zones.
  (b) **Spectral Analysis:** Remote sensing data facilitates spectral analysis to distinguish between natural ecosystems, barren land, and degraded areas by analysing their unique spectral signatures. This approach enhances the assessment of coastal wasteland extent and ecological health.
- **GIS-Based Spatial Analysis**
  (a) **Land Cover Classification:** Geographic Information Systems (GIS) are utilized to classify and map diverse land cover categories within coastal wastelands. This classification provides a detailed understanding of the spatial distribution and ecological characteristics of these areas.
  (b) **Change Detection:** Geospatial tools enable change detection by comparing historical and contemporary satellite imagery. This process monitors shifts in coastal wasteland patterns, such as expansion caused by erosion or human reclamation efforts.
- **Coastal Erosion Modeling**
  (a) **Erosion Risk Assessment:** Geospatial technologies support the development of erosion risk models by incorporating factors such as wave energy,

sea-level rise, and coastal geomorphology. These models identify areas highly vulnerable to coastal erosion.
(b) **Mitigation Planning:** Erosion modelling informs the planning of mitigation strategies, such as beach nourishment, dune stabilization, or coastal vegetation restoration. These insights aid in designing effective measures for coastal protection and sustainability.

- **Habitat Restoration and Conservation**

  (a) **Site Selection:** Geospatial technologies identify optimal locations for habitat restoration by considering factors like soil quality, hydrological patterns, and proximity to intact ecosystems.
  (b) **Restoration Monitoring:** Remote sensing and GIS enable the tracking of habitat restoration progress. Metrics such as vegetation growth, land cover changes, and wildlife return are monitored to evaluate restoration success.

- **Climate Change Adaptation**

  (a) **Sea-Level Rise Modeling:** Geospatial tools model the impacts of sea-level rise on coastal wastelands, offering critical data for adaptation strategies like buffer zone creation, managed retreat, or infrastructure redesign.
  (b) **Carbon Sequestration Estimation:** Monitoring coastal wasteland ecologies provides insights into their carbon sequestration potential, contributing to climate change mitigation by quantifying their role in carbon storage.

In summary, geospatial technologies are vital for monitoring and modelling coastal wasteland ecologies. They support conservation, habitat restoration, erosion management, and climate change adaptation, contributing to the preservation and resilience of these vulnerable ecosystems.

13.3.6 Assessment of Coastal Disaster Monitoring and Management

13.3.6.1 Flood

The assessment of coastal flood disasters using geospatial technologies is crucial for effective disaster management, response, and mitigation. Due to their proximity to the sea and evolving climate patterns, coastal regions are highly prone to flooding. Tools like Geographic Information Systems (GIS), remote sensing, and satellite imagery play an indispensable role in this process. The following is an overview of coastal flood monitoring and management:

- **Early Warning Systems**

  (a) **Precipitation Monitoring:** Remote sensing monitors rainfall patterns in coastal regions, providing critical data for predicting heavy rainfall and potential flooding.

(b) **Sea Level Monitoring:** Satellite altimetry and tide gauge data track sea level rise, enabling real-time early warnings for coastal flooding.

- **Flood Inundation Mapping**

(a) **Flood Extent Maps:** GIS and remote sensing create maps highlighting areas at risk of coastal flooding, aiding in evacuation and resource planning.
(b) **Hydrodynamic Models:** Geospatial tools develop models simulating floodwater flow and spread based on topography, tides, and rainfall.

- **Vulnerability Assessment**

(a) **Land Use and Population Mapping:** GIS identifies high-density and critical infrastructure zones within flood-prone areas, supporting evacuation and resource planning.
(b) **Infrastructure Vulnerability:** Geospatial analysis assesses the flood susceptibility of roads, bridges, and utilities, guiding upgrades and retrofitting efforts.

- **Disaster Response**

(a) **Resource Allocation:** Real-time geospatial data aids in allocating resources and determining safe routes for emergency response and relief efforts.
(b) **Damage Assessment:** Pre- and post-flood satellite imagery facilitates rapid damage evaluation to prioritize recovery and reconstruction.

- **Mitigation and Preparedness**

(a) **Flood Risk Maps:** Geospatial technologies identify high-risk areas, informing land-use planning, building codes, and zoning regulations.
(b) **Scenario Modeling:** GIS-based simulations evaluate the effectiveness of flood mitigation strategies, such as levees, floodwalls, and coastal restoration.

In summary, geospatial technologies are essential for assessing coastal flood disasters, enabling timely warnings, flood mapping, vulnerability analysis, disaster response, and mitigation planning. They play a critical role in reducing the impact of coastal floods on communities and ecosystems, strengthening resilience against sea-level rise and climate change.

13.3.6.2 Storms

Assessing coastal disasters caused by storms and cyclones using geospatial technologies is vital for effective preparedness, response, and mitigation. Given the high vulnerability of coastal regions to these events, accurate monitoring is essential. The following overview of geospatial applications in this context:

- **Cyclone Tracking and Early Warning**
  (a) **Satellite Imagery and Remote Sensing**: Geospatial technologies utilize satellite data to monitor cyclone development, movement, and intensity, enabling early warnings and impact predictions.
  (b) **GIS-Based Tracking**: GIS integrates real-time storm data with coastal topography and population information to identify high-risk areas, aiding in evacuation planning and resource allocation.
- **Storm Surge Modeling**
  (a) **Topographic Data**: High-resolution topographic data is used to model storm surge scenarios, simulating inundation and identifying vulnerable areas for evacuation planning.
  (b) **Hydrodynamic Modeling**: Numerical models incorporating wind speed, tide levels, and coastal geography simulate storm surges, with geospatial tools enhancing their development and visualization.
- **Impact Assessment and Damage Mapping**
  (a) **Post-Storm Satellite Imagery**: Remote sensing compares pre- and post-storm images to assess damage, identify affected areas, and monitor changes in land cover.
  (b) **GIS for Damage Assessment**: GIS creates damage maps to prioritize rescue and recovery efforts, guiding resource allocation effectively.
- **Evacuation and Resource Deployment**
  (a) **Spatial Analysis for Evacuation Routes**: GIS identifies safe evacuation routes based on storm tracking, avoiding areas prone to flooding and storm surges.
  (b) **Resource Allocation**: Geospatial tools provide critical data for efficient deployment of emergency teams and supplies to affected regions.
- **Resilience and Preparedness Planning**
  (a) **Vulnerability Mapping**: GIS-based mapping identifies communities and infrastructure at risk, informing land-use planning and disaster preparedness strategies.
  (b) **Scenario Modeling**: Geospatial technologies simulate storm impacts to guide resilient infrastructure planning and emergency response strategies.

In summary, geospatial technologies are essential for assessing and mitigating the impacts of coastal storms and cyclones. They provide real-time data, early warnings, impact assessments, and support disaster response and preparedness, helping save lives and reduce damage in vulnerable coastal areas.

13.3.6.3 Sea-Level Rise

Assessing the consequences of sea-level rise on coastal areas using geospatial technologies is important for understanding the risks of rising sea levels and formulating effective mitigation and adaptation strategies. Tools like Geographic Information Systems (GIS), remote sensing, and satellite imagery are integral to this process. The following is a brief overview:

- **Sea-Level Rise Monitoring**

  (a) **Satellite Altimetry**: Geospatial technologies use satellite altimetry to measure sea surface height, providing crucial data on sea-level trends.
  (b) **Tide Gauge Networks**: GPS-equipped tide gauges record precise sea-level measurements, which are integrated with satellite data to assess long-term changes.

- **Coastal Vulnerability Mapping**

  (a) **Topographic Data**: High-resolution digital elevation models (DEMs) and LIDAR data in GIS help identify low-lying coastal areas vulnerable to sea-level rise.
  (b) **Flood Inundation Modeling**: Geospatial tools develop models to simulate sea-level rise impacts, predicting flood-prone areas under different scenarios.

- **Impact Assessment**

  (a) **Land Cover Change**: Remote sensing and GIS monitor how sea-level rise alters shorelines, wetlands, and coastal ecosystems.
  (b) **Infrastructure Exposure**: Geospatial analysis identifies critical infrastructure at risk from sea-level rise, helping prioritize adaptation actions.

- **Adaptation and Mitigation Planning**

  (a) **Scenario Modeling**: GIS-based models evaluate potential sea-level rise impacts and adaptation strategies, supporting decision-making and policy development.
  (b) **Coastal Zoning**: GIS aids in creating hazard maps that inform land-use planning and zoning regulations, guiding responsible coastal development.

- **Community Outreach and Education**

  (a) **Public Awareness**: Geospatial tools provide visualizations and maps to raise awareness of sea-level rise impacts on coastal communities.
  (b) **Stakeholder Engagement**: GIS enables interactive maps and models, fostering informed discussions between communities and decision-makers on adaptation strategies.

In summary, geospatial technologies are crucial for assessing and managing the consequences of sea-level rise on coastal regions. They offer valuable data, predictive models, and decision support tools to help communities and governments enhance resilience, protect infrastructure, and ensure the sustainability of vulnerable coastal areas.

13.3.6.4 Coastal Erosion and Others

Assessing coastal disasters, such as coastal erosion, using geospatial technologies is essential for understanding and managing the vulnerabilities of coastal areas to natural hazards. Tools like Geographic Information Systems (GIS), remote sensing, and satellite imagery play a crucial role in evaluating and mitigating the impacts of such disasters. The following is an overview regarding this:

- **Coastal Erosion Assessment**
  - (a) **Remote Sensing for Monitoring**: Satellite and aerial imagery are used to track changes in coastal landforms and erosion rates, identifying vulnerable areas and assessing the severity of erosion.
  - (b) **Beach and Shoreline Mapping**: GIS creates detailed maps of coastal features, aiding in erosion analysis and informing management strategies.
  - (c) **Digital Elevation Models (DEMs)**: High-resolution DEMs are used to detect elevation changes over time, providing insights into coastal erosion dynamics.
- **Hurricane and Storm Surge Modeling**
  - (a) **Meteorological Data Integration**: Geospatial technologies integrate meteorological data with coastal topography to model the impact of hurricanes and storm surges.
  - (b) **Flood Inundation Mapping**: GIS-based models simulate coastal flooding, assisting in vulnerability assessment and disaster response planning.
- **Risk Assessment**
  - (a) **Vulnerability Analysis**: GIS assesses the vulnerability of communities, infrastructure, and ecosystems, considering factors like population density and natural defences.
  - (b) **Scenario Planning**: GIS models various disaster scenarios to support emergency preparedness and risk reduction efforts.
- **Erosion Mitigation and Coastal Protection**
  - (a) **Beach Nourishment Planning**: Geospatial tools identify optimal sand sources and strategies for beach nourishment projects aimed at mitigating erosion.
  - (b) **Structural Solutions**: GIS aids in planning and designing coastal defence structures, such as seawalls and breakwaters, to protect infrastructure.
- **Environmental Impact Assessment**

(a) **Ecosystem Monitoring**: Remote sensing and GIS monitor changes in coastal ecosystems, including wetlands and mangroves, due to erosion.
(b) **Habitat Restoration**: Geospatial technologies guide habitat restoration by identifying critical areas and assessing project success.

In summary, geospatial technologies are essential for assessing and mitigating coastal disasters such as erosion, hurricanes, and storm surges. They offer critical data, predictive models, and decision support tools that assist coastal communities and governments in enhancing resilience, protecting infrastructure, and ensuring the sustainability of vulnerable areas.

13.3.7 Coastal Vulnerability Assessment: Socio-ecological Vulnerability

Assessing coastal vulnerability, particularly socio-ecological vulnerability, through geospatial technologies is crucial for understanding the intricate relationships between natural hazards and human-ecological systems in coastal regions. This assessment identifies high-risk areas and communities, providing critical insights to guide adaptation and mitigation strategies (Fig. 13.5). This is a concise overview:

- **Data Integration:**

(a) **Geospatial Data Sources:** Geospatial technologies utilize remote sensing imagery, GIS databases, and socioeconomic data to build comprehensive datasets for assessing vulnerability.
(b) **Socio-ecological Factors:** These technologies combine social and ecological variables, such as population density, infrastructure, land use, coastal ecosystems, and climate data, into an integrated analysis platform.

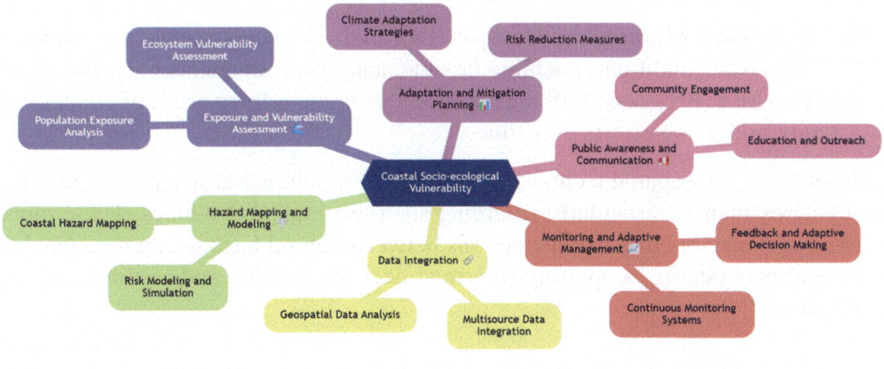

Fig. 13.5 Application of Geospatial Technologies in Coastal socio-ecological vulnerability assessment (Prepared by the authors)

- **Hazard Mapping and Modeling:**
  (a) **Coastal Hazard Identification:** Geospatial tools identify and map hazards like erosion, sea-level rise, storm surges, and flooding, assessing their intensity, frequency, and potential impacts.
  (b) **Sensitivity Analysis:** GIS-based sensitivity analysis examines how socio-ecological factors interact with hazards, identifying the most sensitive areas and communities.
- **Exposure and Vulnerability Assessment:**
  (a) **Exposure Mapping:** Geospatial technologies map exposed areas and assets, including infrastructure, housing, and natural habitats, to coastal hazards.
  (b) **Socio-ecological Vulnerability Index:** Vulnerability indices integrate socio-ecological variables to highlight the most at-risk communities and ecosystems.
- **Adaptation and Mitigation Planning:**
  (a) **Scenario Modeling:** Geospatial technologies facilitate modelling to assess the impacts of adaptation strategies like coastal defences, ecosystem restoration, or managed retreat.
  (b) **Prioritization:** GIS-based analyses help identify and prioritize vulnerable areas requiring urgent adaptation and mitigation measures.
- **Public Awareness and Communication:**
  (a) **Interactive Maps:** Geospatial tools produce interactive vulnerability maps to engage stakeholders, decision-makers, and communities in risk understanding and planning.
  (b) **Community Engagement:** GIS enables participatory processes, incorporating local knowledge into adaptation and mitigation strategies.
- **Monitoring and Adaptive Management:**
  (a) **Real-Time Monitoring:** Geospatial technologies support real-time tracking of coastal conditions, enabling flexible, adaptive management approaches.
  (b) **Tracking Progress:** GIS aids in monitoring the effectiveness of adaptation and mitigation efforts over time.

In summary, geospatial technologies are vital for comprehensive coastal vulnerability assessments, particularly regarding socio-ecological dynamics. These tools offer critical insights into the interactions between natural hazards, human populations, and ecosystems, supporting informed decision-making and resilient coastal management.

13.3.8 Coastal Hydrological Monitoring and Management

Coastal hydrological monitoring and management through geospatial technologies are vital for sustainably managing water resources, ecosystems, and human activities in coastal regions. Tools like GIS, remote sensing, and satellite imagery are integral to this effort. An overview is provided (Fig. 13.6):

- **Water Quality Monitoring:**
  - (a) **Remote Sensing:** Geospatial technologies utilize satellite and aerial imagery to monitor coastal water quality parameters, such as turbidity, chlorophyll-a levels, and pollution. This aids in assessing ecosystem health and water usability.
  - (b) **Temporal Analysis:** Continuous monitoring reveals seasonal trends and identifies pollution sources, supporting targeted remediation efforts.
- **Hydrological Modeling:**
  - (a) **Rainfall and Runoff:** Geospatial tools support hydrological models simulating rainfall-runoff processes in coastal watersheds, predicting freshwater inputs and their impact on water quality and sediment transport.
  - (b) **Flood Modeling:** GIS-based models assess coastal flood risks from heavy rainfall, storm surges, and sea-level rise, aiding disaster preparedness.
- **Tidal and Storm Surge Monitoring:**
  - (a) **Tide Gauges:** GPS-equipped tide gauge networks monitor tidal fluctuations, with geospatial technologies analysing data to predict tides, storm surges, and erosion.

Fig. 13.6 Application of Geospatial Technologies in Coastal Hydrological Monitoring and Management (Prepared by the authors)

(b) **Storm Surge Models:** GIS integrates real-time meteorological and oceanographic data to forecast storm surge impacts, supporting early warnings and coastal defence strategies.

- **Ecosystem Mapping and Conservation:**

  (a) **Habitat Mapping:** Geospatial technologies map ecosystems like mangroves, seagrass beds, and coral reefs, guiding conservation and assessing human impacts.
  (b) **Biodiversity Monitoring:** GIS tracks changes in biodiversity and species distribution, highlighting responses to pollution or habitat loss.

- **Coastal Land-Use Planning:**

  (a) **Zoning and Management:** GIS supports sustainable coastal land-use planning by delineating zones for development, conservation, and recreation, balancing human activities with environmental protection.
  (b) **Beach Nourishment:** GIS identifies optimal locations for nourishment projects to mitigate erosion and preserve ecosystems.

- **Disaster Risk Reduction:**

  (a) **Response Planning:** Real-time geospatial data informs disaster response strategies during coastal events such as storms and floods.
  (b) **Evacuation Planning:** GIS identifies vulnerable communities and develops evacuation routes to enhance safety during emergencies.

In summary, geospatial technologies are vital for coastal hydrological monitoring and management, enabling water quality assessment, hydrological modelling, tidal and storm surge monitoring, ecosystem conservation, sustainable land-use planning, and disaster preparedness. These tools strengthen the resilience of coastal communities and ecosystems while promoting sustainable management practices.

13.3.9 Coastal Zone Mapping and Monitoring

Coastal zone mapping and monitoring through geospatial technologies are essential for the sustainable conservation and management of dynamic, ecologically sensitive coastal regions. Tools like Geographic Information Systems (GIS), remote sensing, and satellite imagery enable comprehensive assessment and management of these areas. Following key aspects include (Fig. 13.7):

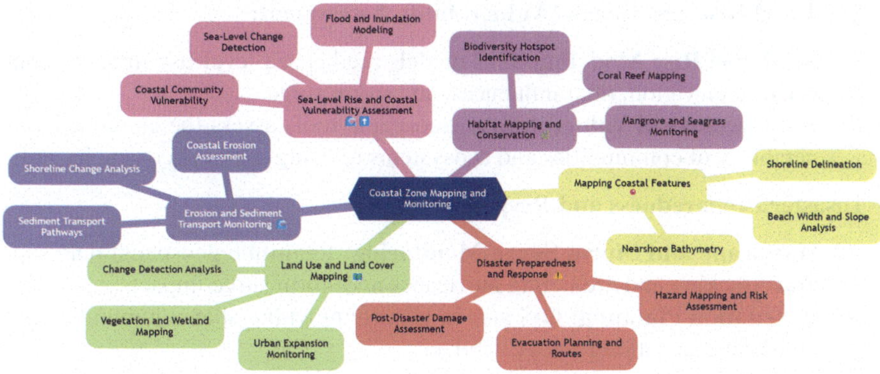

Fig. 13.7 Application of geospatial technologies in coastal zone mapping and monitoring (Prepared by the authors)

- **Mapping Coastal Features:**
  - (a) **Topographic Mapping:** Geospatial technologies produce detailed topographic maps of shorelines, beaches, estuaries, and dunes, providing insights into the physical characteristics of coastal zones.
  - (b) **Bathymetric Mapping:** Remote sensing and bathymetric surveys map seafloor topography and depths, supporting navigation, habitat mapping, and marine resource management.
- **Land Use and Land Cover Mapping:**
  - (a) **Land-Use Classification:** GIS and remote sensing classify and map land use in coastal areas, distinguishing natural habitats, urban zones, agriculture, and wetlands.
  - (b) **Change Detection:** Tracking land-use changes over time helps identify trends and threats to coastal ecosystems.
- **Erosion and Sediment Transport Monitoring:**
  - (a) **Erosion Analysis:** Geospatial tools monitor shoreline changes and beach profiles to assess erosion risks and inform coastal protection strategies.
  - (b) **Sediment Transport Models:** GIS-based models simulate sediment movement along coasts, aiding in understanding sediment dynamics and coastal morphology.
- **Habitat Mapping and Conservation:**
  - (a) **Habitat Identification:** Mapping critical habitats like mangroves, coral reefs, and seagrass beds guides conservation and restoration efforts.
  - (b) **Biodiversity Monitoring:** GIS tracks biodiversity changes and species responses to stressors such as pollution and habitat loss.

- **Sea-Level Rise and Coastal Vulnerability Assessment:**
  (a) **Sea-Level Rise Modeling:** GIS models predict sea-level rise impacts, considering elevation, tidal influences, and storm surge.
  (b) **Socio-ecological Vulnerability:** Geospatial tools assess the combined vulnerability of communities and ecosystems to rising seas and coastal hazards.
- **Disaster Preparedness and Response:**
  (a) **Hurricane and Storm Surge Monitoring:** Real-time geospatial data supports tracking and predicting hurricanes and storm surge impacts.
  (b) **Evacuation Planning:** GIS aids in identifying vulnerable communities and planning evacuation routes effectively.

In summary, geospatial technologies are vital for mapping and monitoring coastal zones, supporting sustainable management, conservation, disaster preparedness, and adaptation to challenges like sea-level rise and erosion. They are essential for protecting the ecological, social, and economic health of coastal regions.

13.3.10 Shoreline Change Detection

Shoreline change detection using geospatial technologies is essential for coastal management and environmental monitoring. By analysing historical and current spatial data, these technologies, such as GIS, remote sensing, and satellite imagery, help track coastline evolution over time. The following is an overview:

- **Data Acquisition:**
  (a) **Historical Imagery:** Geospatial technologies utilize historical satellite and aerial imagery to establish baseline coastal positions over time, providing context for shoreline analysis.
  (b) **Contemporary Imagery:** Current high-resolution satellite and aerial imagery captures up-to-date coastline data, crucial for precise shoreline detection.
- **Shoreline Extraction:**
  (a) **Image Processing:** Remote sensing techniques process imagery to extract shoreline positions for different time periods. Automated algorithms enhance efficiency in this process.
  (b) **Georeferencing:** The extracted shoreline data are georeferenced to align spatially with geographic coordinates, ensuring accurate analysis.
- **Change Detection Analysis:**
  (a) **GIS-Based Comparison:** GIS tools compare historical and contemporary shoreline positions, revealing areas of erosion or accretion.

(b) **Quantitative Assessment:** Shoreline change rates are calculated to assess the magnitude and direction of shifts, identifying regions undergoing rapid erosion or accretion.

- **Identification of Drivers:**

(a) **Natural vs. Anthropogenic Factors:** Detection helps identify the drivers of shoreline changes, including natural forces like wave action and sea-level rise, and human activities such as construction or coastal development.

- **Coastal Management and Planning:**

(a) **Adaptation Strategies:** Data from shoreline change detection inform coastal management, guiding strategies to mitigate erosion, protect infrastructure, and manage development.
(b) **Risk Assessment:** Geospatial technologies evaluate coastal community vulnerability to shoreline changes, supporting disaster preparedness, and land-use planning.

- **Environmental Impact Assessment:**

(a) **Ecosystem Monitoring:** Shoreline change detection evaluates impacts on coastal ecosystems like wetlands and dunes, supporting conservation and restoration efforts.
(b) **Sediment Transport Analysis:** Understanding shoreline changes is essential for analysing sediment dynamics, vital for maintaining healthy coastal environments.

In summary, shoreline change detection through geospatial technologies is an essential tool for effective coastal management, environmental conservation, and disaster risk reduction. By offering insights into coastline dynamics and the underlying factors driving these changes, geospatial technologies support informed decision-making and promote sustainable coastal development.

13.3.11 Oceanographic Data Collection

Oceanographic data collection using geospatial technologies is essential for understanding ocean dynamics, including physical, chemical, and biological processes. Technologies such as satellite remote sensing, GIS, and underwater sensors facilitate the collection, analysis, and visualization of oceanographic data. The following provides an overview of this process:

- **Remote Sensing and Satellite Technology:**
  (a) **Sea Surface Temperature (SST):** Satellites equipped with infrared and microwave sensors gather SST data, critical for studying ocean circulation, climate patterns, marine ecosystems, and hurricane forecasting.
  (b) **Ocean Colour and Chlorophyll Concentration:** Satellites measure ocean surface colour, which correlates with chlorophyll and phytoplankton levels, providing insights into marine productivity, biodiversity, and ecosystem health.
  (c) **Sea Surface Height (SSH):** Satellite altimeters measure changes in SSH, essential for studying ocean circulation, currents, and identifying ocean gyres and eddies.
- **Bathymetric Mapping:**
  (a) **Multibeam Sonar:** Geospatial technologies use multibeam sonar to measure ocean depth and create high-resolution bathymetric maps, crucial for navigation and marine resource exploration.
  (b) **GIS for Ocean Mapping:** GIS helps analyse and visualize bathymetric data, assisting in hazard identification and seafloor topography mapping.
- **Oceanographic Sensors and Buoy Networks:**
  (a) **Oceanographic Sensors:** In-situ sensors deployed on buoys and research vessels collect real-time data on ocean parameters such as temperature, salinity, and nutrient concentrations.
  (b) **Data Transmission:** Geospatial technologies enable real-time data transmission from remote sensors to research centers for climate studies and ecosystem monitoring.
- **Current Profiling and Drifters:**
  (a) **Current Profilers:** Instruments like Acoustic Doppler Current Profilers (ADCPs) measure ocean currents at various depths, aiding in the study of water mass movements and their impact on weather and ecosystems.
  (b) **Drifters:** GPS-equipped surface drifters track ocean current movements, offering valuable data for ocean circulation modelling and climate research.
- **GIS-Based Ocean Modeling:**
  (a) **Numerical Models:** Geospatial technologies support the development of ocean models simulating physical, chemical, and biological processes, helping predict ocean behaviours like temperature and sea level changes.
  (b) **Ecosystem Modeling:** GIS-based ecosystem models predict marine species distribution, including commercially valuable and endangered species.

In summary, geospatial technologies play a key role in oceanographic data collection, offering essential insights into the physical and ecological characteristics of oceans. These technologies support climate research, marine resource management, navigation, disaster forecasting, and ocean conservation efforts.

13.3.12 Coastal Hazard Mapping and Risk Assessment

Coastal hazard mapping and risk assessment with geospatial technologies are vital for identifying and managing the risks associated with natural hazards in coastal regions. Tools like Geographic Information Systems (GIS), remote sensing, and satellite imagery enable effective hazard identification, mapping, and assessment. The following provides an overview:

- **Identification and Mapping of Coastal Hazards:**
  (a) **Storm Surge and Flooding**: Geospatial technologies map areas at risk of storm surges and flooding by analysing topography, tidal patterns, and historical storm data to identify vulnerable zones.
  (b) **Coastal Erosion**: Remote sensing and GIS monitor shoreline changes over time, pinpointing erosion-prone areas that pose risks to infrastructure.
  (c) **Tsunami and Seismic Hazards**: GIS-based seismic hazard maps and tsunami models assess risks from earthquakes and tsunamis along coastlines.
  (d) **Sea-Level Rise**: Geospatial tools model the effects of sea-level rise, identifying areas at risk of flooding.
- **Vulnerability Assessment:**
  (a) **Socioeconomic Vulnerability**: GIS integrates socioeconomic data, such as population density and infrastructure, to identify vulnerable communities and assets.
  (b) **Ecological Vulnerability**: Assessments also evaluate the exposure of critical coastal ecosystems to hazards, guiding conservation efforts.
- **Risk Modeling and Analysis:**
  (a) **Risk Scenarios**: Geospatial technologies combine hazard and vulnerability data to model potential impacts on coastal areas.
  (b) **Quantitative Risk Assessment**: Quantitative analysis estimates the likelihood and potential impact of hazards, guiding risk management decisions.
- **Early Warning Systems:**
  (a) **Monitoring and Alerting**: Geospatial technologies enable real-time hazard monitoring, issuing timely warnings to communities and responders.
  (b) **Evacuation Planning**: GIS supports evacuation route planning, ensuring safe movement of people during coastal emergencies.

- **Adaptation and Mitigation Strategies:**
  (a) **Infrastructure Planning**: GIS aids in designing resilient coastal infrastructure, such as flood barriers and seawalls, to mitigate hazard impacts.
  (b) **Land-Use Planning**: Geospatial tools inform zoning regulations to prevent development in high-risk coastal areas and promote sustainable land-use practices.
  (c) **Ecosystem-Based Adaptation**: GIS identifies opportunities for ecosystem restoration, such as mangrove or dune stabilization, to reduce hazard risks naturally.

In summary, geospatial technologies are vital for coastal hazard mapping and risk assessment, offering critical insights for disaster preparedness, emergency response, land-use planning, and infrastructure development. These technologies enhance resilience and reduce vulnerability to coastal hazards and their impacts.

13.3.13 Coastal Land Use Planning

Coastal land use planning, supported by geospatial technologies, is a vital process for allocating land in coastal areas, balancing environmental, social, and economic considerations. Technologies such as Geographic Information Systems (GIS), remote sensing, and satellite imagery play a key role in ensuring effective and sustainable land use in these dynamic zones. The following provides an overview:

- **Data Integration and Mapping:**
  (a) **Spatial Data Collection:** Geospatial technologies gather data on topography, land cover, habitats, coastal processes, infrastructure, and socioeconomic indicators, integrating these into GIS for comprehensive spatial analysis.
  (b) **Coastal Zoning Maps:** GIS is used to create maps that designate areas for various land uses, such as residential, commercial, industrial, conservation, and recreation, forming the foundation of land-use planning.
- **Environmental Impact Assessment:**
  (a) **Ecosystem Mapping:** Geospatial tools help identify and map coastal ecosystems like wetlands and mangroves, guiding conservation efforts and evaluating land-use impacts.
  (b) **Habitat Protection:** GIS aids in identifying ecologically sensitive areas that need protection from development, preserving biodiversity and ecosystem sustainability.

- **Risk Assessment and Mitigation:**
  (a) **Coastal Hazard Mapping:** Geospatial technologies map hazards such as storm surges, erosion zones, and flood risks, informing land-use decisions to minimize exposure.
  (b) **Resilient Infrastructure Planning:** GIS supports planning resilient infrastructure, including seawalls and flood barriers, to safeguard coastal communities.
- **Climate Adaptation Strategies:**
  (a) **Sea-Level Rise Modeling:** Geospatial technologies model potential sea-level rise impacts, guiding land-use decisions to accommodate future changes.
  (b) **Managed Retreat:** GIS identifies areas for managed retreat, relocating vulnerable infrastructure and communities away from high-risk coastal zones.
- **Public Participation and Decision-Making:**
  (a) **Interactive Maps:** Geospatial tools provide interactive maps to engage the public and stakeholders in land-use planning, allowing for input and informed decision-making.
  (b) **Scenario Modeling:** GIS enables scenario creation to visualize and compare potential impacts of different planning options.
- **Regulatory Compliance and Enforcement:**
  (a) **Monitoring and Compliance:** GIS-based systems ensure adherence to land-use regulations by detecting unauthorized developments and enforcing zoning rules.
  (b) **Permitting Process:** Geospatial technologies streamline the permitting process by providing spatial data for assessing the environmental implications of proposed developments.

In summary, geospatial technologies are essential for coastal land use planning, offering the data and tools needed for informed, sustainable, and resilient decision-making. They help protect coastal ecosystems, reduce disaster risks, and support the development of sustainable coastal communities.

13.3.14 Sustainable Tourism

Sustainable tourism in coastal regions, or coastal ecotourism, focuses on preserving coastal environments and cultures while offering enriching experiences for travellers. Geospatial technologies, such as GIS, remote sensing, and satellite imagery, play a crucial role in managing and promoting sustainable tourism practices in these areas. The following provides an overview:

- **Destination Assessment and Planning:**
  (a) **Geospatial Data for Site Selection:** GIS identifies optimal locations for eco-friendly resorts and recreational activities, considering environmental, accessibility, and cultural factors.
  (b) **Carrying Capacity Analysis:** Geospatial technologies assess the sustainable visitor capacity of destinations, informing tourism management strategies.
- **Environmental Conservation:**
  (a) **Ecosystem Monitoring:** Remote sensing and GIS track environmental changes, including deforestation, coral reef health, and wildlife habitat status.
  (b) **Protected Area Management:** Geospatial tools aid in monitoring and managing protected areas, enforcing regulations, and conducting ecological assessments.
- **Responsible Travel Planning:**
  (a) **Ecotourism Routes:** GIS-based tools design eco-friendly travel itineraries, ensuring minimal environmental impact.
  (b) **Cultural Heritage Preservation:** GIS supports the documentation and responsible visitation of cultural heritage sites.
- **Transportation and Infrastructure:**
  (a) **Sustainable Transport Planning:** Geospatial technologies assist in planning eco-friendly transportation options, such as bike-sharing programs and public transport in coastal areas.
  (b) **Infrastructure Development:** GIS aids in planning sustainable infrastructure with minimal environmental impact, such as eco-lodges.
- **Visitor Management and Education:**
  (a) **Visitor Tracking:** Geospatial technologies manage visitor flows, preventing overcrowding and environmental damage.
  (b) **Interpretive Mapping:** GIS creates interactive maps and apps, providing educational information on the environment and sustainable practices.
- **Community Engagement:**
  (a) **Local Empowerment:** Geospatial tools involve local communities in tourism planning, ensuring economic benefits and active participation in decision-making.
  (b) **Livelihood Diversification:** GIS identifies alternative livelihoods, reducing local dependence on tourism.
- **Monitoring and Reporting:**
  (a) **Impact Assessment:** Geospatial technologies assess the environmental and social impact of tourism, informing management adjustments.

(b) **Transparency and Accountability:** Satellite imagery and GIS enable transparent reporting on sustainability efforts, aiding informed traveller choices.

In summary, geospatial technologies are essential for managing sustainable tourism in coastal regions. They support responsible travel planning, environmental conservation, and community empowerment, while minimizing negative impacts and maximizing tourism's benefits to local economies and ecosystems.

13.3.15 Sustainable Development

Sustainable development in coastal regions is crucial for balancing economic growth, community well-being, and environmental conservation. Geospatial technologies, such as GIS, remote sensing, and satellite imagery, are key tools in promoting and implementing sustainable practices in these areas. The following provides an overview:

- **Coastal Zoning and Land-Use Planning:**
  (a) **Zoning Regulation:** GIS delineates coastal zones for residential, commercial, conservation, and recreational uses, guiding sustainable development and protecting sensitive areas.
  (b) **Sensitivity Analysis:** Geospatial tools identify vulnerable areas, such as those prone to erosion and flooding, supporting informed land-use decisions.
- **Environmental Conservation:**
  (a) **Ecosystem Mapping:** GIS and remote sensing monitor critical ecosystems like mangroves and coral reefs, informing conservation and land management efforts.
  (b) **Habitat Restoration:** Geospatial tools help identify areas for effective ecosystem restoration, enhancing coastal resilience.
- **Climate Change Adaptation:**
  (a) **Sea-Level Rise Modeling:** Geospatial technologies model the impacts of sea-level rise, informing adaptation strategies and infrastructure planning.
  (b) **Managed Retreat:** GIS identifies areas for managed retreat, allowing communities to relocate from vulnerable coastlines.
- **Infrastructure Development:**
  (a) **Resilient Infrastructure:** Geospatial tools aid in planning flood defences, stormwater management, and sustainable transportation networks.
  (b) **Sustainable Energy:** GIS assists in site selection for renewable energy projects like wind and solar, reducing reliance on fossil fuels.

- **Disaster Preparedness and Response:**
  (a) **Early Warning Systems:** Geospatial technologies support early warning systems for coastal hazards, informing disaster response.
  (b) **Evacuation Planning:** GIS identifies safe evacuation routes and locations during coastal disasters, ensuring public safety.
- **Community Engagement and Livelihoods:**
  (a) **Community Involvement:** Geospatial technologies involve local communities in decision-making, ensuring benefits from sustainable development.
  (b) **Alternative Livelihoods:** GIS identifies alternative livelihoods, reducing environmental dependence.
- **Monitoring and Reporting:**
  (a) **Environmental Impact Assessment:** Geospatial tools assess the environmental impacts of development projects, ensuring adherence to sustainability standards.
  (b) **Transparency and Accountability:** Satellite imagery and GIS ensure transparent reporting of sustainable development practices, promoting accountability.

In summary, geospatial technologies play a crucial role in promoting sustainable development in coastal regions. They provide essential data and tools to support informed decisions that balance economic growth, environmental protection, climate resilience, and community well-being, ensuring both prosperity and environmental responsibility.

13.3.16 Others Applications

13.3.16.1 Participatory Decision Making for Man-Environment Conflicts in Coastal Area

Participatory decision-making in addressing man-environment conflicts in coastal areas involves engaging local communities, stakeholders, and experts to balance environmental conservation with human development. Geospatial technologies, including GIS, remote sensing, and participatory mapping, are essential tools in supporting this collaborative process. The following provides an overview:

- **Community Engagement:**
  (a) **Participatory Mapping:** Geospatial tools enable communities to create maps highlighting ecologically important, culturally significant, and conflict-prone areas, reflecting local knowledge and concerns.
  (b) **Stakeholder Involvement:** GIS integrates input from diverse stakeholders, including fishermen, conservationists, and government agencies, ensuring varied perspectives in decision-making.

- **Conflict Identification and Analysis:**
  (a) **Spatial Analysis:** GIS identifies potential conflict zones by analysing the spatial distribution of resources, economic activities, and conservation areas.
  (b) **Data Visualization:** Geospatial technologies simplify the visualization of complex data, aiding stakeholders in understanding conflicts' nature and extent.
- **Scenario Modeling:**
  (a) **GIS-Based Models:** Geospatial tools simulate different development scenarios, helping stakeholders assess trade-offs and make informed decisions.
  (b) **Predictive Analysis:** Remote sensing and GIS can forecast environmental changes (e.g., sea-level rise, erosion), enabling proactive conflict resolution.
- **Resource Management:**
  (a) **Resource Allocation:** GIS assists in distributing resources like fishing quotas and tourism permits fairly, balancing human and environmental needs.
  (b) **Sustainable Practices:** Geospatial tools promote sustainable practices, such as eco-friendly tourism and sustainable fishing.
- **Conflict Resolution Strategies:**
  (a) **Consensus Building:** Geospatial data and maps help facilitate negotiations, providing a common reference to evaluate the impact of decisions.
  (b) **Adaptive Management:** GIS supports adaptive strategies, allowing decisions to evolve based on real-time data and changing conditions.
- **Communication and Education:**
  (a) **Public Awareness:** Geospatial tools enhance public understanding by creating informative maps and interactive visuals of coastal conflicts.
  (b) **Training and Capacity Building:** Communities and stakeholders can be trained to use geospatial technologies for monitoring, data collection, and conflict resolution.

In summary, geospatial technologies are essential in participatory decision-making for resolving man-environment conflicts in coastal areas. They offer the tools and data needed for informed, collaborative solutions that balance community needs, economic development, and environmental conservation.

13.3.16.2 Locate Marine Fishing Grounds

Locating marine fishing grounds using geospatial technologies is vital for identifying optimal fishing areas, supporting sustainable and productive fisheries. The following outlines their application:

- **Satellite-Based Ocean Observation:** Geospatial technologies use satellite data to monitor ocean parameters such as sea surface temperature, chlorophyll concentration, and ocean colour. These observations are crucial for identifying productive fishing grounds by detecting phytoplankton distribution, which forms the base of the marine food chain.
- **Remote Sensing for Fisheries:** Multispectral and hyperspectral sensors capture detailed oceanic data, detecting underwater features like seamounts and reefs that attract fish. Satellite radar imagery identifies surface manifestations of productive fishing zones, such as seagrass beds and upwelling areas.
- **Bathymetric Data:** Bathymetric surveys and sonar systems create detailed seafloor maps, aiding the identification of underwater structures that are favourable to fish. This data is vital for selecting appropriate fishing gear and techniques based on seabed depth and topography.
- **Fisheries Data Integration:** Geospatial technologies integrate diverse data sources such as fisheries records, vessel monitoring systems (VMS), and oceanographic data into one platform. This integration provides a comprehensive view of fishing trends, catch data, and environmental conditions to predict future fishing opportunities.
- **Fisheries Management:** Geospatial technologies assist in managing exclusive economic zones (EEZs) and marine protected areas (MPAs), regulating fishing activities. Vessel tracking systems ensure compliance with designated zones and catch limits.
- **Weather and Oceanographic Data:** Real-time weather and oceanographic data provided by geospatial platforms help fishermen optimize fishing strategies, improve safety, and make informed decisions based on current conditions like temperature and ocean currents.

In summary, geospatial technologies are vital for locating marine fishing grounds by offering detailed insights into ocean conditions, seafloor topography, and marine life distribution. These tools improve fisheries management, support sustainable practices, and protect marine ecosystems.

13.3.16.3 Application of Geospatial Technologies in Climate Migration Policy Making

Geospatial technologies are crucial in climate migration policy-making, offering essential data and tools to assess the consequences of climate change on vulnerable populations and support the creation of effective policies. The following highlights key applications in this context:

- **Vulnerability Assessment:**

  (a) **GIS Mapping:** GIS maps climate hazards (e.g., drought, sea-level rise, extreme weather) and overlays them with socio-economic data to identify vulnerable communities at risk of displacement.

(b) **Data Visualization:** Interactive maps help policymakers understand and communicate the potential effects of climate change on affected areas.

- **Population Tracking:**

(a) **Remote Sensing:** Satellite imagery monitors changes in population density and settlement patterns in at-risk areas, aiding in tracking migration trends.
(b) **Census and Surveys:** Geospatial technologies support population censuses and surveys to collect demographic data on climate-displaced communities.

- **Risk Assessment and Early Warning:**

(a) **Environmental Monitoring:** Real-time environmental monitoring enables early warning systems for climate-related disasters, improving response to displacement events.
(b) **Predictive Modeling:** Climate models integrated with geospatial data predict future scenarios, informing long-term policy planning.

- **Site Selection for Resettlement:**

(a) **GIS-Based Site Selection:** GIS identifies climate-resilient resettlement locations, considering factors like elevation, resources, and infrastructure.
(b) **Infrastructure Planning:** Geospatial tools assist in planning essential infrastructure for resettlement areas, including housing and services.

- **Monitoring and Evaluation:**

(a) **Data Analytics:** Geospatial data analytics and machine learning track the effectiveness of climate migration policies and address displaced populations' needs.
(b) **Impact Assessment:** Geospatial tools support impact assessments, enabling policymakers to adjust migration strategies based on socio-economic and environmental outcomes.

- **Cross-Border Migration:**

(a) **Border Surveillance:** Geospatial technologies monitor cross-border migration, helping governments manage flows and provide humanitarian aid.
(b) **Regional Collaboration:** GIS facilitates regional cooperation to address climate-induced displacement across national borders.

In summary, geospatial technologies are essential for climate migration policymaking, offering critical spatial data and analysis to understand the links between climate change, vulnerability, and migration. They enable governments and organizations to devise strategies that support displaced communities and foster climate-resilient societies.

13.4 Conclusion and Endnotes

In conclusion, geospatial technologies are versatile tools with wide-ranging applications across environmental management, disaster response, urban planning, agriculture, and more. These technologies, including GIS, remote sensing, satellite imagery, and modelling, have revolutionized how we collect, analyse, and visualize spatial data, providing valuable insights for informed decision-making. They play a crucial role in monitoring and managing natural resources, supporting sustainable practices, and aiding disaster preparedness. In agriculture, geospatial tools enable precision farming and crop management. They are also vital for addressing climate change, assessing risks, and supporting climate migration policy. In coastal and marine management, these technologies assist in fisheries assessment, habitat mapping, and protecting ecosystems while supporting local communities. Additionally, they promote sustainable tourism and land use planning in coastal regions. Geospatial technologies are key to resolving man-environment conflicts and fostering participatory decision-making. As they continue to evolve, their integration into policymaking and resource management will be crucial for a sustainable and resilient future. Ongoing research and collaboration will ensure they remain instrumental in solving global challenges.

Further Readings

1. Bahaire, T., & Elliott-White, M. (1999). The application of geographical information systems (GIS) in sustainable tourism planning: A review. *Journal of Sustainable Tourism, 7*(2), 159–174. https://doi.org/10.1080/09669589908667333
2. Bartlett, D., & Celliers, L. (Eds.). (2017). *Geoinformatics for marine and coastal management*. CRC Press (Taylor and Francis Group) ISBN: 9781351722308.
3. Campagna, M. (2018). *GIS for sustainable development*. CRC Press (Taylor and Francis Group) ISBN: 9780367578046.
4. Cenci, L., Disperati, L., Persichillo, M. G., Oliveira, E. R., Alves, F. L., & Phillips, M. (2018). Integrating remote sensing and GIS techniques for monitoring and modeling shoreline evolution to support coastal risk management. *GIScience & Remote Sensing, 55*(3), 355–375. https://doi.org/10.1080/15481603.2017.1376370
5. Chettiyam Thodi, M. F., Gopinath, G., Surendran, U. P., Prem, P., Al-Ansari, N., & Mattar, M. A. (2023). Using RS and GIS techniques to assess and monitor coastal changes of Coastal Islands in the marine environment of a humid tropical region. *Water, 15*(21), 3819. https://doi.org/10.3390/w15213819
6. Durap, A. (2024). Mapping coastal resilience: A Gis-based Bayesian network approach to coastal hazard identification for Queensland's dynamic shorelines. *Anthropocene Coasts, 7*, 23. https://doi.org/10.1007/s44218-024-00060-y
7. Gopinath, G., Thodi, M. F. C., Surendran, U. P., Prem, P., Parambil, J. N., Alataway, A., et al. (2023). Long-term shoreline and islands change detection with digital shoreline analysis using RS data and GIS. *Water, 15*(2), 244. https://doi.org/10.3390/w15020244
8. Güneralp, İ., Castillo, C. R., & Hales, B. U. (2024). Monitoring surface-water connectivity in a coastal lowland river using field-based data loggers. *Journal of Hydrology, 633*, 130972. https://doi.org/10.1016/j.jhydrol.2024.130972

9. Islam, F. R., & Mamun, K. A. (2015). GIS based water quality monitoring system in pacific coastal area: A case study for Fiji. In *2015 2nd Asia-Pacific world congress on computer science and engineering (APWC on CSE)* (pp. 1–7). IEEE. https://doi.org/10.1109/APWCCSE.2015.7476226
10. Liu, K., Li, X., Shi, X., et al. (2008). Monitoring mangrove forest changes using remote sensing and GIS data with decision-tree learning. *Wetlands, 28*, 336–346. https://doi.org/10.1672/06-91.1
11. Marghany, M. (Ed.). (2016). *Applied studies of coastal and marine environments*. IntechOpen ISBN: 978-953-51-2549-5. https://doi.org/10.5772/60743
12. Nageswara Rao, K., Subraelu, P., Naga Kumar, K. C. V., Demudu, G., Hema Malini, B., Ratheesh, R., Rajawat, A. S., & Ajai. (2011). Climate change and sea-level rise: Impact on agriculture along Andhra Pradesh coast—A geomatics analysis. *Journal of the Indian Society of Remote Sensing, 39*(3), 415–422. https://doi.org/10.1007/s12524-011-0120-4
13. Rehman, S., Jahangir, S., & Azhoni, A. (2022). GIS based coastal vulnerability assessment and adaptation barriers to coastal regulations in Dakshina Kannada district, India. *Regional Studies in Marine Science, 55*, 102509. https://doi.org/10.1016/j.rsma.2022.102509
14. Ricketts, P. J. (1992). Current approaches in geographic information systems for coastal management. *Marine Pollution Bulletin, 25*(1–4), 82–87. https://doi.org/10.1016/0025-326X(92)90192-9
15. Samanta, S., & Paul, S. K. (2016). Geospatial analysis of shoreline and land use/land cover changes through remote sensing and GIS techniques. *Modeling Earth Systems and Environment, 2*, 108. https://doi.org/10.1007/s40808-016-0180-0
16. Shalaby, A., & Tateishi, R. (2007). Remote sensing and GIS for mapping and monitoring land cover and land-use changes in the Northwestern coastal zone of Egypt. *Applied Geography, 27*(1), 28–41. https://doi.org/10.1016/j.apgeog.2006.09.004
17. Yang, X. (Ed.). (2009). *Remote sensing and geospatial technologies for coastal ecosystem assessment and management* (Lecture Notes in Geoinformation and Cartography). Springer . ISBN: 978-3-540-88182-7. https://doi.org/10.1007/978-3-540-88183-4
18. Yang, X. (Ed.). (2009). *Remote sensing and geospatial technologies for coastal ecosystem assessment and management*. Springer . ISBN: 978-3-540-88182-7. https://doi.org/10.1007/978-3-540-88183-4
19. Yuan, M. (2021). Geographical information science for the United Nations' 2030 agenda for sustainable development. *International Journal of Geographical Information Science, 35*(1), 1–8. https://doi.org/10.1080/13658816.2020.1766244

Chapter 14
Integrating GIScience in Coastal Policy: A Socio-ecological Perspective

> *"Data is the new science. Big Data holds the answers. Are you asking the right questions?" — Patrick P. Gelsinger.*

Abstract This chapter explores the integration of GIScience in coastal policy-making, focusing on its role in managing coastal socio-ecological systems. It emphasizes the importance of policy in coastal management and how GIScience tools, such as mapping, spatial analysis, data integration, and predictive modelling, enhance decision-making processes. GIScience facilitates multi-disciplinary analyses, offering a comprehensive approach to understanding and addressing the complexities of coastal systems. The chapter highlights real-world examples of GIScience integration, demonstrating the benefits of spatial data for sustainable resource management. It also addresses the challenges of GIScience integration, such as data availability, technical limitations, and stakeholder capacity, while recognizing these challenges as opportunities for innovation and collaboration. Looking ahead, the future of GIScience in coastal policy is promising, with advancements in remote sensing, artificial intelligence, and mobile technology providing new avenues for policy development. The increasing importance of participatory GIS (PGIS) underscores the value of community engagement in policymaking. By incorporating local knowledge, PGIS promotes inclusive, adaptive, and effective coastal management strategies. In summary, the chapter integrating GIScience into coastal policy is crucial for developing resilient, sustainable solutions to the challenges facing coastal regions worldwide.

Keywords Coastal Policy · Socio-ecological systems · GIScience in coastal policy · Participatory GIS · Coastal management · Socio-ecological sustainability

The integration of GIScience into coastal policy-making represents a transformative approach to managing complex socio-ecological systems. This chapter delves into the critical role of GIScience tools, such as spatial analysis, mapping, and predictive

K. D. Malakar et al., *Geospatial Technologies in Coastal Ecologies Monitoring and Management*, Advances in Geographic Information Science,
https://doi.org/10.1007/978-3-031-92017-2_14

modelling, in enhancing decision-making and fostering sustainable resource management. Highlighting real-world applications and addressing challenges like data limitations and stakeholder capacity, it emphasizes innovation and collaboration as pathways forward. With advancements in technologies like GIScience, alongside the growing prominence of participatory GIS, the chapter underscores the necessity of inclusive and adaptive strategies for resilient coastal management in an evolving global context.

Key Points of the Chapter

- The effectiveness of GIScience in coastal policy research.
- Understanding the concept and integration of GIScience with socio-ecological systems.
- The importance and future scope of GIScience and Participatory GIS in coastal policy research.

14.1 Concept and Integration

14.1.1 Coastal Socio-ecological Systems

Coastal socio-ecological systems (SES) are dynamic and complex, emerging from the interactions between social, economic, and ecological components in coastal areas. These systems reflect the intricate relationships between human communities and their coastal surroundings, including resource use, ecosystem services, and the cultural, economic, and social aspects of coastal livelihoods. Various factors shape coastal SES, such as climate change, population growth, urbanization, and natural processes like tidal dynamics, erosion, and sediment deposition. The integration of coastal socio-ecological systems is essential for fostering sustainable development and building resilience. This approach integrates perspectives from natural sciences, social sciences, and policy studies, fostering a comprehensive understanding of human-ecological system interactions for more effective management. Such integration is facilitated through frameworks that evaluate ecosystem services, socio-economic dependencies, governance structures, and adaptive capacity. For instance, the combination of geospatial technologies like GIS and remote sensing with participatory methods such as community-based resource management ensures that both scientific data and local knowledge guide decision-making processes. Effective management of coastal SES requires a careful balance between resource utilization and conservation. Strategies such as ecosystem-based management (EBM), sustainable fisheries practices, and coastal zone planning aim to safeguard biodiversity while supporting the livelihoods of coastal communities. Additionally, climate adaptation strategies such as habitat restoration, mangrove reforestation, and disaster risk reduction are essential for strengthening the resilience of these systems against climate-related challenges like sea-level rise, cyclones, and saltwater intrusion.

In summary, coastal socio-ecological systems highlight the intricate connections between human and natural systems in coastal regions. Their integration is key to addressing contemporary challenges, promoting sustainability, and fostering resilience in vulnerable coastal regions. By prioritizing collaborative governance, inclusive policies, and adaptive management, coastal SES can be effectively managed to benefit both ecosystems and human well-being.

14.1.2 Importance of Policy in Coastal Management

Policy is central to coastal management, providing a structured framework for balancing ecological sustainability, economic growth, and the well-being of coastal communities. Coastal areas frequently face competing demands, including tourism, fisheries, industrial development, and conservation, all of which can place considerable pressure on natural resources and ecosystems. Well-designed policies establish guidelines and regulations to minimize stakeholder conflicts and promote the fair and sustainable management of coastal resources. They also tackle pressing issues such as coastal erosion, habitat degradation, pollution, and climate change impacts, including sea-level rise and extreme weather events. A well-designed coastal management policy integrates scientific knowledge, local expertise, and stakeholder involvement to promote evidence-based decision-making. For example, policies such as integrated coastal zone management (ICZM) aim to align economic activities with environmental protection while accounting for social considerations. Additionally, policies facilitate the enforcement of vital measures, including marine protected areas (MPAs), sustainable fisheries practices, and disaster risk reduction plans, all of which are crucial for preserving biodiversity and strengthening community resilience.

Therefore, coastal policies are also crucial for climate adaptation and disaster preparedness. They offer direction for implementing strategies like mangrove restoration, shoreline protection, and early warning systems to reduce the impacts of natural hazards. By establishing clear objectives, allocating resources, and fostering collaboration among government agencies, communities, and other stakeholders, these policies ensure a unified and comprehensive approach to coastal management. Ultimately, strong and flexible policies are vital for ensuring long-term sustainability and resilience in coastal areas.

14.1.3 GIScience as a Tool for Policy-Making

Geographic Information Science (GIScience) is an essential tool for policy-making, providing spatial insights and data-driven solutions to tackle complex societal and environmental issues. By utilizing Geographic Information Systems (GIS), remote sensing, and spatial analysis, GIScience allows policymakers to visualize, analyse, and interpret data on geography, resources, and human activities. This capability is especially valuable in areas like urban planning, disaster management, environmental conservation, and public health, where spatial patterns are key to making informed decisions (Fig. 14.1).

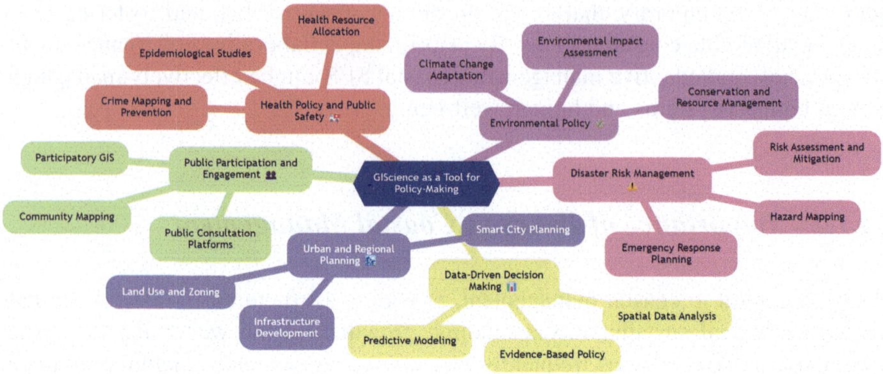

Fig. 14.1 GIScience as a tool for policy-making (Prepared by the authors)

In the policy-making context, GIScience supports evidence-based decisions by integrating diverse datasets, including demographic data, land-use patterns, and environmental indicators. For example, it can identify flood-prone areas, allowing policymakers to prioritize disaster mitigation measures. Similarly, GIS-based analysis can evaluate the consequences of urban development on natural habitats, guiding the creation of sustainable land-use policies. GIScience also enhances participatory planning by enabling stakeholders to engage with and visualize spatial data, promoting transparency and inclusivity in decision-making processes. Furthermore, GIScience aids the monitoring and evaluation of policies by providing tools to track changes over time. Satellite imagery and spatial models, for example, can assess the effectiveness of reforestation programs or monitor the progression of coastal erosion. With the increasing availability of big data and advanced geospatial technologies, including machine learning and real-time analytics, GIScience is evolving as a vital enabler of adaptive and responsive policy-making. In this way, GIScience bridges the gap between scientific research and governance, empowering policymakers to address contemporary challenges with spatial intelligence and precision.

A notable application of GIScience in coastal management is the monitoring of mangrove forests in the Indian Sundarbans. GIS and remote sensing are employed to map and track mangrove coverage, evaluate erosion impacts, and identify areas for potential reforestation. Policymakers use this spatial data to develop strategies for conserving mangroves, which serve as natural buffers against storm surges and provide essential ecosystem services. Additionally, GIS models are utilized to predict the effects of sea-level rise and inform adaptation policies for vulnerable coastal communities in the region. As the availability of big data and advanced geospatial technologies continues to grow, GIScience remains a critical tool for adaptive policy-making. It enables policymakers to make informed decisions, monitor the effectiveness of interventions, and respond to emerging challenges with precision and spatial awareness.

14.2 GIScience in Coastal Policy Integration

Geographic Information Science (GIScience) plays a pivotal role in the integration of coastal policies by providing advanced tools for planning, decision-making, and implementation. Coastal regions face complex challenges such as erosion, habitat loss, sea-level rise, and increasing human pressures, all of which require comprehensive and adaptive policies. GIScience enables policymakers to address these issues by offering spatial data and analytical capabilities that uncover patterns, trends, and interconnections between environmental, social, and economic factors. This spatial intelligence is essential for formulating evidence-based policies that effectively balance development with conservation.

In policy planning, GIScience helps identify high-risk areas, assess resource distribution, and visualize the potential impacts of climate change on coastal systems. For example, spatial modelling can predict the effects of sea-level rise on vulnerable low-lying areas, thereby guiding infrastructure development and disaster preparedness strategies. During the policy-making process, GIScience enhances stakeholder engagement by presenting complex data in user-friendly formats, such as maps and 3D models, fostering inclusivity and transparency. Additionally, GIScience supports scenario analysis, allowing policymakers to evaluate the trade-offs and potential outcomes of various policy options. The implementation of coastal policies is further reinforced by GIScience through real-time monitoring and evaluation. Satellite imagery and remote sensing technologies can track changes in coastal landforms, mangrove coverage, and water quality, providing critical insights to ensure adherence to regulations. GIS-based tools also facilitate adaptive management strategies by updating spatial datasets and models in response to changing conditions. In India's Sundarbans, for example, GIS has been instrumental in integrating policies for mangrove restoration, disaster risk reduction, and sustainable livelihoods, ensuring a holistic approach to coastal management.

Therefore, GIScience expands the scope of coastal policy integration by bridging the gap between scientific research and governance. Its ability to analyse, visualize, and monitor spatial data ensures that policies are not only well-informed but also effectively implemented, thereby contributing to the sustainability and resilience of coastal systems.

14.2.1 Mapping and Spatial Analysis

GIScience offers a crucial foundation for mapping and spatial analysis in the integration of coastal policies. High-resolution maps produced through GIS and remote sensing technologies allow policymakers to identify and visualize key features, including coastal landforms, ecosystems, human settlements, and infrastructure. Spatial analysis methods help uncover patterns, such as areas prone to coastal erosion, regions at risk of flooding, and changes in land use and land cover. These insights inform the planning and prioritization of interventions, such as establishing buffer zones, implementing shoreline protection measures, and conserving vital habitats. By providing detailed spatial datasets, GIScience improves the accuracy and effectiveness of coastal policy-making and management.

14.2.2 Data Integration for Multi-disciplinary Analysis

The effective integration of coastal policies necessitates the synthesis of data from various disciplines, including environmental science, socio-economics, and climate studies. GIScience acts as a platform for combining these diverse datasets, facilitating a comprehensive understanding of the interactions between human activities and coastal ecosystems. For instance, socio-economic data on fishing communities can be merged with environmental data on mangrove health to create policies that strike a balance between supporting livelihoods and promoting conservation. Similarly, GIS enables the integration of real-time weather data with hydrological models to assess flood risks. This interdisciplinary approach promotes collaborative policy planning, ensuring that multiple perspectives and knowledge systems are considered in the decision-making process.

14.2.3 Predictive Modelling and Scenario Planning

The predictive modelling and scenario planning, facilitated by GIScience, are essential for the proactive development of coastal policies. By utilizing geospatial data and advanced modelling techniques, policymakers can simulate future scenarios, such as the effects of sea-level rise, storm surges, or shifts in land use patterns. These models assist in evaluating the long-term consequences of different policy choices and identifying sustainable, resilient strategies. For example, in coastal regions like the Sundarbans, scenario planning through GIS can assess the effectiveness of mangrove restoration in reducing flood risks under varying climate change scenarios. By visualizing potential outcomes, GIScience provides policymakers with the tools to create adaptive, forward-looking coastal policies that anticipate and address future challenges.

14.3 Integrating GIScience in Coastal Policy: The Case of Coastal World

Integrating GIScience in coastal policy improves decision-making by providing spatial data and analytical tools to tackle complex coastal issues. GIS aids policymakers in visualizing key features such as landforms, ecosystems, and infrastructure, helping identify risks like erosion and flooding. It supports evidence-based planning by combining socio-economic, environmental, and climate data, offering a comprehensive view of human-environment interactions. GIScience also enables predictive modelling and scenario planning, informing adaptive policies for sustainability and resilience in vulnerable coastal regions. The following Table 14.1 illustrates GIScience integration across various stages of coastal policy planning and implementation globally.

Table 14.1 Integrating GIScience in coastal policy: The case of coastal world

Stage	GIScience application(s)	Result(s)	Example(s)	Investigated area(s)	References
Policy planning	Mapping and spatial analysis of critical coastal features like erosion hotspots, mangroves, and habitats.	Identification of vulnerable areas and priority zones for intervention.	High-risk zones for storm surges and sea-level rise are mapped, aiding in zoning and evacuation planning.	Bangladesh; and Coastal West Bengal (India)	[6, 11, 12]
Data integration	Combining ecological, socio-economic, and climate data using GIS platforms.	A holistic understanding of human-environment interactions, informing balanced policy frameworks.	Integration of fisheries data with mangrove health to design sustainable fishing policies.	Global perspective; Punta Galeta, Panama	[4, 10, 15, 16]
Scenario planning	Predictive modelling of climate impacts (e.g., sea-level rise, extreme weather events).	Identification of future risks and adaptation measures.	Simulating the effects of a 1-meter sea-level rise on coastal villages to plan for relocation and mitigation.	Coastal Benin (Africa); Indian Sundarbans	[3, 5, 14, 19]
Policy making	Stakeholder engagement through GIS-based participatory tools and decision support systems.	Transparent and inclusive policy-making with community input.	Interactive GIS maps enable local fishermen to suggest no-fishing zones and conservation areas.	Coastal West Bengal (India); Nez Perce Indian Reservation (USA); and Costa Norte, Pacific coastal plain).	[2, 8, 9]
Implementation	Monitoring and evaluation using satellite imagery and real-time geospatial tools.	Effective enforcement and adaptive management of policies.	Tracking mangrove restoration progress using satellite imagery to ensure policy compliance.	Global coast; African coast; Mar Menor; Northern Sri Lanka; and Philippines	[1, 7, 13, 17, 18, 20]

14.4 Challenges and Opportunities

Coastal areas are dynamic socio-ecological systems, where the interplay between human actions and natural processes shapes their sustainability, resilience, and growth. Incorporating Geographic Information Science (GIScience) into coastal policy is crucial for tackling complex issues like climate change, habitat loss, and socio-economic vulnerabilities. While GIScience offers a powerful framework for spatial analysis and decision-making, its integration into coastal policy presents both significant challenges and opportunities. A socio-ecological perspective emphasizes the interconnectedness of these dynamics and highlights the necessity for holistic approaches in policy development.

Challenges

- **Data Availability and Accessibility:** A major challenge in integrating GIScience into coastal policy is the lack of comprehensive, high-quality spatial data, particularly in developing countries. Limited datasets on land use, biodiversity, and climate risks, along with institutional and financial barriers to data access, restrict GIScience's potential in policy formulation.
- **Interdisciplinary Collaboration:** Coastal policy requires multi-disciplinary collaboration, but integrating GIScience across fields such as ecology, economics, and governance is hindered by differences in terminology, methodologies, and goals, making consensus-building difficult.
- **Technical Expertise and Capacity Building:** Effective GIScience use demands specialized skills, which many coastal policymakers and community leaders lack. Limited resources for training further impede widespread GIS adoption, especially in resource-constrained settings.
- **Cost and Infrastructure Constraints:** GIS integration requires significant investment in technology and infrastructure. In low-income coastal regions, insufficient resources, including computing power and internet connectivity, hinder GIS tool effectiveness.
- **Socio-Economic Inequalities:** Coastal communities often experience socio-economic disparities. Policies based on GIScience may inadvertently benefit wealthier groups unless these inequalities are addressed in data collection and analysis.
- **Dynamic Nature of Coastal Systems:** Coastal areas are highly dynamic, influenced by erosion, sea-level rise, and extreme weather. Continuous data updates through advanced remote sensing are necessary to reflect real-time conditions, but these may not always be feasible, limiting policy accuracy.

Opportunities

- **Enhanced Spatial Understanding:** GIScience offers advanced tools for mapping and analysing spatial relationships, allowing policymakers to better understand human-ecological interactions in coastal regions, such as the effects of urbanization on mangroves or fishing pressure, thus guiding informed decisions.

- **Integration of Multi-Source Data:** GIS platforms integrate diverse data sources like satellite imagery, socio-economic surveys, and climate models, enabling a comprehensive view of coastal systems and promoting holistic policy approaches for ecological sustainability and community welfare.
- **Predictive Modeling and Scenario Planning:** GIScience's predictive modelling allows policymakers to simulate future scenarios, such as sea-level rise or land-use changes, helping to evaluate potential policy outcomes and enhance resilience and adaptive capacity.
- **Participatory Policy-Making:** GIScience supports participatory policy-making by enabling community-based mapping and participatory GIS (PGIS), allowing local populations to contribute knowledge, ensuring more inclusive and equitable policy development.
- **Real-Time Monitoring and Adaptive Management:** Advances in remote sensing and geospatial technologies enable real-time monitoring of coastal changes, supporting adaptive management by providing timely data for assessing policy effectiveness and adjusting strategies, such as mangrove restoration progress.
- **Climate Adaptation and Disaster Risk Reduction:** GIScience plays a key role in climate adaptation and disaster risk reduction by identifying vulnerable areas and populations, prioritizing interventions like resilient infrastructure, relocation, or the restoration of natural buffers.
- **Strengthening Governance and Collaboration:** GIScience facilitates collaboration among governments, NGOs, and international organizations, promoting data sharing and improving coordination and transparency, thus strengthening governance and policy implementation.

Therefore, the integrating GIScience into coastal policy offers both challenges and opportunities. Overcoming data and capacity limitations, promoting interdisciplinary collaboration, and including marginalized communities are key to successful integration. A socio-ecological approach ensures policies are both ecologically sustainable and socially equitable. By addressing these challenges, GIScience can significantly contribute to advancing sustainable coastal management and resilience in vulnerable regions globally.

14.5 The Future of GIScience in Coastal Policy

The dynamic and interconnected nature of coastal systems requires adaptive approaches to policymaking. Geographic Information Science (GIScience) has become a transformative tool in coastal management, offering powerful capabilities in spatial analysis, predictive modelling, and participatory planning. Moving forward, GIScience's role in coastal policy will expand, driven by technological advancements, the need for interdisciplinary strategies, and a focus on socio-ecological sustainability. This comprehensive discussion examines the future of

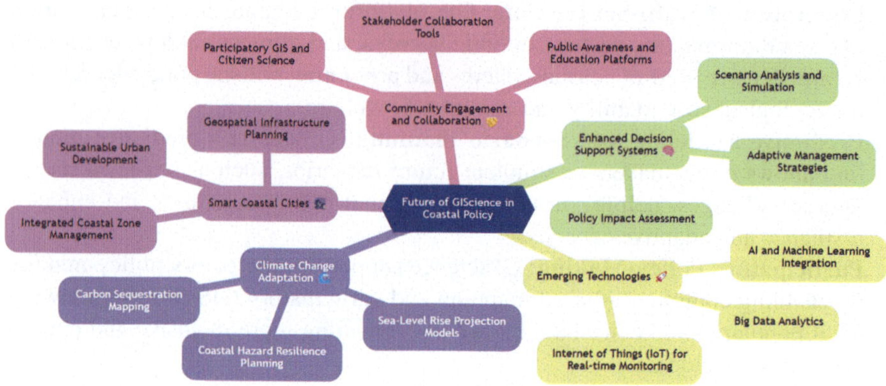

Fig. 14.2 The future of GIScience in coastal policy (Prepared by the authors)

GIScience in coastal policy integration, exploring emerging trends, opportunities, and challenges through a socio-ecological perspective (Fig. 14.2).

The Evolving Role of GIScience in Coastal Policy

- **Technological Innovations in GIScience:** Advances in geospatial technologies will reshape coastal policy development, implementation, and monitoring. Emerging tools like artificial intelligence (AI), machine learning, and big data analytics will enhance GIS applications, allowing policymakers to process large volumes of spatial data in real-time. This will improve prediction accuracy and management strategy responsiveness. For example, AI-driven remote sensing can identify early signs of coastal ecosystem changes, such as mangrove degradation or shoreline erosion, prompting timely interventions. Additionally, the integration of 3D and 4D GIS models will provide dynamic visualizations of coastal processes, such as sea-level rise and storm surges, facilitating more comprehensive scenario planning. These technologies will also make GIS platforms more accessible to local stakeholders, enabling community engagement in decision-making.
- **Incorporating Socio-ecological Data:** Future GIScience applications in coastal policy will emphasize integrating socio-ecological data to address the complex relationship between human and natural systems. Recognizing that sustainable coastal management requires both environmental and social considerations, GIScience will facilitate the inclusion of demographic, traditional, and cultural data alongside ecological information. Tools like participatory GIS (PGIS) will help map community resources, document traditional practices, and identify culturally significant areas. This integration will allow for policies that protect vulnerable populations, promote sustainable livelihoods, and preserve local heritage, while addressing environmental challenges.
- **Strengthening Climate Resilience:** Coastal regions, facing threats from climate change such as extreme weather, sea-level rise, and habitat loss, will benefit from GIScience in enhancing climate resilience. Future GIS applications will focus on

spatial risk assessments, vulnerability mapping, and real-time monitoring to develop climate adaptation strategies. GIS will help identify climate hotspots, prioritize adaptation efforts, and optimize resource allocation. For example, spatial models will evaluate the effectiveness of natural infrastructure, such as mangroves and coral reefs, in mitigating storm surges and coastal flooding. These insights will guide investments in ecosystem-based solutions that support both climate resilience and biodiversity conservation.

Opportunities for Advancing GIScience in Coastal Policy

- **Expanding Citizen Science and Stakeholder Engagement:** The democratization of GIS technologies is opening new avenues for citizen science and stakeholder involvement in coastal policy. With the availability of mobile GIS applications and affordable data collection tools, communities can actively participate in monitoring coastal changes and influencing policy. Crowdsourced data, when combined with scientific datasets, enriches the understanding of coastal dynamics and enhances governance accountability. Participatory approaches also promote social equity by empowering marginalized groups. For example, small-scale fishers and indigenous communities can use GIS tools to document resource use patterns, advocate for their rights, and shape spatial planning decisions.
- **Leveraging Big Data and IoT:** Integrating big data and the Internet of Things (IoT) into GIScience will transform real-time coastal management. Sensors and IoT devices deployed in coastal zones can continuously collect data on water quality, sediment transport, and weather conditions. These real-time datasets, processed through GIS platforms, enable dynamic monitoring and adaptive management of coastal systems. Additionally, big data analytics will improve predictive capabilities, helping policymakers anticipate and mitigate emerging challenges. For instance, machine learning algorithms can analyse historical and live data to predict extreme weather impacts, enhancing early warning systems and disaster preparedness.
- **Fostering International Collaboration:** Global issues like climate change, marine pollution, and biodiversity loss demand coordinated international efforts. GIScience provides a unified framework for data sharing, analysis, and policy alignment across borders. Initiatives such as the Marine Spatial Data Infrastructure (MSDI) and the Global Ocean Observing System (GOOS) highlight how GIScience supports transboundary collaboration in coastal management. By standardizing data formats and methodologies, GIS platforms facilitate knowledge exchange and cross-border cooperation, essential for addressing shared challenges and advancing sustainable coastal development worldwide.

Challenges and Future Directions

Although GIScience holds tremendous potential for advancing coastal policy, several challenges must be overcome to fully harness its benefits:

- **Data Gaps and Inequalities:** Many coastal regions, especially in developing nations, face challenges due to insufficient spatial data and limited access to

advanced technologies. Addressing these disparities requires strategic investments in data collection, infrastructure, and capacity-building initiatives.
- **Interdisciplinary Collaboration:** Coastal policy necessitates collaboration across diverse disciplines, yet differences in methodologies and objectives often impede integration. Promoting interdisciplinary networks and fostering a collaborative culture will be key to bridging these divides.
- **Ethical Challenges:** The application of GIS technologies brings ethical concerns related to data privacy, equity, and representation. Ensuring that GIS-based practices uphold community rights and prioritize social justice is essential for sustainable and inclusive outcomes.

The future of GIScience in coastal policy offers transformative potential for sustainable and equitable socio-ecological management. By leveraging technological advancements, integrating diverse data, and fostering participation, GIScience can help policymakers address complex coastal challenges. Achieving this requires overcoming barriers, building capacity, and prioritizing equity. From enhancing climate resilience to strengthening stakeholder engagement, GIScience is pivotal for reimagining coastal policy. As the field evolves, it will underpin adaptive, inclusive, and science-driven governance, ensuring resilient and sustainable coastal systems for future generations.

14.6 Role of Participatory GIS in Coastal Policy and Its Future

Coastal zones, among the world's most dynamic and vulnerable ecosystems, support livelihoods, economic activities, and biodiversity but face significant pressures from climate change, sea-level rise, urbanization, and industrialization. In regions like the Indian Sundarbans, Participatory Geographic Information Systems (PGIS) have become vital tools for addressing these challenges. By combining community engagement with geospatial technologies, PGIS integrates local knowledge with scientific data, enabling more effective decision-making and promoting sustainable coastal management (Fig. 14.3).

Participatory GIS: An Overview

Participatory GIS (PGIS) is an approach that actively involves local communities and stakeholders in generating, analysing, and interpreting geographic data. It leverages the deep environmental knowledge of local populations to enhance spatial data creation and decision-making processes, diverging from traditional top-down methods that often exclude community input. PGIS typically includes activities like community mapping, participatory data collection through surveys or focus groups, and the application of GIS tools for spatial visualization and analysis. By integrating local expertise with scientific knowledge, PGIS promotes more inclusive, equitable, and effective coastal management practices.

Fig. 14.3 Role of participatory GIS in coastal policy and its future (Prepared by the authors)

The Role of Participatory GIS in Coastal Policy

- **Enhancing Local Engagement and Empowerment:** Participatory GIS (PGIS) significantly enhances local engagement by involving communities, particularly in marginalized regions, in decision-making processes that affect their lives. For instance, traditional marine fishermen in the Sundarbans may use PGIS to map fishing zones, monitor fish population changes, and document habitat conditions. By integrating local knowledge with geospatial tools, PGIS empowers communities to influence policies and strategies, fostering more democratic and inclusive coastal management.
- **Supporting Integrated Coastal Zone Management (ICZM):** PGIS supports Integrated Coastal Zone Management (ICZM) by integrating diverse data sources, such as satellite imagery, climate data, and socio-economic information, with local insights. This comprehensive approach enables policymakers to better understand coastal complexities. For example, PGIS can overlay coastline changes with infrastructure data and community feedback, facilitating targeted interventions for issues like erosion or habitat loss.
- **Disaster Risk Reduction and Climate Change Adaptation:** Coastal areas face increasing risks from natural hazards and climate change. PGIS aids disaster risk reduction and adaptation by mapping hazard-prone zones, identifying vulnerable communities, and informing preparedness plans. For instance, PGIS can create localized flood risk maps based on historical data and sea-level rise projections, enabling communities to develop tailored strategies for disaster resilience.
- **Biodiversity Conservation and Marine Resource Management:** PGIS contributes to biodiversity conservation by providing spatial data on ecosystems, biodiversity hotspots, and resource use patterns. Local stakeholders can share insights on fish stocks and key marine habitats, supporting the design of effective marine protected areas (MPAs) and conservation strategies. This collaborative approach ensures that biodiversity efforts align with local needs and conditions.

- **Monitoring and Enforcement of Coastal Policies:** PGIS is instrumental in monitoring coastal policy effectiveness and ensuring compliance. By tracking changes in land use, habitat health, and policy enforcement, it provides feedback for adaptive management. Community-based monitoring using GPS or mobile applications can also report environmental changes or violations, enabling timely action by authorities.
- **The Future of PGIS in Coastal Policy:** The future of PGIS in coastal policy is shaped by advancing technologies and increasing community-led approaches:

  (a) Integration with Remote Sensing and Big Data: Combining PGIS with remote sensing and big data will improve hazard assessment, climate impact modelling, and resource management.
  (b) Mobile Technologies and Citizen Science: The proliferation of mobile devices will empower communities to collect real-time data, enhancing the participatory nature of GIS.
  (c) Artificial Intelligence and Predictive Analytics: AI integration will enable advanced forecasting, helping policymakers address coastal challenges proactively.
  (d) Innovative Governance Models: PGIS may drive more adaptive, collaborative governance frameworks, replacing rigid hierarchical systems.
  (e) Global Collaboration: International networks can share PGIS tools, data, and strategies to tackle transboundary coastal challenges collectively.

Therefore, the PGIS is pivotal in developing sustainable and inclusive coastal policies by integrating local knowledge with scientific data. It empowers communities, enhances decision-making, and supports effective coastal resource management. With advancing technologies, PGIS offers innovative tools to address the complex challenges of coastal ecosystems and communities. Its growth will be essential for fostering resilient, adaptive, and equitable coastal governance worldwide.

14.7 Conclusion and Endnotes

The integration of GIScience into coastal policy provides a strong framework for managing the complex socio-ecological dynamics of coastal zones. Combining scientific tools with local knowledge, GIScience enables data-driven, informed decision-making to enhance the sustainability of coastal ecosystems and the resilience of dependent communities. Advanced mapping, spatial analysis, and predictive modelling have transformed policy development, allowing for adaptive responses to challenges such as climate change, resource depletion, habitat loss, and others. As well as the example of real-world applications in integrating GIScience in coastal policy, illustrate how GIScience supports socio-ecological sustainability and social equity by engaging diverse stakeholders, including local communities and policymakers. However, challenges such as data gaps, system complexity, and stakeholder capacity highlight the need for further innovation and capacity building. The future of GIScience in coastal policy is promising, with advancements in remote sensing, AI-driven analytics, and participatory tools like PGIS empowering communities to contribute directly to management efforts. These developments will be critical in shaping resilient, inclusive, and sustainable coastal policies globally.

Endnotes

- **Coastal Socio-Ecological Systems and Policy:** Coastal zones are shaped by complex interactions between human and environmental factors, and effective policy is essential for balancing conservation, development, and climate resilience.
- **GIScience in Coastal Management:** GIScience provides crucial tools for spatial analysis, integrating diverse data sources, such as satellite imagery, local knowledge, and environmental data, into coastal management strategies.
- **Data integration and multidisciplinary approaches:** Combining ecological, social, and economic data fosters a comprehensive understanding of coastal systems, enabling the creation of long-term, effective policies.
- **Coastal World Case Study:** This initiative demonstrates GIScience's application in real-world coastal policy, emphasizing spatial planning and community-driven decision-making.
- **Challenges in GIScience Integration:** Challenges in GIScience integration, including data access, infrastructure, and local capacity, must be addressed to enhance policy-making effectiveness.
- **The Future of GIScience and Participatory GIS:** Advancements in AI, remote sensing, and participatory GIS will strengthen coastal governance by enhancing adaptability and community involvement in data collection and policy development.

References

1. Bakirci, M. (2025). Advanced ship detection and ocean monitoring with satellite imagery and deep learning for marine science applications. *Regional Studies in Marine Science, 81*, 103975. https://doi.org/10.1016/j.rsma.2024.103975
2. Bojorquez-Tapia, L. A., Diaz-Mondragon, S., & Ezcurra, E. (2001). GIS-based approach for participatory decision making and land suitability assessment. *International Journal of Geographical Information Science, 15*(2), 129–151. https://doi.org/10.1080/13658810010005534
3. Bolan, S., Padhye, L. P., Jasemizad, T., Govarthanan, M., Karmegam, N., Wijesekara, H., et al. (2024). Impacts of climate change on the fate of contaminants through extreme weather events. *Science of The Total Environment, 909*, 168388. https://doi.org/10.1016/j.scitotenv.2023.168388
4. Chamberland-Fontaine, S., Estrada, G. T., Heckadon-Moreno, S., & Hickey, G. M. (2022). Enhancing the sustainable management of mangrove forests: The case of Punta Galeta, Panama. *Trees, Forests and People, 8*, 100274. https://doi.org/10.1016/j.tfp.2022.100274

5. Deguenon, S. D. D. M., Hounmenou, C. G., Adade, R., Teka, O., Toko, I. I., Aheto, D. W., & Sinsin, B. (2023). Simulation of the impacts of sea-level rise on coastal ecosystems in Benin using a combined approach of machine learning and the sea level affecting marshes model. *Sustainability, 15*(22), 16001. https://doi.org/10.3390/su152216001
6. Gijón Mancheño, A., Herman, P. M., Jonkman, S. N., Kazi, S., Urrutia, I., & van Ledden, M. (2021). Mapping mangrove opportunities with open access data: A case study for Bangladesh. *Sustainability, 13*(15), 8212. https://doi.org/10.3390/su13158212
7. Giménez, J. G., Granero, A., Senent-Aparicio, J., Gómez-Jakobsen, F., Mercado, J. M., Blanco-Gómez, P., et al. (2024). Assessment of oceanographic services for the monitoring of highly anthropised coastal lagoons: The Mar Menor case study. *Ecological Informatics, 81*, 102554. https://doi.org/10.1016/j.ecoinf.2024.102554
8. Jankowski, P. (2009). Towards participatory geographic information systems for community-based environmental decision making. *Journal of Environmental Management, 90*(6), 1966–1971. https://doi.org/10.1016/j.jenvman.2007.08.028
9. Malakar, K. D., & Roy, S. (2024). Mapping community voices in the coastal region of Bengal: Case studies and best practices of participatory GIS. In *Mapping geospatial citizenship*. Springer. https://doi.org/10.1007/978-3-031-63107-8_9
10. McSherry, M., Davis, R. P., Andradi-Brown, D. A., Ahmadia, G. N., Van Kempen, M., & Wingard Brian, S. (2023). Integrated mangrove aquaculture: The sustainable choice for mangroves and aquaculture? *Frontiers in Forests and Global Change, 6*, 1094306. https://doi.org/10.3389/ffgc.2023.1094306
11. Mondal, B., Roy, A., & Saha, A. K. (2022). Vulnerability assessment of mangrove areas in coastal West Bengal, India. *Remote Sensing Applications: Society and Environment, 25*, 100680. https://doi.org/10.1016/j.rsase.2021.100680
12. Polas, A. B., Topp, E., & Plieninger, T. (2024). Mapping landscape values and exploring sense of place in a threatened mangrove biodiversity hotspot: The Sundarbans delta, Bangladesh. *Ecosystems and People, 20*(1), 2370531. https://doi.org/10.1080/26395916.2024.2370531
13. Quevedo, J. M. D., Uchiyama, Y., & Kohsaka, R. (2022). Community perceptions of long-term mangrove cover changes and its drivers from a typhoon-prone province in the Philippines. *Ambio, 51*, 972–989. https://doi.org/10.1007/s13280-021-01608-9
14. Rukhsana, & Hasnine, M. (2024). Delineation of Multi hazards vulnerable zone using the Geospatial approach: A micro-region level study in Indian Coastal Sundarban. *Regional Studies in Marine Science, 71*, 103402. https://doi.org/10.1016/j.rsma.2024.103402
15. Sandilyan, S., & Kathiresan, K. (2012). Mangrove conservation: a global perspective. *Biodiversity and Conservation, 21*, 3523–3542. https://doi.org/10.1007/s10531-012-0388-x
16. Sarkar, U. K., Kumar, T. A., Sood, N., Singh, R. K., Kumar, R., & Tyagi, L. K. (Eds.). (2024). *Sustainable management of fish genetic resources*. Springer . ISBN: 978-981-97-5249-2. https://doi.org/10.1007/978-981-97-5250-8
17. Sathiyamoorthy, S., & Sakurai, T. (2024). Effectiveness of community participation in Mangrove restoration: the evidence from northern Sri Lanka. *Environmental Economics and Policy Studies, 26*, 759–779. https://doi.org/10.1007/s10018-024-00397-1
18. Schmitt, K., & Duke, N. C. (2015). Mangrove management, assessment and monitoring. In M. Köhl & L. Pancel (Eds.), *Tropical forestry handbook*. Springer. https://doi.org/10.1007/978-3-642-41554-8_126-1
19. Sillmann, J., Thorarinsdottir, T., Keenlyside, N., Schaller, N., Alexander, L. V., Hegerl, G., … & Zwiers, F. W. (2017). Understanding, modeling and predicting weather and climate extremes: Challenges and opportunities. Weather and Climate Extremes, 18, 65–74. https://doi.org/10.1016/j.wace.2017.10.003
20. Sunkur, R., Kantamaneni, K., Bokhoree, C., et al. (Eds.). (2024). Mangrove mapping and monitoring using remote sensing techniques towards climate change resilience. *Scientific Reports, 14*, 6949. https://doi.org/10.1038/s41598-024-57563-4

Further Readings

21. Akyarli, A. (1997). GIS as a tool in data management in the coastal zone: An application to Izmir Bay. In N. B. Harmancioglu, M. N. Alpaslan, S. D. Ozkul, & V. P. Singh (Eds.), *Integrated approach to environmental data management systems* (NATO ASI series) (Vol. 31). Springer. https://doi.org/10.1007/978-94-011-5616-5_40
22. Breman, J. (2002). *Marine geography GIS for the oceans and seas*. ESRI Press. ISBN: 978-1589480452.
23. Canessa, R., & Keller, C. P. (2003). User assessment of coastal spatial decision support systems. In D. R. Green & S. D. King (Eds.), *Coastal and marine geo-information systems. Coastal systems and continental margins* (Vol. 4). Springer. https://doi.org/10.1007/0-306-48002-6_31
24. Dhiman, R., Kalbar, P., & Inamdar, A. B. (2018). GIS coupled multiple criteria decision-making approach for classifying urban coastal areas in India. *Habitat International, 71*, 125–134. https://doi.org/10.1016/j.habitatint.2017.12.002
25. Geukes, H. H., van Bodegom, P. M., & van Oudenhoven, A. P. (2024). Setting the stage for decision-making on nature-based solutions for coastal climate adaptation. *Ocean & Coastal Management, 247*, 106916. https://doi.org/10.1016/j.ocecoaman.2023.106916
26. Kousky, C., Fleming, B., & Berger, A. M. (Eds.). (2021). *A blueprint for coastal adaptation: Uniting design, economics, and policy*. Island Press. ISBN: 978-1642831399.
27. National Research Council, Division on Earth, Life Studies, Commission on Geosciences, Committee on Science, & Policy for the Coastal Ocean. (1995). *Science, policy, and the coast: Improving decisionmaking*. National Academies Press. ISBN: 978-0-309-05339-6. https://doi.org/10.17226/4968
28. Olewiler, N., Francisco, H. A., & Ferrer, A. J. G. (Eds.). (2016). *Marine and coastal ecosystem valuation, institutions, and policy in Southeast Asia*. Springer . ISBN: 978-981-10-0139-0. https://doi.org/10.1007/978-981-10-0141-3
29. Rehman, S., Jahangir, S., & Azhoni, A. (2022). GIS based coastal vulnerability assessment and adaptation barriers to coastal regulations in Dakshina Kannada district, India. *Regional Studies in Marine Science, 55*, 102509. https://doi.org/10.1016/j.rsma.2022.102509
30. Scialabba, N. (ed.). (1998). *Integrated coastal area management and agriculture, forestry and fisheries*. FAO guidelines. environment and natural resources service, FAO, Rome. 256 p.
31. Silva, T. S., Rosa, M. L., & Farina, F. (2012). GIS applied to integrated coastal zone and ocean management: Mapping, change detection and spatial modeling for coastal management in southern Brazil. In *Application of geographic information systems* (pp. 219–240). IntechOpen. https://doi.org/10.5772/47840
32. Suresh, R. (Ed.). (2014). *Maritime security of India: The coastal security challenges and policy options*. Vij Books India Pvt. Ltd. ISBN: 978-9384464226.

Part V
Geospatial Techniques for Monitoring and Managing Coastal Ecologies: Field Narratives, Case Studies, and Best Practices of GIScience

Chapter 15
Monitoring and Management of Coastal Ecologies in India: Community Narratives Through PGIS

"The power of geography comes not from seeing the world from a distance, but from understanding it through local voices."—Doreen Massey.

Abstract This chapter investigates the complex relationship between human communities and their coastal environments, emphasizing participatory approaches to sustainable management. Through an exploration of five field narratives, the chapter highlights community-driven initiatives and their integration with scientific methodologies. The first narrative, *Voices from the Shoreline*, highlights collective efforts in mangrove restoration and coastal stewardship, emphasizing the significance of local knowledge and participatory approaches. The second narrative, *Salty Encroachment*, focuses on adaptive strategies to mitigate salinity intrusion, blending traditional practices with scientific tools such as PGIS and ecological assessments. *Crops and Catastrophes*, the third narrative, examines agricultural resilience in response to climatic uncertainties, showcasing innovations such as drought-resistant crops and other adaptive strategies. The fourth narrative, *Nets of Tradition*, underscores the role of indigenous knowledge in managing fishing grounds, demonstrating how participatory mapping and conservation efforts contribute to sustainability. Finally, *Unequal Journeys* addresses the challenges of climate migration and inequality, offering adaptive solutions like eco-villages and skill-building initiatives to enhance community resilience. By integrating these narratives with participatory geographic information systems (PGIS) and other scientific innovations, the chapter highlights the transformative potential of co-creation. It advocates for the fusion of local knowledge with evidence-based approaches to promote ecological balance, equitable resource distribution, and long-term sustainability in coastal ecologies. The narratives illustrate that the solutions to our most pressing challenges are not found in isolation, but through collaboration, where the wisdom of the past informs future innovation and diverse voices contribute to the development of inclusive and effective solutions.

K. D. Malakar et al., *Geospatial Technologies in Coastal Ecologies Monitoring and Management*, Advances in Geographic Information Science,
https://doi.org/10.1007/978-3-031-92017-2_15

Keywords Community narratives · Coastal community · Climate disasters · Management of coastal ecologies · Field narratives and scientific evidence · Applications of participatory GIS · Challenges and sustainability · Socio-ecological relationship · Indian Sundarbans: Coastal West Bengal

This chapter explores the dynamic relationship between coastal communities and their environments, focusing on participatory approaches to sustainable management. By examining five field narratives, it sheds light on community-driven initiatives and their integration with scientific methodologies. These narratives showcase the power of local knowledge, combining traditional practices with innovative tools like PGIS to address challenges such as mangrove restoration, salinity intrusion, agricultural resilience, fishing ground management, and climate migration. Through this exploration, the chapter emphasizes the importance of collaboration, where diverse perspectives and scientific advancements drive inclusive, sustainable solutions.

Key Points of the Chapter

- Examining the diverse perspectives captured in field narratives from coastal communities.
- Explains how humans and nature are interrelated in a socio-ecological context.
- Understanding the contemporary challenges and scope through the field narratives and empowering the following:
  - the strength of community-led approaches;
  - bridging local knowledge and scientific innovation;
  - adapting to climatic and social uncertainty;
  - addressing inequality and ensuring access; and,
  - sustainability as an ongoing commitment.
- Field narratives on the following:
  - mangrove management and coastal stewardship;
  - coastal management in a changing climate;
  - catastrophes, agricultural management, and community;
  - local knowledge in fishing ground management; and,
  - management approaches to climate migration and inequality.

15.1 Introduction

The monitoring and management of coastal ecologies demand a nuanced approach that seamlessly integrates scientific methodologies with community-driven insights. Coastal regions, particularly the Medinipur region, the 24 Parganas districts, and the Indian Sundarbans, are characterized by rich biodiversity and support vulnerable communities whose livelihoods are intricately tied to these fragile ecosystems [7, 13, 24]. Participatory Geographic Information Systems (PGIS) have emerged as a transformative tool, bridging local knowledge with advanced geospatial technologies. By empowering communities to actively engage in ecological monitoring, PGIS facilitates the identification of critical challenges, including coastal erosion, climate change impacts, and biodiversity loss, and promotes the co-creation of sustainable management strategies [21, 35, 38]. The lived experiences and narratives of coastal populations are indispensable for developing resilient and adaptive solutions to safeguard these vital ecosystems [20, 26, 36].

In this context, the chapter examines the coastal region of West Bengal (Fig. 15.1), drawing upon diverse community narratives through different and interrelated stages (Fig. 15.2). These include stories of mangrove management and coastal stewardship, adaptive responses to saline encroachment in a changing climate, agricultural management concerns, the integration of local knowledge in fishing ground management, and thoughts of address climate migration and inequality. Through the application of PGIS, these narratives were explored in collaboration with local coastal communities (Table 15.1). The investigation followed four key stages to ensure an in-depth understanding of the narratives, which are outlined below:

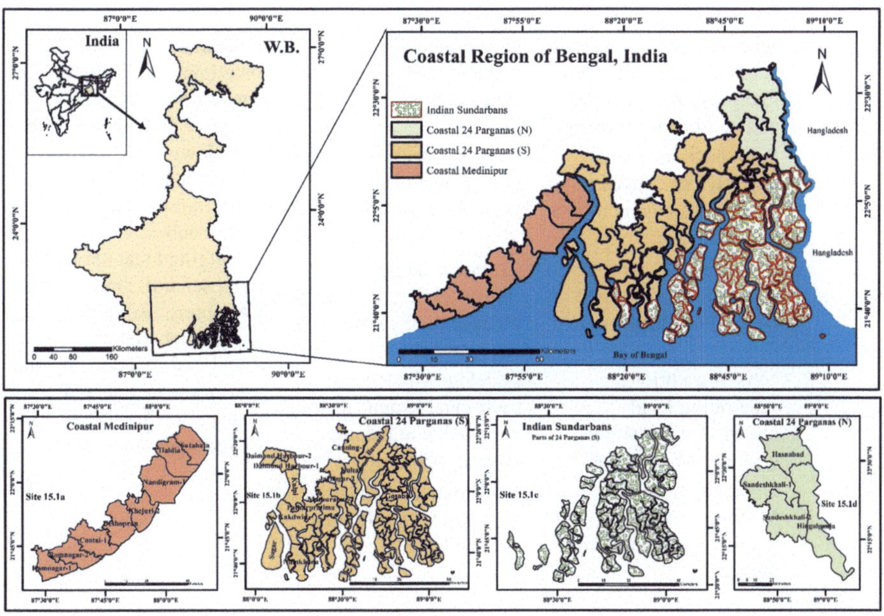

Fig. 15.1 Geographical location of the study area in the coastal region of Bengal, India. **Site 15.1a:** Coastal region of Medinipur; **Site 15.1b:** Coastal South 24 Parganas; **Site 15.1c:** Sundarbans (West Bengal, India); and **Site 15.1d:** Coastal North 24 Parganas. (Prepared by the authors)

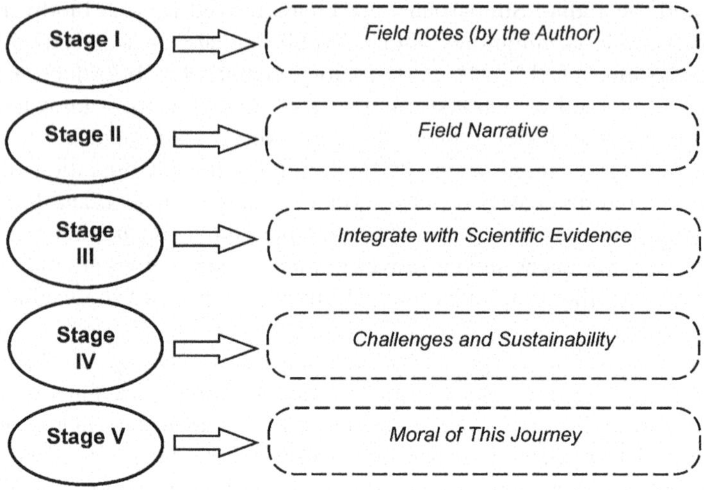

Fig. 15.2 Field Narrative collection and processing stages (Prepared by the authors)

Table 15.1 Narratives collected on the monitoring and management of coastal ecologies using PGIS

Sr. No.	Narrative(s)	Investigation(s): monitoring and management	Integration of key findings, suggestions, and best practices
1	Voices from the shoreline: A collective approach to mangrove management and coastal stewardship	Mangrove conservation, management and coastal stewardship	Understanding the contemporary challenges and scope through the field narratives and empowering the following: The strength of community-led approaches; Bridging local knowledge and scientific innovation; Adapting to climatic and social uncertainty; Addressing inequality and ensuring access; and, Sustainability as an ongoing commitment.
2	Salty encroachment: Adaptive narratives on coastal Management in a changing climate	Salinity intrusion, coastal management in a changing climate	
3	Crops and catastrophes: Community narratives on agricultural management	Climate change, catastrophes, agricultural management, and community	
4	Nets of tradition: Integrating local knowledge in fishing ground management	Local knowledge and traditional technical knowledge in fishing ground management	
5	Unequal journeys: Management approaches to climate migration and inequality	Management approaches to climate migration and inequality	

Credit(s):
Narrative collection: S.R. (November to December 2024 and August to September 2023); K.D.M. (December 2024 and April to June 2022)
Narrative analysis and discussions: K.D.M.; and S.R.
K.D.M.: Kousik Das Malakar; and S.R.: Supriya Roy.

Stage I: Field notes (by the Author)
Stage II: Field Narrative
Stage III: A Process to Integrate with Scientific Evidence
Stage IV: Challenges and Sustainability
Stage V: The Moral of this Journey

15.2 The Power of PGIS Narratives to Understand Challenges and Sustainability: Field Interactions Unveiled

PGIS have become a pivotal tool for addressing complex challenges and advancing sustainability, particularly in regions where local communities are closely interconnected with their natural environments. By facilitating field-based interactions, PGIS uncovers narratives that offer valuable insights into the socio-environmental dynamics, resource utilization, and adaptive strategies employed by these communities. These narratives not only illuminate pressing issues such as environmental degradation, climate vulnerabilities, and socio-economic inequalities but also reveal traditional knowledge and practices vital for sustainable development. By integrating spatial approach with community perspectives, PGIS empowers stakeholders to collaboratively develop solutions that are contextually relevant, equitable, and resilient. The integration of these interactions is outlined as follows:

15.2.1 Voices from the Shoreline: A Collective Approach to Mangrove Management and Coastal Stewardship

The narrative titled *"Voices from the Shoreline"* examined the mangrove region of the Indian Sundarban (Fig. 15.1, Site 15.1c) and employed various PGIS tools and techniques. Table 15.2 provides a brief overview of the narrative collection history

Table 15.2 Voices from the shoreline: functional insights from field observations using participatory methodologies

Location(s)	Indian Sundarban
Time duration	November to December 2024
Participants (The individual's name has been changed)	Mrs. Manashi Mondal (age: 65+) Ms. Radha Mondal (age: 19+) Mr. Surja Das (age: 45+) Others 5 persons (age: 25+)
Used PGIS approaches	HHs; FGDs; CWs; TA
Note	Field-investigated by the author (S. Roy). HHs: Household conversation; FGDs: Focus Group Discussions; CWs: Community Workshops; and TA: Thematic Analysis

and the methods applied. The following discussion presents the author's field notes and the narratives gathered during field investigations and integrates these accounts with scientific evidence. The narrative then addresses the challenges and potential for sustainability, concluding with a reflection on the moral of this journey.

a) **Field notes (by the Author)**

The sun rose gradually over the tidal flats, its early rays reflecting off the waterlogged roots of mangroves sprawling in intricate patterns along the horizon. In the heart of the Indian Sundarban, the voices of the shoreline echoed themes of survival, stewardship, and resilience. *"We are the keepers of the tides,"* declared Mrs. Manashi Mondal, an elder from the local fishing community. Her words carried the profound weight of centuries of wisdom, passed down through generations who had learnt to coexist with the ever-shifting rhythms of the coast.

b) **Field Narrative**

The mangroves, both resilient and vulnerable, have long served as the lifeline of these communities. Ms. Radha, a spirited teenager and Mrs. Manashi's granddaughter, vividly recounted how these trees acted as natural barricades, shielding them from the devastating force of cyclones. *"But now,"* she remarked with a tone of concern, *"we notice gaps where the forest once stood. The waters are encroaching upon us."*

Engagements with the community highlighted their profound connection to the mangroves, which are valued not only as a resource but also as a cultural symbol. Narratives from elders revealed traditional practices such as sustainable fishing and rotational grazing, which upheld a delicate ecological balance. However, younger generations expressed growing unease over diminishing forest cover, increasing salinity, and declining fish stocks. *"We need to protect our mangroves, but we also need to feed our families,"* noted Mr. Surja Das, a local fisherman, encapsulating the tension between environmental preservation and sustaining livelihoods.

c) **A Process to Integrate with Scientific Evidence**

The insights gained from these discussions could provide the foundation for integrating community knowledge with scientific evidence. Through the use of PGIS, community members may actively map their interactions with the mangroves, identifying areas of degradation, zones with high biodiversity, and regions most vulnerable to climate-induced hazards [4, 5, 42]. These community narratives could be further enriched by scientific analyses, including remote sensing data and ecological assessments [17, 23, 27, 37]. This collaborative approach has the potential to facilitate the identification of critical areas where restoration efforts could have the greatest impact.

d) **Challenges and Sustainability**

Despite the collective effort, the journey was fraught with challenges. Tensions arose between immediate needs and long-term objectives as the overharvesting of mangrove timber for fuel and the expansion of aquaculture activities threatened restoration efforts [2, 19, 30]. Climate change further exacerbated these issues, with unpredictable weather patterns eroding both the coastline and the community's confidence [9, 22].

Nevertheless, the process also fostered innovation. Women's self-help groups took the lead in mangrove replantation, combining traditional knowledge of species selection with modern techniques to enhance survival rates [28, 33, 43]. Youth, including Mina, advocated for eco-tourism initiatives that showcased the mangroves' natural beauty while generating sustainable livelihood opportunities. *"If we cannot save the mangroves,"* Ms. Radha remarked at the meeting, *"we will lose ourselves."*

e) **The Moral of this Journey**

The journey of these shoreline communities reveals a profound truth: sustainability is not a fixed endpoint but an ongoing process of negotiation, resilience, and collaboration. The mangroves serve as a living testament to the power of collective stewardship, where the voices of the community and the insights of science converge to sustain both ecosystems and livelihoods. *The mangroves teach us to bend, not break,"* Mrs. Manashi remarked during one of our final field visits. Her words encapsulate the essence of this journey: that when communities and researchers work together, even the most delicate ecologies can find strength, and the most vulnerable shores can stand firm against the encroaching tides.

15.2.2 Salty Encroachment: Adaptive Narratives on Coastal Management in a Changing Climate

The narrative titled *"Salty Encroachment"* explored the issue of salinity intrusion in the land and agricultural areas of Coastal South 24 Parganas (Fig. 15.1, Site 15.1b) and utilized a range of Participatory Geographic Information Systems (PGIS) tools and techniques. Table 15.3 provides an overview of the narrative collection process

Table 15.3 Salty encroachment: functional insights from field observations using participatory methodologies

Location(s)	Coastal South 24 Parganas
Time duration	August to September 2023
Participants (The individual's name has been changed)	Md. Rajesh (age: 61+) Mrs. Rekha Sarkar (age: 56+) Mr. Subal Roy (age: 41+) Others 4 persons (age: 25+)
Used PGIS approaches	HHs; FGDs; CWs; TA
Note	Field-investigated by the author (S. Roy). HHs: Household conversation; FGDs: Focus Group Discussions; CWs: Community Workshops; and TA: Thematic Analysis

and the methodologies employed. The subsequent discussion presents the author's field observations and the narratives documented during field investigations, integrating these accounts with scientific evidence. The narrative further examines the challenges and opportunities for sustainability, concluding with a reflection on the overarching moral of this journey.

a) **Field notes (by the Author)**

The air was thick with the scent of salt as Md. Rajesh gestured toward the edges of his rice field, now a patchwork of cracked soil and brackish water. *"The sea is creeping into our lives,"* he said, his voice resonating with the frustration of years spent struggling against an invisible but unrelenting adversary. In the coastal South 24 Parganas, where rivers merge with the sea, the battle against saltwater intrusion is about more than just land; it is a fight for identity, resilience, and survival.

b) **Field Narrative**

For generations, coastal communities like Md. Rajesh's have relied on agriculture and fishing to sustain their livelihoods. The fertile soils of this delta, enriched by the Ganga rivers, once supported flourishing paddy fields and abundant fishponds. However, the intrusion of saline water, driven by rising sea levels and storm surges, has upended this delicate balance. "*We've seen the soil change,*" commented Mrs. Rekha Sarkar, a farmer and mother of three. *"What once grew enough to feed our families now grows nothing at all."* Salinity has contaminated freshwater resources, compelling families to abandon traditional farming practices. Meanwhile, fishermen like Mr. Subal Roy share accounts of dwindling fish stocks. *"The waters are warmer,"* he explained, *"and the fish are moving farther out to sea, making them harder to reach."*

Amidst these challenges, a spirit of resilience and innovation is emerging. Women's groups are experimenting with salt-tolerant crop varieties, while community elders revive traditional techniques such as constructing earthen embankments to safeguard farmland. The stories gathered from the field capture an ongoing interplay of hardship and determination, illustrating how these communities are adapting to an ever-changing environment.

c) **A Process to Integrate with Scientific Evidence**

Recognizing the significance of lived experiences, future collaborations between researchers and local communities could focus on developing innovative solutions. Using participatory geographic information systems, villagers may actively map areas most vulnerable to salinity intrusion while identifying zones where adaptive practices could prove effective. This participatory approach could be further enriched by scientific insights. Soil testing and satellite imagery could validate the extent of salinization, while advanced climate models might project future trends with greater precision [12, 14, 47]. Community workshops could be organized to share these findings, allowing villagers to contextualize their challenges within broader environmental patterns [3, 10, 29]. *"It would be like assembling pieces of a puzzle."* Such collaborations could pave the way for innovative initiatives, including bio-shields created with mangroves and saline aquaculture projects. By seamlessly integrating scientific rigour with local knowledge, these adaptive strategies could aim to balance ecological conservation with the need to sustain livelihoods.

d) **Challenges and Sustainability**

Despite these advancements, the path to sustainability remains fraught with challenges. While the construction of embankments offers short-term protection, these structures often fail under the strain of increasingly intense storms. Rising temperatures further exacerbate salinity intrusion, jeopardizing even salt-tolerant crops. Additionally, limited access to resources and funding hampers the large-scale implementation of adaptive strategies [40, 41].

Nonetheless, these challenges have not weakened the resolve of the people. Local governance bodies have taken on a more active role, securing microloans for farmers and fishermen to support their transition to sustainable practices. Youth groups have leveraged social media to amplify their narratives, raising awareness about the struggles of coastal communities and advocating for policy reforms. *"We know the sea won't stop,"* Mrs. Rekha stated. *"But neither will we."*

e) **The Moral of this Journey**

The story of salty encroachment is one of resilience, of communities learning to harmonize with the tides rather than struggle against them. It stands as a testament to the power of merging traditional knowledge with scientific innovation, illustrating that adaptation involves not just technology but people. The core lesson of this journey is unity: when communities, researchers, and policymakers collaborate, even the most formidable challenges can be faced with hope, creativity, and strength. As Md. Rajesh poignantly expressed, *"The sea may bring salt, but we will find a way to plant sweetness in its wake."*

15.2.3 Crops and Catastrophes: Community Narratives on Agricultural Management

The narrative titled *"Crops and Catastrophes"* delved into the challenges posed by climate-related catastrophes and the potential for agricultural management in Coastal Medinipur (Fig. 15.1, Site 15.1a), employing a variety of Participatory Geographic Information Systems (PGIS) tools and techniques. Table 15.4 outlines the narrative collection process and the methodologies utilized. The subsequent discussion integrates the author's field observations and the narratives recorded during field investigations with relevant scientific evidence. Furthermore, the narrative critically examines the challenges and opportunities for achieving sustainability, culminating in a reflection on the overarching moral of this journey.

a) **Field notes (by the Author).**

The rhythmic sound of hoeing resonated across the fields as Mrs. Asha Devi worked relentlessly under the scorching sun. Her hands moved with expert

Table 15.4 Crops and catastrophes: functional insights from field observations using participatory methodologies

Location(s)	Coastal Medinipur
Time duration	November to December 2024
Participants (The individual's name has been changed)	Mrs. Asha Bera (age: 51+) Mr. Ramesh Bera (age: 48+) Mr. Ashok Bera (age: 37+) Others 6 persons (age: 25+)
Used PGIS approaches	HHs; FGDs; CWs; TA
Note	Field-investigated by the author (S. Roy). HHs: Household conversation; FGDs: Focus Group Discussions; CWs: Community Workshops; and TA: Thematic Analysis

precision, removing the weeds that threatened her vulnerable crops. *"These fields are our lifeline,"* she remarked, wiping the sweat from her brow. *"But each season now feels like a gamble."* In the heart of the coastal agricultural belt, communities like Kamala's are grappling with an unprecedented set of challenges: erratic weather patterns, declining soil fertility, and the growing threat of climate change.

b) **Field Narrative.**

For decades, farmers in this region relied on predictable monsoons and fertile soil to cultivate staple crops such as rice, jute, and pulses. Their stories once celebrated abundant harvests, communal meals, and festivals that honoured the land's generosity. However, in recent years, these narratives have taken a more sombre turn. *"The rains come too early or too late,"* said Mr. Ramesh, a smallholder farmer. *"Our seeds either rot before they sprout or dry up before they can grow."* His neighbour, Mr. Ashok, shared how a flash flood destroyed an entire season's worth of labour. *"We stood helpless as the water took everything,"* he said, her voice laden with sorrow. Amid these challenges, a spirit of ingenuity is emerging. Farmers are adopting new techniques, such as mulching to conserve soil moisture and planting drought-resistant crop varieties. Women-led collectives are experimenting with organic fertilizers and alternative pest control methods. These grassroots innovations, born out of necessity, reflect the resilience and adaptability of rural communities.

c) **A Process to Integrate with Scientific Evidence.**

Recognizing the value of local knowledge, future collaborations between researchers and farming communities may co-develop innovative solutions to address agricultural challenges. PGIS workshops could empower farmers to map areas susceptible to flooding or drought, uncovering patterns that align with satellite data and predictive models [34, 45, 46]. Therefore, it may be concluded that by combining "their" deep understanding of the land with advanced tools and data, we can enhance "their" resilience. Future initiatives might include the widespread distribution of advanced soil testing kits to monitor pH levels and nutrient deficiencies, alongside drone technology for detailed aerial assessments of crop health [1, 18]. Real-time weather updates, delivered through increasingly accessible mobile platforms, could enable farmers to make more precise decisions about planting and harvesting [15, 31]. These collaborations could extend beyond technology, fostering inclusive spaces for dialogue through regular community meetings. Here, farmers would share their experiences, and researchers would present evolving findings, enabling the co-creation of context-specific strategies that integrate scientific rigour with local wisdom.

d) **Challenges and Sustainability**

Despite these advancements, significant challenges persist. The high cost of modern agricultural tools and seeds often makes them unaffordable for small-scale farmers. Although government subsidies and loans are available, they are often delayed by bureaucratic hurdles. *"By the time the funds arrive, the season is already over,"* Mrs. Asha lamented. Environmental challenges exacerbate these issues. Rising temperatures accelerate soil degradation, while invasive pests and diseases spread with increasing frequency. Market volatility further compounds the uncertainty, as fluctuating prices undermine farmers' financial stability. Nevertheless, the foundation for sustainability is being established. Farmer cooperatives are pooling resources to access machinery and share knowledge. Youth-led initiatives are utilizing social media to advocate for policy reforms and highlight successful practices.

e) **The Moral of this Journey**

The story of crops and catastrophes is ultimately one of resilience. It highlights the power of collective action, where communities, scientists, and policymakers collaborate to address shared challenges. It serves as a reminder that sustainability is not solely about conserving resources but about empowering people to adapt and thrive. As Mrs. Asha Devi reflected, *"We can't control the rain or the sun, but we can control how we respond. Together, we will find a way to keep these fields alive."* Her words encapsulate the essence of this journey, a testament to human determination and the enduring connection between people and the land.

15.2.4 Nets of Tradition: Integrating Local Knowledge in Fishing Ground Management

The narrative titled *"Nets of Tradition"* discussed into the challenges and prospects associated with managing fishing grounds in the coastal region of Bengal (Fig. 15.1, Site 15.1a to 15.1d), employing a variety of Participatory Geographic Information Systems (PGIS) tools and techniques. Table 15.5 outlines the narrative collection process and the methodologies applied. The subsequent discussion presents the author's field observations alongside the narratives documented during field

Table 15.5 Nets of tradition: functional insights from field observations using participatory methodologies

Location(s)	Coastal region of Bengal
Time duration	December 2024
Participants (The individual's name has been changed)	Mr. Suman Jali (age: 62+) Mrs. Madhubala Jali (age: 57+) Mr. Kartik Jali (age: 33+) Others 3 persons (age: 25+)
Used PGIS approaches	HHs; FGDs; CWs; TA
Note	Field-investigated by the author (K. D. Malakar). HHs: Household conversation; FGDs: Focus Group Discussions; CWs: Community Workshops; and TA: Thematic Analysis

investigations, integrating these accounts with scientific evidence. Furthermore, the narrative critically explores the challenges and opportunities for achieving sustainability, culminating in a reflection on the overarching moral of this journey.

a) **Field notes (by the Author)**

The salty breeze carried the scent of the sea as fishermen hauled in their nets along the Sundarban coastline. Amid the rhythmic sound of waves slapping against their boats, Mr. Suman Jali, an experienced fisherman, shared his story. "*Our ancestors taught us the secrets of these waters,*" he said, his voice filled with both pride and concern. "*But now, the tides and the fish no longer behave as they once did.*" It also noted that the traditional marine fishing communities in coastal Bengal are grappling with a delicate balance between preserving their heritage and adapting to the challenges posed by climate change, overfishing, and coastal development, which are disrupting their livelihoods.

b) **Field Narrative**

For generations, fishermen have relied on intricate knowledge passed down orally, encompassing tidal patterns, fish migrations, and the seasonal rhythms of the sea. These stories speak of a time when the waters were abundant with hilsa and prawns, sustaining both livelihoods and traditions. However, the narrative has shifted. Dwindling fish populations and erratic weather patterns have disrupted their way of life. "*We now have to venture farther out to sea,*" said Mrs. Madhubala, a fisherwoman and community leader. "*Sometimes, after days of work, we return empty-handed.*" Her husband, Mr. Kartik, added, "*The mangroves once shielded us, but they're vanishing. Without them, we are vulnerable to the storms.*" In the face of these challenges, the community is showing remarkable resilience. Women's groups have initiated small-scale aquaculture projects, cultivating shrimp and crab in tidal ponds. Meanwhile, fishermen are adopting sustainable practices, such as using larger mesh sizes to allow juvenile fish to escape. These efforts highlight a collective determination to safeguard their traditions while adapting to evolving environmental realities.

c) **A Process to Integrate with Scientific Evidence**

In the future, recognizing the value of local knowledge, scientists, and policy-makers could establish stronger partnerships with fishing communities. Participatory GIS (PGIS) mapping may be utilized to identify critical fishing grounds and assess the health of coastal ecosystems [8, 36]. By integrating traditional knowledge with satellite imagery, researchers could pinpoint areas where conservation efforts would be most urgently needed [6, 16]. So, the fishermen will likely understand these waters better than any map, and their observations could guide research and ensure that proposed solutions are grounded in practicality. Besides that, workshops could be organized to introduce modern techniques, such as artificial reef construction, aimed at restoring fish habitats. Additionally, real-time weather updates and tide predictions delivered via mobile apps could empower fishermen to plan safer and more efficient expeditions. Communities might also embrace mangrove restoration initiatives, planting saplings along eroded coastlines to create natural barriers against storms and provide nurseries for fish and crustaceans. "*We hope to rebuild what we've lost,*" Mrs. Madhubala says. "*It could be our way of giving back to the sea.*" Through these collaborative efforts, a sustainable balance between conservation and livelihoods could be achieved.

d) **Challenges and Sustainability**

The journey remains fraught with challenges. Many families find it difficult to afford the costs of new equipment or participate in conservation programs while grappling with the demands of daily survival. Disputes over fishing rights and the prevalence of illegal trawling further complicate efforts to manage resources sustainably. Although government support exists, it often falls short of addressing the complex and localized issues faced by these communities [25, 32]. Climate change exacerbates these struggles, bringing rising sea levels that threaten coastal villages and ocean warming that disrupts marine ecosystems [9, 44]. "*We're doing everything we can,*" said Mr. Suman, "*but at times, it feels like an uphill battle.*"

Despite these obstacles, hope endures. Younger generations, equipped with education and technology, are stepping up to advocate for their communities and drive change. Collaboration efforts between local leaders, NGOs, and government agencies are shaping a shared vision for sustainable fisheries. Innovations such as solar-powered iceboxes and eco-friendly fishing gear are gaining momentum, helping to reduce both costs and environmental impacts.

e) **The Moral of this Journey**

The narrative of "*Nets of Tradition*" is one of resilience and collective action, emphasizing the power of blending local wisdom with scientific innovation to address intricate challenges. Above all, it showcases the unwavering strength of a community united by a common goal: preserving their heritage while safeguarding their future. As Mrs. Madhubala eloquently expressed, "*The sea has always provided for us, and now it is our responsibility to protect it. Together, we can ensure these waters remain vibrant for generations to come.*" Her words encapsulate the heart of this journey, a profound testament to the enduring connection between people and the natural world.

15.2.5 Unequal Journeys: Management Approaches to Climate Migration and Inequality

The narrative titled *"Unequal Journeys"* delves into the challenges and concern of climate migration and socioeconomic inequality in Coastal 24 Parganas (Fig. 15.1, Site 15.1b and Site 15.1d), employing an array of PGIS tools and techniques. Table 15.6 outlines the narrative collection process and the methodologies applied. The subsequent discussion synthesizes the author's field observations with narratives documented during field investigations, integrating these accounts with scientific evidence. Furthermore, the narrative critically evaluates the challenges and prospects for sustainability, culminating in a reflection on the broader moral and lessons of this journey.

a) **Field notes (by the Author)**

The parched road seemed to stretch endlessly beneath the relentless sun as families made their way, carrying whatever belongings they could salvage. Young children clung to their parents, their eyes brimming with confusion and fear. Such was the plight of the villagers from Sundarban, a coastal settlement claimed by the encroaching seas. *"We had no choice,"* lamented Mrs. Mamtabala, a mother of two. *"The water swallowed our land, our crops, and our home. Now we must begin anew, but where can we possibly go?"*

b) **Field Narrative**

The story of Sundarban-affiliated villages is far from isolated. Across India and beyond, climate-induced migration is becoming an unavoidable reality. As floods, droughts, and extreme weather events displace millions, the burden falls heaviest on marginalized communities. Mrs. Mamtabala's family, like countless others, has joined the growing ranks of climate refugees, individuals forced to abandon their

Table 15.6 Unequal journeys: functional insights from field observations using participatory methodologies

Location(s)	Coastal 24 Parganas
Time duration	April to June 2022
Participants (The individual's name has been changed)	Mrs. Mamtabala Hazra (age: 47+) Mr. Rajsekhar Banik (age: 38+) Others 10 persons (age: 25+)
Used PGIS approaches	HHs; FGDs; CWs; TA
Note	Field-investigated by the author (K. D. Malakar). HHs: Household conversation; FGDs: Focus Group Discussions; CWs: Community Workshops; and TA: Thematic Analysis

homes due to environmental upheaval. *"We used to be farmers,"* shared Mr. Rajsekhar, a former landowner now working as a labourer in the city of Kolkata. *"The land was our identity. Now we are no one."* Migrants often endure discrimination, various climate gaps related issues, loss of livelihoods, and the profound emotional strain of displacement, yet amidst these challenges, resilience shines through. Communities are coming together, forging paths to adapt and support one another in the face of adversity.

c) **A Process to Integrate with Scientific Evidence**

Addressing climate migration will require a comprehensive approach that integrates local knowledge with scientific research. In the future, researchers and NGOs could collaborate more extensively with displaced communities to document their lived experiences and identify emerging patterns. PGIS could be widely utilized to map migration trends, identify vulnerable regions, and develop advanced early warning systems for climate-related disasters [11, 39].

Scientific data will guide the way, but the voices of the people must remain at the centre of the process [20]. Future workshops may serve as platforms for migrants like Mrs. Mamtabala and Mr. Rajsekhar to share their experiences, offering policymakers deeper insights into the social and economic ramifications of displacement. These narratives could be systematically integrated with data on climate variability, infrastructure vulnerabilities, and resource dynamics to inform more inclusive and adaptive management strategies. Innovative pilot programs could pave the way for sustainable resettlement solutions. For example, eco-villages designed with renewable energy systems, advanced water conservation technologies, and cooperative farming practices could become models of sustainable living. Comprehensive skills training programs might empower migrants to participate in urban economies while preserving their cultural identity.

d) **Challenges and Sustainability**

Despite advancements, substantial obstacles persist. Resource allocation tends to favour wealthier regions, leaving marginalized communities underserved and vulnerable. Bureaucratic inefficiencies often delay the delivery of essential aid, while many migrants lack the documentation required to access government support programs. Social tensions further complicate the situation, as host communities frequently view migrants as competitors for jobs, housing, and resources. Climate change intensifies these challenges, perpetuating a cycle of vulnerability. Rising temperatures, erratic weather patterns, and resource scarcity exacerbate existing inequities, while political inertia impedes the implementation of comprehensive, large-scale solutions.

Efforts toward sustainability are shifting the focus to long-term strategies. Investments in renewable energy, climate resilient infrastructure, and education aim to foster a future where migration is an option rather than a necessity. Grassroots movements are also gaining momentum, advocating for climate justice and urging policies that centre on the needs of vulnerable populations. These initiatives represent a collective push for a more equitable and resilient world.

e) **The Moral of this Journey**

The story of *"Unequal Journeys"* exposes a stark reality: climate change deepens existing disparities, yet it also highlights the enduring resilience and collaborative spirit of humanity. By combining traditional wisdom with scientific advancements, communities are finding ways to confront the complexities of climate migration and strive toward a fairer and more sustainable future. According to Mrs. Mamtabala poignantly observed, *"We didn't choose this journey, but we've learnt to walk it together. The road is hard, but with unity and support, we can find our way."* Her words serve as a powerful reminder that, even in the face of immense challenges, it is our shared determination and solidarity that will shape a brighter path forward.

15.3 Conclusion and Arguments

The five field narratives, *Voices from the Shoreline, Salty Encroachment, Crops and Catastrophes, Nets of Tradition,* and *Unequal Journeys*, together underscore the intricate interconnectedness between human communities, their surrounding environments, the complexities and scope of sustainable management. These accounts transcend mere stories of struggle and survival; they serve as testaments to resilience, innovation, and the enduring power of collective action. As a result, the key arguments that emerged from the overall discussion are as follows:

- **The Strength of Community-Led Approaches:** A central theme across all narratives is the transformative power of community-led initiatives. Whether managing mangroves, addressing saline encroachments, sustaining agriculture, preserving fishing grounds, or navigating the complexities of climate migration, local knowledge is the cornerstone of effective management strategies. Communities possess intricate, place-specific understandings of their ecosystems that no external expertise or technology can replicate. By empowering these communities through participatory frameworks, solutions are crafted to be context-specific, pragmatic, and culturally aligned.
- **Bridging Local Knowledge and Scientific Innovation:** Each narrative underscores the significant impact of integrating traditional knowledge with modern scientific advancements. Tools such as Participatory Geographic Information Systems (PGIS), soil testing kits, and renewable energy technologies enhance the ability of local populations to address environmental challenges effectively. This collaboration fosters a co-creative process, where evidence-based science complements local insights, resulting in adaptive solutions that are both innovative and grounded in lived realities.
- **Adapting to Climatic and Social Uncertainty:** Climate change acts as a pervasive disruptor across these narratives, intensifying vulnerabilities and introducing new challenges. Rising sea levels, soil salinization, extreme weather events, and forced migration amplify the uncertainty faced by communities. Nevertheless,

these stories illustrate the remarkable adaptive capacity of human societies. Initiatives such as developing drought-resistant crops, piloting sustainable eco-villages, and fostering collaboration demonstrate that adaptation is achievable through innovation, resilience, and collective effort.

- **Addressing Inequality and Ensuring Access:** Despite the promise of local and scientific approaches, structural inequalities remain a critical obstacle. Marginalized groups, including small-scale farmers, fishermen, and climate migrants, often lack access to essential resources such as funding, technology, and policy support. Bureaucratic delays, social hierarchies, and market inequities further exacerbate their vulnerabilities. These narratives call for systemic reforms to promote equity, streamline resource allocation, and ensure that the needs of the most vulnerable populations are prioritized.
- **Sustainability as an Ongoing Commitment:** The narratives emphasize that sustainability is not a static goal but an ongoing process requiring long-term investment. Whether it involves mangrove restoration, food security, or fisheries management, success depends on sustained efforts in education, infrastructure development, and community engagement. Key components of a sustainable future include renewable energy adoption, cooperative resource management, and youth-driven advocacy. Ultimately, sustainability must achieve a balance among ecological conservation, economic viability, and social justice.

Therefore, the overarching moral of these narratives is that sustainability is a collective responsibility. Local communities, scientific institutions, policymakers, and global stakeholders must collaborate to tackle the complex challenges of environmental degradation and climate change. These narratives demonstrate that the solutions to our most pressing issues are found not in isolation but through cooperation, where the wisdom of the past guides future innovation and diverse voices contribute to shaping inclusive and effective solutions.

References

1. Abdulraheem, M. I., Zhang, W., Li, S., Moshayedi, A. J., Farooque, A. A., & Hu, J. (2023). Advancement of remote sensing for soil measurements and applications: A comprehensive review. *Sustainability, 15*(21), 15444. https://doi.org/10.3390/su152115444
2. Akram, H., Hussain, S., Mazumdar, P., Chua, K. O., Butt, T. E., & Harikrishna, J. A. (2023). Mangrove health: A review of functions, threats, and challenges associated with mangrove management practices. *Forests, 14*(9), 1698. https://doi.org/10.3390/f14091698
3. Allen, K., Castellano, C., & Pessagno, S. (2021). Using dialogue to contextualize culture, ecosystem services, and cultural ecosystem services. *Ecology and Society, 26*(2), 7. https://doi.org/10.5751/ES-12187-260207
4. Ameca, E. I., Nie, Y., Wu, R., Mittermeier, R. A., Foden, W., & Wei, F. (2024). Identifying protected areas in biodiversity hotspots at risk from climate and human-induced compound events for conserving threatened species. *Science of the Total Environment, 938*, 173192. https://doi.org/10.1016/j.scitotenv.2024.173192
5. Antwi, E. K., Boakye-Danquah, J., Owusu, A. B., Loh, S. K., Mensah, R., Boafo, Y. A., & Apronti, P. T. (2015). Community vulnerability assessment index for flood prone savan-

nah agro-ecological zone: A case study of Wa West District, Ghana. *Weather and Climate Extremes, 10*, 56–69. https://doi.org/10.1016/j.wace.2015.10.008

6. Awewomom, J., Dzeble, F., Takyi, Y. D., et al. (2024). Addressing global environmental pollution using environmental control techniques: a focus on environmental policy and preventive environmental management. *Discover Environment, 2*, 8. https://doi.org/10.1007/s44274-024-00033-5
7. Basu, J. P. (2020). *Climate change vulnerability and communities in agro-climatic regions of West Bengal, India: Theory and practice*. Springer . ISBN: 978-3-030-50467-0. https://doi.org/10.1007/978-3-030-50468-7
8. da Silva, C. V., Ortigao, M., Willaert, T., Rosa, R., Nunes, L. C., & Cunha-e-Sa, M. A. (2021). Participatory Geographic Information Systems (PGIS): Alternative approaches to identify potential conflicts and positional accuracy in marine and coastal ecosystem services. *Marine Policy, 131*, 104650. https://doi.org/10.1016/j.marpol.2021.104650
9. Das, A., & Swain, P. K. (2024). Navigating the sea level rise: Exploring the interplay of climate change, sea level rise, and coastal communities in India. *Environmental Monitoring and Assessment, 196*, 1010. https://doi.org/10.1007/s10661-024-13191-z
10. Esmail, N., McPherson, J. M., Abulu, L., Amend, T., Amit, R., Bhatia, S., et al. (2023). What's on the horizon for community-based conservation? Emerging threats and opportunities. *Trends in Ecology & Evolution, 38*(7), 666–680. https://doi.org/10.1016/j.tree.2023.02.008
11. Ghosh, R. C., & Orchiston, C. (2022). A systematic review of climate migration research: Gaps in existing literature. *SN Social Sciences, 2*, 47. https://doi.org/10.1007/s43545-022-00341-8
12. Hassani, A., Azapagic, A., & Shokri, N. (2021). Global predictions of primary soil salinization under changing climate in the 21st century. *Nature Communications, 12*, 6663. https://doi.org/10.1038/s41467-021-26907-3
13. Kantamaneni, K., Panneer, S., Krishnan, A., et al. (2022). Appraisal of climate change and cyclone trends in Indian coastal states: A systematic approach towards climate action. *Arabian Journal of Geosciences, 15*, 814. https://doi.org/10.1007/s12517-022-10076-8
14. Kumar, P., Tiwari, P., Biswas, A., & Srivastava, P. K. (2024). Spatio-temporal assessment of soil salinization utilizing remote sensing derivatives, and prediction modeling: Implications for sustainable development. *Geoscience Frontiers, 15*(6), 101881. https://doi.org/10.1016/j.gsf.2024.101881
15. Kumar, V., Sharma, K. V., Kedam, N., Patel, A., Kate, T. R., & Rathnayake, U. (2024). A comprehensive review on smart and sustainable agriculture using IoT technologies. *Smart Agricultural Technology*, 100487. https://doi.org/10.1016/j.atech.2024.100487
16. Lauer, M., & Aswani, S. (2008). Integrating indigenous ecological knowledge and multispectral image classification for marine habitat mapping in Oceania. *Ocean & Coastal Management, 51*(6), 495–504. https://doi.org/10.1016/j.ocecoaman.2008.04.006
17. Lejano, R. P., Tavares-Reager, J., & Berkes, F. (2013). Climate and narrative: Environmental knowledge in everyday life. *Environmental Science & Policy, 31*, 61–70. https://doi.org/10.1016/j.envsci.2013.02.009
18. Makam, S., Komatineni, B. K., Meena, S. S., et al. (2024). Unmanned aerial vehicles (UAVs): An adoptable technology for precise and smart farming. *Discover Internet of Things, 4*, 12. https://doi.org/10.1007/s43926-024-00066-5
19. Malakar, K. D. (2020). Rural livelihood and mangrove degradation: A case study of Namkhana Block, West Bengal, India. *International Journal of Innovative Science and Research Technology, 5*(1), 721–726.
20. Malakar, K. D., & Roy, S. (2024). Voice mapping of coastal communities: Field narratives from the coastal Medinipur and Sundarbans. In *Mapping geospatial citizenship*. Springer. https://doi.org/10.1007/978-3-031-63107-8_8
21. McCall, M. K., & Minang, P. A. (2005). Assessing participatory GIS for community-based natural resource management: Claiming community forests in Cameroon. *Geographical Journal, 171*(4), 340–356. https://doi.org/10.1111/j.1475-4959.2005.00173.x

22. Mimura, N. (2013). Sea-level rise caused by climate change and its implications for society. *Proceedings of the Japan Academy, Series B, 89*(7), 281–301. https://doi.org/10.2183/pjab.89.281
23. Moezzi, M., Janda, K. B., & Rotmann, S. (2017). Using stories, narratives, and storytelling in energy and climate change research. *Energy Research & Social Science, 31*, 1–10. https://doi.org/10.1016/j.erss.2017.06.034
24. Mondal, B., Roy, A., & Saha, A. K. (2022). Vulnerability assessment of mangrove areas in coastal West Bengal, India. *Remote Sensing Applications: Society and Environment, 25*, 100680. https://doi.org/10.1016/j.rsase.2021.100680
25. Mozumder, M. M. H., Uddin, M. M., Schneider, P., Deb, D., Hasan, M., Saif, S. B., & Nur, A. A. U. (2023). Governance of illegal, unreported, and unregulated (IUU) fishing in Bangladesh: Status, challenges, and potentials. *Frontiers in Marine Science, 10*, 1150213. https://doi.org/10.3389/fmars.2023.1150213
26. Muralikrishnan, L., Manikandan, N., Kumar, J. P. T., Yumnam, A., Seyie, A., Tomar, M., & Venkatesh, P. (2021). Participatory GIS (PGIS) approach for the development of community-based climate smart sustainable agriculture models in the semiarid regions of southern India. *Current Journal of Applied Science and Technology, 40*(5), 1–13. https://doi.org/10.9734/cjast/2021/v40i531300
27. Murthy, M. S. R., Gilani, H., Karky, B. S., Sharma, E., Sandker, M., Koju, U. A., et al. (2017). Synergizing community-based forest monitoring with remote sensing: A path to an effective REDD+ MRV system. *Carbon Balance and Management, 12*, 1–14. https://doi.org/10.1186/s13021-017-0087-8
28. Mylon, V. (2024). Women lead the charge in mangrove restoration in Maharashtra. UNDP India. Report published on July 25, 2024. Retrieved 01 January 2025. Available at: https://www.undp.org/india/stories/women-lead-charge-mangrove-restoration-maharashtra
29. Nash, N., Capstick, S., Whitmarsh, L., Chaudhary, I., & Manandhar, R. (2019). Perceptions of local environmental issues and the relevance of climate change in Nepal's Terai: Perspectives from two communities. *Frontiers in Sociology, 4*, 60. https://doi.org/10.3389/fsoc.2019.00060
30. Onyena, A. P., & Sam, K. (2020). A review of the threat of oil exploitation to mangrove ecosystem: Insights from Niger Delta, Nigeria. *Global Ecology and Conservation, 22*, e00961. https://doi.org/10.1016/j.gecco.2020.e00961
31. Paparrizos, S., Dogbey, R. K., Sutanto, S. J., Gbangou, T., Kranjac-Berisavljevic, G., Gandaa, B. Z., et al. (2023). Hydro-climate information services for smallholder farmers: FarmerSupport app principles, implementation, and evaluation. *Climate Services, 30*, 100387. https://doi.org/10.1016/j.cliser.2023.100387
32. Pham, C. V., Wang, H. C., Chen, S. H., & Lee, J. M. (2023). The threshold effect of overfishing on global fishery outputs: International evidence from a sustainable fishery perspective. *Fishes, 8*(2), 71. https://doi.org/10.3390/fishes8020071
33. Ravaoarinorotsihoarana, L. A., Ratefinjanahary, I., Aina, C., Rakotomahazo, C., Glass, L., Ranivoarivelo, L., & Lavitra, T. (2023). Combining traditional ecological knowledge and scientific observations to support mangrove restoration in Madagascar. *Forests, 14*(7), 1368. https://doi.org/10.3390/f14071368
34. Roncoli, C. (2006). Ethnographic and participatory approaches to research on farmers' responses to climate predictions. *Climate Research, 33*(1), 81–99. https://doi.org/10.3354/cr033081
35. Saadallah, D. M. (2020). Utilizing participatory mapping and PPGIS to examine the activities of local communities. *Alexandria Engineering Journal, 59*(1), 263–274. https://doi.org/10.1016/j.aej.2019.12.038
36. Scully-Engelmeyer, K. M., Granek, E. F., Nielsen-Pincus, M., & Brown, G. (2021). Participatory GIS mapping highlights indirect use and existence values of coastal resources and marine conservation areas. *Ecosystem Services, 50*, 101301. https://doi.org/10.1016/j.ecoser.2021.101301

37. Segarra, J., González-Fernández, A., Osorno-Covarrubias, J., & Couturier, S. (2024). The role of critical remote sensing in environmental justice struggles. *Progress in Environmental Geography, 3*(3), 185–211. https://doi.org/10.1177/27539687241269331
38. Shampa, M. T. A., Shimu, N. J., Chowdhury, K. A., Islam, M. M., & Ahmed, M. K. (2023). A comprehensive review on sustainable coastal zone management in Bangladesh: Present status and the way forward. *Heliyon, 9*, e18190. https://doi.org/10.1016/j.heliyon.2023.e18190
39. Siddiqui, A. R., & Sahay, A. (2022). *Climate change, disaster and adaptations: Contextualising human responses to ecological change*. Springer . ISBN: 978-3-030-91009-9. https://doi.org/10.1007/978-3-030-91010-5
40. Sultan, M. T., Mahmud, U., & Khan, M. Z. (2023). Addressing soil salinity for sustainable agriculture and food security: Innovations and challenges in coastal regions of Bangladesh. *Future Foods*, 100260. https://doi.org/10.1016/j.fufo.2023.100260
41. Tarolli, P., Luo, J., Park, E., Barcaccia, G., & Masin, R. (2024). Soil salinization in agriculture: Mitigation and adaptation strategies combining nature-based solutions and bioengineering. *Iscience, 27*(2), 108830. https://doi.org/10.1016/j.isci.2024.108830
42. Tourinho, L., de Brito Alves, S. M., da Silva, F. B. L., Verdi, M., Roque, N., Conceição, A. A., et al. (2023). A participatory approach to map strategic areas for conservation and restoration at a regional scale. *Perspectives in Ecology and Conservation, 21*(1), 52–61. https://doi.org/10.1016/j.pecon.2022.11.001
43. Uprety, Y., Asselin, H., Bergeron, Y., Doyon, F., & Boucher, J. F. (2012). Contribution of traditional knowledge to ecological restoration: Practices and applications. *Ecoscience, 19*(3), 225–237. https://doi.org/10.2980/19-3-3530
44. Venegas, R. M., Acevedo, J., & Treml, E. A. (2023). Three decades of ocean warming impacts on marine ecosystems: A review and perspective. *Deep Sea Research Part II: Topical Studies in Oceanography, 212*, 105318. https://doi.org/10.1016/j.dsr2.2023.105318
45. Warner, D., Moonsammy, S., & Joseph, J. (2022). Factors that influence the use of climate information services for agriculture: A systematic review. *Climate Services, 28*, 100336. https://doi.org/10.1016/j.cliser.2022.100336
46. White, I., Kingston, R., & Barker, A. (2010). Participatory geographic information systems and public engagement within flood risk management. *Journal of Flood Risk Management, 3*(4), 337–346. https://doi.org/10.1111/j.1753-318X.2010.01083.x
47. Zhang, X., Feng, G., & Sun, X. R. (2024). Advanced technologies of soil moisture monitoring in precision agriculture. *Journal of Agriculture and Food Research*, 101473. https://doi.org/10.1016/j.jafr.2024.101473

Further Readings

48. Aswani, S. (2019). Perspectives in coastal human ecology (CHE) for marine conservation. *Biological Conservation, 236*, 223–235. https://doi.org/10.1016/j.biocon.2019.05.047
49. Damastuti, E., & de Groot, R. (2019). Participatory ecosystem service mapping to enhance community-based mangrove rehabilitation and management in Demak, Indonesia. *Regional Environmental Change, 19*, 65–78. https://doi.org/10.1007/s10113-018-1378-7
50. Levine, A. S., & Feinholz, C. L. (2015). Participatory GIS to inform coral reef ecosystem management: Mapping human coastal and ocean uses in Hawaii. *Applied Geography, 59*, 60–69. https://doi.org/10.1016/j.apgeog.2014.12.004
51. Malakar, K. D., Kumar, M., Anand, S., & Kuzur, G. (2023). Climate change and socio-ecological transformation: Applications and field narratives. In *Climate change and socio-ecological transformation*. Springer Nature. https://doi.org/10.1007/978-981-99-4390-6_7

52. Reis-Filho, J. A., Hatje, V., & Barros, F. (2024). Navigating blue justice: Policy gaps and conflicts in coastal development from small-scale fisher perspectives. *One Earth, 7*(10), 1786–1802. https://doi.org/10.1016/j.oneear.2024.07.014
53. Sulistyawan, B. S., Verweij, P. A., Boot, R. G., Purwanti, B., Rumbiak, W., Wattimena, M. C., et al. (2018). Integrating participatory GIS into spatial planning regulation: The case of Merauke District, Papua, Indonesia. *International Journal of the Commons, 12*(1), 26–59. https://doi.org/10.18352/ijc.759
54. Tait, M., Gaughen, K., Tsang, A., Walton, M., Marcoux, S., Kekoa, L., et al. (2024). Holomua marine initiative: Community-generated socio-cultural principles and indicators for marine conservation and management in Hawai'i. *Ecology and Society, 29*(1), 4. https://doi.org/10.5751/ES-13640-290104

Chapter 16
Geospatial Technologies in Coastal Ecologies Monitoring and Management on the Bengal Coast: Case Studies and Best Practices

"Technology and ecology are not enemies. The key lies in aligning innovation with environmental stewardship."— Adapted from E.O. Wilson.

Abstract Coastal ecosystems are highly dynamic and vulnerable to both natural and anthropogenic pressures, necessitating advanced monitoring and management strategies. This chapter explores the socio-ecological dimensions of coastal ecologies on the Bengal coast in India, emphasizing the application of geospatial technologies for effective coastal monitoring and management. It highlights the integration of participatory approaches to foster sustainable coastal governance, ensuring that scientific assessments align with community-driven priorities. A comprehensive assessment of climate change consequences and adaptation strategies in coastal communities is undertaken through a collection of nine case studies. These case studies illustrate diverse methodologies for monitoring and managing coastal socio-ecological systems. They include a century-long climatic trend analysis (Casebook 1), monitoring and modelling shoreline dynamics (Casebook 2), and assessing mangrove greenness in response to climate change (Casebook 3). Additional case studies explore land use/land cover changes and ecological degradation (Casebook 4), coastal land surface temperature variations (Casebook 5), and mangrove-human conflicts (Casebook 6). The chapter further examines climate adaptation strategies among traditional fishermen (Casebook 7), gaps in climate resilience among displaced populations (Casebook 8), and the spatial dimensions of coastal zone management (Casebook 9). By integrating diverse geospatial methodologies with participatory governance approaches, this chapter presents best practices and innovative insights for enhancing resilience, adaptive capacity, and sustainability in coastal ecosystem management. The findings contribute to informed decision-making for researchers and policymakers, supporting more effective socio-ecological governance in coastal regions.

Keywords Geospatial technologies · Coastal monitoring and management · Climate change · Coastal ecologies · Participatory approaches · Sustainable coastal governance · Socio-ecological pressures · Land use/land cover change · LST and NDVI · Shoreline dynamics · Mangrove greenness · Ecological degradation ·

K. D. Malakar et al., *Geospatial Technologies in Coastal Ecologies Monitoring and Management*, Advances in Geographic Information Science,
https://doi.org/10.1007/978-3-031-92017-2_16

Mangrove-human conflicts · Climate adaptation · Traditional fishermen · Climate migration and climate gaps · Coastal zone management · West Bengal Coast · Indian Sundarban

This chapter examines detailed case studies and best practices for applying geospatial techniques in the monitoring and management of coastal ecologies in the Indian state of West Bengal. The primary objective is to investigate nine selected case studies (casebook), including a century-long climatic trend analysis, shoreline dynamics, mangrove greenness, land use/land cover changes and ecological degradation, coastal land surface temperature and ecological changes, mangrove-human conflicts, climate adaptation strategies of traditional fishermen, climate gaps among climate migrants, and spatial aspects of coastal zone management of the coastal environment.

Key Points of the Chapter

- Socio-ecological perspectives on coastal environments.
- Application of geospatial technologies for coastal monitoring and management.
- Integrating participatory approaches for sustainable coastal governance.
- Assessing climate change impacts and adaptation strategies in coastal communities.
- A collection of casebooks for effective monitoring and management of coastal socio-ecological systems.
  - a century-long climatic trend analysis (casebook 1);
  - monitoring and modelling of shoreline dynamics (casebook 2);
  - modelling and monitoring mangrove greenness (casebook 3);
  - land use/land cover changes and ecological degradation (casebook 4);
  - coastal land surface temperature and ecological changes (casebook 5);
  - mangrove-human conflicts (casebook 6);
  - climate adaptation strategies of traditional fishermen (casebook 7);
  - climate gaps among climate migrants (casebook 8); and,
  - spatial aspects of coastal zone management (casebook 9).

16.1 Introduction

Coastal ecosystems serve as dynamic interfaces between terrestrial and marine environments, playing a crucial role in conserving biodiversity, regulating climate, and sustaining human livelihoods (Mhatre, 2024; Asensio-Montesinos et al., 2024; Zhao et al., 2024; Puthucherril, 2024; Jennings et al., 2024). In regions such as coastal West Bengal, these ecosystems including mangroves, estuaries, and shorelines are particularly vulnerable to climatic changes, human activities, and natural disasters (Das & Swain, 2024; Saengsupavanich et al. 2024; Patra, 2024; Bhattacharyya, 2023; Ghosh & Mistri, 2023; Basu, 2020). To ensure their sustainability and resilience, effective monitoring and management are imperative (Gupta et al. 2024; Sparkes & Werners, 2023; Alibašić, 2022). The *Coastal Ecologies Casebook: A Collection of Effective*

Monitoring and Management seeks to compile comprehensive studies that leverage geospatial tools and participatory approaches to address the challenges confronting coastal ecosystems in West Bengal, while also exploring opportunities of scope for sustainable management (Table 16.1). The following discussions provided a concise

Table 16.1 Coastal ecologies casebook: geospatial technologies in monitoring and management

Casebook(s)	Investigation(s): monitoring and management	Key findings
Casebook 1	A Century-Long Analysis (1901–2022) of Climatic Trends in Coastal West Bengal: An Investigation Using Geospatial Tools and Techniques	Climatic trend analysis and intensity of climate change.
Casebook 2	Geospatial Technologies for Monitoring and Modelling Shoreline Dynamics: A Case Study of the Coastal Region of Bengal, India	Spatio-temporal change assessment of shoreline
Casebook 3	Modelling and Monitoring of Mangrove Ecologies Greenness in the Changing Climate: The Case Study of Indian Sundarbans	Assessment of mangrove greenness and its relation with changing climate.
Casebook 4	Spatio-temporal Analysis of Land Use/Land Cover Changes and Ecological Degradation Using LANDSAT Imagery in Coastal Bengal	Exploring the land use/land cover changes and ecological degradation.
Casebook 5	Assessing Coastal Land Surface Temperature and Ecological Changes through Remote Sensing Data	Examine the land surface temperature and ecological changes.
Casebook 6	Management of Mangrove-Human Conflicts in the Indian Sundarbans, Using Participatory GIS Techniques	Management of mangrove-human conflicts.
Casebook 7	Adaptation Strategies of Traditional Fishermen to Climate Change: A Case Study of Coastal Bengal	Adaptation to climate change and traditional fishermen.
Casebook 8	Management of Climate Gap among Climate Migrants on Sagar Island: A Case Study	Management of climate gap among climate migrants.
Casebook 9	Spatial Consideration of Coastal Zone Management Frameworks: Toward Participatory Policy Recommendations	Spatial consideration of coastal zone management.

Integration of key findings: The study highlights critical aspects of coastal dynamics, emphasizing the intensity of climate change through climatic trend analysis and shoreline spatio-temporal assessments. Changes in mangrove greenness are linked to changing climate, while land use/land cover transformations reveal ongoing ecological degradation. Rising land surface temperatures further exacerbate environmental vulnerabilities. The study also addresses socio-ecological challenges, including mangrove-human conflicts, climate adaptation strategies among traditional fishermen, and the management of climate-induced migration. Further, the spatial considerations in coastal zone management are essential for balancing ecological sustainability and community resilience in the region.

Credit(s):
Data collection: Secondary: K.D.M.; M.K.; and S.R.; primary: S.R. and K.D.M.
Methodologies: K.D.M.; M.K.; and S.R.
Data calculation and mapping: K.D.M.; and S.R.
Result analysis and discussions: K.D.M.; S.R.; and M.K.
K.D.M.: Kousik Das Malakar; M.K.: Manish Kumar; and S.R.: Supriya Roy.

scientific exploration of each casebook topic, emphasizing the context of coastal ecology monitoring and management through the application of geospatial technologies.

Casebook 1
Over the last century, coastal West Bengal has witnessed significant climatic shifts and variations, such as rising temperatures, changes in precipitation patterns, and a higher frequency of extreme weather events. These changes have had a substantial effect on coastal ecosystems, especially the Sundarbans mangrove forest (Subramanian et al. 2023; Kantamaneni et al., 2022; Jabir et al. 2021). Studies have employed geospatial techniques to analyse these climatic trends, providing spatial and temporal insights into their progression and effects (Kumar et al., 2023; Poddar et al., 2023). For instance, research utilizing remote sensing data has revealed significant alterations in land surface temperatures and associated ecological changes in the region. Such analyses are crucial for developing adaptive management strategies to mitigate adverse impacts on these vulnerable ecosystems (Coelho et al., 2024; Azaiez, 2024; Guo et al. 2023; Yu, 2023).

Casebook 2
Shoreline dynamics in coastal regions are shaped by the intricate interaction between natural processes and human interventions (Das & Swain, 2024; Saengsupavanich et al., 2024). The application of geospatial technologies, particularly satellite imagery and GIS modeling, has been pivotal in monitoring and analysing these changes (Harikrishna et al., 2024; Emam & Soliman, 2020). Numerous studies on shoreline extraction and change detection in the region have utilized geospatial techniques to evaluate erosion and accretion patterns, identifying critical areas for coastal management. These tools facilitate the identification of vulnerable zones, support infrastructure planning, and inform conservation strategies to safeguard coastal ecosystems (Colak, 2024; Zambrano-Medina et al., 2023; Ayalke et al., 2023; Mondal et al., 2021; Kermani et al., 2016).

Casebook 3
Mangrove ecosystems are particularly vulnerable to climate variations, making it essential to monitor their health and greenness to assess resilience and carbon sequestration capacity (Saoum & Sarkar, 2024; Mukherjee & Paul, 2021). Remote sensing technologies have been widely used to track temporal changes in mangrove cover and health (Sunkur et al., 2024; Vasquez et al., 2024; Wang et al., 2019). For instance, satellite imagery has revealed changes in species composition and distribution, influenced by rising salinity and sea-level fluctuations. These findings underscore the urgent need for effective conservation strategies to protect mangrove biodiversity and ecological functions (Roy et al., 2025; Zhang et al., 2024a, b, c; Win & Sasaki, 2024; Maurya et al., 2021; Pham et al., 2019).

Casebook 4
Land use and land cover (LULC) changes in coastal regions have resulted in ecological degradation, impacting both terrestrial and aquatic ecosystems (Nath et al.,

2023; Zhu et al., 2022; Kaliraj et al., 2017). Spatio-temporal analyses utilizing LANDSAT imagery have been conducted to map these transformations and identify the underlying drivers of degradation (Islam & Esraz-Ul-Zannat, 2023; Zhao et al., 2021). Studies employing the USGS-LULC classification method with a maximum-likelihood classifier algorithm applied to satellite images have revealed substantial shifts in LULC patterns. The findings indicate a significant expansion of built-up areas at the cost of natural habitats, underscoring the urgent need for sustainable development strategies that balance economic growth with ecological conservation (Sampath & Radhakrishnan, 2023; Islam & Esraz-Ul-Zannat, 2023; Aljenaid et al., 2022; Mukhopadhyay et al., 2018).

Casebook 5

Coastal regions are vital for biodiversity but are increasingly affected by ecological changes driven by land surface temperature (LST) variations. Remote sensing technologies, including Landsat, MODIS, and Sentinel-3, provide critical LST data for monitoring these changes (Patel et al., 2024; García et al., 2023; Yoo et al., 2020). Research indicates that LST fluctuations significantly influence vegetation dynamics and ecological stability, particularly in urbanized coastal areas (Patel et al., 2024; García et al., 2023; Yoo et al., 2020). Land use changes, such as vegetation loss, lead to increased Land Surface Temperature (LST), worsening ecological degradation. Advanced satellite sensors, in conjunction with integrated ecological indicators, enhance monitoring precision and aid in formulating sustainable coastal management strategies. These tools are crucial for mitigating the effects of climate change on coastal ecosystems (Pignatti et al., 2024; Surni et al., 2024; Vadrevu et al., 2023).

Casebook 6

The use of participatory GIS (PGIS) techniques has proven to be a successful approach in addressing human-environment conflicts (Malakar et al., 2023; da Silva et al., 2021; de Carvalho et al., 2021). Engaging local communities in mapping and decision-making processes promotes a sense of ownership and ensures that management strategies are culturally appropriate and practically feasible (Malakar & Roy, 2024a; Fagerholm et al., 2021; Jankowski, 2009). Several research highlights the effectiveness of PGIS in addressing conflicts between mangrove conservation and human activities, promoting more sustainable coexistence. For example, engaging local stakeholders in mapping resource use and identifying conservation priorities has facilitated the development of management plans that balance ecological objectives with community needs (Malakar & Roy, 2024b; Styk & Bogacz, 2022).

Casebook 7

Traditional fishermen in coastal regions are among the communities most vulnerable to climate change, facing challenges such as shifting fish distributions and an increasing frequency of extreme weather events (Islam et al., 2020; Savo et al., 2017). Research on their adaptation strategies has identified a range of responses, including modifications to fishing practices and livelihood diversification. Understanding these strategies offers critical insights for designing support mechanisms and policies that strengthen the resilience of fishing communities. For instance, studies have documented how fishermen adapt by adjusting their gear and targeting different species in response to changing environmental conditions, emphasizing the need for flexible and responsive management approaches (N'Souvi et al., 2024; Mozumder et al., 2023; Shaffril et al., 2017).

Casebook 8
Climate-induced migration is an escalating concern, as rising sea levels and increasing salinity have rendered many areas uninhabitable (Almulhim et al., 2024; Seddiky et al., 2024). Addressing the "climate gap" among migrants requires tackling disparities in vulnerability and access to essential resources. Case studies have examined the socio-economic impacts of climate migration and assessed the effectiveness of various management strategies (Lindegaard et al., 2024; Silchenko & Murray, 2023; Laczko, 2008). Research indicates that participatory approaches, which actively involve migrants in planning and decision-making, are essential for developing equitable solutions that address both immediate needs and long-term sustainability (Bastos et al., 2025; Ross et al., 2015).

Casebook 9
Effective coastal zone management requires spatially informed frameworks that integrate ecological data, human activities, and governance structures. Geospatial technologies play a crucial role in identifying key areas for conservation, sustainable development, and disaster risk reduction (Yang, 2009; Wang et al., 2009; French, 2005). Participatory policy recommendations, developed through inclusive stakeholder engagement, ensure that management plans are context-specific and widely supported by local communities. For example, mapping exercises involving community participation can highlight culturally significant areas or regions with high resource dependency, guiding policies that balance conservation efforts with socio-economic priorities (Ramieri et al., 2019; Soriani et al., 2015; Wu et al., 2012).

Therefore, the integration of geospatial tools and participatory approaches provides a comprehensive framework for monitoring and managing coastal ecologies. The studies compiled in this casebook offer valuable insights and methodologies that can be adapted to similar coastal regions facing environmental and socio-economic challenges. By leveraging technological advancements and fostering community engagement, sustainable coastal ecosystem management can be achieved, ensuring their long-term preservation for future generations. Building on this background context, this chapter critically examines a series of case studies that explore the monitoring and management of Indian coastal ecologies using geospatial tools and participatory approaches.

16.2 Coastal Ecologies Casebook: A Collection of Effective Monitoring and Management

16.2.1 Casebook 1: A Century-Long Analysis (1901–2022) of Climatic Trends in Coastal West Bengal: An Investigation Using Geospatial Tools and Techniques

16.2.1.1 Background and Purpose

Coastal West Bengal (Fig. 16.1), characterized by its complex riverine networks, deltaic formations, and proximity to the Bay of Bengal, has long been highly susceptible to climatic variability (UNESCO, 2023; Panda, 2020). Over the past century, the region has undergone profound environmental transformations driven by both natural and anthropogenic forces, including rising temperatures, erratic precipitation patterns, cyclonic disturbances, and sea-level rise (Das & Swain, 2024; Kantamaneni et al., 2022). Given its ecological significance and socio-economic reliance on climate-sensitive sectors, a comprehensive understanding of long-term climatic trends is essential for ensuring sustainable management and disaster resilience. This study aims to systematically analyse climatic variations in the region by leveraging geospatial techniques, including remote sensing, GIS-based spatial analysis, and statistical modelling. Specifically, it examines historical trends in key climate variables—average, maximum, and minimum temperature; rainfall; potential evapotranspiration; vapor pressure; cloud cover; and wet-day frequency over the past 122 years.

By assessing the magnitude and spatial heterogeneity of climatic changes, this research seeks to contribute to a more nuanced understanding of regional climate dynamics. Additionally, it explores the potential implications of observed trends for

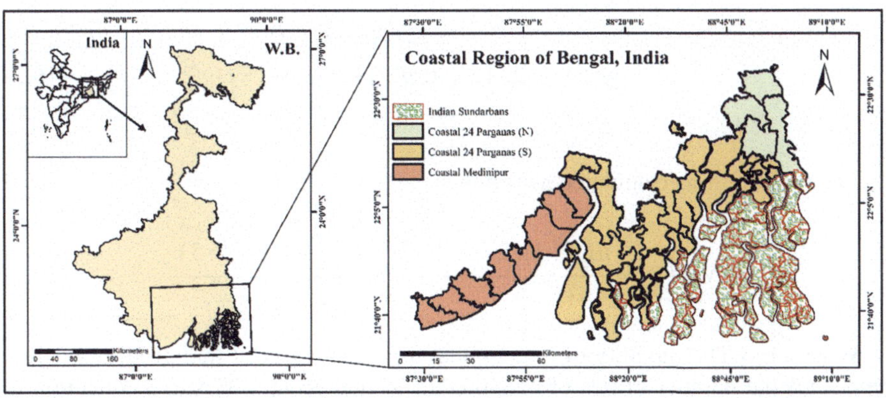

Fig. 16.1 Mapping study area: Coastal region of Bengal, India (Prepared by the authors)

coastal communities, biodiversity conservation, and broader environmental sustainability. The findings will serve as a critical resource for policymakers, researchers, and local stakeholders in designing adaptive strategies to mitigate climate-related risks (Gemeda et al., 2023; Hassan et al., 2023). Through the integration of geospatial methodologies with long-term climate data, this study provides a scientific foundation for strengthening climate resilience planning in one of India's most vulnerable coastal regions (Islam & van Amstel, 2022).

16.2.1.2 Data and Methods

This study utilized 122 years (1901 to 2022) of historical climate data obtained from the Climatic Research Unit and the Met Office (Available at: https://crudata.uea.ac.uk/cru/data/hrg/cru_ts_4.07/). Long-term trends in key climatic variables, including temperature, rainfall, potential evapotranspiration, cloud cover, vapor pressure, and wet day frequency, were analysed using the Mann Kendall test (Eq. 16.4) for trend detection and Sen's Slope estimator (Eq. 16.6) for assessing the magnitude of change (Frimpong et al., 2022; Aditya et al., 2021). Additionally, the spatial distribution of these climate variables across different years within the study region was examined using the Inverse Distance Weighting (IDW) interpolation technique (Fung et al., 2022).

Equations:

The MK test statistic was computed as follows (Kendall, 1955; Mann, 1945);

$$S = \sum_{i=1}^{n-1} \sum_{j=i+1}^{n} \operatorname{sgn}\left(X_j - X_i\right) \tag{16.1}$$

where, n = The cumulative count of data points, where X_i and X_j represent data values in time series i and j (with j being greater than i), respectively, and $\operatorname{sgn}(X_j - X_i)$ denotes the sign function.

$$\operatorname{sgn}\left(X_j - X_i\right) = \begin{cases} +1, if\ Xj - Xi > 0 \\ 0, if\ Xj - Xi = 0 \\ -1, if\ Xj - Xi < 0 \end{cases} \tag{16.2}$$

The variance was calculated as follows:

$$\operatorname{Var}(S) = \frac{n(n+1)(2n+5) - \sum_{i=1}^{m} t_i \left(t_1 - 1\right)\left(2t_1 + 5\right)}{18} \tag{16.3}$$

Where, n = The quantity of data points; m = The count of tied groups, where t_i represents the number of ties of magnitude i. A tied group refers to a collection of sample data sharing identical values.

In scenarios where the sample size exceeds 10 (n > 10), the standard normal test statistic Z_S was determined using the following equation:

$$Z_s = \begin{cases} \frac{S-1}{\sqrt{Var(S)}}, if\ S > 0 \\ 0, if\ S = 0 \\ \frac{S+1}{\sqrt{Var(S)}}, if\ S > 0 \end{cases} \tag{16.4}$$

Positive values of Z_S indicate increasing trends, whereas negative Z_S values indicate decreasing trends.

Sen's slope estimator (Eq. 16.6):

The non-parametric method for estimating the slope of a trend in a sample of N pairs of data was developed by Sen (1968):

$$Q_i = \frac{x_j + x_k}{j-k}\ for\ i = 1,2,3\ldots N \tag{16.5}$$

Here, x_j and x_k represent the data values at times j and k, where j is greater than k.

If there is a single datum in each period, then the eq. $N = \frac{n(n-1)}{2}$ is true, where n represents the number of periods. If there exist many observations within one or more time periods, the value of N can be expressed as $N < \frac{n(n-1)}{2}$, and n is the total number of observations.

The N values of Qi are arranged in ascending order, and the median slope, also known as Sen's slope estimator, was computed as:

$$Q_{med} = \begin{cases} Q\left[\frac{N+1}{2}\right], if\ N\ is\ odd \\ \frac{Q\left[\frac{n}{2}\right] + Q\left[\frac{n+2}{2}\right]}{2}, if\ N\ is\ even \end{cases} \tag{16.6}$$

The Q_{med} sign represents the data trend, and its value denotes the steepness of the trend. To assess whether the median slope is statistically distinct from zero, the confidence interval of Q_{med} at a certain probability is calculated.

Therefore, the calculation of the confidence interval for the time slope was performed as:

$$C\alpha = Z_{1-\alpha/2}\sqrt{Var(S)} \tag{16.7}$$

Here, Var (S) is defined in Eq. 16.3. and $Z_{1-\alpha/2}$ is obtained from the standard normal distribution table.

The confidence interval was calculated in this investigation using a significance threshold of $\alpha = 0.05$.

$$So, M1 = \frac{N - C\alpha}{2} \tag{16.8}$$

The positive numbers in the time series indicate an upward trend, while the negative values show a downward trend.

16.2.1.3 Results, Discussion, and Conclusion

Tables 16.2, 16.3, and 16.4 provide a statistical analysis of long-term trends in key climatic variables across three districts in coastal West Bengal: East Medinipur, South 24 Parganas, and North 24 Parganas. The assessment is based on the Mann Kendall test, which evaluates trend significance, and Sen's Slope estimator, which measures trend magnitude. The analysis spans the period from 1901 to 2022, with a significance level of $\alpha = 0.05$. The key observations are summarized in Table 16.5 and graphically represented in Figs. 16.2, 16.3, and 16.4. Additionally, the spatial distribution (1901–2022) of climatic variables is illustrated in Figs. 16.5, 16.6, 16.7, 16.8, 16.9, 16.10, 16.11 and 16.12.

Implications of Findings

(a) Consistent warming trends across all districts indicate long-term climate change, which may have significant implications for agriculture, water resources, and local livelihoods (Roy et al., 2024a, b; Malhi et al., 2021; Jayaraman & Murari, 2014; Turral et al. 2011).
(b) Increased potential evapotranspiration and vapor pressure suggest higher moisture demand, potentially leading to more frequent drought-like conditions despite stable rainfall trends (Noguera et al., 2023; Zhao et al., 2022; He et al., 2022).
(c) Rising wet day frequency without significant changes in total rainfall could mean more frequent but less intense rainfall events, affecting water availability and flood risks (Breinl et al., 2021; Tabari, 2020).
(d) Uncertain cloud cover trends indicate variability in atmospheric dynamics, which could influence future precipitation patterns (Liu et al., 2023; Zhao et al., 2023).

Table 16.2 East Medinipur: Mann-Kendall and Sen's slope statistic for climatic variables (annual)

Variables (annual average: 1901–2022)	Maximum	Minimum	Mean	Standard deviation	Kendall's tau	S statistic	p-value (two-tailed)	Sen's slope
Average temperature	27.764	25.768	26.513	0.376	0.451	3331	**<0.0001**	0.006270
Maximum temperature	31.588	29.385	30.282	0.429	0.435	3211	**<0.0001**	0.007308
Minimum temperature	24.004	22.092	22.793	0.377	0.375	2767	**<0.0001**	0.005514
Rainfall	187.701	97.082	142.244	19.359	0.101	743	0.1006	0.089651
Potential evapotranspiration	3.885	3.584	3.723	0.060	0.235	1733	**0.0001**	0.000573
Vapour pressure	26.690	24.422	25.332	0.453	0.403	2973	**<0.0001**	0.007374
Cloud coverage	54.067	43.497	48.265	1.857	0.065	479	0.2902	0.004427
Wet day frequency	7.100	4.440	5.635	0.448	0.306	2255	**<0.0001**	0.005559

Note: Significance level alpha = 0.05

Table 16.3 South 24 Parganas: Mann-Kendall and Sen's slope statistic for climatic variables (annual)

Variables (annual average: 1901–2022)	Maximum	Minimum	Mean	Standard deviation	Kendall's tau	S statistic	p-value (two-tailed)	Sen's slope
Average temperature	27.691	25.709	26.464	0.371	0.441	3257	**<0.0001**	0.00611
Maximum temperature	31.442	29.248	30.170	0.421	0.421	3109	**<0.0001**	0.00696
Minimum temperature	23.990	22.126	22.809	0.376	0.366	2705	**<0.0001**	0.00542
Rainfall	201.240	103.367	149.333	20.964	0.078	573	0.2056	0.07691
Potential evapotranspiration	3.787	3.512	3.626	0.058	0.238	1755	**0.0001**	0.00057
Vapour pressure	26.921	24.637	25.598	0.440	0.374	2761	**<0.0001**	0.00678
Cloud coverage	54.183	45.027	49.318	1.671	0.084	619	0.1714	0.00607
Wet day frequency	7.708	4.679	6.007	0.470	0.278	2049	**<0.0001**	0.00497

Note: Significance level alpha = 0.05

Table 16.4 North 24 Parganas: Mann-Kendall and Sen's slope statistic for climatic variables (annual)

Variables (annual average: 1901–2022)	Maximum	Minimum	Mean	Standard deviation	Kendall's tau	S statistic	p-value (two-tailed)	Sen's slope
Average temperature	27.604	25.656	26.427	0.362	0.428	3160	**<0.0001**	0.00576
Maximum temperature	32.081	29.910	30.862	0.408	0.408	3013	**<0.0001**	0.00651
Minimum temperature	23.160	21.369	22.043	0.370	0.352	2599	**<0.0001**	0.00526
Rainfall	196.229	98.102	143.990	20.299	0.031	231	0.6108	0.02765
Potential evapotranspiration	3.721	3.438	3.564	0.056	0.262	1934	**<0.0001**	0.00060
Vapour pressure	26.933	24.352	25.672	0.450	0.281	2075	**<0.0001**	0.00533
Cloud coverage	54.021	45.919	50.003	1.552	0.099	733	0.1053	0.00656
Wet day frequency	8.117	4.942	6.256	0.504	0.239	1767	**<0.0001**	0.00488

Note: Significance level alpha = 0.05

Table 16.5 Key observations: trend analysis of climatic variables from 1901 to 2022 using Mann-Kendall test and Sen's slope estimator

Sr. No.	Climatic variables	Key observations
1	Temperature trends	Average, maximum, and minimum temperatures exhibit a statistically significant increasing trend across all three districts ($p < 0.0001$). Sen's slope values indicate a similar rate of warming, with maximum temperatures rising at the highest rate, followed by average and minimum temperatures. East Medinipur shows the highest rate of increase in maximum temperature (0.007308 °C per year), while north 24 Parganas has the lowest (0.00651 °C per year).
2	Rainfall trends	Unlike temperature, rainfall trends are not statistically significant in any district ($p > 0.05$). Sen's slope values are positive, suggesting a slight increase in rainfall over time, but the lack of statistical significance implies high interannual variability. East Medinipur has the highest Sen's slope (0.089651 mm per year), but it remains statistically insignificant ($p = 0.1006$).
3	Potential evapotranspiration (PET) trends	PET shows a significant increasing trend in all three districts ($p \leq 0.0001$), indicating higher atmospheric demand for moisture over time. North 24 Parganas has the highest Sen's slope (0.00060 mm per year), slightly higher than the other two districts.
4	Vapor pressure trends	Vapor pressure has a strong and significant increasing trend in all three districts ($p < 0.0001$). The highest Sen's slope is observed in East Medinipur (0.007374 hPa per year), indicating an increase in atmospheric moisture content.
5	Cloud cover trends	Cloud cover does not show a significant trend in any district ($p > 0.05$). Although Sen's slope values suggest a slight increase, the high p-values indicate that these trends are not strong enough to be considered statistically significant.
6	Wet day frequency trends	Wet day frequency (days with measurable rainfall) shows a statistically significant increasing trend in all three districts ($p < 0.0001$). The rate of increase is highest in East Medinipur (0.005559 days per year) and lowest in North 24 Parganas (0.00488 days per year).

Source: Table 16.2, 16.3, and 16.4

The findings reveal a pronounced warming trend accompanied by changes in atmospheric moisture and evapotranspiration, while no definitive pattern is observed in total rainfall or cloud cover. These trends underscore the necessity for adaptive climate strategies to mitigate the effects of rising temperatures, evolving rainfall patterns, and their broader socio-environmental implications in coastal West Bengal.

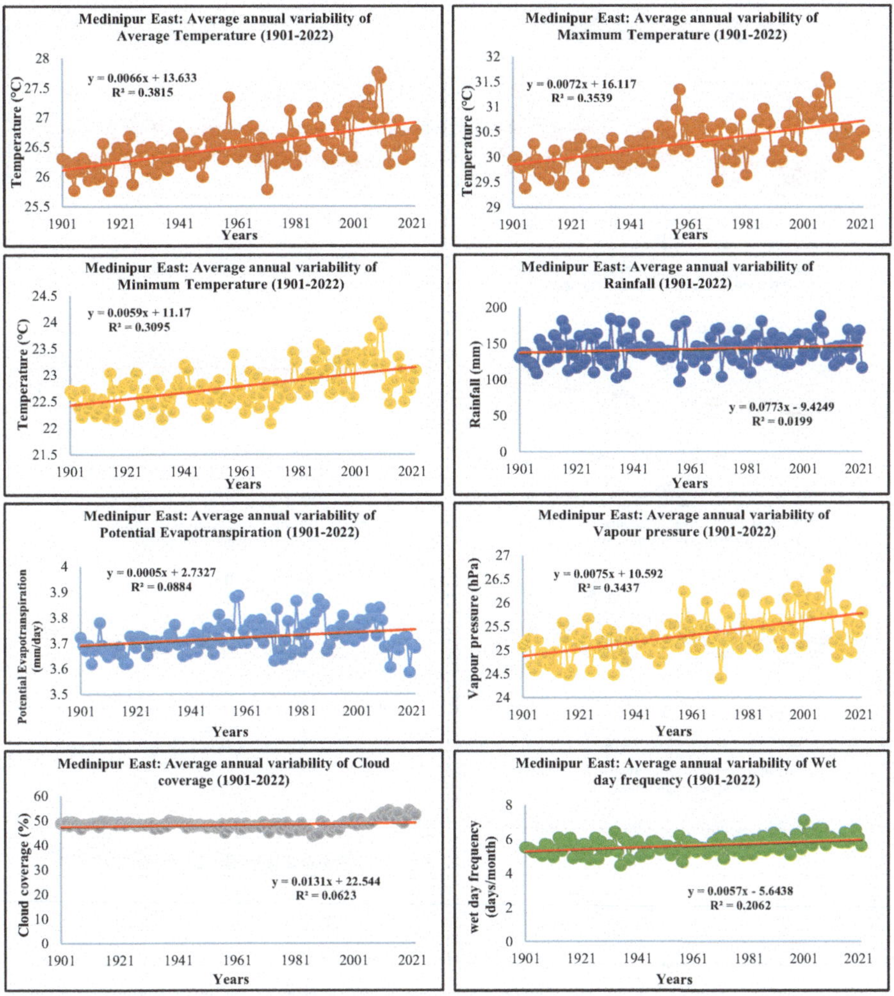

Fig. 16.2 Average annual variability of climatic variables (1901–2022) in East Medinipur/Midnapur East. (Prepared by the authors)

16.2.2 Casebook 2: Geospatial Technologies for Modelling and Monitoring Shoreline Dynamics: A Case Study of the Coastal Region of Bengal, India

16.2.2.1 Background and Purpose

The coastal region of Bengal, India (Fig. 16.13), is characterized by significant dynamism, shaped by both natural processes and anthropogenic factors. Cyclonic disturbances, sea-level rise, tidal fluctuations, and sediment transport exert

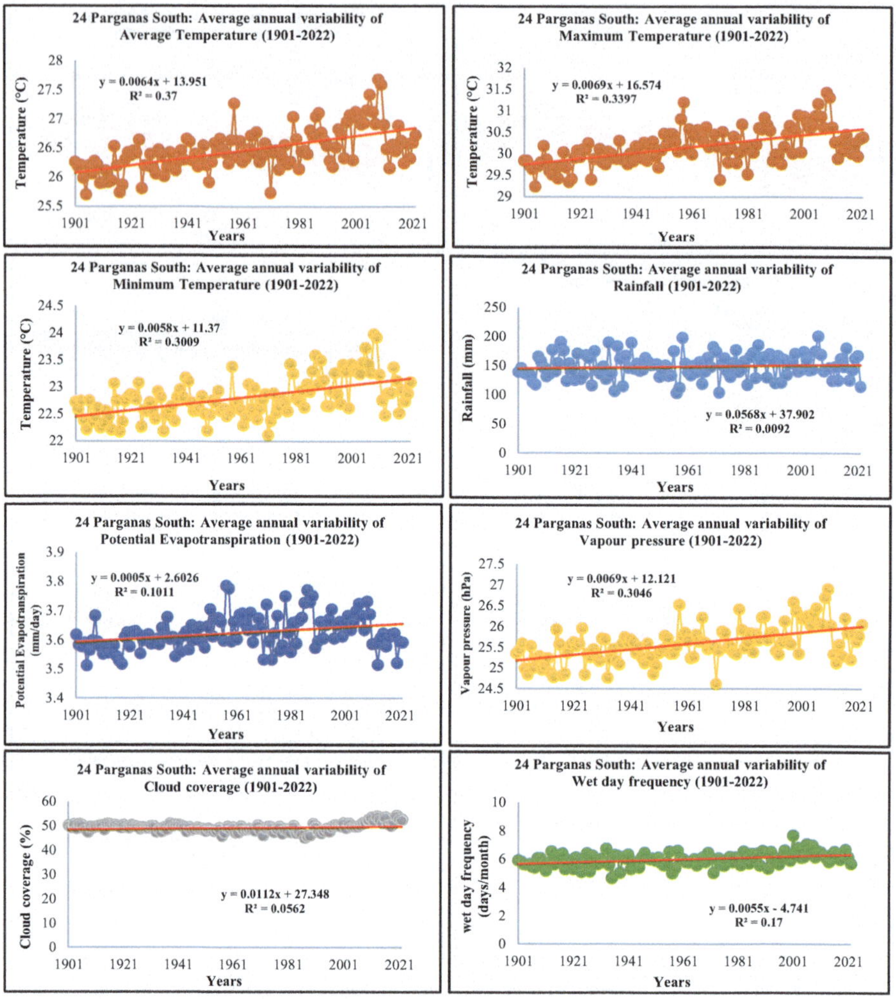

Fig. 16.3 Average annual variability of climatic variables (1901–2022) in 24 Parganas South/ South 24 Parganas (Prepared by the authors)

considerable influence on shoreline stability, resulting in patterns of erosion and accretion at various locations (Zhang et al., 2024a, b, c; Dong et al., 2024; Pradeep et al., 2022; Mohamed Rashidi et al., 2021). In addition, human activities such as construction, land reclamation, and modifications to embankments further alter the region's natural coastal processes (Hastuti et al., 2024; Wang et al., 2023). Understanding the dynamics of shorelines is vital for effective coastal management, disaster preparedness, and ecosystem conservation (Benhur et al., 2024; Shampa et al., 2023). However, traditional field-based techniques often fail to capture long-term changes and spatial variations in shoreline dynamics at a detailed scale (Zhang et al., 2024a, b, c; Fromard et al., 2004).

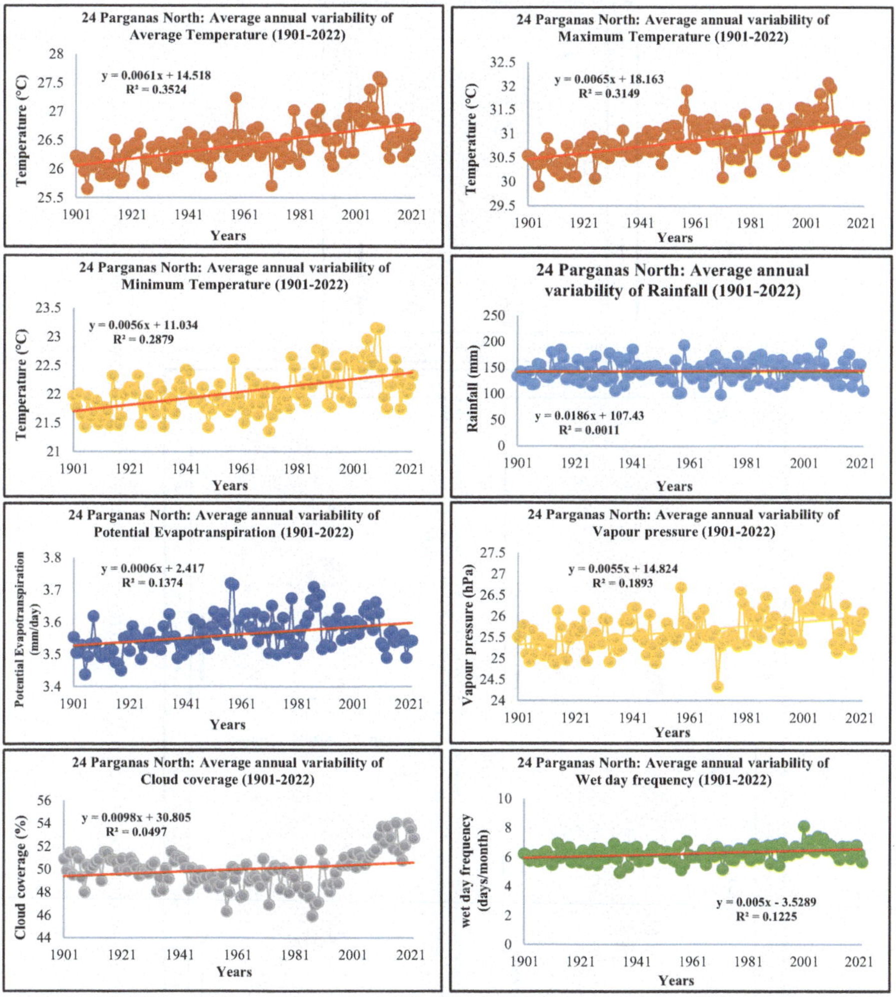

Fig. 16.4 Average annual variability of climatic variables (1901–2022) in 24 Parganas North/ North 24 Parganas (Prepared by the authors)

This study aims to leverage geospatial technologies, including remote sensing and Geographic Information Systems (GIS), to monitor and model shoreline dynamics along the Bengal coast. Advanced geospatial methods such as the Digital Shoreline Analysis System (DSAS), change detection algorithms, and statistical modelling were used to analyse trends in erosion and accretion (Kundu & Mandal, 2024; Dobrică et al., 2022; Barik et al., 2021). The outcomes of this research will provide critical insights into the underlying factors driving coastal changes, facilitating the development of adaptive strategies for enhancing coastal resilience. By integrating geospatial methodologies with long-term shoreline data, this study aims to provide a scientific basis for evidence-based coastal planning and climate

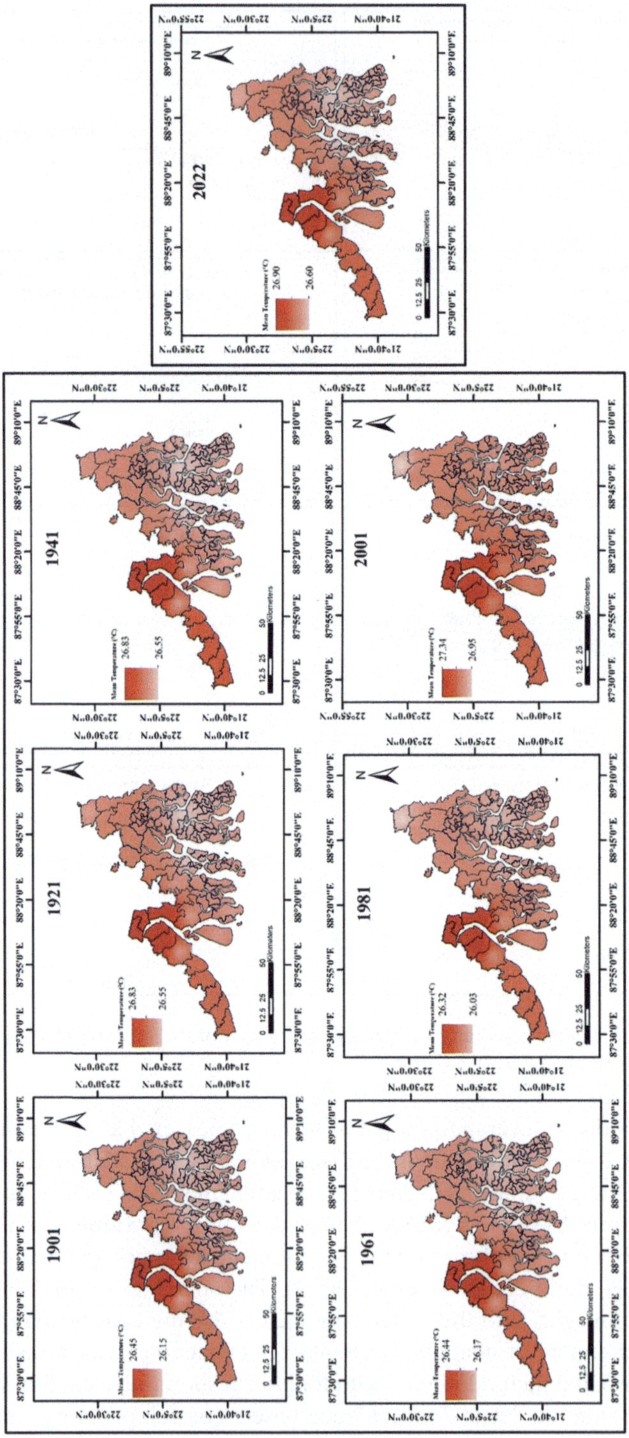

Fig. 16.5 Spatial distribution (1901–2022) of annual average temperature in the coastal region of Bengal, India (Prepared by the authors)

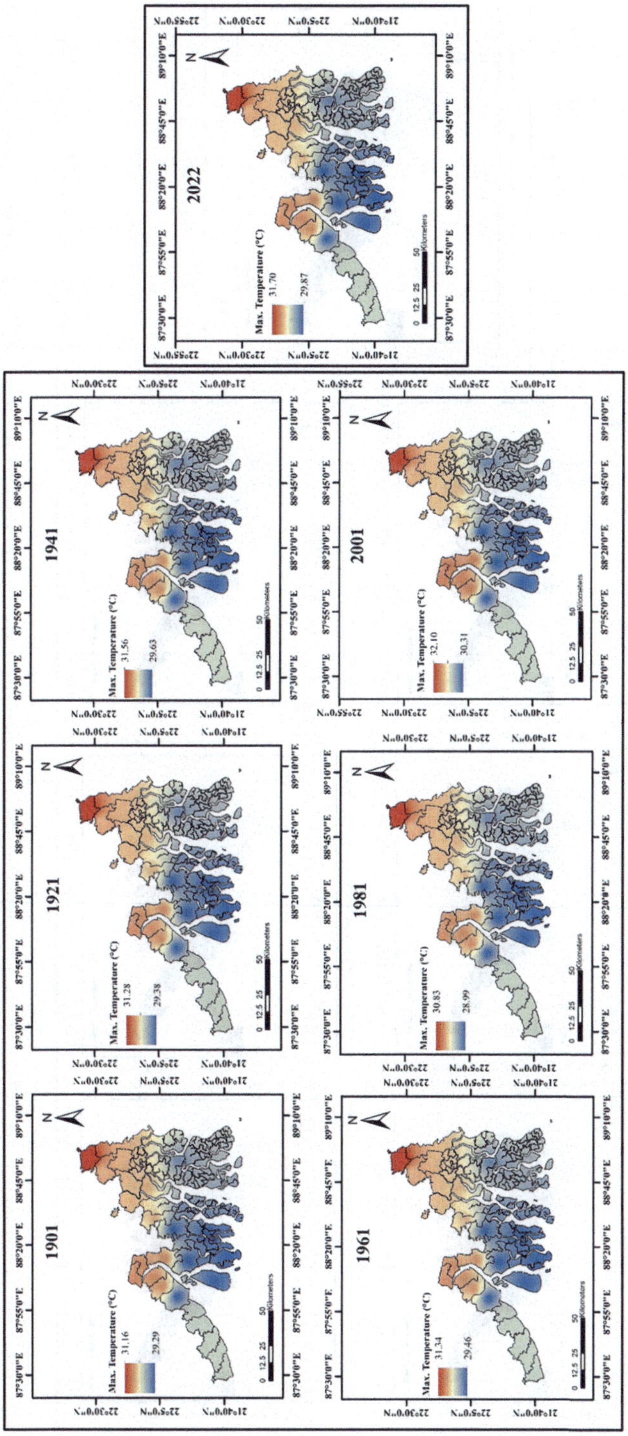

Fig. 16.6 Spatial distribution (1901–2022) of annual average maximum temperature in the coastal region of Bengal, India (Prepared by the authors)

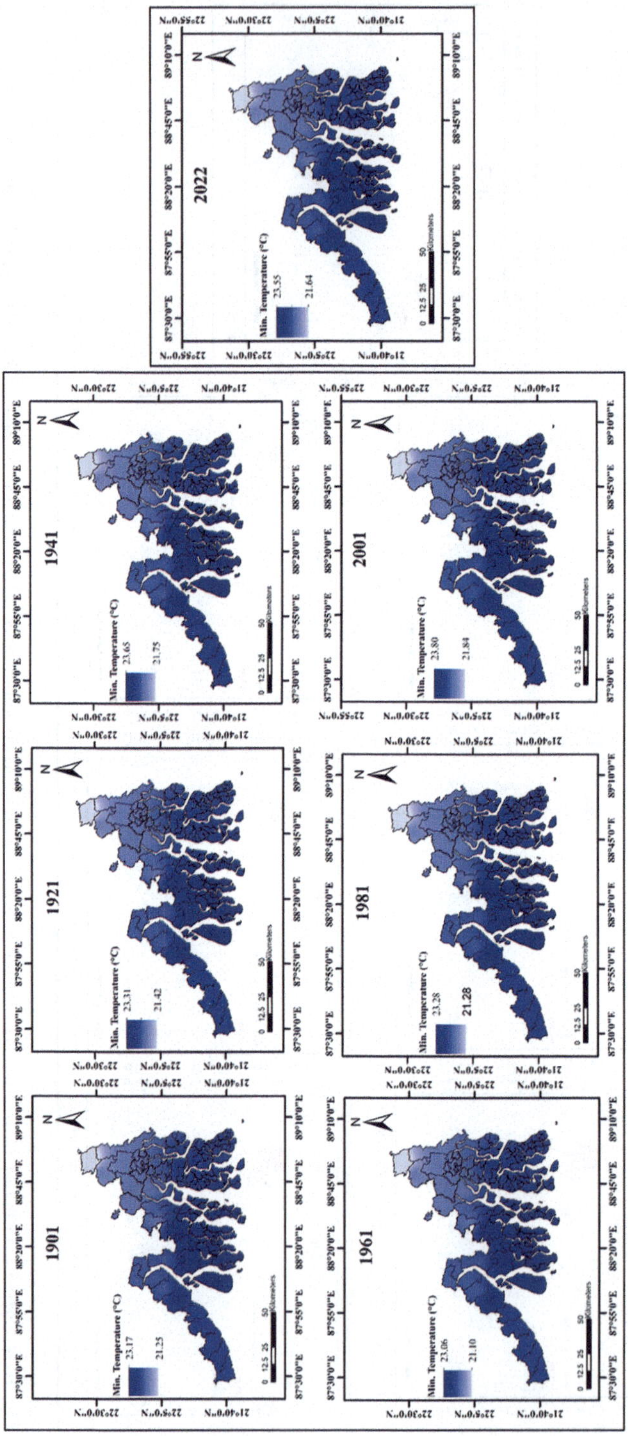

Fig. 16.7 Spatial distribution (1901–2022) of annual average minimum temperature in the coastal region of Bengal, India (Prepared by the authors)

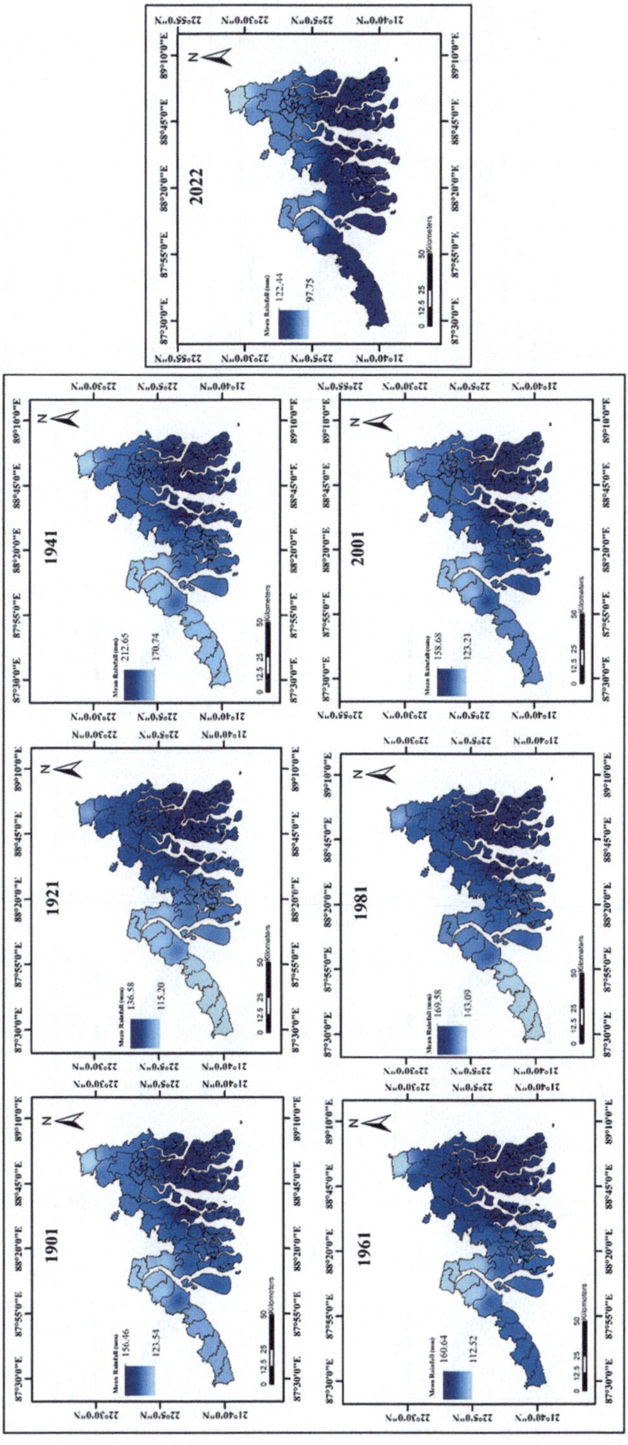

Fig. 16.8 Spatial distribution (1901–2022) of annual average rainfall in the coastal region of Bengal, India (Prepared by the authors)

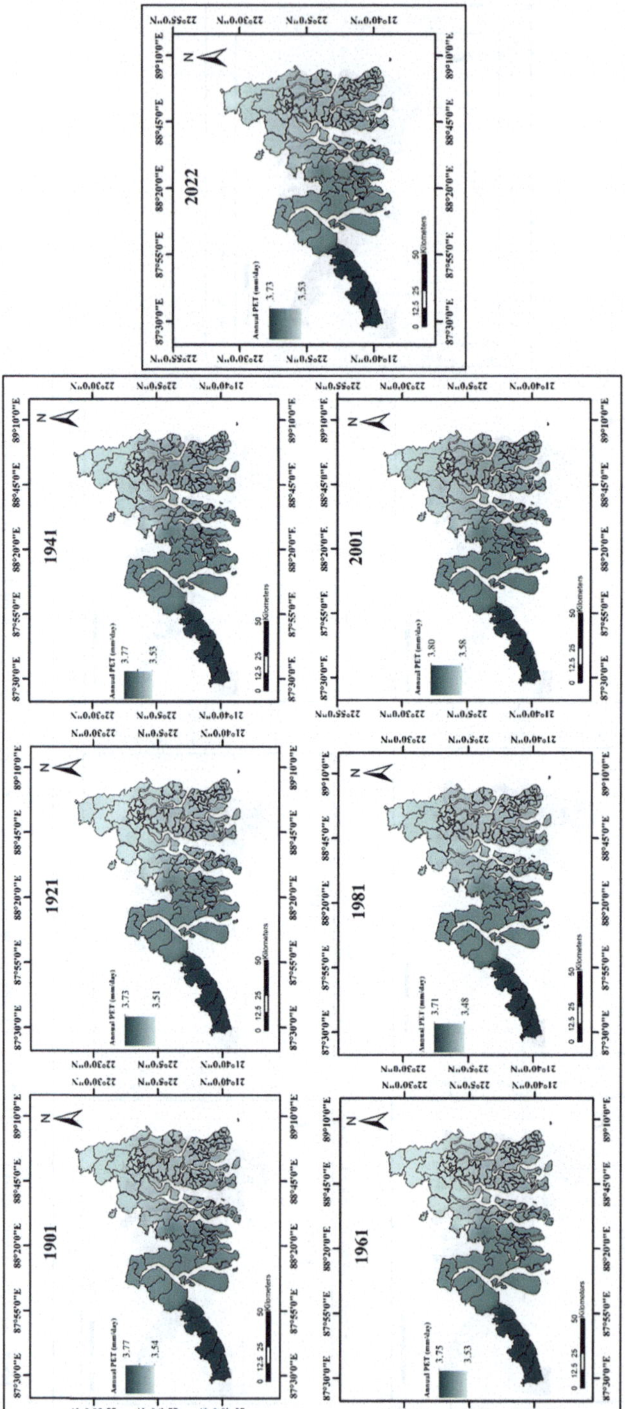

Fig. 16.9 Spatial distribution (1901–2022) of annual average potential evapotranspiration in the coastal region of Bengal, India (Prepared by the authors)

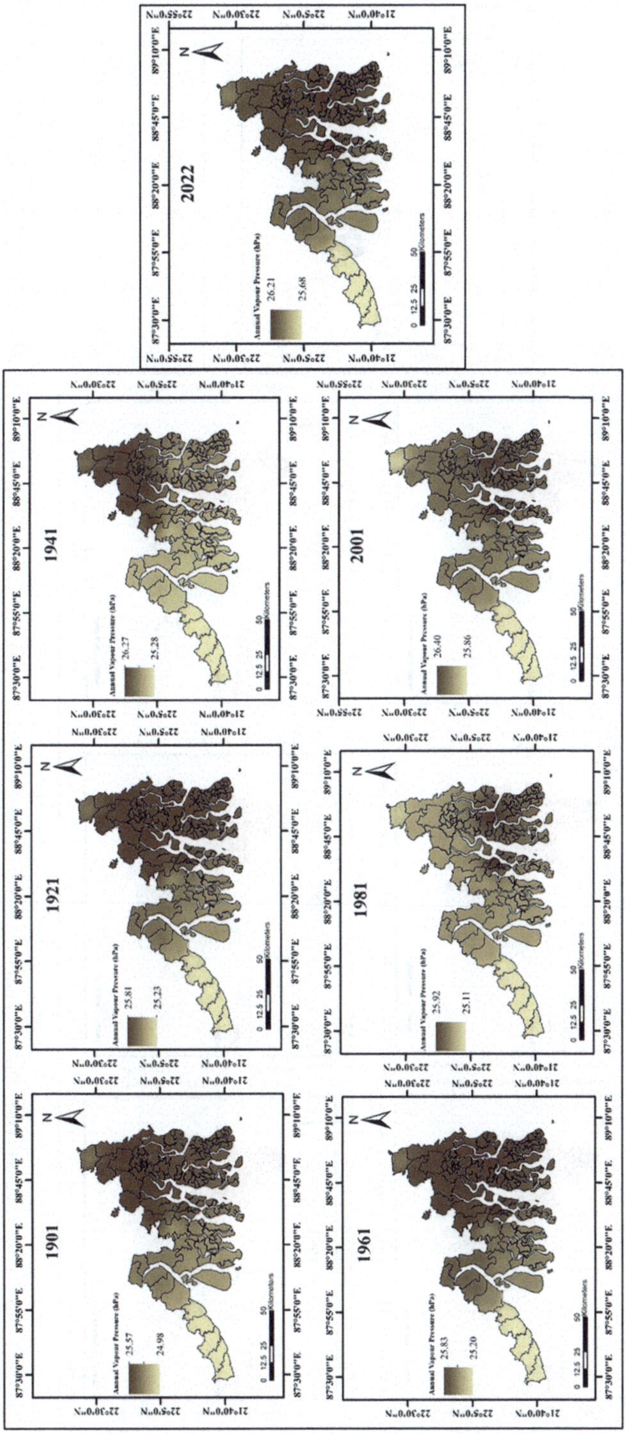

Fig. 16.10 Spatial distribution (1901–2022) of annual average vapour pressure in the coastal region of Bengal, India (Prepared by the authors)

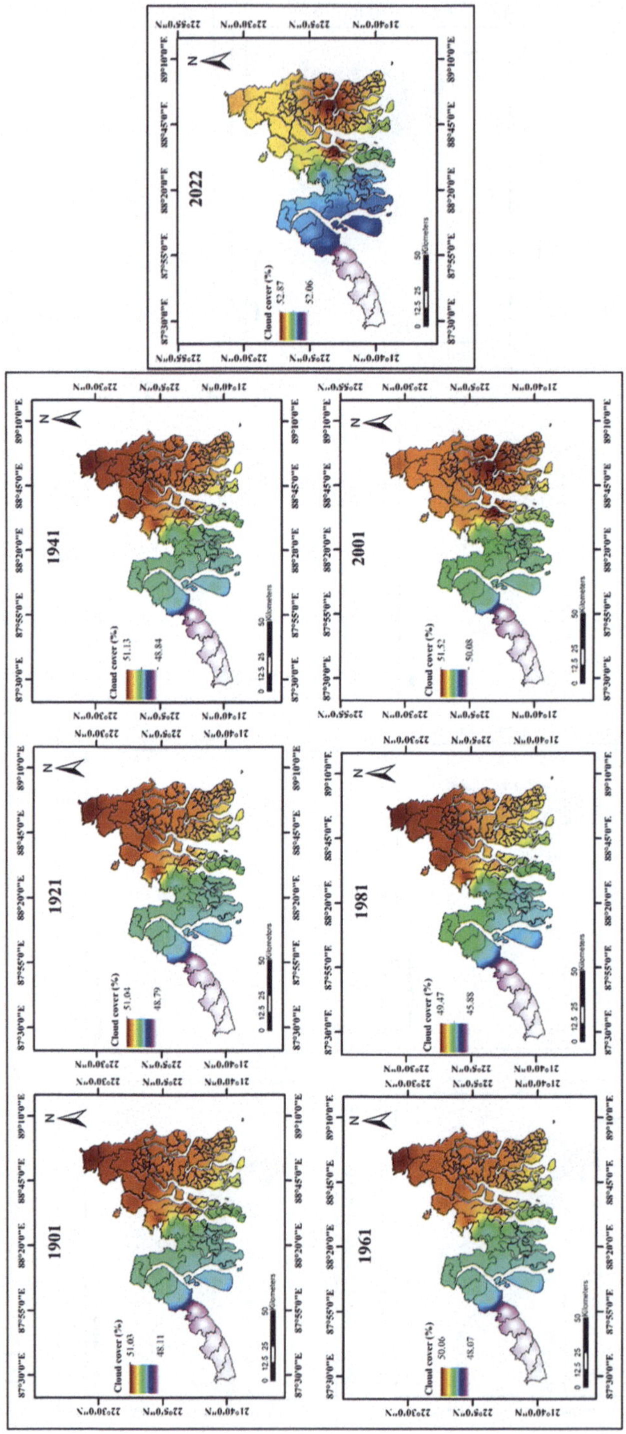

Fig. 16.11 Spatial distribution (1901–2022) of annual average cloud coverage in the coastal region of Bengal, India (Prepared by the authors)

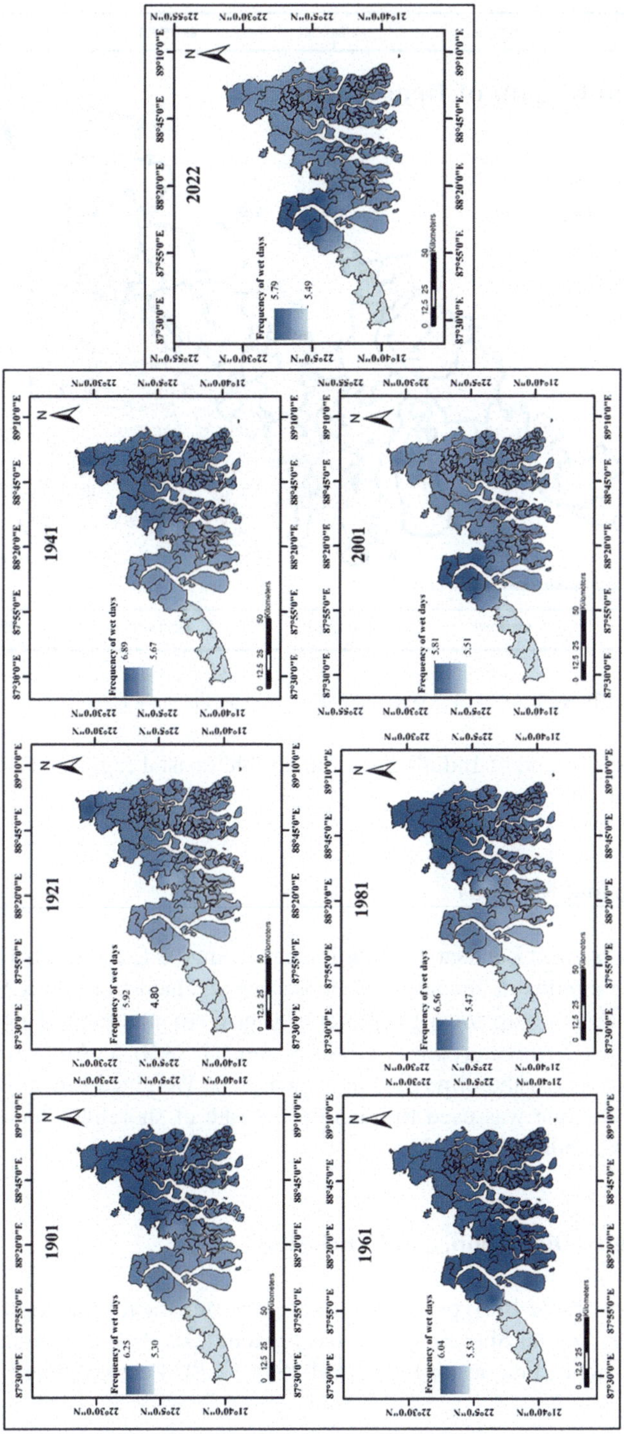

Fig. 16.12 Spatial distribution (1901–2022) of annual average wet day frequency in the coastal region of Bengal, India (Prepared by the authors)

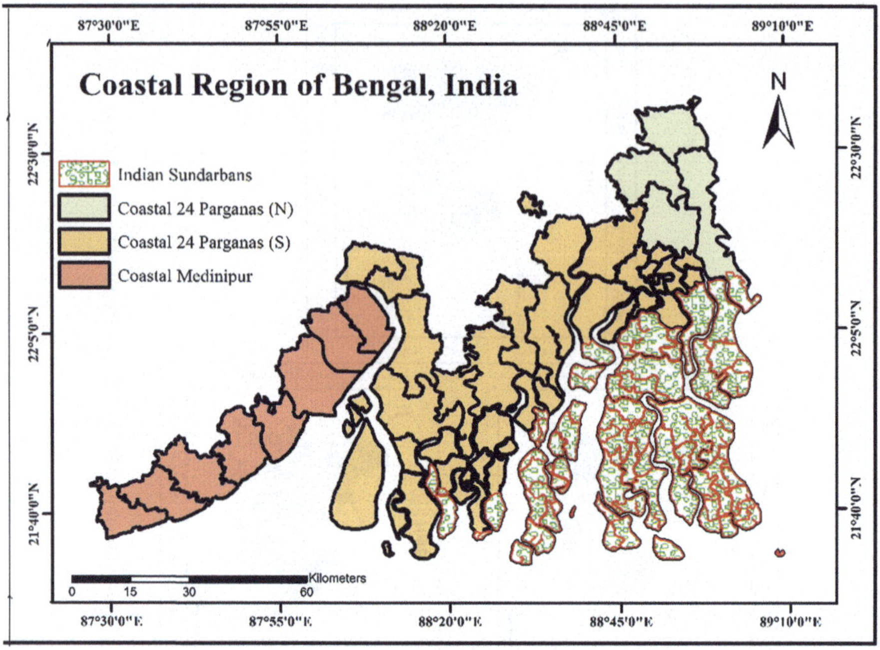

Fig. 16.13 Coastal Bengal: study area overview (Prepared by the authors)

adaptation efforts in one of India's most vulnerable coastal regions (Pereira et al., 2022; Song et al., 2021).

16.2.2.2 Data and Methods

This case study utilized Landsat satellite images from 1994, 2004, 2014, and 2024 (captured in the months of January and February) obtained from the USGS Earth Explorer (Available at: https://earthexplorer.usgs.gov/) to monitor and model shoreline changes. The Digital Shoreline Analysis System (DSAS), an ArcGIS extension, was employed to assess shoreline dynamics, while the Weighted Linear Regression Rate (WLR) technique was used to quantify the rate of shoreline change (Thinh, 2022; Sheik & Chandrasekar, 2011.

16.2.2.3 Results, Discussion, and Conclusion

The key findings (Table 16.6) presents a descriptive statistical analysis of shoreline change using the Digital Shoreline Analysis System (DSAS) and Weighted Linear Regression (WLR) technique for the period 1994 to 2024 across three study areas

Table 16.6 Descriptive statistics of DSAS shoreline change analysis and WLR values (1994 to 2024)

	CRB	CM	CS24
Total transects	14,693	472	14,221
Total erosional transects	5078	80	4998
Total Accretional transects	6359	55	6304
Total stable transects	23	0	23
Erosional transects (%)	34.56068	16.94915	35.14521
Accretional transects (%)	43.27911	11.65254	44.32881
Stable transects (%)	0.156537	0	0.161733
Mean shoreline change (m/year)	0.435412	−1.32837	0.456437
Max shoreline change (m/year)	49.3	14.92	49.3
Min shoreline change (m/year)	−49.93	−14.64	−49.93
Mean erosion (m/year)	−1.02375	−1.21551	−1.01739
SD erosion (m/year)	2.560194	3.147097	2.538276
Mean accretion (m/year)	1.363357	0.835572	1.380874
SD accretion (m/year)	2.945581	2.955898	2.94372

Note: CRB: Coastal Region of Bengal; CM: Coastal Medinipur; and CS24: Coastal 24 Parganas (South and North)

namely, the Coastal Region of Bengal (CRB), Coastal Medinipur (CM), and Coastal 24 Parganas (CS24). The analysis provides insights into shoreline stability, erosion, and accretion trends based on transect-based assessments.

A total of 14,693 transects were analysed in CRB, 472 in CM, and 14,221 in CS24. The proportion of transects experiencing erosion was 34.56% in CRB, 16.95% in CM, and 35.14% in CS24, while accretional transects accounted for 43.28% in CRB, 11.65% in CM, and 44.33% in CS24. Stable transects constituted a negligible percentage across all regions, with no stable transects observed in CM. The mean shoreline changes rates varied across the study areas, with a positive trend of 0.435 m/year in CRB and 0.456 m/year in CS24, indicating a net accretional tendency. In contrast, CM exhibited a negative mean shoreline change of −1.328 m/year, suggesting dominant erosion. The maximum shoreline change recorded was 49.3 m/year in both CRB and CS24, while CM showed a lower maximum of 14.92 m/year. The minimum shoreline changes values indicated substantial erosion, with −49.93 m/year in CRB and CS24 and −14.64 m/year in CM. Erosion and accretion statistics further highlight regional variations. The mean erosion rate was highest in CM (−1.215 m/year), followed by CRB (−1.023 m/year) and CS24 (−1.017 m/year). The mean accretion rate was highest in CS24 (1.38 m/year), followed by CRB (1.36 m/year), while CM exhibited the lowest accretion rate (0.83 m/year). The standard deviations for both erosion and accretion indicate high variability in shoreline change, particularly in CRB and CS24. Overall, the results suggest that while CRB and CS24 exhibit a net accretional trend, CM is undergoing significant erosion, necessitating targeted coastal management interventions to mitigate shoreline instability (Figs. 16.14 and 16.15).

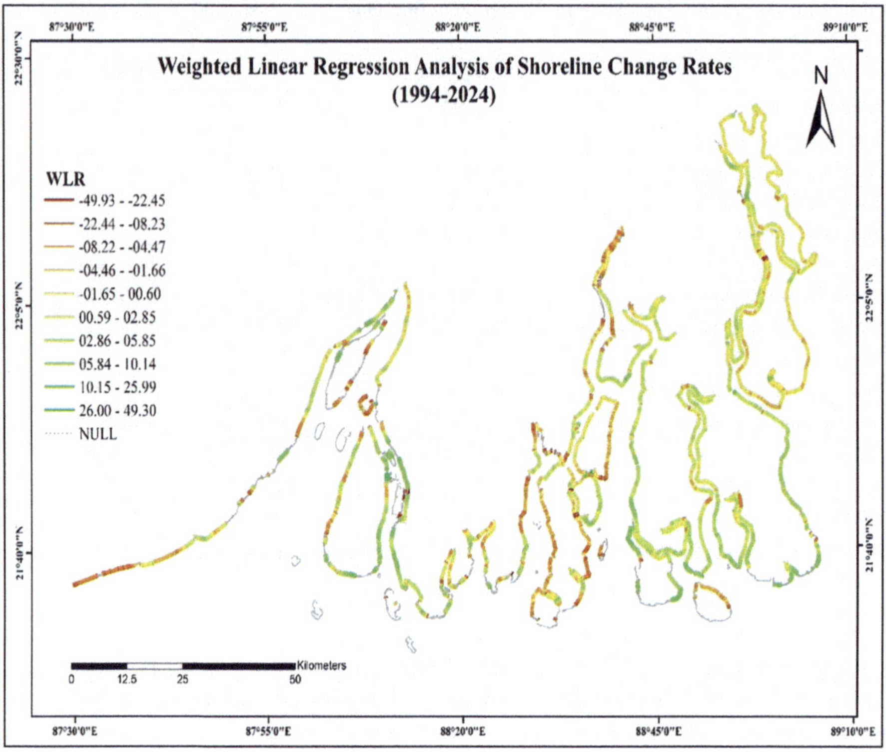

Fig. 16.14 Weighted Linear Regression (WLR) analysis of coastal region of Bengal during 1994 to 2024 (Prepared by the authors)

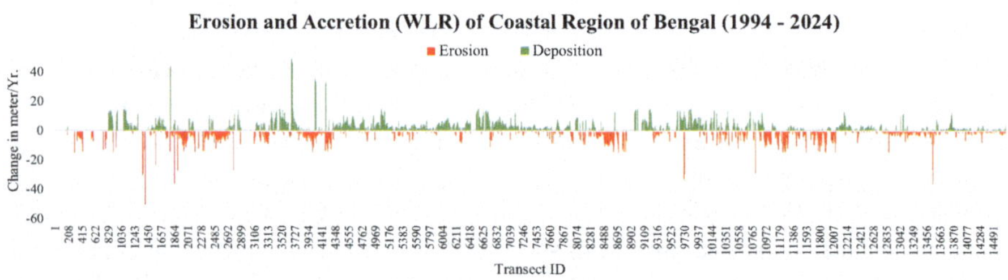

Fig. 16.15 Temporal variations in erosion and accretion rates in the coastal region of Bengal (1994–2024) (Prepared by the authors)

Implications of Findings

The results of this shoreline change analysis have critical implications for coastal management, infrastructure planning, and climate adaptation strategies in Bengal's coastal regions (Zambrano-Medina et al., 2023; Narayana, 2016). The varying rates of erosion and accretion across the three study areas highlight the combined influence of natural processes and human interventions on shoreline stability. Coastal

Medinipur (CM), which exhibits the highest erosion rates, faces significant challenges, including land loss, habitat degradation, and an increased risk of saltwater intrusion. The pronounced negative shoreline change trend in CM indicates heightened vulnerability for coastal communities, agricultural lands, and infrastructure, necessitating immediate interventions such as sustainable shoreline protection measures, sediment management, and the restoration of coastal buffer zones. In contrast, the Coastal Region of Bengal (CRB) and Coastal 24 Parganas (CS24) demonstrate a net accretional trend, suggesting dynamic sediment deposition that may provide natural coastal protection. However, localized erosion persists in these areas, requiring continuous monitoring to ensure that accretion does not obscure long-term coastal vulnerabilities. The significant variability in shoreline changes, as reflected in the standard deviations, underscores the need for site-specific management strategies rather than a uniform approach (Hastuti et al., 2024; Boussetta et al., 2022). Furthermore, these findings are essential for policymakers, urban planners, and environmental agencies in designing evidence-based coastal adaptation strategies. The integration of geospatial technologies in shoreline monitoring facilitates the early identification of high-risk zones, allowing for proactive measures such as sustainable embankment designs, eco-engineering solutions, and community-driven adaptation initiatives. Additionally, the results emphasize the urgency of regulating unplanned coastal development, which exacerbates shoreline instability and increases long-term risks for coastal settlements.

The analysis of shoreline dynamics using DSAS and WLR techniques reveals long-term trends of erosion and accretion along the Bengal coast. While CRB and CS24 show net accretion, CM experiences severe erosion, underscoring the need for region-specific coastal management. The high variability in shoreline changes highlights the necessity of continuous monitoring and adaptive strategies to mitigate climate-induced risks. Here, the socio-economic and ecological importance of Bengal's coastal regions, proactive resilience planning is crucial for safeguarding livelihoods, ecosystems, and infrastructure. Future research should integrate hydrodynamic modelling, sediment transport analysis, and socio-environmental assessments to develop comprehensive coastal sustainability frameworks. By leveraging geospatial technologies and long-term monitoring, this study provides a scientific basis for informed decision-making in one of India's most vulnerable coastal landscapes.

16.2.3 Casebook 3: Modelling and Monitoring of Mangrove Ecologies Greenness in the Changing Climate: The Case Study of Indian Sundarbans

16.2.3.1 Background and Purpose

The Indian Sundarbans (Fig. 16.16), a crucial coastal ecosystem and UNESCO World Heritage Site, contain one of the world's most extensive and ecologically rich mangrove forests (UNESCO, 1984). These mangroves play a vital role in coastal

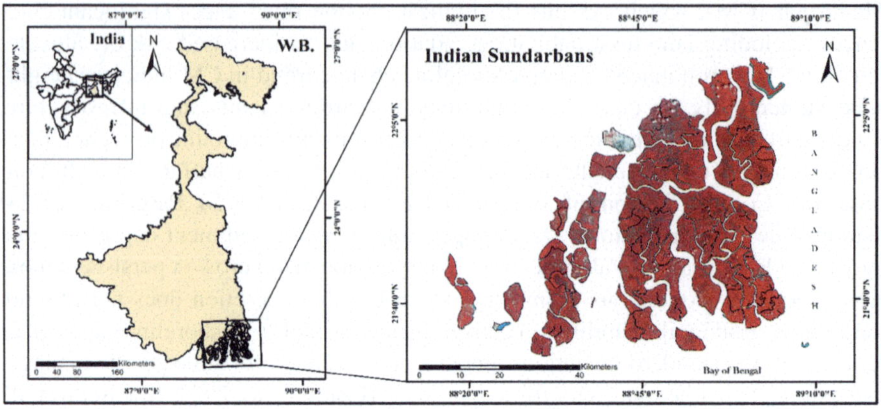

Fig. 16.16 Spatial extent of the study: The Indian Sundarbans (Prepared by the authors)

defence, carbon storage, and sustaining local communities (Mitra, 2013). However, the ecosystem is increasingly threatened by climate change, rising sea levels, salinity intrusion, and human activities such as deforestation and aquaculture expansion. These pressures adversely impact mangrove health and greenness, underscoring the importance of continuous monitoring for effective conservation and sustainable management (Paul et al., 2023; Malakar et al., 2023; Ghosh & Roy, 2022; Ghosh et al., 2015). This case study seeks to model and monitor the greenness dynamics of mangrove ecologies in the Indian Sundarbans under evolving climatic conditions. Utilizing remote sensing and geospatial techniques, the study examines temporal variations in mangrove greenness from 1990 to 2023. Vegetation indices derived from satellite data, including the Ratio Vegetation Index (RVI) (Tran et al., 2022; Zhou et al., 2016), Normalized Difference Vegetation Index (NDVI) (Tran et al., 2022; Rouse et al., 1974), Atmospherically Resistant Vegetation Index (ARVI) (Yuan et al., 2016), and Enhanced Vegetation Index (EVI) (Prananda et al., 2020; Hunt et al., 2011), were employed to quantify greenness trends and detect spatial-temporal variations. Additionally, climate variables such as temperature and rainfall were analysed to assess their influence on mangrove health.

The study's findings will offer critical insights into the resilience and vulnerability of the Sundarbans' mangrove ecosystems. By integrating geospatial methodologies with ecological assessments, the research aims to assist policymakers, conservationists, and local stakeholders in developing evidence-based strategies for mangrove conservation. The research highlights the importance of adaptive management strategies in addressing climate-related challenges and safeguarding the long-term sustainability of this ecologically vital region.

16.2.3.2 Data and Methods

This study relies on secondary datasets for mangrove greenness analysis. Table 16.7 provides an overview of the satellite data used in the investigation. These datasets

Table 16.7 Details of the satellite (Landsat) dataset used in the present study over the Indian Sundarbans (1990 to 2023)

Year	Spacecraft	Sensor	Scene ID	Acquisition date	Time (GMT)	Others	
2023	Landsat 8/9	OLI and TIRS	LC91380452023065LGN00	2023-03-06	04:31:36.23	Cloud cover (%): 0.09 Station ID: LGN Image quality OLI: 9 Image quality TIRS: 9 Sun azimuth: 133.64458351 Sun elevation: 52.24437839 Earth Sun distance: 0.9918821	WRS type: 2 WRS path: 138 WRS row: 045 Map projection: UTM UTM zone: 45 Datum: WGS84 Ellipsoid: WGS84
2015	Landsat 8	OLI and TIRS	LC81380452015035LGN02	2015-02-04	04:31:20.07	Cloud cover (%): 2.23 Station ID: LNG Image quality OLI: 9 Image quality TIRS: 9 Sun azimuth: 142.81434425 Sun elevation: 43.62606508 Earth Sun distance: 0.9856985	
1990	Landsat 5	TM	LT51380451990030BKT00	1990-01-30	03:53:25.15	Cloud cover (%): 0.00 Station ID: BKT Image quality: 9 Sun azimuth: 134.94627028 Sun elevation: 36.76515610 Earth Sun distance: 0.9850764	

Source: USGS Earth Explorer (https://earthexplorer.usgs.gov/)

were further integrated with gridded climate data from the Climatic Research Unit (CRU TS Version 4.07), which has a spatial resolution of 0.5° (accessible at https://crudata.uea.ac.uk/cru/data/hrg/cru_ts_4.07). The analysis primarily focuses on temperature and precipitation trends from 1989 to 2022 as a climate variable.

The analysis was conducted using the ArcGIS 10.8 application on pre-processed Landsat satellite imagery of the Indian Sundarban region, which included atmospheric correction and incorporated various vegetation indices (VIs), as detailed in Table 16.8. Additionally, the study applied the Mann-Kendall (MK) test to assess trends in mean temperature and rainfall (Frimpong et al., 2022; Gadedjisso-Tossou et al., 2021). The MK test operates under the null hypothesis (H0), which assumes no trend, meaning the data are independent and randomly distributed, and it is compared against the alternative hypothesis (Ha), which suggests the presence of a trend (Koudahe et al., 2018). The MK test statistic was computed as follows;

$$S = \sum_{i=1}^{n-1} \sum_{j=i+1}^{n} \operatorname{sgn}\left(X_j - X_i\right) \tag{16.9}$$

Here, n = The cumulative count of data points, where Xi and Xj represent data values in time series i and j (with j being greater than i), respectively, and sgn(Xj − Xi) denotes the sign function.

$$\operatorname{sgn}\left(X_j - X_i\right) = \begin{cases} +1, if\ Xj - Xi > 0 \\ 0, if\ Xj - Xi = 0 \\ -1, if\ Xj - Xi < 0 \end{cases} \tag{16.10}$$

Table 16.8 Modelling equations for vegetation indices

Vegetation indices	Equation	References	Equation No.
Ratio vegetation index (RVI)	$\frac{NIR}{RED}$	Tran et al. (2022); Zhou et al. (2016)	16.13
Normalized difference vegetation index (NDVI)	$\frac{(NIR - RED)}{(NIR + RED)}$	Tran et al. (2022); Rouse et al. (1974)	16.14
Atmospherically resistant vegetation index (ARVI)	$\frac{(NIR - (2 \times RED) + BLUE)}{(NIR + (2 \times RED) + BLUE)}$	Yuan et al. (2016)	16.15
Enhanced vegetation index (EVI)	$2.5\frac{(NIR - RED)}{(NIR + 6RED - 7.5BLUE) + 1}$	Prananda et al. (2020); Hunt et al. (2011)	16.16

Note: NIR: Reflectance value in the near-infrared band; RED: Reflectance value in the red band; and BLUE: Reflectance value in the blue band

The variance was calculated as follows:

$$Var(S) = \frac{n(n+1)(2n+5) - \sum_{i=1}^{m} t_i (t_1 - 1)(2t_1 + 5)}{18} \tag{16.11}$$

Here, n = The quantity of data points, m = The count of tied groups, where t_i represents the number of ties of magnitude i. A tied group refers to a collection of sample data sharing identical values.

In scenarios where the sample size exceeds 10 (n > 10), the standard normal test statistic Z_S was determined using the following equation:

$$Z_s = \begin{cases} \frac{S-1}{\sqrt{Var(S)}}, if\ S > 0 \\ 0, if\ S = 0 \\ \frac{S+1}{\sqrt{Var(S)}}, if\ S > 0 \end{cases} \tag{16.12}$$

Positive values of Z_S indicate increasing trends, whereas negative Z_S values indicate decreasing trends.

Furthermore, this study employed spatial statistical analysis (Herrera Estrella et al., 2021; Hou et al., 2015; Roberts et al., 1997), specifically utilizing the Ordinary Least Squares (OLS) technique (Eq. 16.17). This approach facilitates an in-depth examination of the relationship between vegetation indices (VIs) and climatic parameters, including temperature and precipitation, over the period 1990–2023. Additionally, regression analysis was conducted to investigate spatio-temporal variations in mean temperature, rainfall, and each vegetation index.

$$\hat{Y} = b_0 + b_1 x_1 + b_2 x_2 + \ldots + b_p x_p \tag{16.17}$$

Here, $\hat{Y}$ is the estimated or predicted value of the dependent variable; b_0 is the y-intercept; b_1, b_2, … b_p are the coefficients for the independent variables x_1, x_2, … x_p are the independent variables. In a general context, when the slope is greater than zero, it indicates a rising trend in the dependent variable, whereas when the slope is less than zero, it signifies a declining trend in the dependent variable.

Table 16.9 Calculated p-values (two-tailed) of Mann-Kendall (MK) test

Stations/locations (N/E)		Station 1 (22.152192 /89.004106)	Station 2 (21.674547 /89.00721)	Station 3 (21.847734 /88.716955)	Station 4 (21.703789 /88.544289)	Station 5 (21.691271 /88.325971)
p-values of M-K test	Temperature (1989–2022)	0.804	0.963	0.988	0.988	0.951
	Rainfall (1989–2022)	0.258	0.486	0.566	0.566	0.914
Remarks	There is no statistically significant trend (in the case of temperature and rainfall). Significance level alpha = 0.05					

16.2.3.3 Results, Discussion, and Conclusion

The analysis of climatic data, specifically temperature and rainfall in the Indian Sundarbans, using the Mann-Kendall (MK) test indicated no statistically significant trends between 1989 and 2022 (Table 16.9). Since the p-values for all selected data points exceeded the significance threshold of $\alpha = 0.05$, the null hypothesis (H_0) could not be rejected (Koudahe et al., 2018). Additionally, this study presented the graphical distribution of these climatic variables (Fig. 16.17), illustrating variations over time. These fluctuations in climate variables have influenced the mangrove ecosystems within the study area (Fanous et al., 2023; Ghosh et al., 2017).

Modeling Greenness Contents

Modeling the greenness content within mangrove ecosystems necessitates a comprehensive grasp of the variables that influence mangrove greenness. This understanding is pivotal in guiding conservation and management strategies (Sun et al., 2024). During the course of the study, the temporal fluctuations in the calculated RVI, NDVI, ARVI, and EVI values across three distinct time periods were depicted in Figs. 16.18 and 16.19. It was observed that the RVI values were 1.943 in 1990, 1.593 in 2015, and 1.640 in 2023. This resulted in an overall decline in pixel values of −0.30 from 1990 to 2023, indicating a temporal decrease in mangrove greenness. This trend suggests potential stress and reduced density in mangrove coverage over time within the study area. Such analysis aids in understanding the dynamics of vegetation density and health (Pamungkas, 2023; Tran et al., 2022). Regarding NDVI, there was a decrease in mean pixel values of −0.08 from 1990 to 2015, followed by a slight increase of +0.01 from 2015 to 2023. This cumulative change of −0.07 from 1990 to 2023 indicates a moderate to sparse vegetated condition (Rutkay et al., 2020). Conversely, the application of the ARVI index to mitigate the impact of atmospheric aerosols revealed a continuous decrease in mean pixel values, with a decrease of −0.29 observed in 2023 (0.131), indicating sparse and stressed vegetation (Yuan et al., 2016). In contrast, EVI demonstrated a consistent increase (+0.03) throughout the study period. Thus, the temporal assessment of mean pixel values (Table 16.10) predominantly suggests that vegetation greenness content experienced sparse and stressed conditions in the study area.

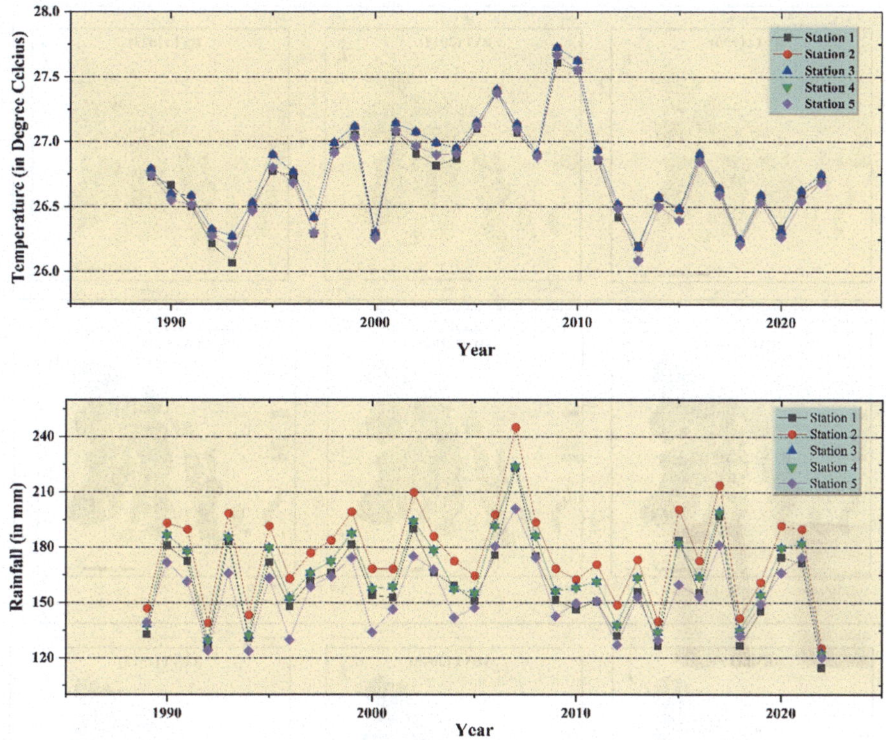

Fig. 16.17 Distribution of climate data in the Indian Sundarbans from 1989 to 2022. (Prepared by the authors)

This study extended one step further, establishing a link between greenness content and changing climate conditions in the study area. The standard deviation of residuals for temperature (Fig. 16.20) and rainfall (Fig. 16.21) in relation to RVI, NDVI, ARVI, and EVI from 1990 to 2023 was spatially mapped, highlighting the distribution of selected sample pixel points within the study area, and Fig. 16.22 depicts the relationships between these variables. Through this analysis, it becomes evident that the variability in climatic conditions within the study area significantly influences the modelling of greenness content. While an increase in temperature has a positive impact on the greenness content of the mangroves in some instances (Ghebrezgabher et al., 2020), it is essential to note that rising temperatures, a key driver of global warming, contribute to changing climate conditions (Ahmed, 2022). These changes, including a continuous decrease in rainfall, affect various environmental factors and subsequently lead to reductions in greenness content (Duke et al., 2017; Ward et al., 2016), thereby increasing stress on the mangrove ecosystems of the study area. A notable change to these analyses unfolded during the content analysis in 2015. This change was characterized by a decline in climatic factors, particularly a decrease in temperature compared to 1989 and 2022. In this

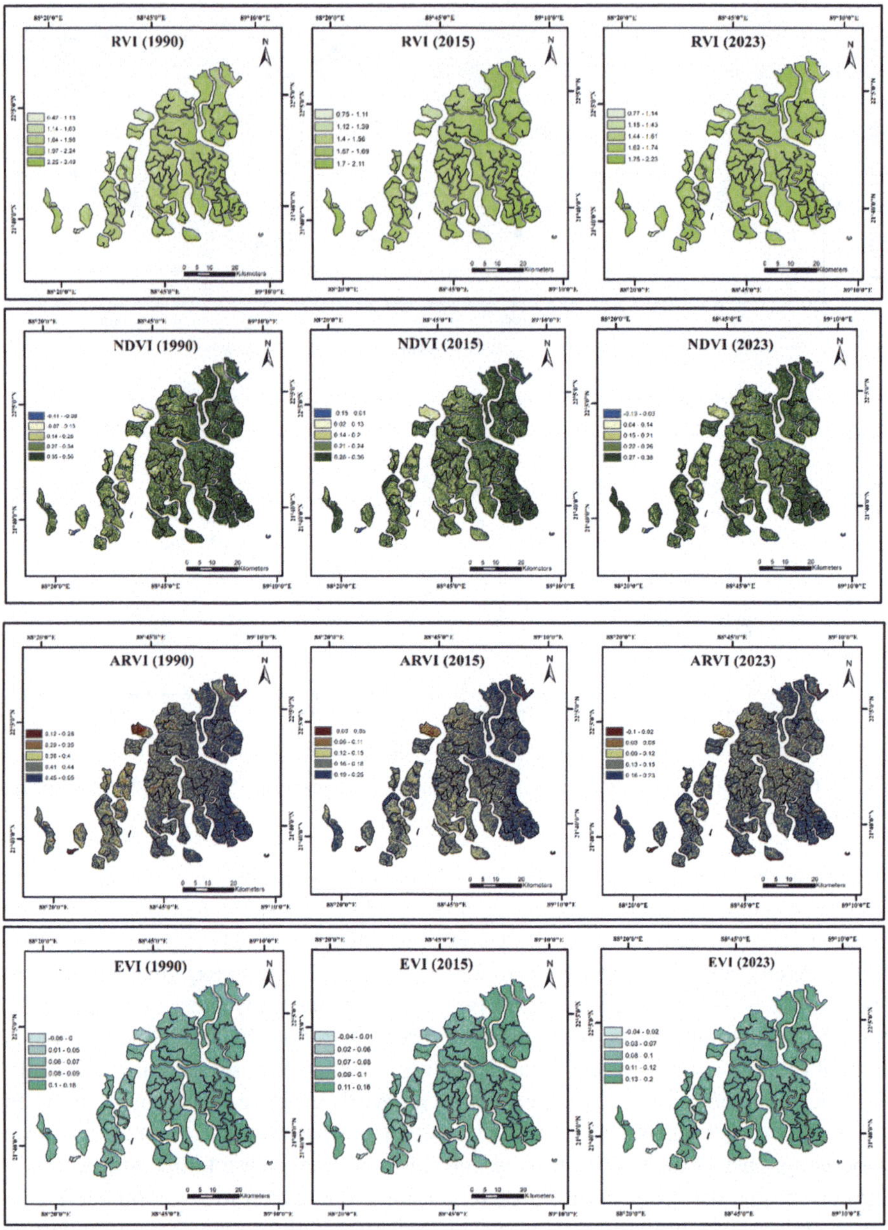

Fig. 16.18 Spatio-temporal mapping of RVI, NDVI, ARVI, and EVI spectral values (1990–2023) (Prepared by the authors)

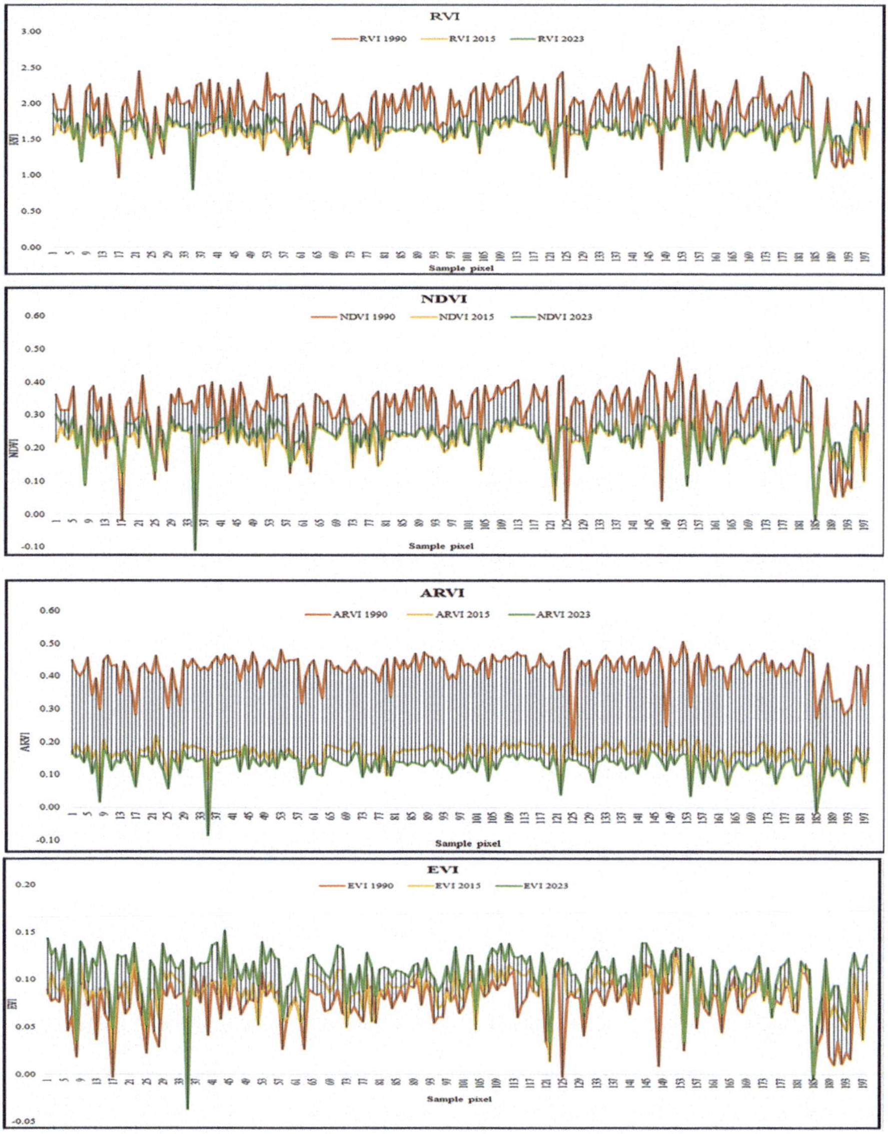

Fig. 16.19 Temporal trends in spectral values of RVI, NDVI, ARVI, and EVI (1990–2023) (Prepared by the authors)

Table 16.10 Statistics of the greenness contents modelling

Indices	RVI			NDVI			ARVI			EVI			Temperature (°C)			Precipitation (mm)		
Time periods	1990	2015	2023	1990	2015	2023	1990	2015	2023	1990	2015	2023	1989	2014	2022	1989	2014	2022
Max.	2.800	1.885	1.935	0.474	0.307	0.319	0.506	0.220	0.188	0.134	0.129	0.152	26.73	26.55	26.70	1714.03	1519.70	1471.72
Min.	0.966	0.996	0.802	−0.018	−0.002	−0.110	0.163	0.048	−0.086	−0.003	−0.001	−0.037	26.56	26.32	26.53	1597.73	1433.45	1305.96
Mean	1.943	1.593	1.640	0.310	0.226	0.239	0.421	0.167	0.131	0.076	0.088	0.107	26.64	26.44	26.63	1651.04	1478.61	1403.61
Sample variance	0.110	0.020	0.020	0.008	0.002	0.002	0.002	0.001	0.002	0.001	0.000	0.001						
Standard deviation	0.330	0.140	0.150	0.090	0.040	0.050	0.050	0.020	0.030	0.020	0.020	0.020						
Standard error	0.020	0.010	0.010	0.006	0.003	0.003	0.003	0.002	0.002	0.001	0.001	0.001						
R^2 (with tem.)	0.0380	0.0530	0.0027	0.0010	0.0457	0.0028	0.2420	0.0516	0.0018	0.0289	0.0518	0.0030						
R^2 (with pre.)	0.0021	0.0608	0.0002	0.0242	0.0550	0.0016	0.0002	0.0631	0.0008	0.0012	0.0594	0.0005						
Changeability																		
1990 to 2015	−0.35			−0.08			−0.25			0.01			1989 to 2014	−0.21		−172.42		
2015 to 2023	0.04			0.01			−0.04			0.02			2014to 2022	0.19		−75.01		
1990 to 2023	−0.30			−0.07			−0.29			0.03			1989 to 2022	−0.01		−247.43		

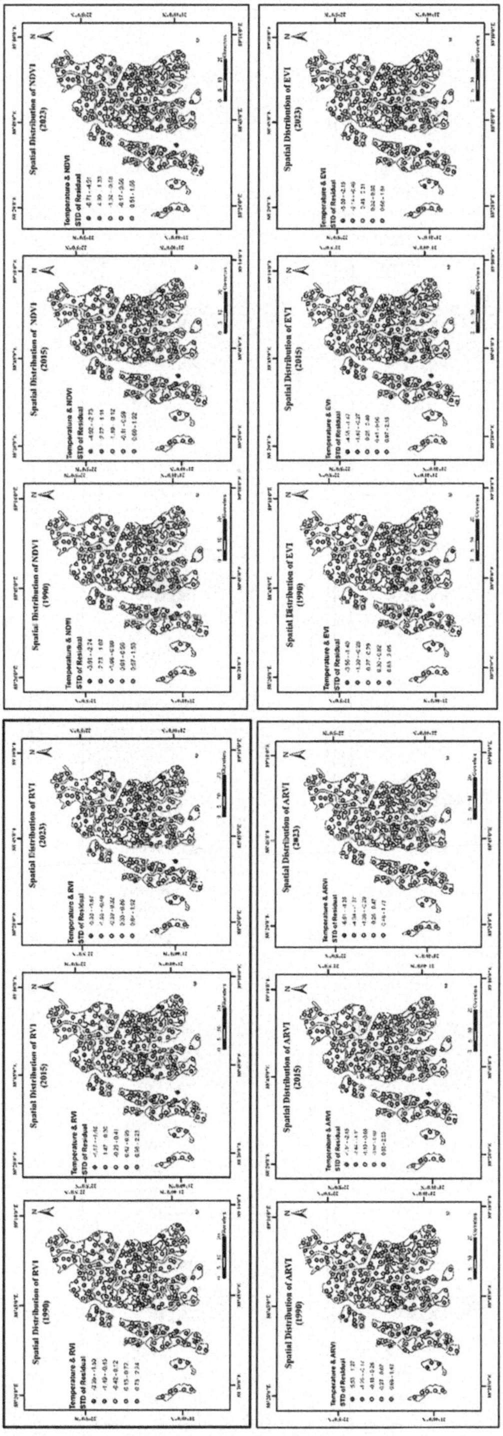

Fig. 16.20 Spatial distribution of relationships between climatic factors (temperature) and vegetation indices (RVI, NDVI, ARVI, and EVI) (Prepared by the authors)

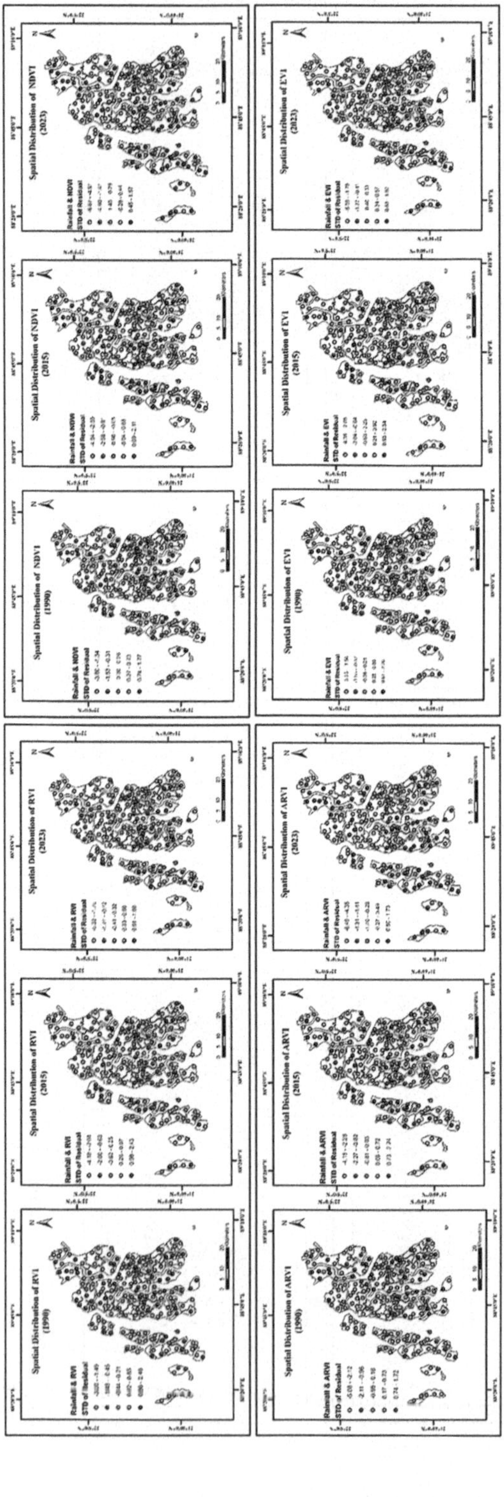

Fig. 16.21 Spatial distribution of relationships between climatic factors (rainfall) and vegetation indices (RVI, NDVI, ARVI, and EVI) (Prepared by the authors)

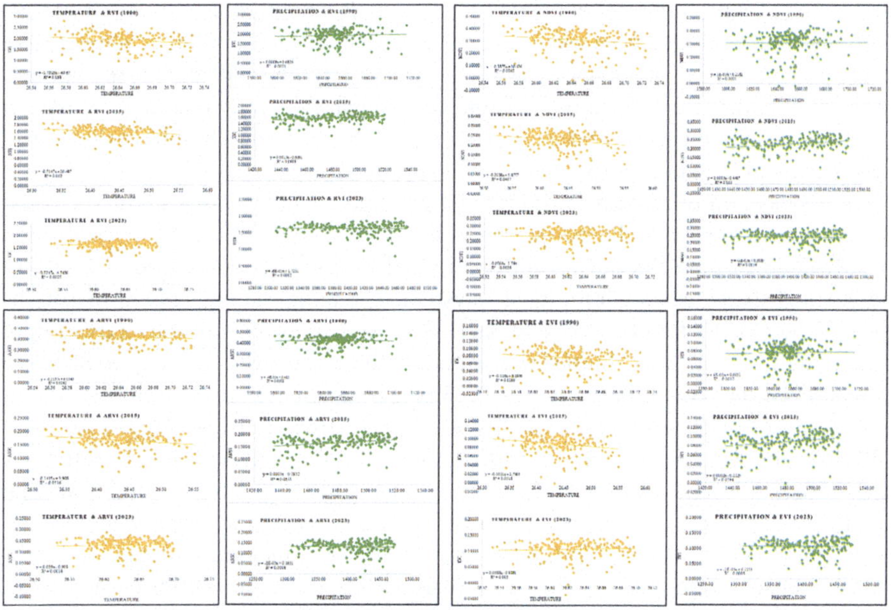

Fig. 16.22 Relational trend between Climatic Factors (Temperature and Rainfall) and Vegetation Indices (RVI, NDVI, ARVI, and EVI) (1990–2023) (Prepared by the authors)

scenario, the statistical modelling of greenness content was also affected (Pan et al., 2019). Consequently, it can be inferred that the overall shifts in climate conditions or changes in climate variability exerted an influence on the greenness of the Sundarbans mangroves. This resulted in the transition from healthier and denser conditions to situations characterized by reduced greenness and heightened stress levels.

Implications of Findings

This study explores both temporal and spatial variations in the responses of vegetation indices (VIs) to climate factors (CFs) within the Indian Sundarbans. Over the past 33 years, climate data has provided scientific evidence of changing climatic conditions in the region. The observed data amplitudes indicate significant spatial variations in climate parameters between 1989 and 2022, sometimes with considerable fluctuations. Due to its coastal setting, the study also highlights the impacts of global climate change on the Indian Sundarbans, including rising sea levels (Kanan et al., 2023), extreme climatic events (Mondal et al., 2022), increasing sea salinity (Das et al., 2022), and rising temperatures (Jabir et al., 2021; Sharadi et al., 2020).

In light of evolving climate variability conditions, the study explores the status of health risk and stress levels in mangrove vegetation using various vegetation indices. A comprehensive analysis of summary statistics revealed significant changes that were closely related to climate change and variability. Moreover, the study underscored these pivotal findings throughout the results discussions, as well as in the analysis pertaining to the year 2015, which observed analogous patterns. Notably, climate indicators experienced a decline in 2014, with a temperature decrease of −0.21 °C since 1989, alongside ongoing decreases in rainfall. Nonetheless, the decrease in rainfall and subsequent rise in temperature indicate a continual escalation up to the present. As a result, these changes highlight the impact on all vegetative indices throughout this period.

In essence, an evaluation of mangrove greenness suggests that climate change has a negative impact on the mangroves in the study region. This impact has resulted in a decline in their overall health com-pared to previous conditions, leading to heightened stress levels within the ecosystem. The temporal changes further highlight an increasing trend of stressful environmental conditions in the study region. Additionally, the growing human intrusion into the mangrove forest has contributed to an escalation in disturbance levels within the natural environment, as observed in studies by Malakar et al. (2023), Rasyid et al. (2016), and DasGupta and Shaw (2015). Building on these findings, this study strongly advocates for continuous monitoring of long-term trends in mangrove vegetation responses to climate change in the Indian Sundarbans. Additionally, it recommends the implementation of nature-based solutions (NbS) to tackle key challenges, including mangrove biodiversity conservation (Key et al., 2022), carbon sequestration (Hilmi et al., 2023), coastal flood mitigation (Van Hespen et al., 2023; Gijsman et al., 2021), and the promotion of sustainable resource management through active local community involvement (Wibisono et al., 2023; Kusmana, 2015).

16.2.4 Casebook 4: Spatio-temporal Analysis of Land Use/ Land Cover Changes and Ecological Degradation Using LANDSAT Imagery in Coastal Bengal

16.2.4.1 Background and Purpose

The coastal region of Bengal (Fig. 16.23) has experienced substantial land use and land cover (LULC) transformations driven by both natural processes and anthropogenic activities. Rapid urbanization, agricultural expansion, aquaculture, and infrastructure development have significantly altered land cover patterns, adversely affecting local ecosystems and biodiversity (Nath et al., 2023; Bera et al., 2022a, b; Datta & Deb, 2012). Furthermore, climate-induced stressors, including coastal erosion, sea-level rise, and extreme weather events, have intensified ecological degradation, posing additional challenges to environmental sustainability (Saengsupavanich et al., 2024; Kantamaneni et al., 2022; Mondal et al., 2022; Jabir et al., 2021; Swapna et al., 2020). Understanding these spatio-temporal dynamics is crucial for effective environmental management and sustainable development planning. However, conventional field-based approaches often fail to capture large-scale and long-term trends, underscoring the necessity of remote sensing for

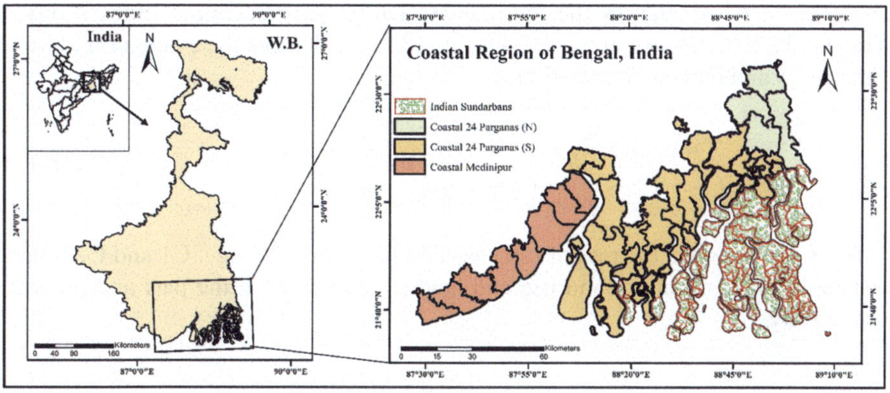

Fig. 16.23 Study site: Coastal Bengal and its dynamic landscape (Prepared by the authors)

comprehensive LULC monitoring in this vulnerable coastal landscape (Annan et al., 2024; Picuno et al., 2015).

This study aims to investigate LULC changes and their implications for ecological degradation in Coastal Bengal using LANDSAT satellite imagery. Employing geospatial techniques such as supervised classification, change detection analysis, and statistical modelling, it examines the extent and patterns of land cover transitions over the past few decades. Furthermore, incorporating vegetation indices such as the Normalized Difference Vegetation Index (NDVI) enhances the evaluation of vegetation loss and habitat degradation. The study's findings offer valuable insights for policymakers, environmental planners, and local stakeholders in formulating data-driven strategies to strengthen coastal resilience. By highlighting the spatial and temporal dynamics of land use change, this research underscores the critical need for sustainable land management practices to curb environmental degradation and support long-term ecological stability in Coastal Bengal.

16.2.4.2 Data and Methods

The present case study employed LANDSAT satellite imagery from 1994, 2004, 2014, and 2024 (captured in the months of January and February) acquired from the USGS Earth Explorer (Available at: https://earthexplorer.usgs.gov) to analyse spatio-temporal land use and land cover (LULC) changes and ecological degradation.

LULC changes in the study area were assessed using supervised classification, applying the maximum likelihood algorithm to images from the specified years (Nath et al., 2023). The classification accuracy was evaluated using the kappa coefficient to ensure the reliability of the results (Cohen, 1960). Additionally, change detection techniques were utilized to quantify LULC transitions over time, following the methodology outlined by Puyravaud (2003) and employing the corresponding equation (Eq. 16.18). This analytical approach offers a comprehensive perspective on the extent and patterns of land cover changes, providing crucial insights into the region's environmental dynamics. Furthermore, the integration of vegetation indices, such as the Normalized Difference Vegetation Index (NDVI) (Tran et al., 2022; Rouse et al., 1974) (Eq. 16.19), enhances the evaluation of vegetation loss and habitat degradation.

$$R = \left[\frac{1}{t_1 - t_2}\right] \times \left[\ln\left(\frac{C_1}{C_2}\right)\right] \tag{16.18}$$

Here, R represents the rate of land use and land cover change, C1 and C2 represents areas have different land use and land cover classes in the past and present, respectively.

$$NDVI = \frac{(NIR - RED)}{(NIR + RED)} \tag{16.19}$$

Here, NIR: Reflectance value in the near-infrared band; RED: Reflectance value in the red band.

16.2.4.3 Results, Discussion, and Conclusion

Spatio-Temporal Analysis of LULC: Key Observations
The spatio-temporal analysis of land use and land cover (LULC) changes in Coastal Bengal over the past three decades (1994–2024) highlights substantial transformations influenced by both anthropogenic interventions and environmental dynamics (Fig. 16.24). The findings (Table 16.11) indicate a marked reduction in agricultural land, mangrove cover, and natural vegetation, while concurrently, there has been a significant expansion of settlements and shrimp farming, reflecting shifting land use patterns driven by socio-economic and ecological factors (Fig. 16.25 and Table 16.12).

Implications of Findings
The findings reveal rapid land use changes in Coastal Bengal, driven by expanding settlements and shrimp farming at the cost of agricultural land, mangroves, and vegetation (Thakur et al., 2021; Samanta & Paul, 2016; Jana et al., 2016). Despite some mangrove recovery, the overall decline in natural vegetation and water bodies signals ecological challenges, including biodiversity loss and increased climate vulnerability (Kayal & Chowdhury, 2022; Ghosh & Roy, 2022). This study emphasizes the need for sustainable land management, integrated coastal zone management (ICZM), and policy interventions to balance development and conservation. Continuous geospatial monitoring is essential for informed decision-making and adaptive strategies.

Spatio-Temporal Analysis of LULC: Changes
The analysis of LULC changes in Coastal Bengal from 1994 to 2024 provides valuable insights into long-term landscape transformations. The findings (Table 16.13)

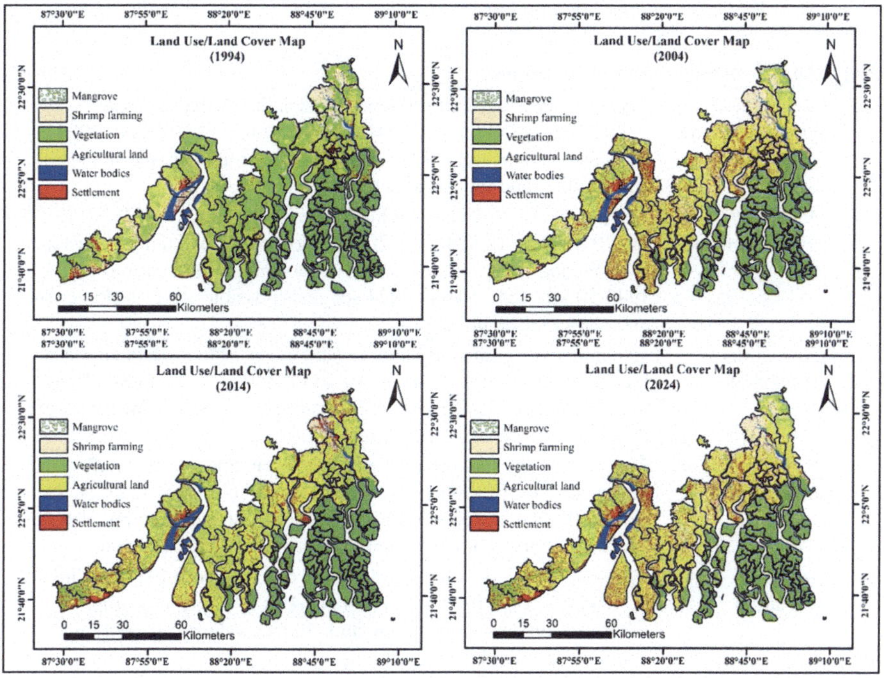

Fig. 16.24 Land use/land cover maps of coastal region of Bengal, 1994–2024. (Prepared by the authors)

reveal a consistent decline in agricultural land and vegetation, degradation of mangroves, and the expansion of human settlements and shrimp farming. These shifts are driven by a complex interplay of anthropogenic pressures, climate-induced stressors, and economic transitions toward aquaculture and construction development. The changes outlined in Table 16.14 are summarized as follows:

Implications of Findings

The findings reveal growing anthropogenic pressure on Coastal Bengal's fragile ecosystems, marked by declining agricultural land, mangroves, and vegetation, raising concerns about biodiversity loss and climate vulnerability (Ghosh & Roy, 2022). The expansion of shrimp farming and settlements reflects economic and demographic shifts, necessitating integrated coastal management (Roy et al., 2024a, b; Majumder, 2024; Giri et al., 2022; Pitchaikani, 2020; Chopra et al., 2009). Sustainable land use planning, conservation policies, and adaptive strategies are crucial to balancing development with ecological preservation. Continuous remote sensing and geospatial analysis will support evidence-based decision-making and resilience planning in this dynamic coastal landscape.

Table 16.11 Key observations: spatio-temporal analysis of LULC in Coastal Bengal (India)

Sr. No.	LULC features	Key observations
1	Decline in agricultural land	The area under agricultural land has steadily decreased from 2788.55 km^2 (38.00%) in 1994 to 2637.30 km^2 (35.94%) in 2024. This reduction is indicative of land conversion for urban expansion, aquaculture, and industrial activities.
2	Mangrove degradation and slight recovery	The mangrove cover has declined from 2129.07 km^2 (29.02%) in 1994 to 1705.52 km^2 (23.24%) in 2014, reflecting deforestation, coastal erosion, and rising salinity. However, a marginal increase to 1743.50 km^2 (23.76%) in 2024 suggests possible afforestation efforts and conservation initiatives.
3	Expansion of settlements	Housing has significantly altered land cover, with settlement areas increasing from 364.76 km^2 (4.97%) in 1994 to 763.64 km^2 (10.41%) in 2024. This rise is attributed to population growth, infrastructure development, and coastal migration.
4	Rapid growth of shrimp farming	Shrimp farming has emerged as a dominant land use, expanding from 407.97 km^2 (5.56%) in 1994 to 1063.06 km^2 (14.49%) in 2024. This shift highlights the economic importance of aquaculture but also raises concerns about ecological degradation and salinity intrusion.
5	Decline in vegetation cover	The vegetation cover has significantly reduced from 1312.86 km^2 (17.89%) in 1994 to 854.23 km^2 (11.64%) in 2024, indicating deforestation, land conversion, and climate-induced stressors.
6	Reduction in water bodies	The area occupied by water bodies has slightly decreased from 334.51 km^2 (4.56%) in 1994 to 275.99 km^2 (3.76%) in 2024, possibly due to land reclamation and sedimentation processes.

Source: Fig. 16.24 and 16.25; Table 16.12

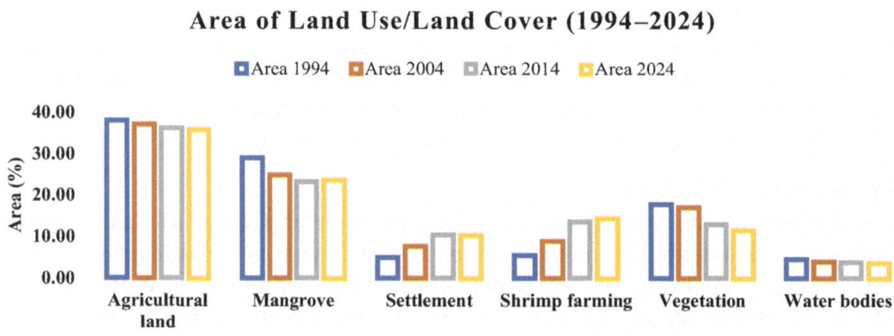

Fig. 16.25 Land use/land cover area in coastal region of Bengal (1994–2024). (Prepared by the authors)

Spatio-Temporal Analysis of Vegetation Loss and Habitat Degradation: Key Observations

The analysis of Normalized Difference Vegetation Index (NDVI) values from 1994 to 2024 offers critical insights into vegetation loss and habitat degradation in Coastal Bengal (Fig. 16.26). As a key indicator of vegetation health and density, NDVI

Table 16.12 Land use/land cover area in coastal region of Bengal (1994–2024)

	1994		2004		2014		2024	
LU/LC classes	Area (sq. km)	Area (%)	Area (sq. km)	Area (%)	Area (sq. km)	Area (%)	Area (sq. km)	Area (%)
Agricultural land	2788.55	38.00	2728.28	37.18	2652.29	36.15	2637.30	35.94
Mangrove	2129.07	29.02	1831.23	24.96	1705.52	23.24	1743.50	23.76
Settlement	364.76	4.97	564.46	7.69	760.33	10.36	763.64	10.41
Shrimp farming	407.97	5.56	655.53	8.93	992.51	13.53	1063.06	14.49
Vegetation	1312.86	17.89	1263.99	17.23	946.69	12.90	854.23	11.64
Water bodies	334.51	4.56	294.22	4.01	280.36	3.82	275.99	3.76

Source: Calculated by the authors. Note: Data for January/February was calculated over the period from 1994 to 2024

Table 16.13 Changes in land use/land cover area in coastal Bengal (1994–2024)

	1990–2004	2004–2014	2014–2024	1994–2024
LU/LC classes	Area (sq.km)	Area (sq.km)	Area (sq.km)	Area (sq.km)
Agricultural land	−60.27	−75.99	−14.99	−151.25
Mangrove	−297.84	−125.71	37.98	−385.57
Settlement	199.70	195.87	3.31	398.88
Shrimp farming	247.56	336.98	70.55	655.09
Vegetation	−48.88	−317.30	−92.46	−458.64
Water bodies	−40.29	−13.86	−4.37	−58.52

Source: Calculated by the authors. Note: Data for January/February was calculated over the period from 1994 to 2024

ranges from −1 to +1, with higher values representing dense vegetation and lower values indicating barren land, degraded vegetation, or water bodies. The observed decline in NDVI over time reflects increasing environmental degradation, ecosystem stress, and land cover transformations driven by both anthropogenic activities and climatic influences. Table 16.15 provides a comprehensive overview of these trends.

Implications of Findings

This NDVI-based spatio-temporal assessment emphasizes the need for integrated ecological monitoring and targeted policy interventions to restore degraded landscapes and strengthen the resilience of Coastal Bengal's vulnerable ecosystems against climate change and anthropogenic pressures (Das & Sarkar, 2025; Das & Das, 2025; Chauhan, 2023). The following points further outline the broader implications of these findings:

- Biodiversity Loss: The decline in vegetation cover threatens mangrove-dependent species and disrupts ecosystem stability (Saoum & Sarkar, 2024; Biswas & Biswas, 2019).

Table 16.14 Changes: spatio-temporal analysis of LULC in Coastal Bengal (India)

Sr. No.	LULC features	Key observations
1	Declining agricultural land	Agricultural land decreased by 151.25 km^2 over the 30-year period, with the highest loss occurring between 2004 and 2014 (−75.99 sq. km). The decline is attributed to land conversion for settlements and aquaculture, affecting local food security and traditional livelihoods.
2	Mangrove degradation with partial recovery	The mangrove cover experienced a net loss of 385.57 km^2, with the highest reduction between 1994 and 2004 (−297.84 km^2). However, between 2014 and 2024, a slight increase (+37.98 km^2) suggests afforestation efforts and conservation measures. The long-term decline underscores deforestation, rising salinity, and erosion, affecting biodiversity and coastal resilience.
3	Settlements expansion	Settlements increased significantly by 398.88 km^2, with the most rapid expansion occurring in 1994–2014. This growth reflects population increase, infrastructural development, and economic diversification, but also raises concerns over habitat loss and environmental stress.
4	Rapid shrimp farming expansion	Shrimp farming grew substantially by 655.09 km^2, with the highest expansion between 2004 and 2014 (+336.98 km^2). The shift towards aquaculture is driven by economic incentives, yet it has contributed to mangrove degradation, increased salinity intrusion, and ecosystem imbalances.
5	Significant vegetation loss	Vegetation cover saw the most extensive decline, losing 458.64 km^2 over 30 years, with the most severe loss between 2004 and 2014 (−317.30 km^2). This loss is linked to land conversion for settlements and aquaculture, as well as climatic pressures such as rising sea levels and extreme weather events.
6	Reduction in water bodies	Water bodies decreased by 58.52 km^2, with the highest decline between 1994 and 2004 (−40.29 km^2). This reduction could result from land reclamation, sedimentation, and water extraction for agriculture and urban use.

Source: Table 16.13

- Coastal Vulnerability: The degradation of natural vegetation increases the region's exposure to cyclones, storm surges, and coastal erosion (Das & Das, 2025; Mohammed et al., 2024; Rodrigues et al. 2024).
- Need for Restoration: The findings highlight the urgent need for reforestation, conservation policies, and sustainable land management practices to mitigate long-term environmental impacts (Gayen, 2025; Barua et al., 2024; Sandilyan & Kathiresan, 2012).

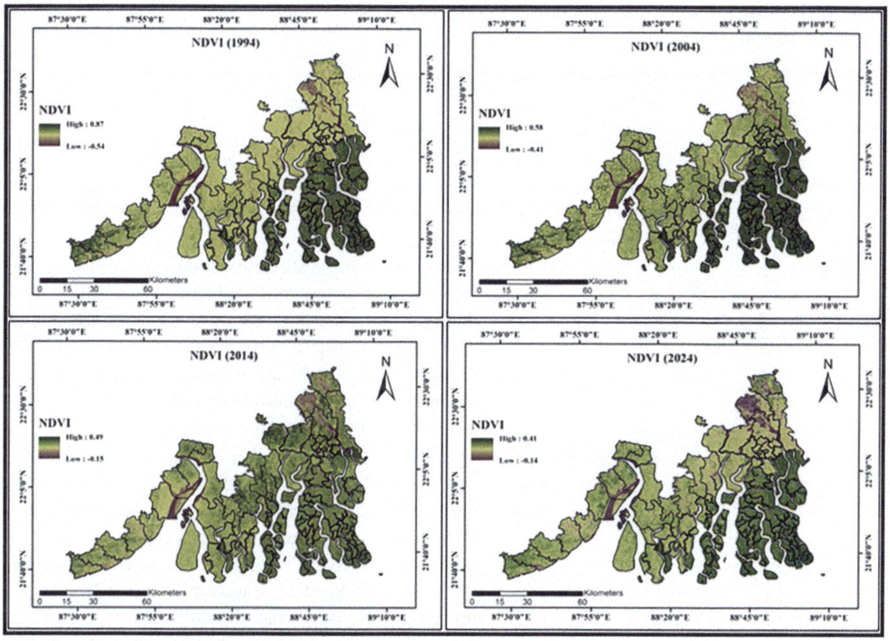

Fig. 16.26 Spatio-temporal distribution of NDVI in coastal region of Bengal during 1994, 2004, 2014, and 2024. (Prepared by the authors)

16.2.5 Casebook 5: Assessing Coastal Land Surface Temperature and Ecological Changes Through Remote Sensing Data

16.2.5.1 Background and Purpose

Coastal regions are highly dynamic environments where fluctuations in land surface temperature (LST) and ecological changes result from both natural processes and human activities (García et al., 2023; Naserikia et al., 2023). Key drivers such as global warming, urban expansion, deforestation, and land use changes contribute to significant thermal variations, impacting coastal ecosystems, biodiversity, and local communities (García et al., 2023; Prabhu Gaonkar et al., 2022; Arulbalaji et al., 2020). The rising incidence of heat stress, habitat degradation, and microclimatic shifts further exacerbates environmental challenges in these vulnerable areas (Patra et al., 2025; Agbelade, 2025; Salam et al., 2024; Stark et al., 2023). In this context, the present case study examines coastal LST variations and associated ecological transformations using remote sensing data in the coastal region of Bengal, India (Fig. 16.27). Utilizing Landsat satellite imagery and geospatial analysis, the study explores multi-decadal LST trends and their relationship with land cover changes. Vegetation indices, including the Normalized Difference Vegetation Index (NDVI),

Table 16.15 Key observations: spatio-temporal analysis of vegetation loss and habitat degradation in Coastal Bengal (India)

Sr. No.	NDVI trends	Vegetation loss and habitat degradation	Key observations
1	Overall decline in NDVI high values	Vegetation health and density reduction	The highest NDVI value decreased from 0.87 (1994) to 0.41 (2024), indicating a substantial reduction in vegetation density and overall greenness. This decline suggests deforestation, conversion of vegetated areas to settlements and aquaculture, and degradation of natural habitats. The sharpest drop occurred between 1994 and 2004 (0.87 to 0.58), likely due to increased land use changes, including mangrove deforestation and agricultural intensification.
2	Increasing degradation in low NDVI values	Habitat loss and land degradation	The lowest NDVI value improved slightly from −0.54 (1994) to −0.14 (2024), indicating a relative decline in barren land or water-covered areas. However, this change does not necessarily imply ecosystem recovery but rather land conversion into alternative uses, such as aquaculture ponds, infrastructure, or degraded shrubland. The sharpest improvement occurred between 2014 and 2024 (−0.15 to −0.14), suggesting stabilization in the conversion process but not a significant ecological recovery.
3	Low NDVI values	Progressive vegetation loss and land use transitions	The declining trend in NDVI values correlates with the loss of mangroves, shrinking natural vegetation, and agricultural transformation. The reduction in high NDVI values, alongside a stabilization of low values, suggests that dense forest areas are being cleared and replaced with lower-quality land cover types (e.g., degraded vegetation, settlements, or aquaculture farms).

Source: Fig. 16.26

Enhanced Vegetation Index (EVI), Normalized Difference Moisture Index (NDMI), and Modified Normalized Difference Water Index (MNDWI), were used to assess ecological responses to temperature fluctuations. The study's findings will offer valuable insights into the interaction between LST dynamics and ecological stability in coastal regions. By integrating remote sensing techniques with statistical modelling, this research aims to inform policymakers, environmental planners, and conservationists in developing adaptive strategies to mitigate climate-induced thermal stress and ecological degradation. Ultimately, the study emphasizes the necessity of sustainable coastal management practices to enhance resilience and ensure long-term ecological balance.

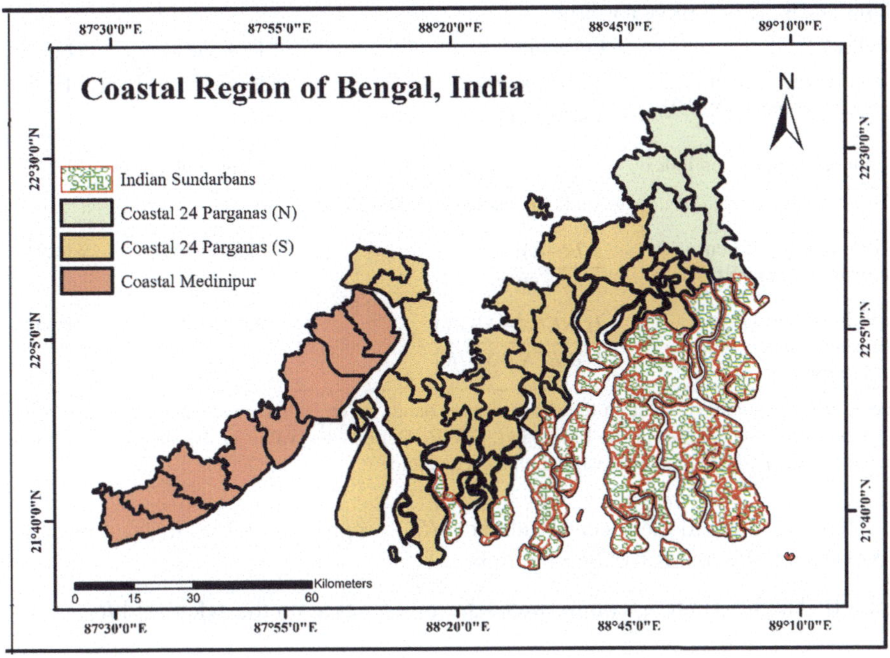

Fig. 16.27 Geographical context of the study area: Coastal Bengal (Prepared by the authors)

16.2.5.2 Data and Methods

The present case study used LANDSAT satellite imagery from 1994, 2004, 2014, and 2024 acquired from the USGS Earth Explorer (Available at: https://earthexplorer.usgs.gov) to analyse the land surface temperature (LST) and ecological changes.

LSTs were retrieved from the thermal infrared band (for Landsat TM data: band 6 and Landsat OLI/TIRS band: 10) using the Spectral Radiance model (Eq. 16.21) and the MWA model (Mono Window Algorithm model) (Eq. 16.23), respectively (Wan Mohd Jaafar et al., 2020; Sahana et al., 2016). Surface ecological conditions were monitored using the Normalized Difference Vegetation Index (NDVI), Enhanced Vegetation Index (EVI), Normalized Difference Moisture Index (NDMI), and Modified Normalized Difference Water Index (MNDWI), which established a relationship with the retrieved LST in the study area. Table 16.16 provides an overview of the spectral indices that have been chosen.

Equations:

Spectral Radiance Model

Step-I: Estimation of Spectral Radiance (R_S):

$$R_S = R_{S\,MIN} + \left(R_{S\,MAX} - R_{S\,MIN}\right) \times {}^{DN}\!/_{255} \tag{16.20}$$

Table 16.16 Various spectral indices

Indices	Equation	References	Eq. No.
Normalized difference vegetation index (NDVI)	$\frac{(NIR - RED)}{(NIR + RED)}$	Rouse et al. (1974)	16.29
Enhanced vegetation index (EVI)	$2.5\frac{(NIR - RED)}{(NIR + 6RED - 7.5BLUE) + 1}$	Hunt et al. (2011)	16.30
Normalized difference moisture index (NDMI)	$\frac{NIR - SWIR}{NIR + SWIR}$	Gao, (1996)	16.31
Modified normalized difference water index (MNDWI)	$\frac{(GREEN - SWIR)}{(GREEN + SWIR)}$	Xu (2006)	16.32

Here, NIR: Reflectance value in the near-infrared band; RED: Reflectance value in the red band; BLUE: Reflectance value in the blue band; SWIR: Reflectance value in the shortwave infrared band; and GREEN: Reflectance value in the green band

Here, R_S refers to the spectral radiance, $R_{S\,MIN}$ represents 1.238, $R_{S\,MAX}$ represents 15.600, and *DN* stands for digital number.

Step-II: The Kelvin temperature was determined based on the calculated R_S.

$$LST_A = \left[K_2 / \left\{\ln\left(K_1 / R_S + 1\right)\right\}\right] \quad (16.21)$$

Here, LST_A refers to the Land Surface Temperature, K_2 represents calibration constant 2 with a value of 1260.56, K_1 represents calibration constant 1 with a value of 607.76, and R_S stands for spectral radiance.

Step-III: Converted values from Kelvin to degree Celsius using the LST_A formula:

$$T(\text{degree celsius}) = T(\text{Kelvin}) - 273.15 \quad (16.22)$$

Where, T (Kelvin) = LST_A *OR* LST_B

MWA (Mono Window Algorithm) Model

Step-I: LST Calculation:

$$LST_B = T / 1 + w \times (T / p) \times \ln(e) \quad (16.23)$$

Where, LST_B represents the land surface temperature in kelvin, *T* represents the at satellite brightness temperature in kelvin, *w* is wavelength of emitted radiance (11.5 μm), e means land surface emissivity, and p = (h × c/s) means h = Plank's Constant (6.625×10^{-34} Js), s = Boltzmann Constant (1.38×10^{-23} J/K), c = Velocity of Light (2.998×10^{8} m/s), Therefore, p = 14,380.

Step-II: Other associate calculations 1:

$$T = K_2 / \ln\left[\left\{\left(K_1 / L\lambda\right) + 1\right\} - 273.15\right] \tag{16.24}$$

Where, T represents At satellite Brightness Temperature in degree Celsius, K_2 = Band-specific constant (calculate from metadata), K_1 = Band-specific constant (calculate from metadata), L_λ = TOA spectral Radiance (Watts/ (m^2*srad*μm)), and 273.15 = convert kelvin to degree Celsius.

$$\text{TOA spectral Radiance}\left(L\lambda\right) = M_L Q_{cal} + A_L \tag{16.25}$$

Where, M_L = Band-specific multiplicative rescaling factor (calculate from metadata), Q_{cal} = Digital Numbers (DN), and A_L = Band-specific additive rescaling factor (calculate from metadata).

Step-II: Other associate calculations 2:

$$\text{Land Surface Emissivity}\left(\text{e}\right) = 0.004\, P_V + 0.986 \tag{16.26}$$

Here, P_V means Proportion of Vegetation.

$$P_V = \left(NDVI - NDVI_{MIN}\right) / \left(NDVI_{MAX} - NDVI_{MIN}\right)^2 \tag{16.27}$$

Here, NDVI means Normalized Difference Vegetation Index.

$$NDVI = \left(NIR - RED\right) / \left(NIR + RED\right) \tag{16.28}$$

Here, NIR: Reflectance value in the near-infrared band; RED: Reflectance value in the red band.

Spectral Indices

The surface ecological conditions of the study area have been assessed using the spectral indices listed in Table 16.3.

16.2.5.3 Results, Discussion, and Conclusion

Spatio-Temporal Trend Analysis of Land Surface Temperature

The spatio-temporal analysis of land surface temperature (LST) in the coastal region of Bengal demonstrates a pronounced warming trend over the past three decades (Fig. 16.28). The findings (Table 16.17) indicate a consistent rise in both the minimum and maximum recorded temperatures over time.

The overarching trend underscores a continuous warming trajectory in the coastal region of Bengal, carrying significant implications for ecosystem transformations, heightened coastal vulnerability, and shifting local climate patterns. The findings align closely with broader global climate change trends, highlighting the urgency of

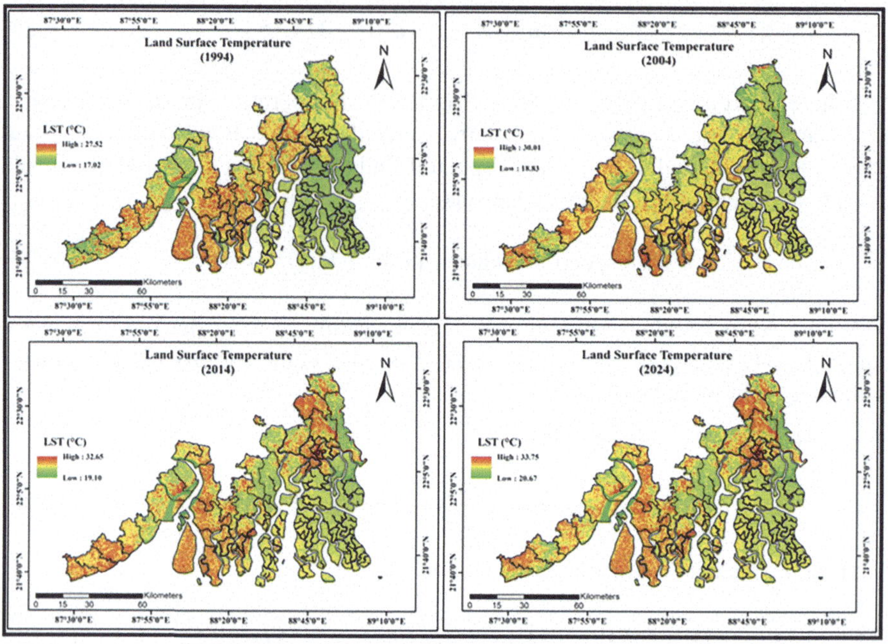

Fig. 16.28 Spatio-temporal distribution of LST in coastal region of Bengal during 1994, 2004, 2014, and 2024. (Prepared by the authors)

implementing sustainable coastal management, urban heat mitigation measures, and climate adaptation strategies in the region.

Spatio-Temporal Analysis of Vegetation and Water Indices

The spatio-temporal analysis of vegetation and water indices namely, NDVI (Fig. 16.29), EVI (Fig. 16.30), NDMI (Fig. 16.31), and MNDWI (Fig. 16.32) in the coastal region of Bengal from 1994 to 2024 reveals significant environmental transformations. These changes reflect patterns of vegetation loss, water stress, and coastal ecosystem alterations due to climatic and anthropogenic factors (Table 16.18).

The long-term analysis of NDVI, EVI, NDMI, and MNDWI in the coastal region of Bengal indicates a worsening ecological condition, marked by vegetation loss (declining NDVI and EVI), reduced soil moisture (declining NDMI), and decreasing surface water availability (dropping MNDWI). These changes are likely driven by climate change, rising temperatures, sea-level rise, land-use transformations, urban expansion, and hydrological alterations. To mitigate these adverse trends, implementing conservation measures, afforestation programs, and sustainable coastal management strategies is essential.

Relationship Between LST and Vegetation and Water Indices in the Coastal Region of Bengal

The relationship between Land Surface Temperature (LST) and vegetation and water indices, namely NDVI, EVI, NDMI, and MNDWI, offers valuable insights

Table 16.17 Spatio-temporal trend analysis of land surface temperature (LST) in the coastal region of Bengal (India)

Key observation	Details
Temporal trend (over the years)	From 1994 to 2024, there has been a gradual rise in both high and low LST values. The highest LST increased from 27.52 °C in 1994 to 33.75 °C in 2024, marking a rise of 6.23 °C over 30 years. Similarly, the lowest LST increased from 17.02 °C in 1994 to 20.67 °C in 2024, showing a rise of 3.65 °C over the same period. The rate of increase appears non-linear, with more pronounced warming after 2004.
Spatial implications (coastal influence)	Coastal regions are highly sensitive to climate variability, influenced by sea surface temperatures, monsoonal variations, land-use changes, and urbanization. The increasing LST suggests intensified warming due to climate change, possibly linked to reduced vegetative cover, built-up expansion, and changes in coastal dynamics. The rise in the low-temperature values signifies a decline in nighttime cooling, which may indicate changes in humidity levels, increased heat retention, and changing wind patterns.
Decadal comparisons and climatic indications	The 1994–2004 decade saw a moderate rise in high LST (2.49 °C) and low LST (1.81 °C). The 2004–2014 decade recorded a sharper increase in high LST (2.64 °C) but a relatively smaller rise in low LST (0.27 °C). The 2014–2024 decade shows a continued warming trend, though the rate of high LST rise (1.10 °C) slightly slowed, while low LST increased significantly (1.57 °C), indicating potential changes in atmospheric moisture retention and cloud cover patterns.

Source: Fig. 16.28

into environmental transformations in the coastal region of Bengal. The coefficient of determination (R (Adeleke et al., 2021)) values highlights the extent to which LST is correlated with these indices over time (1994 to 2024), reflecting dynamic changes in vegetation cover, moisture availability, and surface water distribution (Table 16.19).

Implications of Findings

- **Weakening Role of Vegetation in Temperature Regulation:** The correlations between NDVI and EVI with LST have nearly disappeared in 2024, suggesting that vegetation no longer plays a significant role in regulating surface temperature. This decline in correlation points to widespread deforestation, degradation, and urban expansion in the coastal region, where vegetation cover has been substantially reduced, diminishing its influence on local temperature regulation (Gloria & Gnanasekaran, 2024; Salam et al., 2024; Putri et al., 2024).
- **Moisture and Water Indices Show a Stronger Relationship with LST:** The strong correlation between LST and NDMI/MNDWI suggests that temperature variations are increasingly influencing water availability and soil moisture content. This relationship highlights evident drying trends, rising heat stress, and the

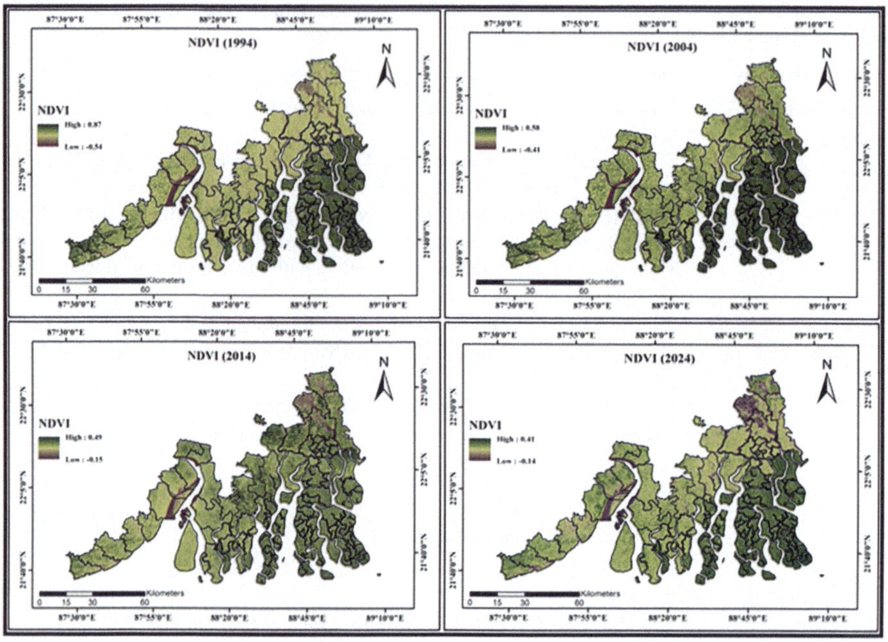

Fig. 16.29 Spatio-temporal distribution of NDVI in coastal region of Bengal during 1994, 2004, 2014, and 2024. (Prepared by the authors)

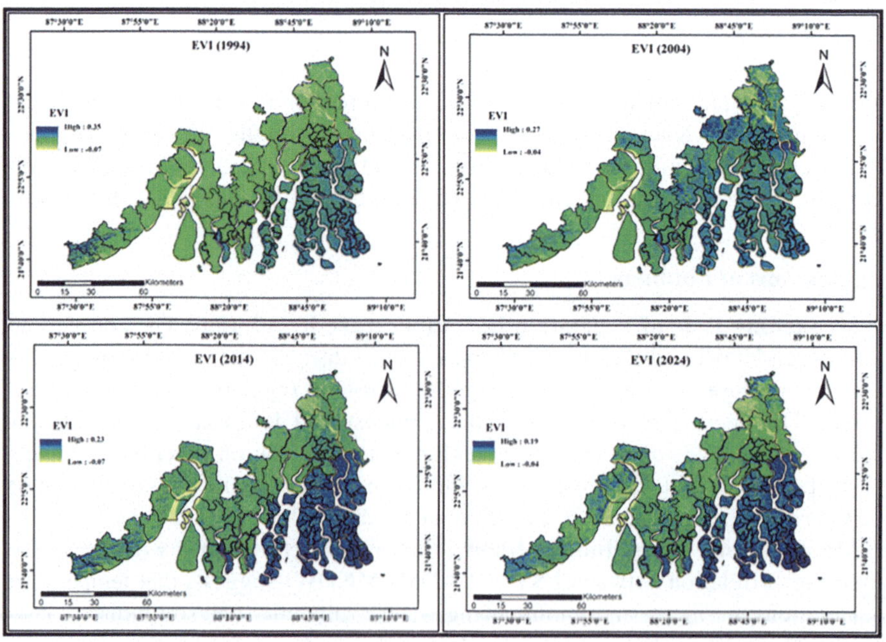

Fig. 16.30 Spatio-temporal distribution of EVI in coastal region of Bengal during 1994, 2004, 2014, and 2024. (Prepared by the authors)

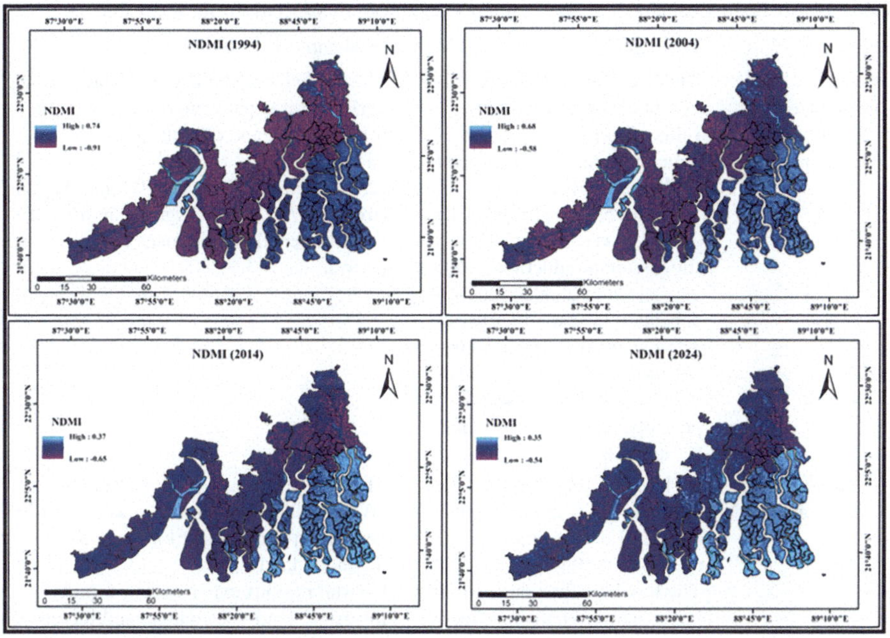

Fig. 16.31 Spatio-temporal distribution of NDMI in coastal region of Bengal during 1994, 2004, 2014, and 2024. (Prepared by the authors)

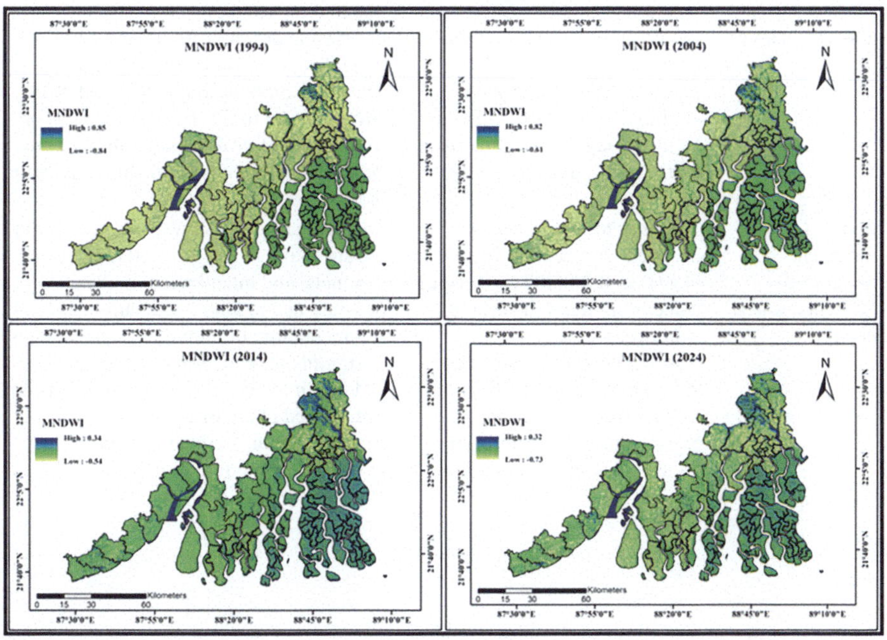

Fig. 16.32 Spatio-temporal distribution of MNDWI in coastal region of Bengal during 1994, 2004, 2014, and 2024. (Prepared by the authors)

Table 16.18 Ecological changes: spatio-temporal analysis of vegetation and water indices

Indices	Trend analysis	Implications
NDVI (normalized difference vegetation index)	NDVI values show a declining trend in both high and low values from 1994 to 2024, indicating decreasing vegetation health and cover. The high NDVI dropped from 0.87 (1994) to 0.41 (2024), suggesting a reduction in dense vegetation or conversion of vegetated land to other uses. The low NDVI increased from −0.54 to −0.14, indicating a reduction in barren land or possibly expansion of human settlements.	The decline may be due to deforestation, agricultural expansion, coastal erosion, urbanization, and climate-induced changes. The drastic drop between 1994 and 2004 (from 0.87 to 0.58) suggests significant environmental disturbances during this period.
EVI (enhanced vegetation index)	EVI values also show a continuous decline, with the high EVI decreasing from 0.35 (1994) to 0.19 (2024). The low EVI fluctuates slightly but remains negative, reflecting persistently poor vegetation health in certain areas.	The EVI decline, in conjunction with NDVI reduction, suggests loss of healthy vegetation and a transition towards degraded landscapes. Coastal regions experiencing salinity intrusion, land-use change, and increased human activities may be contributing to this trend.
NDMI (normalized difference moisture index)	The high NDMI decreased from 0.74 (1994) to 0.35 (2024), indicating a significant reduction in vegetation moisture content. The low NDMI improved slightly from −0.91 (1994) to −0.54 (2024), suggesting that extreme dry areas have slightly recovered.	The overall decline points to reduced water availability for vegetation, likely due to climate change impacts, decreasing groundwater levels, or increasing land surface temperatures. The steep drop between 2004 and 2014 (from 0.68 to 0.37) suggests an event or trend (such as prolonged droughts, land degradation, or coastal erosion) leading to drastic moisture reduction.
MNDWI (modified normalized difference water index)	The high MNDWI values dropped significantly from 0.85 (1994) to 0.32 (2024), showing a declining trend in surface water presence. The low MNDWI fluctuated, with a notable dip in 2024 (−0.73), suggesting an expansion of dry or built-up areas replacing water bodies.	The decrease in MNDWI values suggests the shrinking of coastal wetlands, river channels, and inland water bodies, possibly due to land reclamation, sedimentation, or climate-driven changes. A significant reduction between 2004 and 2014 (from 0.82 to 0.34) may indicate an accelerated phase of coastal transformation, wetland loss, or hydrological modifications.

Source: Figs. 16.29, 16.30, 16.31 and 16.32

Table 16.19 Relationship between LST and vegetation and water indices in the coastal region of Bengal

Relationship	Trend analysis	Implications
LST and NDVI	The correlation between LST and NDVI is weak and varies over time, showing fluctuations across the study period (Fig. 16.33). In 1994, the correlation was 0.1275, increasing to a peak of 0.1713 in 2004. However, it declined to 0.1195 in 2014 and dropped drastically to 0.0033 in 2024, indicating a near-total loss of association between land surface temperature and vegetation cover.	Initially, there was a moderate negative correlation, indicating that higher temperatures were associated with lower vegetation cover. The sharp decline in 2024 (0.0033) suggests that NDVI is no longer strongly linked to LST, possibly due to urban expansion, loss of dense vegetation, or changes in land-use patterns. This could indicate a shift towards built-up areas, degraded land, or non-vegetated surfaces, which do not respond to temperature variations in the same way as healthy vegetation.
LST and EVI	The correlation between LST and EVI follows a pattern similar to that of NDVI but remains generally weaker (Fig. 16.34). In 1994, the correlation stood at 0.1652, reaching its peak at 0.2165 in 2004. However, it declined to 0.1355 in 2014 and dropped sharply to 8E-05 in 2024, indicating an almost negligible relationship between LST and enhanced vegetation index in recent years.	The initial moderate correlation in 2004 suggests that vegetation productivity played a role in regulating temperature. However, the drastic drop to near zero in 2024 suggests that EVI is no longer a key factor influencing LST. This could indicate loss of dense vegetation, increased urbanization, or shifts in coastal ecosystem composition. The disappearance of correlation implies that land surface heating is now being driven by factors other than vegetation health, such as soil exposure, built-up surfaces, and changing coastal processes.

(continued)

Table 16.19 (continued)

Relationship	Trend analysis	Implications
LST and NDMI	The correlation between LST and NDMI remains strong but fluctuates over time (Fig. 16.35). In 1994, the correlation was 0.6032, increasing slightly to 0.6117 in 2004 and reaching its peak at 0.6519 in 2014. However, by 2024, it had declined to 0.4868, suggesting a weakening correlation between land surface temperature and moisture availability, likely driven by increasing aridity and shifting hydrological conditions.	The peak in 2014 (0.6519) suggests that moisture content in vegetation was strongly influenced by temperature at that time. The drop in 2024 (0.4868) indicates that this relationship is weakening, possibly due to increased land dryness, declining soil moisture, and reduced vegetation cover. This could be linked to higher evaporation rates, declining groundwater levels, or shifts in rainfall patterns in the region.
LST and MNDWI	The correlation between LST and MNDWI remains significant and shows an increasing trend over time (Fig. 16.36). In 1994, the correlation was 0.5112, slightly decreasing to 0.487 in 2004. However, it rose to 0.6425 in 2014 before declining to 0.5524 in 2024, indicating a sustained but variable relationship between LST and surface water availability, likely influenced by changing hydrological dynamics and increasing thermal stress on water bodies.	The rising correlation between LST and MNDWI suggests that temperature changes are increasingly affecting surface water availability. The strong correlation in 2014 and 2024 indicates that as LST rises, wetland and water body coverage decreases. This may be due to higher evaporation rates, reduced precipitation, and increased salinity intrusion in coastal water bodies. The slight drop from 2014 to 2024 suggests possible interventions such as water conservation efforts or changing coastal hydrodynamics that have temporarily stabilized the trend.

Source: Figs. 16.33, 16.34, 16.35 and 16.36

decline of wetland areas, underscoring the growing impact of temperature fluctuations on coastal hydrological conditions (Beninato et al., 2024; Pandey et al., 2023; Pablos et al. 2016).

- **Coastal Climate Vulnerability Is Increasing:** The strengthening relationship between LST and water stress indicates escalating climate change impacts, including higher evaporation rates, reduced groundwater recharge, and increased saline water intrusion. These changes pose significant risks, leading to agricultural losses, freshwater scarcity, and heightened vulnerability of coastal ecosystems (Debnath & Alamdari, 2025; Guild et al., 2024; Chowdhury et al., 2024).

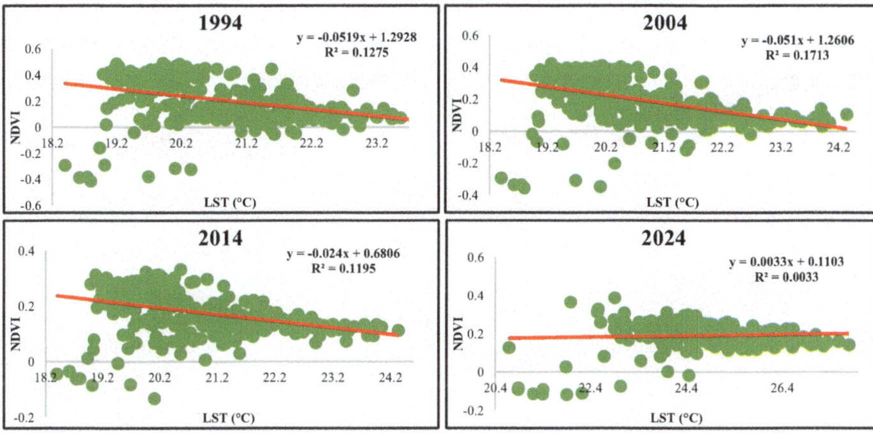

Fig. 16.33 Relationship between LST and NDVI (Prepared by the authors)

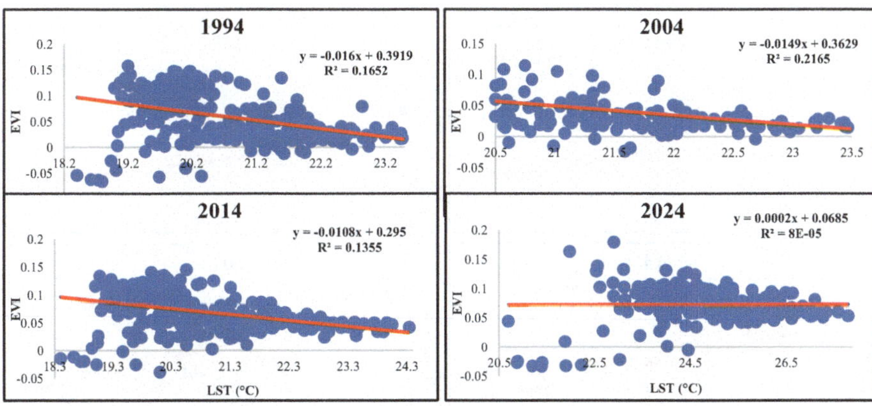

Fig. 16.34 Relationship between LST and EVI (Prepared by the authors)

- **Need for Sustainable Land and Water Management:** Restoring vegetation cover through afforestation and agroforestry could help re-establish the relationship between LST and vegetation, while improved water conservation and wetland protection are crucial for mitigating heat stress and maintaining ecological balance. Additionally, sustainable urban planning should prioritize reducing impervious surfaces and integrating green infrastructure to minimize heat buildup and enhance environmental resilience (Zouari & Hachicha, 2024; Veeravalli, 2024; Van Stan et al., 2018; Grant et al., 2013).

In conclusion, the coastal region of Bengal is experiencing a significant transformation, with the diminishing role of vegetation in temperature regulation and the increasing prominence of moisture stress and water body shrinkage as primary climate concerns. Addressing these challenges requires urgent interventions in climate adaptation, land-use planning, and water resource management to mitigate the intensifying impacts of rising land surface temperatures and ecosystem degradation.

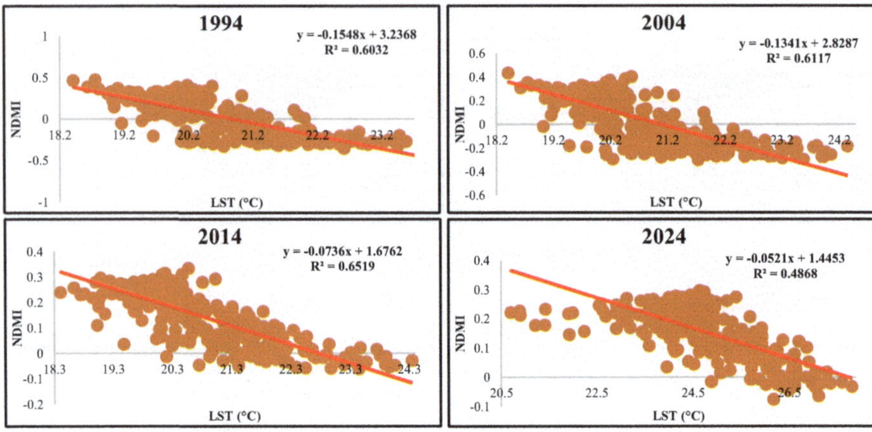

Fig. 16.35 Relationship between LST and NDMI (Prepared by the authors)

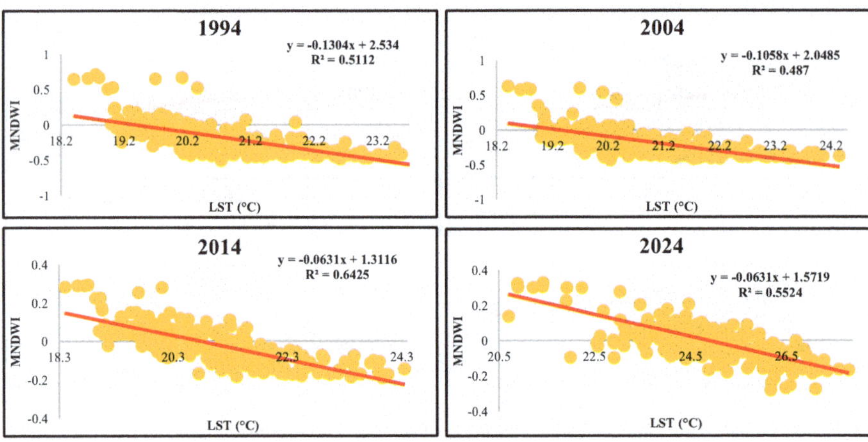

Fig. 16.36 Relationship between LST and MNDWI (Prepared by the authors)

16.2.6 Casebook 6: Management of Mangrove-Human Conflicts in the Indian Sundarbans, Using Participatory GIS Techniques

16.2.6.1 Background and Purpose

The Indian Sundarbans (Fig. 16.37), designated as a UNESCO World Heritage Site, represent a unique yet ecologically vulnerable region where mangrove ecosystems and human settlements coexist (UNESCO, 1984). However, this coexistence has increasingly given rise to conflicts between conservation imperatives and the livelihood needs of local communities, particularly traditional marine fishermen (Malakar et al., 2023; Upadhyay et al., 2002). The intensification of these conflicts is driven by rapid land-use transformations, coastal erosion, climate-induced stressors, and

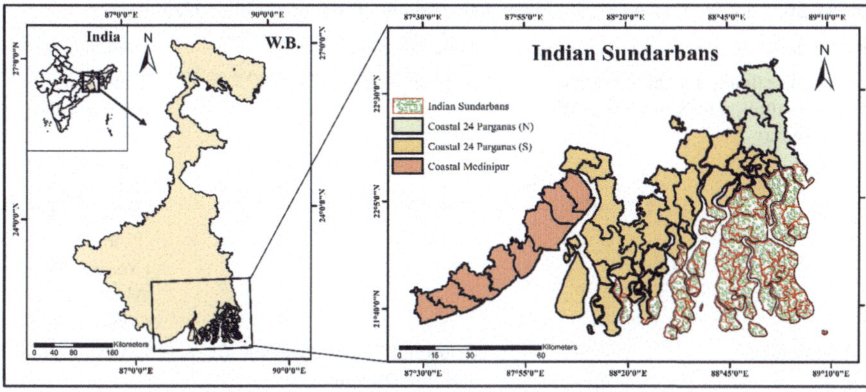

Fig. 16.37 Location and extent of the study area: The Indian Sundarbans (Prepared by the authors)

economic pressures, underscoring the necessity for an integrated management approach. Mangroves play a crucial role as natural buffers against coastal hazards, sustaining biodiversity and providing essential resources for local populations. However, their degradation due to encroachment, deforestation, and overexploitation poses significant threats to both environmental sustainability and human well-being (Saoum & Sarkar, 2024; Sunkur et al., 2023; Mitra, 2013). This case study aims to assess the potential of Participatory GIS (PGIS) techniques in managing mangrove-human conflicts in the Indian Sundarbans.

PGIS combines geospatial technologies with community-driven knowledge, fostering a bottom-up approach to environmental management. By actively involving local stakeholders in mapping resource distribution, identifying conflict zones, and assessing vulnerabilities, this study seeks to develop inclusive, data-driven strategies for sustainable mangrove conservation (Dhyani et al., 2023; Damastuti & de Groot, 2019; DasGupta & Shaw, 2017; Adeel & Pomeroy, 2002). The findings will inform policy recommendations, advocating for collaborative governance, environmentally sustainable livelihood alternatives, and adaptive land-use planning. Ultimately, this research aims to harmonize ecological conservation with socio-economic development, fostering the long-term resilience of both mangrove ecosystems and human communities in the Sundarbans.

16.2.6.2 Data and Methods

This case study utilizes primary field survey data from the Indian Sundarbans and integrates participatory methodologies, including GIS-based approaches. Table 16.20 outlines the comprehensive participatory methods employed to document and analyse field observations in detail. In additions, the present study employed qualitative approaches to explore the conflicts and management opportunities within the study area.

Table 16.20 Participatory methodologies: management of mangrove-human conflicts

Survey sites	Time duration	Participants[a]
Coastal Hingalganj, Gosaba, Pathar Pratima, Kultali, and Namkhana (West Bengal, India)	August to September 2023	Total respondents: 25. Age groups: Below 25 (Years): 11 25 to 35 (Years): 08 Above 35 (Years): 06
Methodologies	Households' conversation, focus group discussion, and thematic analysis; Application of Google map location and Google earth (for data collection and location identification);	

Note: Field data collected by the author (S. Roy)
[a]Unwilling to disclose the respondent's name

16.2.6.3 Results, Discussion, and Conclusion

Mangrove-Human Conflicts

According to the collective perspectives of the respondents, the interaction between mangrove ecosystems and human activities in the Indian Sundarbans has resulted in multiple conflicts, largely driven by resource dependency and environmental degradation. The key conflicts identified in the study area are as follows:

- **Livelihood Dependency:** Traditional fishermen and local communities rely on mangroves for fishing, honey collection, and wood harvesting. However, unsustainable resource extraction leads to ecological degradation, creating conflicts between livelihood security and conservation efforts.
- **Fuelwood Extraction:** Heavy dependence on mangrove wood for cooking and heating contributes to deforestation, weakening natural coastal defences and accelerating habitat loss.
- **Traditional Medicine Harvesting:** Many mangrove species are utilized for medicinal purposes, but excessive and unregulated harvesting threatens biodiversity and depletes local healthcare resources.
- **Use of Mangroves for Housing:** Mangrove timber is extensively used for construction, increasing deforestation and reducing the region's resilience to coastal hazards such as storms and tidal surges.
- **Impact of Shrimp Farming:** The expansion of shrimp aquaculture often involves clearing mangrove forests, leading to habitat destruction, biodiversity loss, and saline water intrusion in agricultural lands.
- **Agricultural Encroachment:** The conversion of mangrove areas into farmland disrupts the hydrological balance, depletes soil fertility, and increases erosion risks, further threatening the ecosystem.

- **Settlement Expansion:** Population growth and infrastructure development continue to encroach upon mangrove forests, intensifying conflicts between conservation objectives and human settlement needs.
- **Unregulated Tourism:** Rapid and poorly managed tourism contributes to pollution, waste accumulation, and the disturbance of sensitive mangrove habitats, further straining ecological stability.
- **Climate-Induced Challenges:** Rising sea levels, increased coastal erosion, and extreme weather events accelerate mangrove degradation, exacerbating conflicts between conservation priorities and human survival needs.

Managing Conflicts Through Participatory Approaches:
A sustainable resolution to mangrove-human conflicts requires active community engagement through participatory approaches, fostering a balance between ecological conservation and livelihood security. The following management strategies have been developed in consultation with the local community in the study area.

- **Community-Based Resource Management:** Actively involving local communities in resource monitoring and controlled harvesting practices fosters a balance between conservation efforts and economic sustenance (Viswanathan, 2016; On-prom, 2014).
- **Sustainable Fuel Alternatives:** Promoting the use of alternative energy sources, including biogas, solar cooking, and afforestation programs, can help decrease dependence on mangrove wood and curb deforestation (Bekere & Megerssa, 2020; Kuhe et al., 2017).
- **Ethnobotanical Conservation:** Promoting the cultivation of medicinal plants outside mangrove ecosystems helps preserve biodiversity while sustaining traditional medicine practices (Patil et al., 2024; Alum, 2024).
- **Eco-friendly Housing Solutions:** Advocating for sustainable construction materials and policies that discourage the use of mangrove timber can help mitigate habitat destruction (Anggraeni, 2024; Ashton, 2022).
- **Regulated Shrimp Farming:** Implementing environmentally responsible aquaculture practices, such as integrated mangrove-shrimp farming, minimizes deforestation while ensuring economic viability (Hieu et al., 2025; Widigdo et al., 2024; Anh et al., 2024).
- **Agroforestry and Sustainable Agriculture:** Integrating mangroves into agricultural systems through agroforestry can enhance soil health, reduce erosion, and promote sustainable food production (Sekhar et al., 2024; Verma et al., 2024; VijayKumar et al., 2024).
- **Planned Settlement Expansion:** Establishing land-use regulations and designated conservation zones can help manage human settlements while ensuring mangrove protection (Anggraeni, 2024; Prasita et al., 2016).
- **Sustainable Tourism Initiatives:** Enforcing eco-tourism policies, waste management strategies, and controlled visitor access can reduce environmental disturbances and support conservation efforts (Nasution et al., 2025; Daneshwar & Revaty, 2024).

- **Climate Adaptation Strategies:** Strengthening coastal defences, restoring degraded mangrove ecosystems, and integrating traditional knowledge with scientific approaches can enhance resilience to erosion and sea-level rise (Bastien-Olvera et al., 2024; Marquez & Olavides, 2024; Kospa, 2024).

In conclusion, achieving a sustainable balance between mangrove conservation and human livelihoods in the Sundarbans requires community-driven, participatory approaches. Effective conflict mitigation depends on stakeholder engagement, sustainable resource management, and climate adaptation, ensuring both ecological resilience and local well-being.

16.2.7 Casebook 7: Adaptation Strategies of Traditional Fishermen to Climate Change: A Case Study of Coastal Bengal

16.2.7.1 Background and Purpose

The coastal region of Bengal (Fig. 16.38), home to a substantial population of traditional marine fishermen, faces escalating vulnerabilities due to climate change-induced challenges such as extreme weather events, rising sea levels, coastal erosion, and shifting fishery resources (Debnath et al., 2024; Das & Swain, 2024;

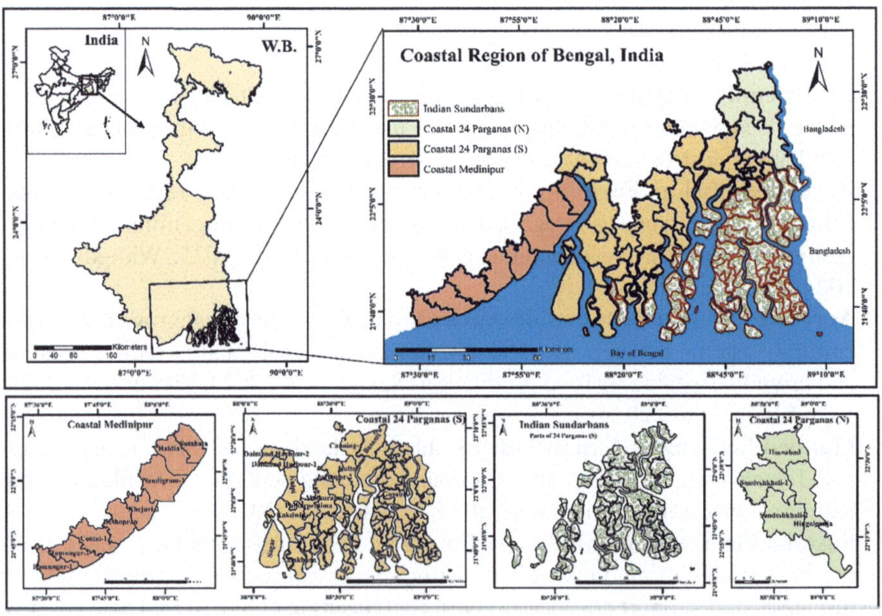

Fig. 16.38 Coastal Bengal, India: study area overview (Prepared by the authors)

Bera et al., 2021). These environmental stressors pose significant threats to both the livelihoods and cultural heritage of fishing communities, which have long depended on marine and estuarine ecosystems for sustenance (Sarkar et al., 2024; Etongo & Arrisol, 2021). The interplay between climate change and socio-economic vulnerabilities underscores the need for an urgent assessment of the adaptive strategies employed by traditional fishermen to navigate these evolving challenges (Handayati et al., 2025; Kasri et al., 2024; Pattanaik, 2021).

This case study examines the climatic challenges and adaptive responses of traditional fishermen in coastal Bengal, emphasizing their indigenous knowledge, resource management strategies, and socio-economic adjustments. Through field-based participatory approaches, the study explores how fishermen modify their fishing techniques, diversify income sources, and integrate traditional wisdom with modern technologies to sustain their livelihoods. By documenting these adaptation strategies, the study enhances understanding of grassroots-level responses to climate change and provides valuable insights for policy recommendations. Ultimately, the findings seek to inform climate adaptation planning and coastal resource governance, ensuring that traditional fishing communities can navigate environmental uncertainties while preserving their cultural identity and economic resilience.

16.2.7.2 Data and Methods

This case study utilized climate data from the Climatic Research Unit (available at: https://crudata.uea.ac.uk/cru/data/hrg/cru_ts_4.07) to assess the intensity of climate change from 1901 to 2022. Additionally, sea-level data were obtained from the Permanent Service for Mean Sea Level (PSMSL) at the National Oceanographic Centre (available at: https://psmsl.org/data), alongside primary survey data from 480 fishing households (using simple random sampling). The study examined fishermen's climate perceptions, narratives, local livelihoods, and adaptation strategies. Climate data were analysed using the Mann-Kendall (MK) test (Casebook 1 | Eq. 16.4) to identify long-term trends, while sea-level rise patterns were assessed through scatter plot visualizations. The quantitative and qualitative aspects, including fishermen's perceptions, narratives, and adaptation strategies, were examined using a mixed-method approach that incorporated participatory insights. Furthermore, fishermen's perspectives on key climate variables and observable climate change impacts in coastal Bengal were analysed using the Garrett ranking technique (Eq. 16.33) (Garrett, 1979), providing a structured understanding of local climate challenges and adaptive responses.

$$\text{Percent position} = \frac{100\left(R_{ij} - 0.5\right)}{N_j} \tag{16.33}$$

Here, R_{ij} is the rank given for the ith ith factors by the jth respondents, and N_j is the number of factors ranked by the jth respondents.

16.2.7.3 Results, Discussion, and Conclusion

Climate Change: Examining Scientific Data

The Mann-Kendall trend analysis for annual climatic variables reveals significant patterns in temperature and rainfall across three coastal districts of Bengal: East Medinipur, South 24 Parganas, and North 24 Parganas. The R^2 values indicate the strength of trends observed in average annual temperature and mean annual rainfall over time (Fig. 16.39). The analysis shows that temperature trends are more pronounced across all three districts, with East Medinipur ($R^2 = 0.3815$) exhibiting the strongest increasing trend, followed by South 24 Parganas ($R^2 = 0.37$) and North 24 Parganas ($R^2 = 0.3524$). These values suggest a consistent rise in temperature, aligning with broader global warming patterns and increasing heat stress in the region. In contrast, rainfall trends are relatively weak and inconsistent. East Medinipur ($R^2 = 0.0199$) shows a slightly stronger positive trend, indicating a marginal increase in annual rainfall, whereas South 24 Parganas ($R^2 = 0.0092$) and North 24 Parganas ($R^2 = 0.0011$) exhibit even weaker trends, suggesting rainfall variability with no clear significant increase. The low R^2 values for rainfall indicate high inter-annual fluctuations, which could pose challenges for water resource management,

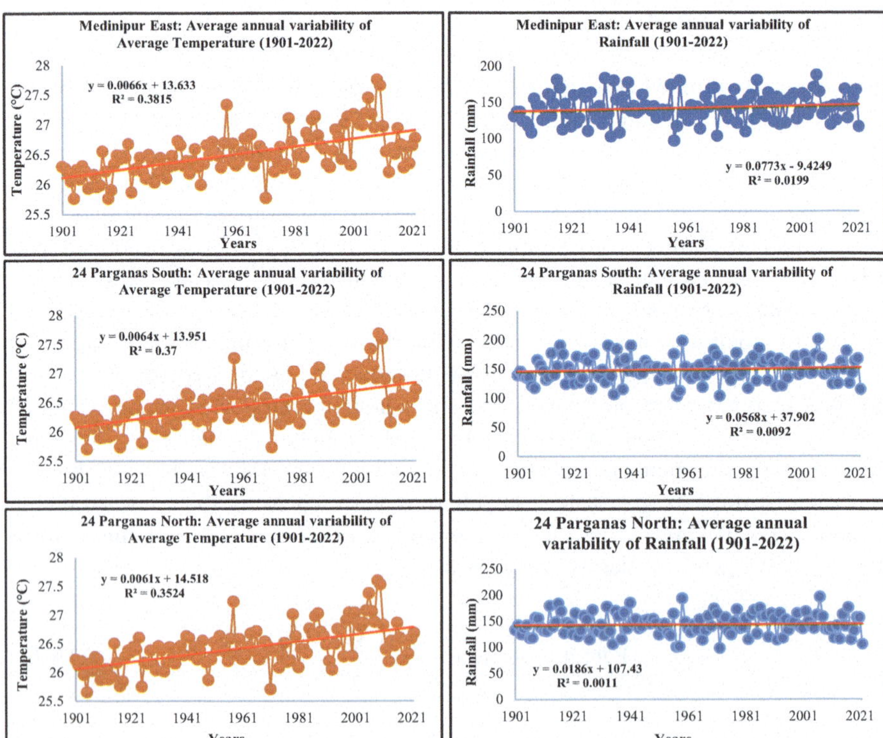

Fig. 16.39 Climate trend analysis in coastal region of Bengal (Prepared by the authors)

agriculture, and local livelihoods. Overall, the findings highlight a notable warming trend across all districts, while rainfall patterns remain uncertain and less predictable, emphasizing the need for adaptive climate strategies, particularly in water resource planning and heat stress mitigation in coastal Bengal.

Climate Change Evidence: Examining Sea-Level Data

The trend analysis of sea-level rise data, based on R^2 values, highlights varying degrees of sea-level changes across three coastal locations in Bengal: Gangra, Haldia, and Diamond Harbour. The R^2 values indicate the strength of the observed trends over time, with higher values suggesting more pronounced and consistent changes (Fig. 16.40). Among the three locations, Diamond Harbour ($R^2 = 0.7069$) exhibits the strongest increasing trend, indicating a significant and consistent rise in sea level. This suggests that Diamond Harbour is experiencing the most severe impacts of sea-level rise, likely due to factors such as coastal subsidence, sediment deposition, and climate-induced oceanic expansion. Haldia ($R^2 = 0.5573$) also shows a strong positive trend, reflecting a substantial rise in sea level, though slightly less pronounced than in Diamond Harbour. As a major port and industrial hub, rising sea levels in Haldia may pose risks to coastal infrastructure, port operations, and surrounding settlements. In contrast, Gangra ($R^2 = 0.1078$) has the weakest trend, indicating a relatively lower rate of sea-level rise or greater variability in data trends over time. This could suggest that localized geological or hydrodynamic factors influence sea-level changes in this area, leading to less consistent patterns. Overall, the analysis suggests that sea-level rise is a pressing concern in the coastal Bengal region, with Diamond Harbour and Haldia experiencing more rapid and pronounced changes compared to Gangra. These trends highlight the urgent need for coastal adaptation strategies, including shoreline protection, sustainable

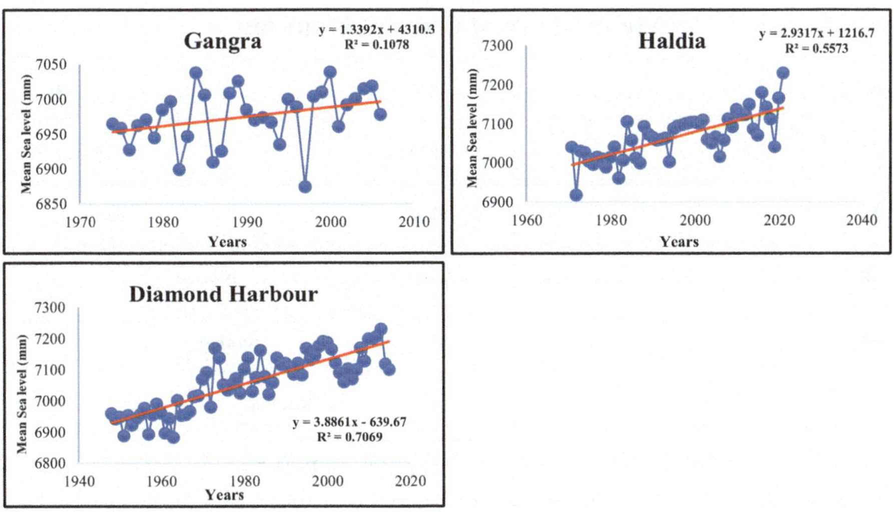

Fig. 16.40 Changes in sea level at Gangra, Haldia, and Diamond Harbour. (Prepared by the authors)

land-use planning, and climate-resilient infrastructure to mitigate the impacts of rising sea levels.

Climate Change Evidence: Fishermen's Perceptions

The Garrett ranking analysis of fishermen's perceptions regarding climate change in coastal Bengal provides critical insights into the environmental variables that they consider most impactful. The Garrett Mean Scores (Table 16.21) and Garrett Ranks (Fig. 16.41) highlight the relative significance of different climate parameters based on field survey responses.

Key Findings from Fishermen's Perspectives

- **Air Temperature (AR) Increase (Rank 1, Score: 70.17):** Fishermen perceive rising air temperature as the most significant indicator of climate change. Higher temperatures affect fish availability, fishing conditions, and overall marine ecosystem health, making it the most concerning factor.

Table 16.21 Fishermen's (TMFC) perspectives on key variables/ parameters and evidence of climate change in the coastal region of Bengal

Variables/parameters	AR	RF	CS	WS	SR	SST	SSS
Garrett mean score	70.17	62.95	66.88	31.98	43.73	40.27	31.02
Garrett rank	1	3	2	6	4	5	7

Note: AR: Air temperature (increase); RF: Rainfall; CS: Frequency of cyclones (increase); WS: Wind speed (change); SR: Sea-level rise (increase); SST: Sea surface temperature (increase); and SSS: Sea surface salinity (increase)

Data source: TMFC perception of field survey (N: 480), 2022 (by the researcher)

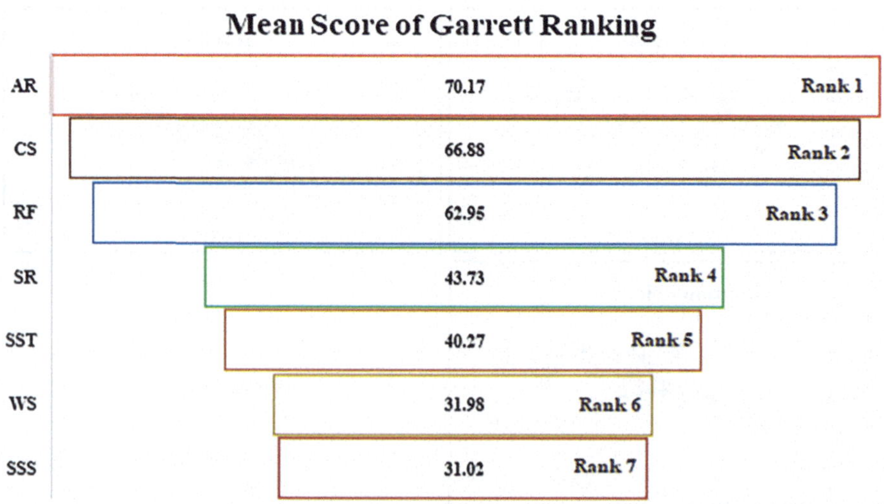

Fig. 16.41 Plotted the Garrett ranking scores for key variables/ parameters and evidence of climate change as identified by fishermen. (Prepared by the authors)

- **Cyclone Frequency (CS) Increase (Rank 2, Score: 66.88):** The increasing frequency of cyclones is a major threat, disrupting fishing activities, damaging boats and settlements, and endangering livelihoods. Cyclonic storms have become more intense and frequent, heightening the vulnerability of fishing communities.
- **Rainfall (RF) Variability (Rank 3, Score: 62.95):** Fishermen consider shifting rainfall patterns, whether heavy rains or extended dry periods, a major concern. Irregular rainfall disrupts coastal and estuarine ecosystems, affecting fish breeding and availability.
- **Sea-Level Rise (SR) (Rank 4, Score: 43.73):** Perceived as a growing but moderate threat, rising sea levels contribute to coastal erosion, saltwater intrusion, and loss of fishing grounds, gradually affecting local livelihoods.
- **Sea Surface Temperature (SST) Increase (Rank 5, Score: 40.27):** Warmer Sea surface temperatures alter fish migration patterns and reduce fish stocks, making it harder for fishermen to sustain their traditional practices.
- **Wind Speed (WS) Changes (Rank 6, Score: 31.98):** Although shifting wind patterns affect navigation and fishing conditions, fishermen consider this a less immediate concern compared to other climate-related impacts.
- **Sea Surface Salinity (SSS) Increase (Rank 7, Score: 31.02):** The increase in sea surface salinity due to reduced freshwater inflow and rising evaporation rates is the least prioritized issue, though it still impacts estuarine fish species and overall marine biodiversity.

The analysis underscores that rising air temperature, increasing cyclonic activity, and changing rainfall patterns are the most pressing climate-related challenges for traditional fishermen in coastal Bengal. These perceptions align with broader climatic trends, emphasizing the urgent need for adaptive strategies, resilient infrastructure, and climate-responsive fisheries management to safeguard livelihoods and marine ecosystems.

Climate Change Evidence: Fishermen's Field Narratives

Traditional marine fishermen (three respondents) in coastal West Bengal have historically depended on their extensive knowledge of the sea, weather patterns, and fish behaviour. However, climate change has significantly disrupted these established patterns, impacting their fishing practices and livelihoods. Fishermen have observed shifts in monsoon cycles, irregular rainfall, rising sea levels, and increasing sea surface temperatures, all of which influence fish availability. Many report declining fish stocks and unpredictable catch seasons, necessitating adaptations such as altering fishing zones, modifying gear, or seeking alternative livelihoods. Additionally, cyclones and coastal erosion pose severe threats, causing damage to boats, nets, and homes. Despite these adversities, fishermen continue to integrate traditional ecological knowledge with modern technology, utilizing ancestral wisdom to predict fish movements while incorporating GPS and advanced fishing equipment. Their narratives underscore both resilience and deep concerns regarding the long-term sustainability of their livelihoods in the face of persistent environmental changes. [Respondents: A. Khatun (Age: 28 Years); P. Roy (Age: 46 Years); and R. Jana (Age: 51+ Years). | N.B.: Name changed. | Data collection: August to September 2023 by the author of S. Roy.]

Climate Impacts and Traditional Fishermen: Lives and Livelihoods
The respondents faced various climate-related challenges, with cyclones severely impacting 83.75% of households and floods affecting 59.17%. Riverbank erosion (42.29%) and lowland inundation (56.67%) also posed risks. Salinity intrusion, land degradation, and mangrove loss further threatened livelihoods. Crop production declined for 50% of households, while fish and honey production fell for 82.71%. Additionally, 67.08% reported increased livestock diseases, highlighting the widespread impact of environmental changes (Fig. 16.42).

Sea Surface Temperature
Figure 16.43 illustrates fishermen's ecological perceptions of sea surface temperature (SST) and fish abundance in coastal Bengal. When SST is high, 67.8% of fishermen report a decline in fish availability, while only 4.16% observe an increase.

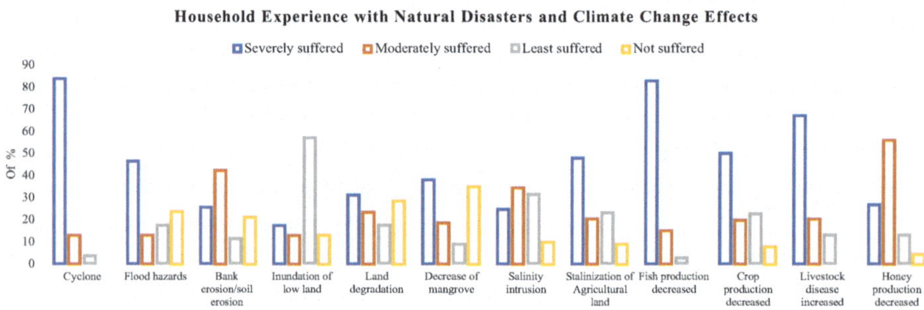

Fig. 16.42 Household experience with natural disasters and climate change effects. (Prepared by the authors)

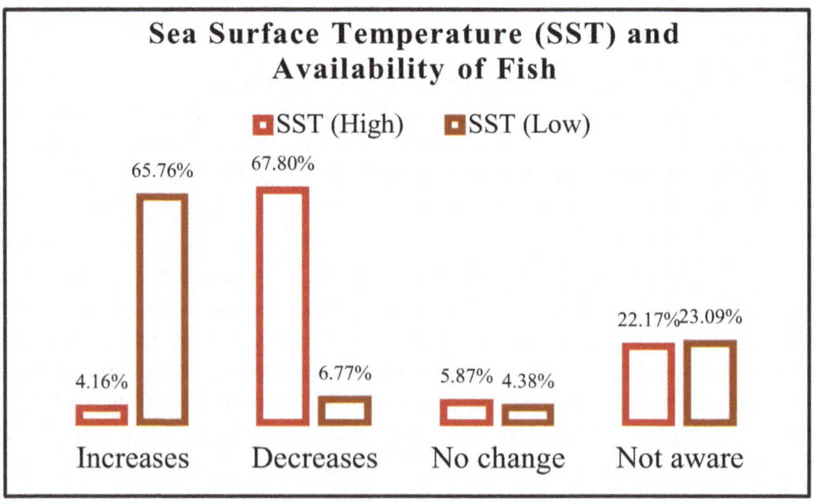

Fig. 16.43 Ecological perceptions of fishermen on sea surface temperature (SST) and availability of fish (Prepared by the authors)

Around 5.87% notice no change, and 22.17% are unaware of SST's impact. Most fishermen associate rising SST with reduced fish populations, reflecting concerns over warming oceans. Conversely, 65.76% of fishermen link low SST to increased fish abundance, while 6.77% believe it decreases availability. Around 4.38% see no change, and 23.09% are unaware of SST's influence. These perceptions align with broader ecological findings, emphasizing SST fluctuations' impact on fish populations and the coastal fishing sector.

Impact of Rainfall

Figure 16.44 illustrates fishermen's perceptions of rainfall's impact on fish abundance in coastal Bengal. When rainfall is high, 66.8% of fishermen observe an increase in fish availability, attributing it to enhanced nutrient runoff and improved fish habitats. Around 11.02% report a decline, 1.89% see no change, and 20.29% are unaware of its effects. In contrast, 57.9% of fishermen associate low rainfall with declining fish populations, likely due to reduced nutrient runoff and lower water levels. Only 14.87% believe fish abundance increases with less rainfall, while 7.68% see no change, and 19.55% are unaware of its impact. These findings underscore fishermen's ecological perceptions and highlight the potential consequences of changing rainfall patterns on coastal fisheries.

Sea-Surface Salinity

Figure 16.45 illustrates fishermen's perceptions of sea-surface salinity (SSS) and fish abundance in coastal Bengal. When SSS is high, 34.18% observe a decline in fish availability, linking excessive salinity to unfavourable marine conditions. Only 10.22% see no change, while 43.19% are unaware of its effects, highlighting a knowledge gap in understanding salinity's impact on fisheries. Conversely, 30.44%

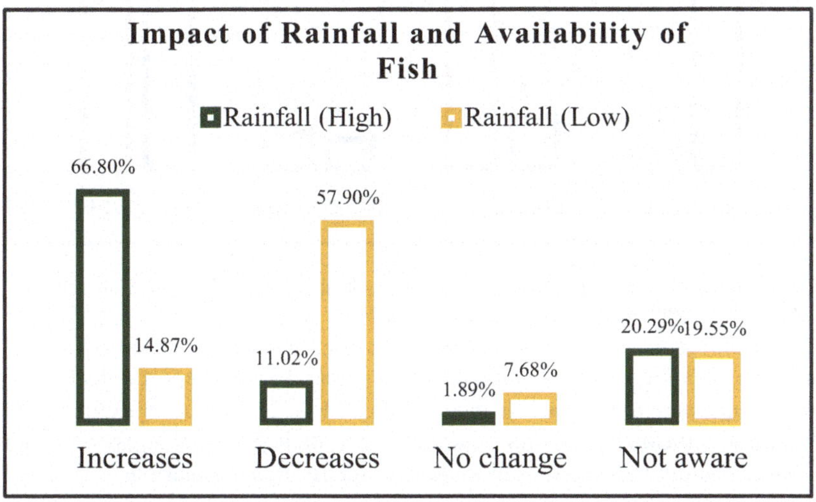

Fig. 16.44 Ecological perceptions of fishermen on impact of rainfall and availability of fish (Prepared by the authors)

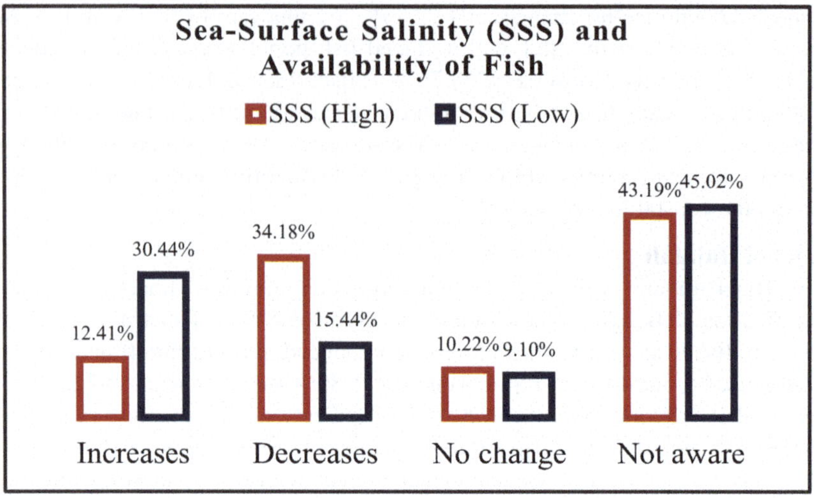

Fig. 16.45 Ecological perceptions of fishermen on sea-surface salinity (SSS) and availability of fish (Prepared by the authors)

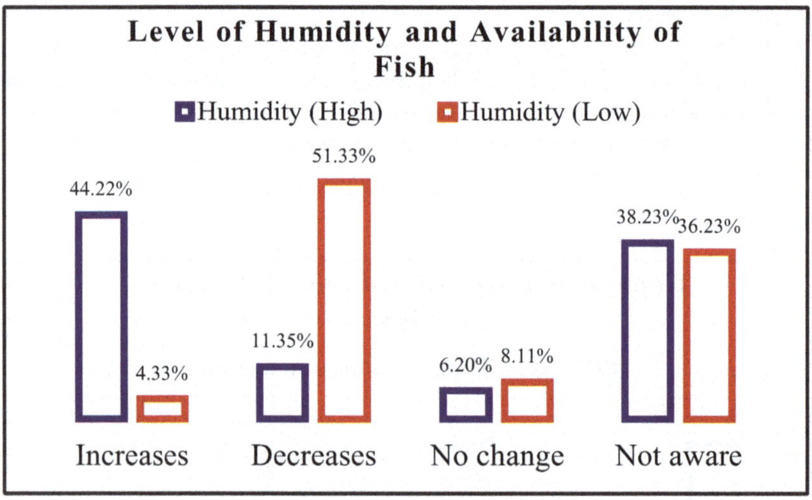

Fig. 16.46 Ecological perceptions of fishermen on level of humidity and availability of fish (Prepared by the authors)

believe low SSS increases fish abundance, whereas 15.44% report a decline, 9.1% see no change, and 45.02% are unaware of its effects. These findings emphasize the ecological significance of salinity variations and their potential stress on fisheries, with lower salinity perceived as more favourable for fish populations.

Humidity

Figure 16.46 highlights fishermen's perceptions of humidity's impact on fish availability in coastal Bengal. When humidity is high, 44.22% observe an increase in fish abundance, linking it to favourable conditions, while 11.35% perceive a decline, 6.2% see no change, and 38.23% are unaware of its effects, indicating a knowledge gap. Conversely, low humidity is viewed negatively, with 51.33% reporting a decline in fish availability, while only 4.33% observe an increase, 8.11% see no change, and 36.23% are unaware of its influence. These findings suggest that fishermen generally associate high humidity with better fish abundance, aligning with ecological insights that humidity affects marine ecosystems and fish behaviour.

Coastal Upwelling
Figure 16.47 illustrates fishermen's perceptions of coastal upwelling's impact on fish abundance in coastal Bengal. A majority (77.12%) believe upwelling increases fish availability, aligning with ecological principles that nutrient-rich deep waters enhance marine productivity. This process supports phytoplankton growth, attracting fish and boosting populations. Conversely, 6.47% perceive a decline in fish supply, 4.19% notice no change, and 12.22% are unaware of upwelling's impact. These findings suggest that most fishermen recognize upwelling as beneficial, reflecting broader ecological understanding that nutrient enrichment sustains marine biodiversity and fish stocks.

Frequency of Cyclone
Figure 16.48 highlights fishermen's perceptions of cyclones' impact on fish availability in coastal Bengal. A significant majority (80.18%) believe cyclones reduce fish supply, citing habitat destruction, fish displacement, and ecosystem disruptions. In contrast, 4.77% perceive an increase, likely due to temporary nutrient upwelling, though its benefits are short-lived. A small fraction (2.88%) reports no change, possibly due to localized resilience, while 12.17% are unaware of any impact. These findings emphasize the severe challenges fishermen face due to cyclonic

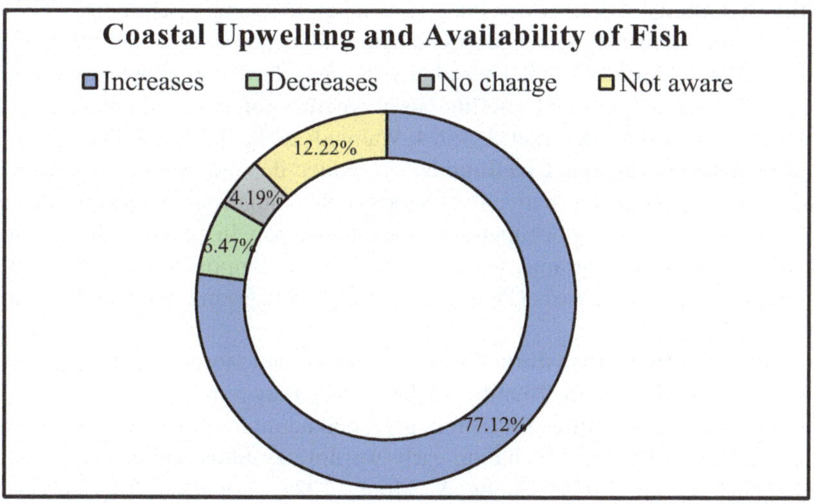

Fig. 16.47 Ecological perceptions of fishermen on coastal upwelling and availability of fish (Prepared by the authors)

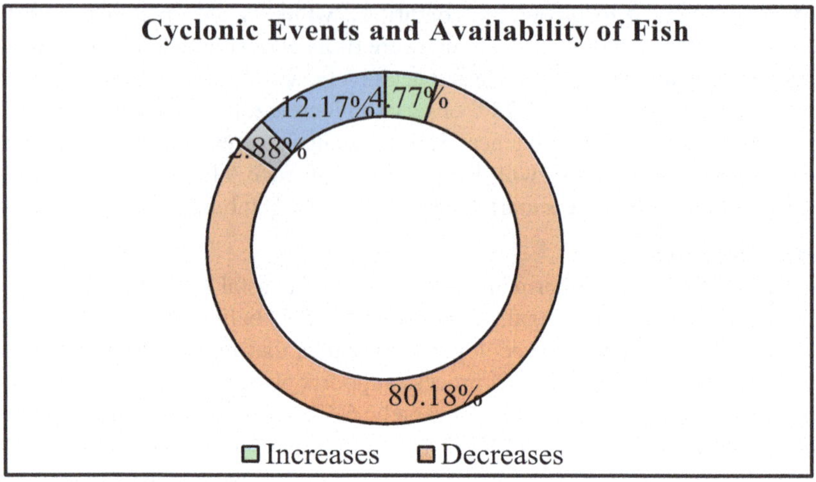

Fig. 16.48 Ecological perceptions of fishermen on cyclonic events and availability of fish (Prepared by the authors)

disturbances, reinforcing the broader ecological concern about extreme weather events affecting marine livelihoods.

Adaptation Strategies of Traditional Fishermen to Climate Change

Traditional fishermen in coastal Bengal employ various adaptation strategies to cope with the growing challenges posed by climate change. These strategies encompass modifications in fishing techniques, resource diversification, and knowledge integration to sustain their livelihoods amid environmental uncertainties.

- **Climate Change Adaptation:** Fishermen adapt to changing climatic conditions by adjusting their fishing schedules, shifting to deeper waters, and utilizing traditional knowledge to predict weather patterns. They also adopt modern technologies, such as GPS and satellite-based weather forecasts, to enhance fishing efficiency and safety (Kasri et al., 2024; Wahyudi et al., 2024; Rahim et al., 2024).
- **Fishermen's Lives and Livelihoods:** To reduce dependence on fishing, many fishermen engage in alternative livelihoods such as small-scale aquaculture, agriculture, and non-fishing occupations. Some participate in government-sponsored skill development programs, while others migrate temporarily to seek employment in nearby urban areas (Pical & Rahman, 2025; Anang et al., 2024; Diartho et al., 2024).
- **Impact of Natural Disasters:** Cyclones, floods, and storm surges pose significant threats to fishermen. To mitigate these risks, they reinforce their boats, build elevated storage structures for fishing gear, and adopt community-based disaster preparedness initiatives, including early warning systems and evacuation drills (Lacheheb et al., 2025; Chong & Moey, 2024; Tripathy, 2023; Castro & Magnaye, 2023).

- **Sea Surface Temperature (SST):** Rising SST affects fish availability, leading fishermen to shift target species, adopt selective fishing practices, and use traditional ecological knowledge to track seasonal fish migration patterns. Some invest in cold storage facilities to reduce post-harvest losses (Ojea et al., 2020; Ho et al., 2020).
- **Impact of Rainfall:** Variability in rainfall influences fish breeding and availability. Fishermen adjust their fishing efforts based on rainfall patterns, choosing locations where nutrient-rich runoff enhances fish productivity. They also engage in rainwater harvesting for domestic and small-scale aquaculture use (Nugraha et al., 2023; Farquhar et al., 2022; Adeleke et al., 2021).
- **Sea-Surface Salinity (SSS):** Changes in salinity levels impact fish habitats. Fishermen respond by targeting species that thrive under varying salinity conditions and modifying net designs to optimize catches in fluctuating environments. Some communities construct artificial breeding areas to support fish population sustainability (Blondeau-Bidet et al., 2024; Chung et al., 2024; Röthig et al. 2023).
- **Humidity Adaptation:** Humidity affects fish drying and preservation methods. Fishermen adapt by modifying fish drying techniques, such as using solar dryers or indoor drying facilities, to maintain product quality and reduce spoilage (Hasan et al., 2025; Reza & Hossain, 2024; Babiker et al., 2024).
- **Coastal Upwelling Utilization:** Coastal upwelling enhances fish abundance by bringing nutrient-rich waters to the surface. Fishermen strategically time their fishing activities to coincide with upwelling events, improving catch rates and reducing fishing effort (Zaba et al., 2021; Jacox et al., 2018).
- **Frequency of Cyclones:** The increasing occurrence of cyclones forces fishermen to adopt cyclone-resistant infrastructure, secure financial support through microfinance schemes, and participate in cooperative societies for mutual aid. Insurance coverage for boats and fishing gear is also gaining importance (Rout, 2024; Goyal, 2024; Gills et al., 2024).

These adaptation strategies reflect the resilience and resourcefulness of traditional fishermen in coping with climate change. While local knowledge remains a key asset, integrating scientific innovations and policy support can further enhance their capacity to sustain livelihoods and marine ecosystems in the face of ongoing environmental changes.

16.2.8 Casebook 8: Management of Climate Gap Among Climate Migrants on Sagar Island: A Case Study

16.2.8.1 Background and Purpose

Sagar Island (Fig. 16.49), the largest island in the Sundarbans delta of West Bengal, India, faces significant climate change vulnerabilities due to rising sea levels, coastal erosion, and extreme weather events such as storm surges and cyclones(Ghosh et al., 2024; Bera et al., 2022a, b). Over time, these environmental stressors have

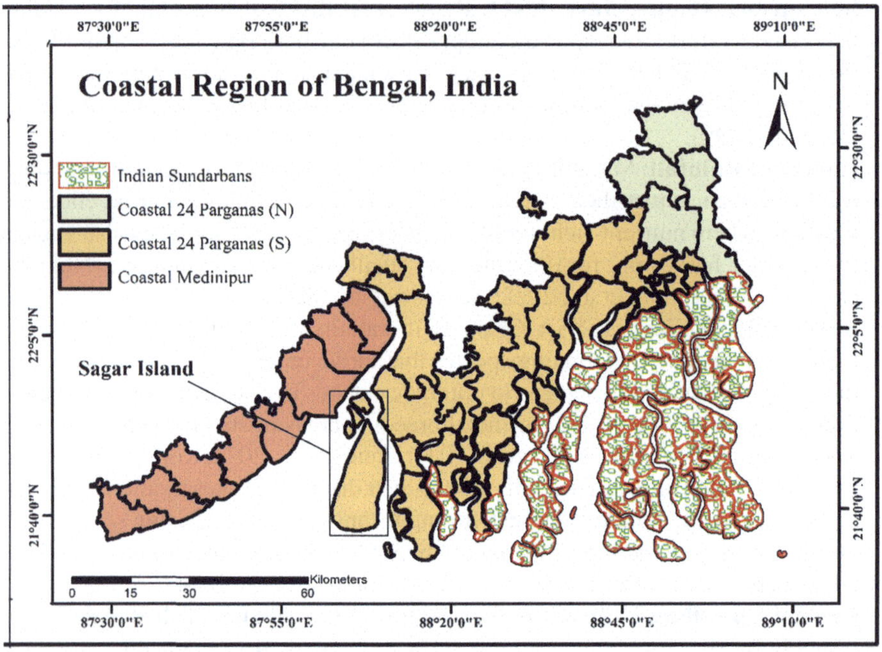

Fig. 16.49 Mapping the Study Region: Sagar Island, West Bengal (Prepared by the authors)

compelled many islanders to migrate internally, resulting in the emergence of climate migrants, individuals displaced due to climate-induced habitat loss and livelihood disruptions (Sinha, 2022; Bera et al., 2021; Bera, 2020). The increasing frequency of such displacements has exacerbated the "climate gap," referring to disparities in the capacity of affected communities to adapt and recover from climate-induced adversities. This case study examines the management of the climate gap among climate migrants on Sagar Island, with a focus on their socio-economic vulnerabilities, adaptation strategies, and relevant policy interventions. By analysing the lived experiences of climate migrants, the study aims to contribute to a broader discourse on climate justice, emphasizing the need for equitable adaptation measures that address both immediate relief and long-term sustainability. The findings seek to inform policymakers and stakeholders about effective strategies for minimizing the climate gap and enhancing the resilience of displaced communities on the island. The study employs a qualitative approach to explore narratives of climate migration and associated challenges. It draws on data collected through 20 focus group discussions (FGDs) conducted (August to September 2023) across Sagar Island, utilizing participatory methodologies to gain deeper insights into the migrants' experiences and adaptation mechanisms.

16.2.8.2 Results, Discussion, and Conclusion

Climate Gap Among Climate Migrants

The climate gap encompasses disparities in exposure, vulnerability, and adaptive capacity among climate migrants, particularly those displaced by environmental changes. In regions such as Sagar Island, climate-induced migration is primarily driven by sea-level rise, coastal erosion, and extreme weather events. Migrants frequently encounter socio-economic hardships, including loss of livelihoods, inadequate housing, and restricted access to essential resources. The inability to secure stable employment or restore traditional occupations further exacerbates their vulnerability. Additionally, deficiencies in infrastructure, education, and healthcare limit their capacity to adapt to new environments. Addressing these disparities necessitates targeted policy interventions, community-driven solutions, and sustainable adaptation strategies.

The climate gap among migrants' manifests across economic, social, and institutional dimensions. Economic disparities arise from declining agricultural productivity, shrinking fishing zones, and limited alternative livelihood opportunities. Social inequalities are reflected in displacement-induced psychological distress, erosion of cultural identity, and challenges in community integration. Institutional gaps stem from inadequate governmental policies, a lack of climate-resilient infrastructure, and insufficient financial support for affected populations. A comprehensive understanding of these gaps requires both qualitative and quantitative approaches, including field observations, interviews, and statistical analyses. Identifying these disparities is essential for developing inclusive adaptation strategies that address the specific needs of climate migrants and enhance their long-term resilience.

Management of Climate Gap Using Participatory Approaches

Effective management of the climate gap among migrants necessitates inclusive, community-driven strategies that integrate local knowledge into adaptation efforts. Participatory approaches enhance resilience by ensuring affected communities actively contribute to decision-making and adaptation planning.

Community-based adaptation (CBA) is a key strategy where migrants and local stakeholders collaboratively identify risks, develop solutions, and implement climate-resilient practices. Training in climate-resilient agriculture and sustainable fishing helps reduce economic vulnerabilities, while participatory mapping and risk assessments guide relocation and adaptation strategies (Tanwattana & Toyoda, 2024; Oino & Musau, 2024). Engaging local governance is essential in bridging institutional gaps. Community climate action groups facilitate dialogue between migrants and policymakers, ensuring their needs inform adaptation policies. Participatory budgeting further enables communities to influence resource allocation for infrastructure, healthcare, and disaster preparedness (Igbokwe et al., 2024; Wagner & Lima, 2023; and Restrepo-Mieth et al., 2023). Further, the social cohesion and knowledge-sharing strengthen adaptive capacity. Peer-to-peer learning fosters integration of traditional ecological knowledge with scientific methods, while inclusive participation of women and youth enhance innovation in adaptation strategies (Wannewitz & Garschagen, 2024; Samsuddin et al. 2024; McMichael et al., 2024; Lloyd & Lewis, 2024).

Therefore, incorporating participatory approaches, climate gap management becomes more effective, fostering long-term resilience and equitable adaptation for climate migrants on Sagar Island and other vulnerable coastal regions.

16.2.9 Casebook 9: Spatial Consideration of Coastal Zone Management Frameworks: Toward Participatory Policy Recommendations

16.2.9.1 Background and Purpose

Coastal zones represent dynamic yet vulnerable landscapes that necessitate comprehensive management frameworks to harmonize ecological sustainability, economic development, and community resilience (Berdenov et al., 2024; Aykut & Tezcan, 2024; Xu et al., 2024; Shetty et al., 2019). This study critically examines the influence of spatial dimensions on coastal governance and decision-making within the coastal region of Bengal, India (Fig. 16.50). These areas are increasingly exposed to

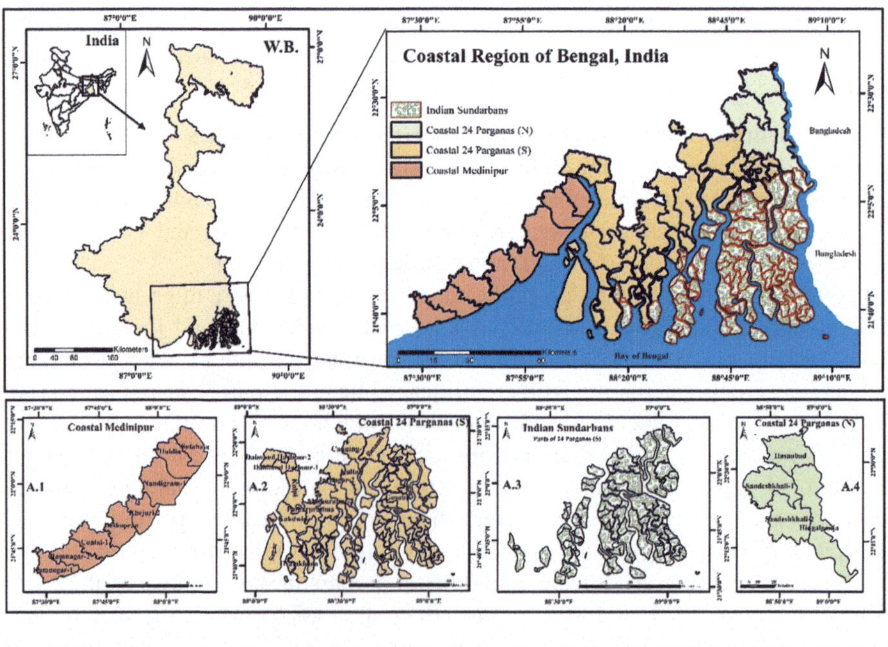

Fig. 16.50 Geographical Context of the Study Area: Coastal Bengal, India (A.1: Coastal East Medinipur; A.2: Coastal South 24 Parganas; A.3: Indian Sundarbans; and A.4: Coastal North 24 Parganas) (Prepared by the authors)

climate change, rising sea levels, and anthropogenic pressures, highlighting the need for an integrated framework that combines spatial analysis with participatory methodologies (Das & Swain, 2024; Samanta et al., 2023; Roy et al., 2023). Conventional management strategies frequently fail to account for the spatial heterogeneity of coastal ecosystems and the diverse priorities of local stakeholders, often resulting in fragmented policies and suboptimal interventions (Lukambagire et al., 2024; Mirli et al., 2024; Zamani et al., 2024).

This case study investigates the significance of spatial considerations in coastal zone management (CZM) and the extent to which participatory approaches can enhance policy efficiency. The research emphasizes the necessity of incorporating local communities, policymakers, and experts into the decision-making process to develop sustainable and context-specific management strategies. Ultimately, this study aims to formulate spatially informed policy recommendations that bridge the divide between scientific inquiry and real-world applications. By integrating spatial analysis with participatory governance, the research aspires to foster a more resilient and equitable coastal management paradigm.

Building on this background, the case study applied participatory approaches to gather data through 25 focus group discussions (FGDs), facilitating an in-depth exploration of community perspectives on coastal management. These discussions provided valuable qualitative insights into local knowledge, perceptions of environmental risks, and spatial considerations influencing coastal governance. The participatory nature of the study ensured that the voices of diverse stakeholders, including traditional marine fishermen, policymakers, and environmental experts, were incorporated into the analysis. The qualitative data obtained from the FGDs were systematically analysed to identify recurring themes, concerns, and spatial priorities within the community. This analysis highlighted the interconnections between socio-economic vulnerabilities, environmental changes, and existing coastal management practices. Additionally, the study assessed how spatially embedded local knowledge could contribute to more inclusive and adaptive policy frameworks. By integrating qualitative insights with spatial considerations, the research underscores the necessity of context-specific interventions that align with both scientific assessments and community-driven priorities. Through this approach, the study bridges the gap between top-down policy formulation and grassroots realities, ultimately contributing to more resilient and participatory coastal zone management strategies in the study area.

16.2.9.2 Results, Discussion, and Conclusion

Contemporary Issues of Coastal Bengal

Based on insights from focus group discussions (FGDs) held in the coastal region of Bengal between August and September 2023, it is clear that West Bengal's coastal areas are grappling with numerous contemporary challenges driven by both natural and human-induced factors. Climate change-induced sea-level rise, coastal erosion, and the increasing frequency of cyclones pose significant threats to the region's fragile ecosystems and local livelihoods. The Sundarbans, in particular, remains highly vulnerable, as rising salinity levels continue to impact agriculture and freshwater availability. Furthermore, unplanned settlement expansion and unsustainable aquaculture practices contribute to habitat degradation and biodiversity loss. Socio-economic vulnerabilities are further exacerbated by displacement, the decline of traditional livelihoods, and inadequate disaster preparedness. Additionally, gaps in policy implementation and limited community participation hinder effective coastal zone management. These concerns are further corroborated by recent scientific research conducted by Mukherjee and Siddique (2024), Malakar et al. (2023), Ghosh and Roy (2022), and Pattanaik (2021), reinforcing the urgency of addressing these pressing issues through integrated and participatory management strategies.

Community Engagement, Perception, and Spatial Consideration: Policy Recommendations

Community engagement plays a crucial role in shaping effective coastal management strategies in West Bengal, where local populations face increasing socio-economic and environmental challenges. The perceptions of coastal communities are deeply informed by their lived experiences, encompassing concerns about climate-induced risks, livelihood insecurities, and resource accessibility. This analysis highlights the intricate linkages between socio-economic vulnerabilities, environmental changes, and existing coastal management practices, underscoring the necessity for more inclusive governance mechanisms. Despite possessing extensive knowledge of their local environment, coastal communities are often marginalized in traditional top-down management frameworks, resulting in ineffective policy interventions (Viswanathan et al., 2024; Sele & Mukundi, 2024; Al Falah et al., 2024; Hoffmann, 2022). Furthermore, this study evaluates the role of spatially embedded local knowledge in fostering inclusive and adaptive policy frameworks. By integrating community perspectives with scientific data, spatial considerations help identify high-risk zones, resource-use patterns, and socio-ecological dynamics essential for sustainable management. The research emphasizes the need for context-specific spatial interventions that align both scientific assessments and community-driven priorities. In this regard, Table 16.22 presents key participatory policy recommendations that inform a more comprehensive coastal zone management strategy.

By adopting this approach, the study bridges the divide between top-down policy formulation and grassroots realities, ultimately fostering more resilient and participatory coastal management strategies in the region. Strengthening community engagement ensures that policies are not only scientifically rigorous but also socially relevant, contributing to the long-term sustainability of coastal West Bengal.

16.3 Conclusion and Arguments

The findings from the selected case studies emphasize the crucial role of geospatial technologies in monitoring and managing coastal ecosystems along the Bengal

Table 16.22 Coastal zone management frameworks: participatory policy recommendations

Policy recommendations	Development of coastal zone management suggestions
Bridging the gap between policy and ground realities	Through this approach, the study bridges the gap between top-down policy formulation and grassroots realities. Strengthening participatory governance ensures resilient and inclusive coastal zone management strategies.
Integration of socio-economic and environmental factors	This analysis highlighted the interconnections between socio-economic vulnerabilities, environmental changes, and existing coastal management practices. Policies should address climate-induced risks, livelihood insecurities, and resource management challenges faced by coastal communities.
Utilization of spatially embedded local knowledge	The study assessed how spatially embedded local knowledge could contribute to more inclusive and adaptive policy frameworks. Engaging communities in data collection and spatial mapping enhances policy relevance and effectiveness.
Context-specific and adaptive management approaches	By integrating qualitative insights with spatial considerations, the research underscores the necessity of context-specific interventions. Policies should align scientific assessments with community-driven priorities to ensure sustainability.
Policy implementation strategies	Strengthen local governance structures to enhance community participation. Promote climate-resilient infrastructure and sustainable livelihood options. Improve disaster preparedness by integrating local knowledge with scientific risk assessments.

Source: FGDs in the coastal region of West Bengal, India (August and September, 2023)

coast. Climatic trend analyses highlight the growing impacts of climate change, while spatio-temporal assessments of shoreline dynamics reveal persistent patterns of erosion and accretion. Evaluations of mangrove greenness indicate its vulnerability to shifting climate conditions, reinforcing the urgency of targeted conservation efforts. Furthermore, analyses of land use and land cover changes expose ongoing ecological degradation driven by both human activities and environmental transformation. Furthermore, the observed rise in land surface temperatures correlates with broader ecological transformations, necessitating adaptive and proactive management approaches. The studies also underscore the complexities of mangrove-human conflicts, emphasizing the importance of sustainable resource governance. While traditional fishermen exhibit resilience in adapting to climate change, unresolved challenges in climate migration management continue to exacerbate socio-economic vulnerabilities. A central argument emerging from these case studies is the imperative for spatially informed and participatory coastal zone management frameworks. Geospatial technologies provide essential data-driven insights for policymakers, facilitating evidence-based decision-making. These case studies advocate for an integrated, multi-scalar, and interdisciplinary approach to coastal governance, reinforcing the need for adaptive policies and community engagement to address climate-induced vulnerabilities effectively.

References

Adeel, Z., & Pomeroy, R. (2002). Assessment and management of mangrove ecosystems in developing countries. *Trees, 16*, 235–238. https://doi.org/10.1007/s00468-002-0168-4

Adeleke, L., Jerry, J. V., Ayal, D., Oluwatobi, A. J., Sunday, A. I., & Amos, A. I. (2021). Climate variability on fishing activities in inland waters: Case of Owena River in Ondo and Osun States, Nigeria. In N. Oguge, D. Ayal, L. Adeleke, & I. da Silva (Eds.), *African handbook of climate change adaptation*. Springer. https://doi.org/10.1007/978-3-030-45106-6_91

Aditya, F., Gusmayanti, E., & Sudrajat, J. (2021). Rainfall trend analysis using Mann-Kendall and Sen's slope estimator test in West Kalimantan. In *IOP conference series: Earth and environmental science* (Vol. 893, p. 012006). IOP Publishing. https://doi.org/10.1088/1755-1315/893/1/012006

Agbelade, A. D. (2025). Analyzing the influence of urban vegetation cover on land surface temperature in Southwestern Nigeria. *Discover Environment, 3*, 10. https://doi.org/10.1007/s44274-025-00186-x

Ahmed, M. (2022). The science of climate change. In *The Palgrave handbook of global sustainability* (pp. 1–28). Springer. https://doi.org/10.1007/978-3-030-38948-2_22-1

Al Falah, P., Fian, S., Saputra, A., Ramadhan, B. S., & Likuwatan, P. (2024). Peran masyarakat lokal dalam pengelolaan sumber daya alam. *Paradigma Mandiri, 2*(02), 94–107. https://doi.org/10.37949/pm22173

Alibašić, H. (2022). *Strategic resilience and sustainability planning: Management strategies for sustainable and climate-resilient communities and organizations*. Springer . ISBN: 978-3-030-91158-4. https://doi.org/10.1007/978-3-030-91159-1

Aljenaid, S. S., Kadhem, G. R., AlKhuzaei, M. F., et al. (2022). Detecting and assessing the Spatio-temporal land use land cover changes of Bahrain Island during 1986–2020 using remote sensing and GIS. *Earth Systems and Environment, 6*, 787–802. https://doi.org/10.1007/s41748-022-00315-z

Almulhim, A. I., Alverio, G. N., Sharifi, A., et al. (2024). Climate-induced migration in the Global South: An in depth analysis. *npj Climate Action, 3*, 47. https://doi.org/10.1038/s44168-024-00133-1

Alum, E. U. (2024). Sustainable harvesting of medicinal plants: Balancing therapeutic benefits with environmental conservation. *Agroecology and Sustainable Food Systems*, 1–6. https://doi.org/10.1080/21683565.2024.2421948

Anang, S. A., Otchere-Peprah, K., Aidoo-Mensah, D., Mensah, N. O., Nakuja, T., & Boateng-Gyambiby, D. (2024). Livelihood diversification and coping strategies: Evidence from small-scale fish farmers in Ghana. *Cogent Social Sciences, 10*(1). https://doi.org/10.1080/23311886.2024.2433712

Anggraeni, I. D. (2024). The role of regulations in supporting the restoration of mangrove ecosystems for the achievement of sustainable development goals: A literature review. *Eduvest, 4*(11), 10361–10378. https://doi.org/10.59188/eduvest.v4i11.43690

Anh, H. H., Beaulieu, A., Hanh, T. M. D., & Tru, L. C. (2024). Economic effects of a controlled mangrove-to-pond coverage ratio policy on mangrove conservation and shrimp farming: A case study in vietnam using instrumental regression analysis. *Trees, Forests and People, 16*, 100579. https://doi.org/10.1016/j.tfp.2024.100579

Annan, E., Amponsah, W., Adjei, K. A., Disse, M., Hounkpè, J., Biney, E., et al. (2024). Spatio-temporal land use and land cover change assessment: Insights from the Ouémé River Basin. *Scientific African, 25*, e02262. https://doi.org/10.1016/j.sciaf.2024.e02262

Arulbalaji, P., Padmalal, D., & Maya, K. (2020). Impact of urbanization and land surface temperature changes in a coastal town in Kerala, India. *Environmental Earth Sciences, 79*, 400. https://doi.org/10.1007/s12665-020-09120-1

Asensio-Montesinos, F., Molina, R., Anfuso, G., Manno, G., & Lo Re, C. (2024). Natural and human impacts on coastal areas. *Journal of Marine Science and Engineering, 12*(11), 2017. https://doi.org/10.3390/jmse12112017

Ashton, E. C. (2022). Threats to mangroves and conservation strategies. In S. C. Das, A. Pullaiah, & E.C. (Eds.), *Mangroves: Biodiversity, livelihoods and conservation*. Springer. https://doi.org/10.1007/978-981-19-0519-3_10

Ayalke, Z. G., Şişman, A., & Akpinar, K. (2023). Shoreline extraction and analyzing the effect of coastal structures on shoreline changing with remote sensing and geographic information system: Case of Samsun, Turkey. *Regional Studies in Marine Science, 61*, 102883. https://doi.org/10.1016/j.rsma.2023.102883

Aykut, F., & Tezcan, D. (2024). Evaluating sea level rise impacts on the Southeastern Türkiye coastline: A coastal vulnerability perspective. *PFG, 92*, 335–352. https://doi.org/10.1007/s41064-024-00284-0

Azaiez, N. (2024). The contribution of remote sensing in highlighting the resilience of rangelands through indicators of land surface temperature and the normalized difference vegetation index in the tihama alluvial plain. *Geographia Technica*, 139–151. https://doi.org/10.21163/gt_2024.192.11

Babiker, A. M. O. A., et al. (2024). Effect of solar dryer and direct open sun drying methods on drying air temperatures and moisture content of Fish Bayad (Bagrus bayad). *European Journal of Ecology, Biology and Agriculture, 1*(2), 3–13. https://doi.org/10.59324/ejeba.2024.1(2).01

Barik, G., Guru, B., & Sangma, F. (2021). Shoreline changes analysis and forecast using digital shoreline assessment system 5.0: Evidences from parts of East Coast of India. *Journal of the Indian Society of Remote Sensing, 49*, 2815–2830. https://doi.org/10.1007/s12524-021-01424-4

Barua, P., Chowdhury, S. S., & Mitra, A. (2024). Review on conservation approach study of largest mangrove ecosystem as the mitigation of climate change. *Journal of Climate Change, 10*(3), 31–42. https://doi.org/10.3233/jcc240021

Bastien-Olvera, B. A., et al. (2024). Mangrove preservation could have significantly reduced damages from hurricane Otis on the coast of Guerrero, Mexico. *Science of the Total Environment, 957*, 177822. https://doi.org/10.1016/j.scitotenv.2024.177822

Bastos, M. F. L., Kneipp, J. M., Gomes, C. M., Perlin, A. P., & Bichueti, R. S. (2025). Management strategies for climate change mitigation and adaptation in coastal regions: A systematic literature review. *Coasts, 5*(1), 5. https://doi.org/10.3390/coasts5010005

Basu, J. P. (2020). *Climate change vulnerability and communities in agro-climatic regions of West Bengal, India: Theory and practice*. Springer . ISBN: 978-3-030-50467-0. https://doi.org/10.1007/978-3-030-50468-7

Bekere, Y., & Megerssa, G. (2020). Role of biogas technology adoption in forest conservations: Evidence from Ethiopia. *Journal of Degraded and Mining Lands Management, 7*(2), 2035–2039. https://doi.org/10.15243/JDMLM.2020.072.2035

Benhur, J., Vendhan, M., Kumar, P., & Janagiraman, R. (2024). Coastal resilience and shoreline dynamics: Assessing the impact of a hybrid beach restoration strategy in Puducherry, India. *Frontiers in Marine Science, 11*, 1426627. https://doi.org/10.3389/fmars.2024.1426627

Beninato, S., Holzman, M., & Rivas, R. E. (2024). *Sensitivity of the land surface temperature and NDWI index to soil moisture on wheat and barley*. https://doi.org/10.21203/rs.3.rs-4401715/v1

Bera, A. (2020). Review article on climate induced migration: Risk assessment and management: A case study of Sagar Island of Indian Sundarban. *SGVU Journal of Climate Change and Water, 7*, 47–55.

Bera, A., Taloor, A. K., Meraj, G., Kanga, S., Singh, S. K., Đurin, B., & Anand, S. (2021). Climate vulnerability and economic determinants: Linkages and risk reduction in Sagar Island, India: A geospatial approach. *Quaternary Science Advances, 4*, 100038. https://doi.org/10.1016/j.qsa.2021.100038

Bera, A., Meraj, G., Kanga, S., Farooq, M., Singh, S. K., Sahu, N., & Kumar, P. (2022a). Vulnerability and risk assessment to climate change in Sagar Island, India. *Water, 14*(5), 823. https://doi.org/10.3390/w14050823

Bera, P., Modak, B. K., Mandi, S., Modak, P., & Mandal, M. (2022b). Chronological variation of land use and land cover of coastal West Bengal. *Journal of Scientific Enquiry, 2*(2), 6–9. https://doi.org/10.54280/jse.222214

Berdenov, Z., Yeginbayeva, A. Y., Zinabdin, N., Beketova, A., Mendybayeva, G., Assylbekova, A., & Önal, H. (2024). Recreational and functional zoning of territories with technogenic impact for the purpose of sustainable development of the region. *Geojournal of Tourism and Geosites, 55*(3), 1354–1363. https://doi.org/10.30892/gtg.55336-1308

Bhattacharyya, R. (2023). Community vulnerability and disaster risk reduction at Sundarbans. In A. Singh (Ed.), *International handbook of disaster research.* Springer. https://doi.org/10.1007/978-981-19-8388-7_65

Biswas, P. L., & Biswas, S. R. (2019). Mangrove forests: Ecology, management, and threats. In W. Leal Filho, A. Azul, L. Brandli, P. Özuyar, & T. Wall (Eds.), *Life on land. Encyclopedia of the UN sustainable development goals.* Springer. https://doi.org/10.1007/978-3-319-71065-5_26-1

Blondeau-Bidet, E., Tine, M., Gonzalez, A.-A., Guinand, B., & Lorin-Nebel, C. (2024). Coping with salinity extremes: Gill transcriptome profiling in the black-chinned tilapia (Sarotherodon melanotheron). *Science of the Total Environment*, 172620. https://doi.org/10.1016/j.scitotenv.2024.172620

Boussetta, A., Niculescu, S., Bengoufa, S., & Zagrarni, M. F. (2022). Spatio-temporal analysis of shoreline changes and erosion risk assessment along Jerba Island (Tunisia) based on remote-sensing data and geospatial tools. *Regional Studies in Marine Science, 55*, 102564. https://doi.org/10.1016/j.rsma.2022.102564

Breinl, K., Lun, D., Müller-Thomy, H., & Blöschl, G. (2021). Understanding the relationship between rainfall and flood probabilities through combined intensity-duration-frequency analysis. *Journal of Hydrology, 602*, 126759. https://doi.org/10.1016/j.jhydrol.2021.126759

Castro, A. G. and Magnaye, D. (2023). *Coastal hazards and the challenges for planning fishing communities in the Philippines.* EGU General Assembly 2023, Vienna, Austria, 24–28 Apr 2023, EGU23-885, https://doi.org/10.5194/egusphere-egu23-885

Chauhan, T. (2023). *Dynamics of Sundarbans mangroves under climate extremes and changing soil nutrient composition.* EGU General Assembly 2023, Vienna, Austria, 24–28 Apr 2023, EGU23-13968. https://doi.org/10.5194/egusphere-egu23-13968

Chong, F., & Moey, S. (2024). *Coastal resilience planning: Harnessing data insights to assess and mitigate future tropical cyclones impacts in vulnerable coastal communities* (pp. 604–608). IEEE. https://doi.org/10.1109/igarss53475.2024.10641424

Chopra, K., Kapuria, P., & Kumar, P. (2009). *Biodiversity land use change and human well being: A study of aquaculture in the Indian Sundarbans* (Research papers in economics). Oxford University Press. ISBN: 9780198060215. https://econpapers.repec.org/RePEc:oxp:obooks:9780198060215

Chowdhury, K. J., Ali, M., Chowdhury, M. A., & Islam, S. L. U. (2024). Climate change induced risks assessment of a coastal area: A "socioeconomic and livelihood vulnerability index" based study in coastal Bangladesh. *Natural Hazards Research, 5*, 75. https://doi.org/10.1016/j.nhres.2024.06.005

Chung, M.-H. J., Noble, D. W. A., Fox, R. J., Harrison, L. M., & Jennions, M. D. (2024). Fluctuating salinity during development impacts fish productivity. *bioRxiv.* https://doi.org/10.1101/2024.02.01.578515

Coelho, P., Camargo Campos, C. G., Sá, É. S., Biffi, L., & da Silveira, P. C. (2024). Application of remote sensing for climatic vulnerability analysis of the urban area of Lages/SC through the use of radiometric index and surface temperature. *Revista Brasileira de Geografia Física., 17*, 785. https://doi.org/10.26848/rbgf.v17.2.p785-803

Cohen, J. A. (1960). Coefficient of agreement for nominal scales. *Educational and Psychological Measurement, 1960*(20), 37–46.
Colak, A. T. I. (2024). Geospatial analysis of shoreline changes in the Oman coastal region (2000-2022) using GIS and remote sensing techniques. *Frontiers in Marine Science, 11*, 1305283. https://doi.org/10.3389/fmars.2024.1305283
da Silva, C. V., Ortigao, M., Willaert, T., Rosa, R., Nunes, L. C., & Cunha-e-Sa, M. A. (2021). Participatory Geographic Information Systems (PGIS): Alternative approaches to identify potential conflicts and positional accuracy in marine and coastal ecosystem services. *Marine Policy, 131*, 104650. https://doi.org/10.1016/j.marpol.2021.104650
Damastuti, E., & de Groot, R. (2019). Participatory ecosystem service mapping to enhance community-based mangrove rehabilitation and management in Demak, Indonesia. *Regional Environmental Change, 19*, 65–78. https://doi.org/10.1007/s10113-018-1378-7
Daneshwar, D., & Revaty, E. V. (2024). A paradigm shift towards environmental responsibility in sustainable green tourism and hospitality. *International Journal of Social Science and Human Research, 7*(5), 2795–2798. https://doi.org/10.47191/ijsshr/v7-i05-33
Das, K., & Das, K. (2025). Chapter 10: Climate resilience or impeding vulnerabilities in the light of mangroves of Indian Sundarban. In *Surviving sundarbans: Tragedy and hope* (1st ed., pp. 169–185. ISBN: 978-93-48859-73-0). BP International. https://doi.org/10.9734/bpi/mono/978-93-48859-73-0/ch10
Das, C. S., & Sarkar, P. (2025). Chapter 5: The transboundary Sundarban mangroves: A struggling ecosystem. In *Surviving sundarbans: Tragedy and hope* (1st ed., pp. 80–103. ISBN: 978-93-48859-73-0). BP International. https://doi.org/10.9734/bpi/mono/978-93-48859-73-0/ch5
Das, A., & Swain, P. K. (2024). Navigating the sea level rise: Exploring the interplay of climate change, sea level rise, and coastal communities in india. *Environmental Monitoring and Assessment, 196*, 1010. https://doi.org/10.1007/s10661-024-13191-z
Das, S., Chanda, A., & Ghosh, T. (Eds.). (2022). *Pond ecosystems of the Indian Sundarbans: An overview* (Water science and technology library). Springer. https://doi.org/10.1007/978-3-030-86786-7
DasGupta, R., & Shaw, R. (2015). An indicator based approach to assess coastal communities' resilience against climate related disasters in Indian Sundarbans. *Journal of Coastal Conservation, 19*(1), 85–101. https://doi.org/10.1007/s11852-014-0369-1
DasGupta, R., & Shaw, R. (Eds.). (2017). *Participatory mangrove management in a changing climate: Perspectives from the Asia-Pacific*. Springer . ISBN: 978-4-431-56481-2. https://doi.org/10.1007/978-4-431-56481-2
Datta, D., & Deb, S. (2012). Analysis of coastal land use/land cover changes in the Indian Sunderbans using remotely sensed data. *Geo-spatial Information Science, 15*(4), 241–250. https://doi.org/10.1080/10095020.2012.714104
Datta, P., Behera, B., & Rahut, D. B. (2024). Climate change and water-related threats in the Indian Sundarbans: Food security and management implications. *International Journal of Water Resources Development, 40*(3), 323–344. https://doi.org/10.1080/07900627.2023.2224459
de Carvalho, C. M., Luiz Giatti, L., Fagerholm, N., Bedran-Martins, A. M., & Kytta, M. (2021). Participatory geographic information systems (PGIS) to assess water, energy and food availability in a vulnerable community in Guarulhos (Brazil). *International Journal of Urban Sustainable Development, 13*(3), 516–529. https://doi.org/10.1080/19463138.2021.2019041
Debnath, M., & Alamdari, N. (2025). Impacts of climate change and sea-level rise on groundwater levels and saltwater intrusion in coastal aquifers. *SSRN*. https://doi.org/10.2139/ssrn.5085447
Debnath, S., Sarkar, U. K., Kumari, S., et al. (2024). Exploring the vulnerability of the coastal wetlands of India to the changing climate and their adaptation strategies. *International Journal of Biometeorology, 68*, 749–760. https://doi.org/10.1007/s00484-024-02620-w
Dhyani, S., Shukla, J., Kadaverugu, R., Dasgupta, R., Panda, M., Kundu, S. K., Santhanam, H., Pujari, P. R., Kumar, P., & Hashimoto, S. (2023). Participatory stakeholder assessment for driv-

ers of mangrove loss to prioritize evidence-based conservation and restoration in Bhitarkanika and Mahanadi Delta, India. *Sustainability, 15*(2), 963. https://doi.org/10.3390/su15020963

Diartho, H. C., Elan, M. C., & Jumiati, A. (2024). Sustainable livelihood approach for supporting aquaculture households amid structural shifts in Sidoarjo regency. *Society, 12*(2), 140–154. https://doi.org/10.33019/society.v12i2.691

Dobrică, G., Cerneagă, C., & Maftei, C. (2022). Shoreline change analysis using digital shoreline analysis system along southeast Romanian coast. In A. Çiner et al. (Eds.), *Recent research on geomorphology, sedimentology, marine geosciences and geochemistry. CAJG 2019. Advances in science, technology & innovation*. Springer. https://doi.org/10.1007/978-3-030-72547-1_18

Dong, W. S., Ismailluddin, A., Yun, L. S., Ariffin, E. H., Saengsupavanich, C., Maulud, K. N. A., et al. (2024). The impact of climate change on coastal erosion in Southeast Asia and the compelling need to establish robust adaptation strategies. *Heliyon, 10*(4), e25609. https://doi.org/10.1016/j.heliyon.2024.e25609

Duke, N. C., Kovacs, J. M., Griffiths, A. D., Preece, L., Hill, D. J., Van Oosterzee, P., et al. (2017). Large-scale dieback of mangroves in Australia's Gulf of Carpentaria: A severe ecosystem response, coincidental with an unusually extreme weather event. *Marine and Freshwater Research, 68*(10), 1816–1829.

Emam, W. W. M., & Soliman, K. M. (2020). Applying geospatial technology in quantifying spatiotemporal shoreline dynamics along Marina El-Alamein Resort, Egypt. *Environmental Monitoring and Assessment, 192*, 459. https://doi.org/10.1007/s10661-020-08432-w

Etongo, D., & Arrisol, L. (2021). Vulnerability of fishery-based livelihoods to climate variability and change in a tropical island: Insights from small-scale fishers in Seychelles. *Discover Sustainability, 2*, 48. https://doi.org/10.1007/s43621-021-00057-4

Fagerholm, N., Raymond, C. M., Olafsson, A. S., Brown, G., Rinne, T., Hasanzadeh, K., et al. (2021). A methodological framework for analysis of participatory mapping data in research, planning, and management. *International Journal of Geographical Information Science, 35*(9), 1848–1875. https://doi.org/10.1080/13658816.2020.1869747

Fanous, M., Eden, J. M., Remesan, R., & Daneshkhah, A. (2023). Challenges and prospects of climate change impact assessment on mangrove environments through mathematical models. *Environmental Modelling & Software, 162*, 105658. https://doi.org/10.1016/j.envsoft.2023.105658

Farquhar, S. D., Nirindrainy, A. F., Heck, N., & Saldarriaga, M. G. (2022). The impacts of long-term changes in weather on small-scale fishers' available fishing hours in Nosy Barren, Madagascar. *Frontiers in Marine Science, 9*. https://doi.org/10.3389/fmars.2022.841048

French, P. W. (2005). Coastal zone management. In M. L. Schwartz (Ed.), *Encyclopedia of coastal science* (Encyclopedia of earth science series). Springer. https://doi.org/10.1007/1-4020-3880-1_92

Frimpong, B. F., Koranteng, A., & Molkenthin, F. (2022). Analysis of temperature variability utilising Mann–Kendall and Sen's Slope estimator tests in the Accra and Kumasi metropolises in Ghana. *Environmental Systems Research, 11*(1), 24. https://doi.org/10.1186/s40068-022-00269-1

Fromard, F., Vega, C., & Proisy, C. (2004). Half a century of dynamic coastal change affecting mangrove shorelines of French Guiana. A case study based on remote sensing data analyses and field surveys. *Marine Geology, 208*, 265–280. https://doi.org/10.1016/j.margeo.2004.04.018

Fung, K. F., Chew, K. S., Huang, Y. F., Ahmed, A. N., Teo, F. Y., Ng, J. L., & Elshafie, A. (2022). Evaluation of spatial interpolation methods and spatiotemporal modeling of rainfall distribution in Peninsular Malaysia. *Ain Shams Engineering Journal, 13*, 101571. https://doi.org/10.1016/j.asej.2021.09.001

Gadedjisso-Tossou, A., Adjegan, K. I., & Kablan, A. K. M. (2021). Rainfall and temperature trend analysis by Mann–Kendall test and significance for rainfed cereal yields in Northern Togo. *Science, 3*(1), 17. https://doi.org/10.3390/sci3010017

Gao, B. C. (1996). NDWI—A normalized difference water index for remote sensing of vegetation liquid water from space. *Remote Sensing of Environment, 58*(3), 257–266. https://doi.org/10.1016/S0034-4257(96)00067-3

García, D. H., Riza, M., & Díaz, J. A. (2023). Land surface temperature relationship with the land use/land cover indices leading to thermal field variation in the Turkish Republic of Northern Cyprus. *Earth Systems and Environment, 7*, 561–580. https://doi.org/10.1007/s41748-023-00341-5

Garrett, H. E. (1979). *Statistics in psychology and education* (p. 329). Vakils, Feffer and Simons Ltd.. (9th Indian Reprnt).

Gayen, M. (2025). Chapter 7: Blue carbon dynamics in the Sundarban biosphere reserve (SBR): Impacts of climate and anthropogenic factors. In *Surviving sundarbans: Tragedy and hope* (1st ed., pp. 114–137. ISBN: 978-93-48859-73-0). BP International. https://doi.org/10.9734/bpi/mono/978-93-48859-73-0/ch7

Gemeda, D. O., Korecha, D., & Garedew, W. (2023). Determinants of climate change adaptation strategies and existing barriers in southwestern parts of Ethiopia. *Climate Services, 30*, 100376. https://doi.org/10.1016/j.cliser.2023.100376

Ghebrezgabher, M. G., Yang, T., Yang, X., & Eyassu Sereke, T. (2020). Assessment of NDVI variations in responses to climate change in the horn of Africa. *Egyptian Journal of Remote Sensing and Space Science, 23*(3), 249–261. https://doi.org/10.1016/j.ejrs.2020.08.003

Ghosh, S., & Mistri, B. (2023). Cyclone-induced coastal vulnerability, livelihood challenges and mitigation measures of Matla–Bidya inter-estuarine area, Indian Sundarban. *Natural Hazards, 116*, 3857–3878. https://doi.org/10.1007/s11069-023-05840-2

Ghosh, S., & Roy, S. (2022). Climate change, ecological stress and livelihood choices in Indian Sundarban. In A. K. E. Haque, P. Mukhopadhyay, M. Nepal, & M. R. Shammin (Eds.), *Climate change and community resilience* (pp. 399–413). Singapore. https://doi.org/10.1007/978-981-16-0680-9_26

Ghosh, A., Schmidt, S., Fickert, T., & Nüsser, M. (2015). The Indian Sundarban mangrove forests: History, utilization, conservation strategies and local perception. *Diversity, 7*(2), 149–169. https://doi.org/10.3390/d7020149

Ghosh, M. K., Kumar, L., & Roy, C. (2017). Climate variability and mangrove cover dynamics at species level in the Sundarbans, Bangladesh. *Sustainability, 9*(5), 805. https://doi.org/10.3390/su9050805

Ghosh, A., Baidya, S., & Gupta, A. K. (2024). Climate change impact on Landuse and livelihood in Sundarbans: A case study of Sagar Island. In A. K. Gupta, A. Gupta, & P. Acharya (Eds.), *Disaster risk and management under climate change. Disaster resilience and green growth*. Springer. https://doi.org/10.1007/978-981-99-4105-6_20

Gijsman, R., Horstman, E. M., van der Wal, D., Friess, D. A., Swales, A., & Wijnberg, K. M. (2021). Nature-based engineering: A review on reducing coastal flood risk with mangroves. *Frontiers in Marine Science, 8*, 702412. https://doi.org/10.3389/fmars.2021.702412

Gills, R., Sulochanan, B., Padua, S., Nakhawa, D. A., Azeez, P. A., Edward, L., Jayasankar, R., Shettigar, V., Thakurdas, T., Pilli, S. K., & Ambrose, T. (2024). Impact of cyclone Tauktae on fishermen's livelihood in Karnataka: A climate change perspective analysis. *Indian Journal of Fisheries, 71*(3). https://doi.org/10.21077/ijf.2024.71.3.136688-16

Giri, S., Daw, T. M., Hazra, S., Troell, M., Samanta, S., Basu, O., Marcinko, C. L. J., & Chanda, A. (2022). Economic incentives drive the conversion of agriculture to aquaculture in the Indian Sundarbans: Livelihood and environmental implications of different aquaculture types. *AMBIO: A Journal of the Human Environment, 51*(9), 1963–1977. https://doi.org/10.1007/s13280-022-01720-4

Gloria, S. J., & Gnanasekaran, S. P. (2024). Impact of urban vegetation loss on urban Heat Islands: A case study of Chennai metropolitan area. *Indian Journal of Science and Technology, 17*(2), 134–141. https://doi.org/10.17485/ijst/v17i2.1071

Goyal, P. (2024). *Cyclone disaster mitigation and management in India: An overview* (pp. 121–147). Springer Nature. https://doi.org/10.1007/978-981-99-4105-6_7

Grant, G. E., Tague, C. L., & Allen, C. D. (2013). Watering the forest for the trees: An emerging priority for managing water in forest landscapes. *Frontiers in Ecology and the Environment, 11*(6), 314–321. https://doi.org/10.1890/120209

Guild, R., Wang, X., & Quijón, P. A. (2024). Climate change impacts on coastal ecosystems. *Environmental Research, 3*, 042006. https://doi.org/10.1088/2752-5295/ad9f90

Guo, Z. C., Xie, Y., Guo, H., Zhang, X., Wenguo, H., Bie, Q., Xi, G., & Ma, C. (2023). Do the ecosystems of Gansu Province in Western China's crucial ecological security barrier remain vulnerable? Evidence from remote sensing based on geospatial analysis. *Journal of Cleaner Production, 402*, 136740. https://doi.org/10.1016/j.jclepro.2023.136740

Gupta, A. K., Gupta, A., & Acharya, P. (Eds.). (2024). *Disaster risk and management under climate change*. Springer . ISBN: 978-981-99-4104-9. https://doi.org/10.1007/978-981-99-4105-6

Handayati, P., Nasih, A. M., Susilowati, I., Idris, I., Nayak, P. K., & Narmaditya, B. S. (2025). From vulnerable to resilience: An assessment of small-scale fisheries livelihood in South Malang of Indonesia. *Discover Sustainability, 6*(1). https://doi.org/10.1007/s43621-025-00810-z

Harikrishna, S., Gurugnanam, B., Bairavi, S., & Shekar, S. (2024). Geospatial technology based shoreline change analysis (1992-2022) and predictive forecast modeling for 2032 and 2042, utilising DSAS, along the eastern coast of India. *Kuwait Journal of Science*, 100272. https://doi.org/10.1016/j.kjs.2024.100272

Hasan, M., Sarkar, F. R., Aktar, S., & Islam, K. (2025). Modification and performance evaluation of IoT-enabled solar dryer for fish drying. *GSC Advanced Research and Reviews, 22*, 001. https://doi.org/10.30574/gscarr.2025.22.1.0519

Hassan, W. H., Nile, B. K., Kadhim, Z. K., et al. (2023). Trends, forecasting and adaptation strategies of climate change in the middle and west regions of Iraq. *SN Applied Sciences, 5*, 312. https://doi.org/10.1007/s42452-023-05544-z

Hastuti, A. W., Nagai, M., Ismail, N. P., Priyono, B., Suniada, K. I., & Wijaya, A. (2024). Spatiotemporal analysis of shoreline change trends and adaptation in Bali Province, Indonesia. *Regional Studies in Marine Science, 76*, 103598. https://doi.org/10.1016/j.rsma.2024.103598

He, Q.-L., Xiao, J.-L., & Shi, W.-Y. (2022). Responses of terrestrial evapotranspiration to extreme drought: A review. *Water, 14*(23), 3847. https://doi.org/10.3390/w14233847

Herrera Estrella, E., Stoeth, A., Krakauer, N. Y., & Devineni, N. (2021). Quantifying vegetation response to environmental changes on the Galapagos Islands, Ecuador using the normalized difference vegetation index (NDVI). *Environmental Research Communications, 3*(6), 065003. https://doi.org/10.1088/2515-7620/ac0bd1

Hieu, H. H., Chau, D. M., Thuy, N. T. T., & Nguyen, M.-L. (2025). A comparative analysis of livelihood sustainability among farmers in Ca Mau Province, Vietnam: A case study of organic mangrove–shrimp and rice–shrimp farming systems. Cogent. *Social Sciences, 11*(1). https://doi.org/10.1080/23311886.2025.2451736

Hilmi, N., Benitez Carranco, M. B., Broussard, D., Mathew, M., Djoundourian, S., Cassotta, S., et al. (2023). Tropical blue carbon: Solutions and perspectives for valuations of carbon sequestration. *Frontiers in Climate, 5*, 1169663. https://doi.org/10.3389/fclim.2023.1169663

Ho, C.-H., Yagi, N., & Tian, Y. (2020). An impact and adaptation assessment of changing coastal fishing grounds and fishery industry under global change. *Mitigation and Adaptation Strategies for Global Change, 25*(6), 1073–1102. https://doi.org/10.1007/S11027-020-09922-5

Hoffmann, S. (2022). Challenges and opportunities of area-based conservation in reaching biodiversity and sustainability goals. *Biodiversity and Conservation, 31*, 325–352. https://doi.org/10.1007/s10531-021-02340-2

Hou, W., Gao, J., Wu, S., & Dai, E. (2015). Interannual variations in growing-season NDVI and its correlation with climate variables in the southwestern karst region of China. *Remote Sensing, 7*(9), 11105–11124. https://doi.org/10.3390/rs70911105

Hunt, E. R., Daughtry, C. S. T., Eitel, J. U. H., & Long, D. S. (2011). Remote sensing leaf chlorophyll content using a visible band index. *Agronomy Journal, 103*(4), 1090–1099. https://doi.org/10.2134/agronj2010.0395

Igbokwe, J. C., et al. (2024). Community engagement in local governance: A review of USA and African strategies. *World Journal of Advanced Research and Reviews, 21*(2), 105–112. https://doi.org/10.30574/wjarr.2024.21.2.0421

Islam, M. R., & Esraz-Ul-Zannat, M. (2023). Remote sensing based investigation of coastal LULC dynamics in the coastal region of Bangladesh. *Remote Sensing Applications: Society and Environment, 31*, 100982. https://doi.org/10.1016/j.rsase.2023.100982

Islam, M. N., & van Amstel, A. (Eds.). (2022). *India: Climate change impacts, mitigation and adaptation in developing countries*. Springer. . ISBN: 978-3-030-67979-8. https://doi.org/10.1007/978-3-030-67865-4

Islam, M. M., Islam, N., Habib, A., & Mozumder, M. M. H. (2020). Climate change impacts on a tropical fishery ecosystem: Implications and societal responses. *Sustainability, 12*(19), 7970. https://doi.org/10.3390/su12197970

Jabir, A. A., Hasan, G. J., & Anam, M. M. (2021). Correlation between temperature, sea level rise and land loss: An assessment along the Sundarbans coast. *Journal of King Saud University-Engineering Sciences*. https://doi.org/10.1016/j.jksues.2021.07.012

Jacox, M. G., Jacox, M. G., Edwards, C. A., Hazen, E. L., & Bograd, S. J. (2018). Coastal upwelling revisited: Ekman, Bakun, and improved upwelling indices for the U.S. West Coast. *Journal of Geophysical Research, 123*(10), 7332–7350. https://doi.org/10.1029/2018JC014187

Jana, A., Maiti, S., & Biswas, A. (2016). Seasonal change monitoring and mapping of coastal vegetation types along Midnapur-Balasore Coast, Bay of Bengal using multi-temporal landsat data. *Modeling Earth Systems and Environment, 2*, 7. https://doi.org/10.1007/s40808-015-0062-x

Jankowski, P. (2009). Towards participatory geographic information systems for community-based environmental decision making. *Journal of Environmental Management, 90*(6), 1966–1971. https://doi.org/10.1016/j.jenvman.2007.08.028

Jayaraman, T., & Murari, K. (2014). Climate change and agriculture: Current and future trends, and implications for India. *Review of Agrarian Studies, 4*(1), 1–49. https://doi.org/10.25003/RAS.04.01.0001

Jennings, V., San Antonio, K. M., Brown, M. J., Choice, L., Simpson, Q., Ford, I., Solis, P., Lacey, A., & Robinson, D. (2024). Place-based conservation in coastal and marine ecosystems: The importance of engagement with underrepresented communities. *Sustainability, 16*. https://doi.org/10.3390/su16229965

Kaliraj, S., Chandrasekar, N., Ramachandran, K. K., Srinivas, Y., & Saravanan, S. (2017). Coastal landuse and land cover change and transformations of Kanyakumari coast, India using remote sensing and GIS. *The Egyptian Journal of Remote Sensing and Space Science, 20*(2), 169–185. https://doi.org/10.1016/j.ejrs.2017.04.003

Kanan, A. H., Pirotti, F., Masiero, M., & Rahman, M. M. (2023). Mapping inundation from sea level rise and its interaction with land cover in the Sundarbans mangrove Forest. *Climatic Change, 176*(8), 104. https://doi.org/10.1007/s10584-023-03574-5

Kantamaneni, K., Panneer, S., Krishnan, A., et al. (2022). Appraisal of climate change and cyclone trends in Indian coastal states: A systematic approach towards climate action. *Arabian Journal of Geosciences, 15*, 814. https://doi.org/10.1007/s12517-022-10076-8

Kasri, K., Hasani, M. C., Baso, A., & Amiluddin, A. (2024). Perceptions of small-scale fishermen on the impact of climate change in coastal and Small Island areas of Pangkajene and islands regency. *PONGGAWA: Journal of Fisheries Socio-Economic*, 92–107. https://doi.org/10.35911/ponggawa.v4i2.42105

Kayal, P., & Chowdhury, I. R. (2022). Impact of frequent cyclonic storms on land use and land cover changes along the coastal areas of Indian Sundarban. In M. N. Islam & A. V. Amstel (Eds.), *India II: Climate change impacts, mitigation and adaptation in developing countries* (Springer climate). Springer. https://doi.org/10.1007/978-3-030-94395-0_2

Kendall, M. G. (1955). *Rank correlation methods*. Griffin.

Kermani, S., Boutiba, M., Guendouz, M., Guettouche, M. S., & Khelfani, D. (2016). Detection and analysis of shoreline changes using geospatial tools and automatic computation: Case of jijelian sandy coast (East Algeria). *Ocean & Coastal Management, 132*, 46–58. https://doi.org/10.1016/j.ocecoaman.2016.08.010

Key, I. B., Smith, A. C., Turner, B., Chausson, A., Girardin, C. A., Macgillivray, M., & Seddon, N. (2022). Biodiversity outcomes of nature-based solutions for climate change adaptation: Characterising the evidence base. *Frontiers in Environmental Science, 10*, 905767. https://doi.org/10.3389/fenvs.2022.905767

Kospa, H. S. D. (2024). Systematic literature review and future study directions to coastal resilience. *Jurnal Tekno Global UIGM Fakultas Teknik, 13*(02), 66–73. https://doi.org/10.36982/jtg.v13i02.4849

Koudahe, K., Koffi, D., Kayode, J., Awokola, S., & Adebola, A. (2018). Impact of climate variability on crop yields in Southern Togo. *Environment Pollution and Climate Change, 02*(01). https://doi.org/10.4172/2573-458X.1000148

Kuhe, A., Bisu, D. Y., & Iortyer, H. A. (2017). Optimization of cooking energy mix, an alternative strategy to reduce deforestation: An example of households and restaurants in the Bauchi Metropolis, Nigeria. *African Journal of Science, Technology, Innovation and Development, 9*(2), 207–213. https://doi.org/10.1080/20421338.2017.1305653

Kumar, A., Sankalp, S., & Remesan, R. (2023). Spatiotemporal rainfall variability and trend analysis over all the districts of West Bengal during 1980–2021. *Developments in Environmental Science, 14*, 1–15. https://doi.org/10.1016/B978-0-443-18640-0.00004-3

Kundu, K., & Mandal, J. K. (2024). Shoreline change detection and future prediction of Sundarban delta using remote sensing data and digital shoreline analysis system. *Journal of the Indian Society of Remote Sensing, 52*, 485–503. https://doi.org/10.1007/s12524-024-01833-1

Kusmana, C. (2015). Integrated sustainable mangrove forest management. *Journal of Natural Resources and Environmental Management, 5*(1), 1–6.

Lacheheb, M., Noy, I., & Pundit, M. (2025). Impact of tropical cyclones on fishing activity in the Philippines. *Fisheries Research, 281*, 107221. https://doi.org/10.1016/j.fishres.2024.107221

Laczko, F. (2008). *Migration and climate change*. IOM (International Organization for Migration), Geneva, No. 31. ISBN: 978-92-9068-405-3.

Lindegaard, L. S., Jarawura, F. X., & Kleist, N. (2024). Climate migration management? Contrasting international policy approaches with evidence from Ghana. *Climate Policy*, 1–16. https://doi.org/10.1080/14693062.2024.2411326

Liu, H., Koren, I., Altaratz, O., & Chekroun, M. D. (2023). Opposing trends of cloud coverage over land and ocean under global warming. *Atmospheric Chemistry and Physics, 23*(11), 6559–6569. https://doi.org/10.5194/acp-23-6559-2023

Lloyd, N., & Lewis, P. R. (2024). *A socio-technical perspective on fostering sustainable community development via social capital and social cohesion* (pp. 61–66). https://doi.org/10.1109/acsos-c63493.2024.00032

Lukambagire, I., Matovu, B., Manianga, A., Bhavani, R. R., & Anjana, S. (2024). Towards a collaborative stakeholder engagement pathway to increase ocean sustainability related to marine spatial planning in developing coastal states. *Environmental Challenges, 15*, 100954. https://doi.org/10.1016/j.envc.2024.100954

Majumder, S. (2024). Livelihood & socio-economic study on sundarban area: A brief research review. *IOSR Journal of Economics and Finance, 16*(6), 23–29. https://doi.org/10.9790/5933-1506022329

Malakar, K. D., & Roy, S. (2024a). *Mapping geospatial citizenship: The power of participatory GIS*. Springer Nature. ISBN: 978-3-031-63106-1.

Malakar, K. D., & Roy, S. (2024b). Mapping community voices in the coastal region of Bengal: Case studies and best practices of participatory GIS. In *Mapping geospatial citizenship*. Springer. https://doi.org/10.1007/978-3-031-63107-8_9

Malakar, K. D., Kumar, M., Anand, S., & Kuzur, G. (2023). Climate change and socio-ecological transformation: Applications and field narratives. In *Climate change and socio-ecological transformation; advances in geographical and environmental sciences* (pp. 295–374). Springer Nature. https://doi.org/10.1007/978-981-99-4390-6_7

Malhi, G. S., Kaur, M., & Kaushik, P. (2021). Impact of climate change on agriculture and its mitigation strategies: A review. *Sustainability, 13*(3), 1318. https://doi.org/10.3390/su13031318

Mann, H. B. (1945). Nonparametric tests against trend. *Econometrica, 13*(3), 245–259. https://doi.org/10.2307/1907187

Marquez, G. P. B., & Olavides, R. D. (2024). Integrating science-based and local ecological knowledge: A case study of mangrove restoration and rehabilitation projects in the Philippines. *Disasters*, e12630. https://doi.org/10.1111/disa.12630

Maurya, K., Mahajan, S., & Chaube, N. (2021). Remote sensing techniques: Mapping and monitoring of mangrove ecosystem—A review. *Complex & Intelligent Systems, 7*, 2797–2818. https://doi.org/10.1007/s40747-021-00457-z

McMichael, C., Yee, M., Cornish, G., Lutu, A., & McNamara, K. E. (2024). *Adaptive capacity in Pacific Islands: Responding to coastal and climatic change in Nagigi village, Fiji*. https://doi.org/10.31223/x53q6v.

Mhatre, M. A. (2024). Understanding key stressors for marine and coastal biodiversity loss: An overview. *Uttar Pradesh Journal of Zoology, 45*(15), 330–336. https://doi.org/10.56557/upjoz/2024/v45i154248

Mirli, A., Bakas, T., Latinopoulos, D., Kagalou, I., & Spiliotis, M. (2024). Participatory Management of a Mediterranean Lagoon Complex Social-Ecological System Using Intuitionistic Fuzzy TOPSIS. *Sustainability, 16*(23), 10647. https://doi.org/10.3390/su162310647

Mitra, A. (2013). *Sensitivity of mangrove ecosystem to changing climate*. New Delhi. https://doi.org/10.1007/978-81-322-1509-7

Mohamed Rashidi, A. H., Jamal, M. H., Hassan, M. Z., Mohd Sendek, S. S., Mohd Sopie, S. L., & Abd Hamid, M. R. (2021). Coastal structures as beach erosion control and sea level rise adaptation in Malaysia: A review. *Water, 13*(13), 1741. https://doi.org/10.3390/w13131741

Mohammed, M., Sultana, F., Khan, A., Ahammed, S., Reza Saimun, M. S., Bhuiyan, M. S., Srivastava, S. K., Mukul, S. A., & Khan, M. A. S. A. (2024). Assessing vulnerability to cyclone hazards in the world's largest mangrove forest, the sundarbans: A geospatial analysis. *Forests, 15*(10), 1722. https://doi.org/10.3390/f15101722

Mondal, B., Saha, A. K., & Roy, A. (2021). Shoreline extraction and change estimation using geospatial techniques: A study of coastal West Bengal, India. *Proceedings of the Indian National Science Academy, 87*, 595–612. https://doi.org/10.1007/s43538-021-00059-w

Mondal, M., Biswas, A., Haldar, S., Mandal, S., Mandal, P., Bhattacharya, S., & Paul, S. (2022). Climate change, multi-hazards and society: An empirical study on the coastal community of Indian Sundarban. *Natural Hazards Research, 2*(2), 84–96. https://doi.org/10.1016/j.nhres.2022.04.002

Mozumder, M. M. H., Schneider, P., Islam, M. M., Deb, D., Hasan, M., Monzer, M. A., & Nur, A. A. U. (2023). Climate change adaptation strategies for small-scale Hilsa fishers in the coastal area of Bangladesh: Social, economic, and ecological perspectives. *Frontiers in Marine Science, 10*, 1151875. https://doi.org/10.3389/fmars.2023.1151875

Mukherjee, D., & Paul, A. K. (2021). Mangrove sensitivities to climate change and its impacts in the Sundarbans: A case study in the Patibania Island of south western Sundarbans, India. In *Modern cartography series* (Vol. 10, pp. 353–385). Academic Press. https://doi.org/10.1016/B978-0-12-823895-0.00030-0

Mukherjee, N., & Siddique, G. (2024). *Impact of climate change in the Indian Sundarbans region*. Springer . ISBN: 978-3-031-54237-4. https://doi.org/10.1007/978-3-031-54238-1

Mukhopadhyay, A., Hornby, D. D., Hutton, C. W., Lázár, A. N., Amoako Johnson, F., & Ghosh, T. (2018). Land cover and land use analysis in coastal Bangladesh. In R. Nicholls, C. Hutton, W. Adger, S. Hanson, M. Rahman, & M. Salehin (Eds.), *Ecosystem services for well-being in deltas*. Palgrave Macmillan. https://doi.org/10.1007/978-3-319-71093-8_20

N'Souvi, K., Adjakpenou, A., Sun, C., & Ayisi, C. L. (2024). Climate change perceptions, impacts on the catches, and adaptation practices of the small-scale fishermen in Togo's coastal area. *Environmental Development, 49*, 100957. https://doi.org/10.1016/j.envdev.2023.100957

Narayana, A. C. (2016). Shoreline changes. In M. J. Kennish (Ed.), *Encyclopedia of estuaries. Encyclopedia of earth sciences series*. Springer. https://doi.org/10.1007/978-94-017-8801-4_118

Naserikia, M., Hart, M. A., Nazarian, N., Bechtel, B., Lipson, M., & Nice, K. A. (2023). Land surface and air temperature dynamics: The role of urban form and seasonality. *Science of the Total Environment, 905*, 167306. https://doi.org/10.1016/j.scitotenv.2023.167306

Nasution, M., Hussin, H., Rusli, Z., Mayarni, M., Mashur, D., Zulkarnaini, Z., & Ananda, F. (2025). Integration of environmental policies in sustainable Management of Mangrove Ecotourism. *E3S Web of Conferences, 611*, 04003. https://doi.org/10.1051/e3sconf/202561104003

Nath, A., Koley, B., Choudhury, T., Saraswati, S., Ray, B. C., Um, J. S., & Sharma, A. (2023). Assessing coastal land-use and land-cover change dynamics using geospatial techniques. *Sustainability, 15*(9), 7398. https://doi.org/10.3390/su15097398

Noguera, I., Vicente-Serrano, S. M., Peña-Angulo, D., Domínguez-Castro, F., Juez, C., Tomás-Burguera, M., et al. (2023). Assessment of vapor pressure deficit variability and trends in Spain and possible connections with soil moisture. *Atmospheric Research, 285*, 106666. https://doi.org/10.1016/j.atmosres.2023.106666

Nugraha, R., Wildan, D. M., Rahmadya, A., Setiawan, F., & Ridwansyah, I. (2023). *Effect of rainfall on water condition and catch results of eel (Anguilla spp.) in Cicatih River, Sukabumi, West Java. 1260*, 012020. https://doi.org/10.1088/1755-1315/1260/1/012020

Oino, P. G., & Musau, E. G. (2024). Community engagement in climate change and adaptation in Kenya: A socio-anthropological and linguistic perspective. *African Journal of Climate Change and Resource Sustainability, 3*(1), 387–404. https://doi.org/10.37284/ajccrs.3.1.2444

Ojea, E., Lester, S. E., & Salgueiro-Otero, D. (2020). Adaptation of fishing communities to climate-driven shifts in target species. *One Earth, 2*(6), 544–556. https://doi.org/10.1016/J.ONEEAR.2020.05.012

On-prom, S. (2014). Community-based mangrove Forest Management in Thailand: Key lesson learned for environmental risk management. In N. Kaneko, S. Yoshiura, & M. Kobayashi (Eds.), *Sustainable living with environmental risks*. Springer. https://doi.org/10.1007/978-4-431-54804-1_8

Pablos, M., Martínez-Fernández, J., Piles, M., Sánchez, N., Vall-llossera, M., & Camps, A. (2016). Multi-temporal evaluation of soil moisture and land surface temperature dynamics using in situ and satellite observations. *Remote Sensing, 8*(7), 587. https://doi.org/10.3390/RS8070587

Pamungkas, S. (2023). Analysis of vegetation index for Ndvi, Evi-2, and Savi for mangrove Forest density using Google Earth engine in Lembar Bay, Lombok Island. *IOP Conference Series: Earth and Environmental Science, 1127*(1), 012034. https://doi.org/10.1088/1755-1315/1127/1/012034

Pan, S., Zhao, X., & Yue, Y. (2019). Spatiotemporal changes of NDVI and correlation with meteorological factors in Northern China from 1985-2015. *E3S Web of Conferences, 131*, 01040. https://doi.org/10.1051/e3sconf/201913101040

Panda, A. (2020). Climate change, displacement, and managed retreat in coastal India. *Migration Information. The Online Journal of the Migration Policy Institute*. Accessed 14 Feb 2023 https://www.migrationpolicy.org/article/climate-change-displacementmanaged-retreat-india

Pandey, A. G., Mondal, A., Guha, S., Upadhyay, P. K., & Singh, D. M. (2023). A long-term analysis of the dependency of land surface temperature on land surface indices. *Papers in Applied Geography*, 1–16. https://doi.org/10.1080/23754931.2023.2187314

Patel, S., Indraganti, M., & Jawarneh, R. N. (2024). Land surface temperature responses to land use dynamics in urban areas of Doha, Qatar. *Sustainable Cities and Society, 104*, 105273. https://doi.org/10.1016/j.scs.2024.105273

Patil, P., Rajah, A. R., Sivakumar, B., Bora, N. R., Vasanth, V., Sri, S. D., Vas, M. A., Rajaram, R., Prathyusha, K., Datta, P. L., & Nath, I. (2024). A comprehensive review on sustainable cultivation practices of medicinal plants. *International Journal of Plant and Soil Science, 36*(9), 536–544. https://doi.org/10.9734/ijpss/2024/v36i95002

Patra, T. (2024). Impacts of climate change on the sundarbans mangrove ecosystem: A comprehensive analysis. *International Research Journal of Modernization in Engineering Technology and Science, 6*(8), 1171–1179. https://doi.org/10.56726/irjmets60970

Patra, P. K., Behera, D., Chettry, V., et al. (2025). Geospatial analysis of unplanned urbanization: impact on land surface temperature and habitat suitability in Cuttack, India. *Discover Sustainability, 6*, 118. https://doi.org/10.1007/s43621-025-00920-8

Pattanaik, A. (2021). *Mangrove-dependent small-scale fisher (SSF) communities in the Sundarbans-vulnerable yet viable*. Thesis submitted to University of Waterloo, Ontario, Canada.

Paul, S., Das, C. S., & Chaudhuri, S. (2023). Erosion-accretion pattern in the tidal-Rivers of Indian Sundarban in relation to estuarine hydrodynamics. *Continental Shelf Research, 252*, 104865.

Pereira, G. F., Gurugnanam, B., Goswami, S., et al. (2022). Application of geospatial techniques to determine coastal erosion and accretion along the Ramanathapuram shore, Tamil Nadu, India. *Journal of the Geological Society of India, 98*, 1261–1270. https://doi.org/10.1007/s12594-022-2161-4

Pham, T. D., Yokoya, N., Bui, D. T., Yoshino, K., & Friess, D. A. (2019). Remote sensing approaches for monitoring mangrove species, structure, and biomass: Opportunities and challenges. *Remote Sensing, 11*(3), 230. https://doi.org/10.3390/rs11030230

Pical, V. J., & Rahman, R. (2025). Alternative livelihoods of traditional fishermen as the impact of climate change in Ambon City. *Egyptian Journal of Aquatic Biology and Fisheries, 29*(1), 243–256. https://doi.org/10.21608/ejabf.2025.404165

Picuno, P., Sica, C., Dimitrijevic, A., Tortora, A., Capobianco, R. L., & Statuto, D. (2015). New technologies for the sustainable management and planning of rural land and environment. In A. Vastola (Ed.), *The sustainability of agro-food and natural resource systems in the mediterranean basin*. Springer. https://doi.org/10.1007/978-3-319-16357-4_21

Pignatti, S., Carfora, M. F., Coluzzi, R., D'Amato, L., De Feis, I., Mora, D. F., Laneve, G., Imbrenda, V., Lanfredi, M., Mirzaei, S., Palombo, A., Pascucci, S., Rossi, F., Santini, F., Simoniello, T., & Vanguri, R. (2024). *Detection of critical areas prone to land degradation using Prisma: The Metaponto coastal area in South Italy test case* (pp. 1063–1066). IEEE. https://doi.org/10.1109/igarss53475.2024.10641499

Pitchaikani, J. S. (2020). *Integrated coastal zone management practices for Sundarbans, India*. NISCAIR-CSIR, India. http://nopr.niscair.res.in/handle/123456789/54432

Poddar, D., Mandal, T., & Das, J. (2023). Spatio-temporal changes of rainfall pattern under changing climate in West Bengal, India. In *Climate change, agriculture and society: Approaches toward sustainability* (pp. 15–36). Springer. https://doi.org/10.1007/978-3-031-28251-5_2

Prabhu Gaonkar, V. G., Nadaf, F. M., BalajiraoKapale, V., Gaonkar, S., Shetkar, S., & D'Silva, M. (2022). Analysing Spatio-temporal changes in land surface temperature of coastal Goa using LANDSAT satellite data. In U. Chatterjee, A. O. Akanwa, S. Kumar, S. K. Singh, & A. Dutta Roy (Eds.), *Ecological footprints of climate change* (Springer climate). Springer. https://doi.org/10.1007/978-3-031-15501-7_20

Pradeep, J., Shaji, E., Chandran, S., Ajas, H., Chandra, S. V., Dev, S. D., & Babu, D. S. (2022). Assessment of coastal variations due to climate change using remote sensing and machine learning techniques: A case study from west coast of India. *Estuarine, Coastal and Shelf Science, 275*, 107968. https://doi.org/10.1016/j.ecss.2022.107968

Prananda, A. R. A., Kamal, M., & Wijaya Kusuma, D. (2020). The effect of using different vegetation indices for mangrove leaf area index modelling. *IOP Conference Series: Earth and Environmental Science, 500*(1), 012006. https://doi.org/10.1088/1755-1315/500/1/012006

Prasita, V. D., Subianto, A., & Asbar, A. (2016). *Evaluation of zonation of the mangrove conservation areas (MCA) in Pamurbaya*. https://dspace.hangtuah.ac.id/xmlui/handle/dx/1074.

Puthucherril, T. G. (2024). *Protecting the coastal and marine environment* (pp. 443–462). https://doi.org/10.1093/oxfordhb/9780198884682.013.26

Putri, D. O. A., Ermawati, M., & Yanto, N. P. (2024). Analisis dampak perubahan penggunan lahan perkotaan terhadap urban heat island di kecamatan kedaton. *Jurnal Perencanaan Dan Pengembangan Kebijakan, 4*(1), 27. https://doi.org/10.35472/jppk.v4i1.1464

Puyravaud, J. P. (2003). Standardizing the calculation of the annual rate of deforestation. *Forest Ecology and Management, 177*(1), 593–596.

Rahim, A., Hastuti, D. R. D., Ahmadin, A., Sari, D. A. P., & Afandi, M. (2024). Survival decisions and adaptation strategies of small-scale fishers in the face of extreme weather impacts in coastal areas. *Journal of Marine and Island Cultures, 13*(3). https://doi.org/10.21463/jmic.2024.13.3.05

Ramieri, E., Bocci, M., & Markovic, M. (2019). Linking integrated coastal zone management to maritime spatial planning: The Mediterranean experience. In J. Zaucha & K. Gee (Eds.), *Maritime spatial planning*. Palgrave Macmillan. https://doi.org/10.1007/978-3-319-98696-8_12

Rasyid, A., As, M. A., Nurdin, N., & Jaya, I. (2016). Ibrahim. Impact of human interventions on mangrove ecosystem in spatial perspective. *IOP Conference Series: Earth and Environmental Science, 47*, 012041. https://doi.org/10.1088/1755-1315/47/1/012041

Restrepo-Mieth, A., Perry, J., Garnick, J., & Weisberg, M. (2023). Community-based participatory climate action. *Global Sustainability (Cambridge University Press), 6*. https://doi.org/10.1017/sus.2023.12

Reza, M. S., & Hossain, A. (2024). Development and performance assessment of a hybrid solar cabinet dryer for fish drying. *Energy and Thermofluids Engineering, 3*. https://doi.org/10.38208/ete.v3.758

Roberts, D. A., Green, R. O., & Adams, J. B. (1997). Temporal and spatial patterns in vegetation and atmospheric properties from AVIRIS. *Remote Sensing of Environment, 62*(3), 223–240. https://doi.org/10.1016/S0034-4257(97)00092-8

Rodrigues, J. N., Bhattacharya, S., Sachdev, B. K., & Cabete, D. C. R. (2024). Disaster management in the Sundarbans delta ecosystem through environmental governance and sustainable planning. *Revista de Direito Econômico e Socioambiental., 15*, e268. https://doi.org/10.7213/revdireconsoc.v15i2.31364

Ross, H., Shaw, S., Rissik, D., et al. (2015). A participatory systems approach to understanding climate adaptation needs. *Climatic Change, 129*, 27–42. https://doi.org/10.1007/s10584-014-1318-6

Röthig, T., Trevathan-Tackett, S. M., Voolstra, C. R., Ross, C., Chaffron, S., Durack, P. J., & Sweet, M. (2023). Human-induced salinity changes impact marine organisms and ecosystems. *Global Change Biology, 29*, 4731. https://doi.org/10.1111/gcb.16859

Rouse, J. W., Haas, R. H., Schell, J. A., Deering, D. W., et al. (1974). Monitoring vegetation systems in the Great plains with ERTS. *NASA Special Publications, 351*(1), 309.

Rout, P. R. (2024). Planning for resilience: Examining gaps in cyclone preparedness and impacts on marginalized communities in India's east coast cities. *Indian Scientific Journal of Research in Engineering and Management, 08*(05), 1–5. https://doi.org/10.55041/ijsrem32478

Roy, P., Pal, S. C., Chakrabortty, R., Chowdhuri, I., Saha, A., & Shit, M. (2023). Effects of climate change and sea-level rise on coastal habitat: Vulnerability assessment, adaptation strategies and policy recommendations. *Journal of Environmental Management, 330*, 117187. https://doi.org/10.1016/j.jenvman.2022.117187

Roy, A., Kumar, S., & Rahaman, M. (2024a). Exploring climate change impacts on rural livelihoods and adaptation strategies: Reflections from marginalized communities in India. *Environmental Development, 49*, 100937. https://doi.org/10.1016/j.envdev.2023.100937

Roy, S. S., Ghosh, T., & Hewavithana, D. K. (2024b). Transformation of coastal wetlands in the Sundarban Delta (1999–2020). *Environmental Monitoring and Assessment, 196*, 758. https://doi.org/10.1007/s10661-024-12901-x

Roy, S. K., Mojumder, P., Chowdhury, A. A., & Hasan, M. (2025). Evaluating mangrove forest dynamics and fragmentation in Sundarbans, Bangladesh via high-resolution satellite images (Sentinel-2). *Global Ecology and Conservation*, e03493. https://doi.org/10.1016/j.gecco.2025.e03493

Rutkay, A., Kalkan, K., & Gürsoy, Ö. (2020). Determining the Forest fire risk with sentinel 2 images. *Turkish Journal of Geosciences, 1*(1), 22–26.

Saengsupavanich, C., Agarwala, N., Magdalena, I., et al. (2024). Impacts of a growing population on the coastal environment of the Bay of Bengal. *Anthropocene Coasts, 7*, 24. https://doi.org/10.1007/s44218-024-00055-9

Sahana, M., Ahmed, R., & Sajjad, H. (2016). Analyzing land surface temperature distribution in response to land use/land cover change using split window algorithm and spectral radiance model in Sundarban Biosphere Reserve, India. *Modeling Earth Systems and Environment, 2*, 81. https://doi.org/10.1007/s40808-016-0135-5

Salam, M., Islam, M. K., Jahan, I., et al. (2024). Assessing the impacts of vegetation loss and land surface temperature on surface urban Heat Island (SUHI) in Gazipur District, Bangladesh. *Computational Urban Science, 4*, 24. https://doi.org/10.1007/s43762-024-00136-y

Samanta, S., & Paul, S. K. (2016). Geospatial analysis of shoreline and land use/land cover changes through remote sensing and GIS techniques. *Modeling Earth Systems and Environment, 2*, 108. https://doi.org/10.1007/s40808-016-0180-0

Samanta, S., Hazra, S., French, J. R., Nicholls, R. J., & Mondal, P. P. (2023). Exploratory modelling of the impacts of sea-level rise on the Sundarbans mangrove forest, West Bengal, India. *Science of the Total Environment, 903*, 166624. https://doi.org/10.1016/j.scitotenv.2023.166624

Sampath, V. K., & Radhakrishnan, N. (2023). A comparative study of LULC classifiers for analysing the cover management factor and support practice factor in RUSLE model. *Earth Science Informatics, 16*, 733–751. https://doi.org/10.1007/s12145-022-00911-7

Samsuddin, N. S., Wan Talaat, W. I. A., Saputra, J., & Salleh, H. S. (2024). The utilization of traditional knowledge on the impact of climate change towards island communities' social well-being: The mediating effect of adaptive capacity. *Planning Malaysia Journal, 22*. https://doi.org/10.21837/pm.v22i34.1626

Sandilyan, S., & Kathiresan, K. (2012). Mangrove conservation: A global perspective. *Biodiversity and Conservation, 21*, 3523–3542. https://doi.org/10.1007/s10531-012-0388-x

Saoum, M. R., & Sarkar, S. K. (2024). Monitoring mangrove forest change and its impacts on the environment. *Ecological Indicators, 159*, 111666. https://doi.org/10.1016/j.ecolind.2024.111666

Sarkar, M., Paul, S., & Garai, J. (2024). Climate-induced coastal occupational vulnerability and livelihood insecurity: Insights from coastal Bangladesh. *Progress in Disaster Science, 24*, 100382. https://doi.org/10.1016/j.pdisas.2024.100382

Sartorius, J. V., Geddes, A., Gagnon, A. S., & Burnett, K. A. (2024). Participation and co-production in climate adaptation: Scope and limits identified from a meta-method review of research with European coastal communities. *Wiley Interdisciplinary Reviews: Climate Change, 15*(3), e880. https://doi.org/10.1002/wcc.880

Savo, V., Morton, C., & Lepofsky, D. (2017). Impacts of climate change for coastal fishers and implications for fisheries. *Fish and Fisheries, 18*(5), 877–889. https://doi.org/10.1111/faf.12212

Seddiky, M. A., Ara, E., & Karim, A. (2024). Climate change-induced hazard risks and migration in Bangladesh: A case study. *Current Research in Environmental Sustainability, 7*, 100253. https://doi.org/10.1016/j.crsust.2024.100253

Sekhar, M., Lallawmkimi, M. C., Vijaya Rani, D., Upadhyay, L., Tiwari, P., Verma, U., & Rajput, A. (2024). Agroforestry practices and their impact on soil health and fertility: A review. *Journal of Experimental Agriculture International, 46*(9), 511–528. https://doi.org/10.9734/jeai/2024/v46i92849

Sele, J. P., & Mukundi, M. B. (2024). Community-based approaches to environmental conservation: Empowering local initiatives. *Greener Journal of Social Sciences, 14*(2), 289–299. https://doi.org/10.15580/gjss.2024.2.122024211

Sen, P. K. (1968). Estimates of the regression coefficient based on Kendall's tau. *Journal of the American Statistical Association, 63*, 1379. https://doi.org/10.1080/01621459.1968.10480934

Shaffril, H. A. M., Samah, A. A., & D'Silva, J. L. (2017). Climate change: Social adaptation strategies for fishermen. *Marine Policy, 81*, 256–261. https://doi.org/10.1016/j.marpol.2017.03.031

Shampa, M. T. A., Shimu, N. J., Chowdhury, K. A., Islam, M. M., & Ahmed, M. K. (2023). A comprehensive review on sustainable coastal zone management in Bangladesh: Present status and the way forward. *Heliyon, 9*(8), e18190. https://doi.org/10.1016/j.heliyon.2023.e18190

Sharadi, P., Pal, N., Mitra, A., Pramanick, P., Zaman, S., & Mitra, A. (2020). *Climate change in Indian Sundarbans using air temperature as proxy*. Konnagar Municipality, Govt. of West Bengal (Project: Climate Changel A, B, C of Environment), Article 6.

Sheik, M., & Chandrasekar. (2011). A shoreline change analysis along the coast between Kanyakumari and Tuticorin, India, using digital shoreline analysis system. *Geo-Spatial Information Science, 14*, 282–293. https://doi.org/10.1007/s11806-011-0551-7

Shetty, A., Jayappa, K. S., Ramakrishnan, R., et al. (2019). Shoreline dynamics and vulnerability assessment along the Karnataka coast, India: A geo-statistical approach. *Journal of the Indian Society of Remote Sensing, 47*, 1223–1234. https://doi.org/10.1007/s12524-019-00980-0

Silchenko, D., & Murray, U. (2023). Migration and climate change–the role of social protection. *Climate Risk Management, 39*, 100472. https://doi.org/10.1016/j.crm.2022.100472

Sinha, D. (2022). Sundarbans' climate refugees face an uncertain future. *DW (Nature and Environment, India)*. Retrieved 31 January 2025. Available at: https://www.dw.com/en/forgotten-people-sundarbans-climate-refugees-forced-to-move-again/a-61162969

Song, Y., Shen, Y., Xie, R., et al. (2021). A DSAS-based study of central shoreline change in Jiangsu over 45 years. *Anthropocene Coasts, 4*, 115–128. https://doi.org/10.1139/anc-2020-0001

Soriani, S., Buono, F., Tonino, M., & Camuffo, M. (2015). Participation in ICZM initiatives: Critical aspects and lessons learnt from the Mediterranean and Black Sea experiences. *Marine Pollution Bulletin, 92*(1–2), 143–148. https://doi.org/10.1016/j.marpolbul.2014.12.045

Sparkes, E., & Werners, S. E. (2023). Monitoring, evaluation and learning requirements for climate-resilient development pathways. *Current Opinion in Environmental Sustainability, 64*, 101329. https://doi.org/10.1016/j.cosust.2023.101329

Stark, G., Ma, L., Zeng, Z. G., Du, W. G., & Levy, O. (2023). Cool shade and not-so-cool shade: How habitat loss may accelerate thermal stress under current and future climate. *Global Change Biology, 29*(22), 6201–6216. https://doi.org/10.1111/gcb.16802

Styk, K., & Bogacz, P. (2022). A method for stakeholder mapping in connection with the implementation of a development project. *Energies, 15*(4), 1592. https://doi.org/10.3390/en15041592

Subramanian, A., Nagarajan, A. M., Vinod, S., Chakraborty, S., Sivagami, K., Theodore, T., et al. (2023). Long-term impacts of climate change on coastal and transitional eco-systems in India: An overview of its current status, future projections, solutions, and policies. *RSC Advances, 13*(18), 12204–12228. https://doi.org/10.1039/D2RA07448F

Sun, Z., An, Y., Kong, J., Zhao, J., Cui, W., Nie, T., et al. (2024). Exploring the spatio-temporal patterns of global mangrove gross primary production and quantifying the factors affecting its estimation, 1996–2020. *Science of the Total Environment, 908*, 168262. https://doi.org/10.1016/j.scitotenv.2023.168262

Sunkur, R., Kantamaneni, K., Bokhoree, C., & Ravan, S. (2023). Mangroves' role in supporting ecosystem-based techniques to reduce disaster risk and adapt to climate change: A review. *Journal of Sea Research, 196*, 102449. https://doi.org/10.1016/j.seares.2023.102449

Sunkur, R., Kantamaneni, K., Bokhoree, C., et al. (2024). Mangrove mapping and monitoring using remote sensing techniques towards climate change resilience. *Scientific Reports, 14*, 6949. https://doi.org/10.1038/s41598-024-57563-4

Surni, N. H., Harsuko, R., Leksono, A. S., & Baja, S. (2024). The impact of land use change on land surface temperature 1990-2020 for environmental monitoring in Kapota Island, Wakatobi regency, Indonesia. *International Review for Spatial Planning and Sustainable Development, 12*, 43. https://doi.org/10.14246/irspsd.12.2_43

Swapna, P., et al. (2020). Sea-level rise. In R. Krishnan, J. Sanjay, C. Gnanaseelan, M. Mujumdar, A. Kulkarni, & S. Chakraborty (Eds.), *Assessment of climate change over the Indian region*. Springer. https://doi.org/10.1007/978-981-15-4327-2_9

Tabari, H. (2020). Climate change impact on flood and extreme precipitation increases with water availability. *Scientific Reports, 10*, 13768. https://doi.org/10.1038/s41598-020-70816-2

Tanwattana, P., & Toyoda, Y. (2024). Climate resilience action to policy: Embodied community-based climate risk resilience (CBCRR) in drought prone community in Thailand's economic corridor. *Climate Policy*, 1–18. https://doi.org/10.1080/14693062.2024.2430682

Thakur, S., Maity, D., Mondal, I., et al. (2021). Assessment of changes in land use, land cover, and land surface temperature in the mangrove forest of Sundarbans, northeast coast of India. *Environment, Development and Sustainability, 23*, 1917–1943. https://doi.org/10.1007/s10668-020-00656-7

Thinh, N. A. (2022). Studying shoreline change in Ky Anh coastal area of Ha Tinh Province during 1989–2013 based on the digital shoreline analysis system (DSAS). In *Global changes and sustainable development in Asian emerging market economies, 2*. Springer. https://doi.org/10.1007/9783-030-81443-4_33

Tran, T. V., Reef, R., & Zhu, X. (2022). A review of spectral indices for mangrove remote sensing. *Remote Sensing, 14*(19), 4868. https://doi.org/10.3390/rs14194868

Tripathy, S. (2023). *Fishers, community resilience, and disaster management: Learning from the grassroots of Odisha, India* (pp. 355–368). https://doi.org/10.1007/978-981-19-8388-7_204

Turral, H., Burke, J., & Faurès, J. (2011). *Climate change, water and food security*. Food and Agriculture Organization of the United Nations. ISBN: 978-92-5-106795-6.

UNESCO. (1984). *Sundarbans National Park: World heritage*. United Nations Educational, Scientific and Cultural Organization. Retrieved 6 November 2023. https://whc.unesco.org/en/list/452

UNESCO. (2023). *The Sundarban*. United Nations Educational. Accessed 30 Dec 2023 https://whc.unesco.org/en/decisions/8287

Upadhyay, V. P., Ranjan, R., & Singh, J. S. (2002). Human–mangrove conflicts: The way out. *Current Science, 83*(11), 1328–1336. http://repository.ias.ac.in/72920/

Vadrevu, K. P., Gutman, G., Tsuneo, M., & Justice, C. (2023). Land use changes, degradation, and impact on ecosystem services in Asia and Southeast Asia. *LDD, 35*, 919. https://doi.org/10.1002/ldr.4971

Van Hespen, R., Hu, Z., Borsje, B., De Dominicis, M., Friess, D. A., Jevrejeva, S., et al. (2023). Mangrove forests as a nature-based solution for coastal flood protection: Biophysical and ecological considerations. *Water Science and Engineering, 16*(1), 1–13. https://doi.org/10.1016/j.wse.2022.10.004

Van Stan, J. T., Underwood, S. J., & Friesen, J. (2018). *Urban forestry: An underutilized tool in water management* (Vol. 3, pp. 35–61). https://doi.org/10.1016/BS.APMP.2018.04.003

Vasquez, J., Acevedo-Barrios, R., Miranda-Castro, W., et al. (2024). Determining changes in mangrove cover using remote sensing with landsat images: A review. *Water, Air, and Soil Pollution, 235*, 18. https://doi.org/10.1007/s11270-023-06788-6

Veeravalli, S. G. (2024). Assessing the impact of vegetation and its spatial configuration on land surface temperature in selected Indian cities. In *Proceedings of SPIE 12977, eighth geoinformation science symposium 2023: Geoinformation science for sustainable planet* (p. 129771N). https://doi.org/10.1117/12.3009564

Verma, K., Sharma, P., Bhardwaj, D. R., Sharma, V., & Thakur, P. (2024). *Towards a greener future: Scaling up agroforestry for global sustainability*. https://doi.org/10.47852/bonviewglce42022481

VijayKumar, R., Tiwari, P., Daniel, S., Kumar, K. R., Mishra, I., Ks, A., & Shah, D. G. (2024). Agroforestry systems: A pathway to resilient and productive landscapes. *International Journal of Enviornment and Climate Change, 14*(12), 177–193. https://doi.org/10.9734/ijecc/2024/v14i124617

Viswanathan, P. (2016). Multifunctional benefits of community-based mangrove restoration in Gujarat: An analysis. In N. Ghosh, P. Mukhopadhyay, A. Shah, & M. Panda (Eds.), *Nature, economy and society*. Springer. https://doi.org/10.1007/978-81-322-2404-4_12

Viswanathan, M., Sreekumar, A., Sridharan, S., et al. (2024). Addressing grand challenges through the bottom-up marketing approach: Lessons from subsistence marketplaces and marketplace literacy. *Journal of the Academy of Marketing Science, 52*, 1279–1300. https://doi.org/10.1007/s11747-024-01022-z

Wagner, P. M., & Lima, V. (2023). The interests, ideas, and institutions shaping public participation in local climate change governance in Ireland. *Local Environment*, 1–15. https://doi.org/10.1080/13549839.2023.2284946

Wahyudi, A., Deffinika, I., Arif, M., Setiawan, A. R., Priadmodjo, A., & Erwin, M. A. (2024). Climate change impact and adaptation of traditional fishermen in Rangas Village. *Majene Regency, 1406*, 012003. https://doi.org/10.1088/1755-1315/1406/1/012003

Wan Mohd Jaafar, W. S., Abdul Maulud, K. N., et al. (2020). The influence of deforestation on land surface temperature: A case study of and Kedah, Malaysia. *Forests, 11*(6), 670. https://doi.org/10.3390/f11060670

Wang, J., Zhou, L., & Yang, X. (2009). Geographic information systems and spatial analysis for coastal ecosystem research and management. In X. Yang (Ed.), *Remote sensing and geospatial technologies for coastal ecosystem assessment and management* (Lecture notes in geoinformation and cartography). Springer. https://doi.org/10.1007/978-3-540-88183-4_3

Wang, L., Jia, M., Yin, D., & Tian, J. (2019). A review of remote sensing for mangrove forests: 1956–2018. *Remote Sensing of Environment, 231*, 111223. https://doi.org/10.1016/j.rse.2019.111223

Wang, Q., Ma, Y., Cheng, Z., & Du, Y. (2023). Coastline changes under natural and anthropogenic drivers in a macro-tidal estuary between 2000-2020. *Frontiers in Marine Science, 10*, 1335064. https://doi.org/10.3389/fmars.2023.1335064

Wannewitz, M., & Garschagen, M. (2024). The role of social identities for collective adaptation capacities– General considerations and lessons from Jakarta, Indonesia. *International Journal of Disaster Risk Reduction, 100*, 104194. https://doi.org/10.1016/j.ijdrr.2023.104194

Ward, R. D., Friess, D. A., Day, R. H., & Mackenzie, R. A. (2016). Impacts of climate change on mangrove ecosystems: A region by region overview. *Ecosystem Health and Sustainability, 2*(4), e01211. https://doi.org/10.1002/ehs2.1211

Wibisono, G., Sardjono, M., Suhardiman, A., et al. (2023). Community-based mangrove management in Berau Regency. In *IOP conference series: Earth and environmental science* (Vol. 1192, p. 012019). IOP Publishing.

Widigdo, B., Ilman, M., Santoso, N., Kusmana, C., & Krisanti, M. (2024). Performance of integrated mangrove-aquaculture (IMA) in District of Sidoarjo, East Java Indonesia. *BIO Web of Conferences, 92*, 01022. https://doi.org/10.1051/bioconf/20249201022

Win, K. S., & Sasaki, J. (2024). The change detection of mangrove forests using deep learning with medium-resolution satellite imagery: A case study of Wunbaik Mangrove Forest in Myanmar. *Remote Sensing, 16*(21), 4077. https://doi.org/10.3390/rs16214077

Wu, X. Q., Gao, M., Wang, D., Wang, Y., Lu, Q. S., & Zhang, Z. D. (2012). Framework and practice of integrated coastal zone management in Shandong Province, China. *Ocean & Coastal Management, 69*, 58–67. https://doi.org/10.1016/j.ocecoaman.2012.07.030

Xu, H. (2006). Modification of normalised difference water index (NDWI) to enhance open water features in remotely sensed imagery. *International Journal of Remote Sensing, 27*, 3025–3033. https://doi.org/10.1080/01431160600589179

Xu, D., Wang, Y., & Wang, J. (2024). A review of social-ecological system vulnerability in desertified regions: Assessment, simulation, and sustainable management. *Science of the Total Environment*, 172604. https://doi.org/10.1016/j.scitotenv.2024.172604

Yang, X. (Ed.). (2009). *Remote sensing and geospatial technologies for coastal ecosystem Assessment and management.* Springer . ISBN: 978-3-540-88182-7. https://doi.org/10.1007/978-3-540-88183-4

Yoo, C., Im, J., Cho, D., Yokoya, N., Xia, J., & Bechtel, B. (2020). Estimation of all-weather 1 km MODIS land surface temperature for humid summer days. *Remote Sensing, 12*(9), 1398. https://doi.org/10.3390/rs12091398

Yu, H. (2023). *Research on ecological environmental vulnerability assessment and monitoring based on remote sensing images* (pp. 436–441). IEEE. https://doi.org/10.1109/icpsitiags59213.2023.10527731

Yuan, L., Pu, R., Zhang, J., Wang, J., & Yang, H. (2016). Using high spatial resolution satellite imagery for mapping powdery mildew at a regional scale. *Precision Agriculture, 17*, 332–348. https://doi.org/10.1007/s11119-015-9421-x

Zaba, K. D., Franks, P., & Ohman, M. D. (2021). The California undercurrent as a source of upwelled waters in a coastal filament. *Journal of Geophysical Research, 126*(2). https://doi.org/10.1029/2020JC016602

Zamani, N. P., Rahimah, I., Lestarina, P. M., Harahap, Z. A., & Paembonan, R. E. (2024). Moving from an integrated coastal management approach to a transdisciplinary coastal management approach: A review of case study in Indonesia. *BIO Web of Conferences, 106*, 02011. https://doi.org/10.1051/bioconf/202410602011

Zambrano-Medina, Y. G., Plata-Rocha, W., Monjardin-Armenta, S. A., & Franco-Ochoa, C. (2023). Assessment and forecast of shoreline change using geo-spatial techniques in the Gulf of California. *Land, 12*(4), 782. https://doi.org/10.3390/land12040782

Zhang, B., Zhang, L., Chen, B., Deng, L., Fu, B., Yan, M., & Ji, C. (2024a). Assessment of mangrove health based on pressure–state–response framework in Guangxi Beibu Gulf, China. *Ecological Indicators, 167*, 112685. https://doi.org/10.1016/j.ecolind.2024.112685

Zhang, Y., Li, H., Li, D., Hou, X., Guo, P., & Guo, J. (2024b). Evolutionary dynamics of island shoreline in the context of climate change: Insights from extensive empirical evidence. *International Journal of Digital Earth, 17*(1). https://doi.org/10.1080/17538947.2024.2329816

Zhang, Z., Wang, Z., Liang, B., Leng, X., Yang, B., & Shi, L. (2024c). Shoreline change analysis in the estuarine area of Rizhao based on remote sensing images and numerical simulation. *Frontiers in Marine Science, 11*, 1488577. https://doi.org/10.3389/fmars.2024.1488577

Zhao, Y., An, R., Xiong, N., Ou, D., & Jiang, C. (2021). Spatio-temporal land-use/land-cover change dynamics in coastal plains in Hangzhou Bay Area, China from 2009 to 2020 using Google Earth engine. *Land, 10*(11), 1149. https://doi.org/10.3390/land10111149

Zhao, M., et al. (2022). Evapotranspiration frequently increases during droughts. *Nature Climate Change, 12*, 1024–1030. https://doi.org/10.1038/s41558-022-01505-3

Zhao, K., Zhu, G., Sang, L., Liu, J., Wang, L., Liu, Y., et al. (2023). Temporal and spatial variation of cloud cover in arid regions of Central Asia in the last 40 years. *Research in Cold and Arid Regions, 15*(2), 66–72. https://doi.org/10.1016/j.rcar.2023.06.002

Zhao, Y., Han, Z., Zhang, C.-R., Wang, Y., Zhong, J., & Gao, M. (2024). Coastal cultural ecosystem services: A bridge between the natural ecosystem and social ecosystem for sustainable development. *Land, 13*(9), 1352. https://doi.org/10.3390/land13091352

Zhou, L., Chen, N., Chen, Z., & Xing, C. (2016). ROSCC: An efficient remote sensing observation-sharing method based on cloud computing for soil moisture mapping in precision agriculture. *IEEE Journal of Selected Topics in Applied Earth Observations and Remote Sensing, 9*(12), 5588–5598. https://doi.org/10.1109/JSTARS.2016.2574810

Zhu, L., Song, R., Sun, S., Li, Y., & Hu, K. (2022). Land use/land cover change and its impact on ecosystem carbon storage in coastal areas of China from 1980 to 2050. *Ecological Indicators, 142*, 109178. https://doi.org/10.1016/j.ecolind.2022.109178

Zouari, M., & Hachicha, M. (2024). Agroforestry as a sustainable solution for mitigating climate change impacts and addressing water scarcity in Tunisia: A review. *Journal of Oasis Agriculture and Sustainable Development, 6*(02), 22–30. https://doi.org/10.56027/joasd.182024

Part VI
Transforming Coastal Research: Geographic Information Science in Transdisciplinary Solutions and Strategic Decision Making

Chapter 17
Advancing Coastal Research: Integration of GIScience for Transdisciplinary Solutions and Strategic Decision-Making

"In today's world, transdisciplinary collaboration is the compass to navigate complex challenges, and geospatial science is our map." — by K. D. Malakar (Author)

Abstract This chapter explores the role of GIScience in advancing coastal research, emphasizing transdisciplinary approaches and strategic decision-making for sustainable coastal management. It highlights the role of GIS technologies in providing spatially explicit data and analytical tools that facilitate informed decision-making, adaptive management, and multi-stakeholder collaboration. The chapter discusses key themes such as the integration of GIS into coastal policy frameworks, the use of AI and machine learning for predictive modelling, and the importance of adaptive management approaches for dynamic coastal systems. The chapter addresses challenges such as data gaps, technological limitations, and governance issues, alongside ethical considerations in participatory research. It also explores the potential of remote sensing technologies, including LiDAR and UAVs, as well as cloud-based GIS platforms, in enhancing the monitoring and evaluation of coastal policies. The chapter concludes by highlighting the importance of ongoing capacity building, ethical practices, and inclusive participation to ensure GIS-based coastal research fosters long-term environmental sustainability and community resilience.

Keywords Advances in coastal research · GIScience in coastal research · Disciplinary principles in coastal research · Coastal management · Glocal approaches · Strategic decision-making for coastal sustainability · Coastal ecologies · Transdisciplinary solutions · Tools and techniques of GIScience

This chapter delves into the transformative potential of GIScience in coastal research, emphasizing its role in advancing sustainable management through transdisciplinary collaboration and strategic decision-making. It explores how GIS technologies provide essential spatial data and analytical tools for informed

K. D. Malakar et al., *Geospatial Technologies in Coastal Ecologies Monitoring and Management*, Advances in Geographic Information Science,
https://doi.org/10.1007/978-3-031-92017-2_17

decision-making, adaptive management, and multi-stakeholder involvement. The chapter addresses emerging themes such as AI-driven predictive modelling, the integration of GIS into policy frameworks, and the importance of adaptive management in dynamic coastal environments. It also tackles challenges like data gaps, technological barriers, and governance issues, while emphasizing ethical considerations and the need for continuous capacity building.

Key Points of the Chapter

- Exploring disciplinary principles in advancing coastal research.
- Various Tools and Techniques of GIScience in Coastal Research.
- Understanding the integration of GIScience for transdisciplinary solutions and strategic decision-making.

17.1 Advancing Coastal Research

17.1.1 History of Coastal Research

The study of coastal environments has a rich history, reflecting humanity's enduring connection with coastlines. Early research, rooted in natural history and maritime exploration, focused on coastlines for navigation, trade, and settlement. Ancient civilizations, such as Greece, Rome, and India, documented coastal geography and tides for commercial and defensive purposes. The nineteenth century marked a shift with the adoption of modern scientific methods. Researchers, influenced by pioneers like Charles Lyell, systematically explored coastal geomorphology, tidal processes, and sediment transport, laying the groundwork for theories on erosion, deposition, and delta formation. In the twentieth century, advancements such as aerial photography, satellite imagery, and GIS propelled coastal research, enabling detailed studies of wave action, longshore drift, and human impacts like urbanization. The field became increasingly interdisciplinary, integrating geology, biology, sociology, and climatology. Today, coastal research addresses global challenges like sea-level rise, climate change, and disaster risk. Coastal zones are examined as complex socio-ecological systems, emphasizing sustainability and resilience. Collaborative approaches and advanced modelling continue to enhance understanding in this critical field.

17.1.2 Disciplinary Principles and Advancing Coastal Research: Intra-discipline, Multidisciplinary, Interdisciplinary, and Transdisciplinary Approaches

Coastal research has evolved significantly as a discipline, shaped by the complexity and diversity of coastal environments and the pressing challenges they encounter. These dynamic socio-ecological systems represent the intersection of natural processes and human activities. Addressing the multifaceted issues of coastal areas has necessitated the adoption of diverse methodological frameworks, ranging from intra-disciplinary and multidisciplinary to interdisciplinary and transdisciplinary approaches (Fig. 17.1). Each of these pathways offers unique avenues for generating knowledge and driving innovation in coastal studies. There are such as;

(a) **Intra-disciplinary Approach:** The intra-disciplinary approach focuses on studying coastal systems within the framework of a single academic discipline. This specialization enables researchers to explore specific aspects of coastal environments in depth. For instance, coastal geomorphologists examine shoreline dynamics, sediment transport, and erosion patterns, while marine ecologists concentrate on biodiversity and habitat conservation within coastal zones.

    This approach significantly advances foundational knowledge and discipline-specific theories and methodologies. For example, understanding wave energy and its role in sediment deposition is essential for coastal engineering, while research in coastal climatology on wind patterns and storm surges informs risk assessments and disaster preparedness. However, while the intra-disciplinary approach provides critical depth, it often lacks the integrative perspective required to address the multifaceted and interconnected challenges of coastal systems, which demand insights from diverse fields.

(b) **Multidisciplinary Approach:** The multidisciplinary approach brings together experts from various disciplines, each addressing coastal issues through their distinct lenses. While disciplines remain independent, their collective contributions provide a broader understanding of coastal systems. For instance, a multidisciplinary investigation into coastal erosion might involve geologists studying physical processes, economists evaluating financial impacts, and sociologists examining community responses. A key strength of this approach lies in its ability to aggregate diverse perspectives, fostering a more comprehensive understanding of complex coastal challenges. However, it often encounters dif-

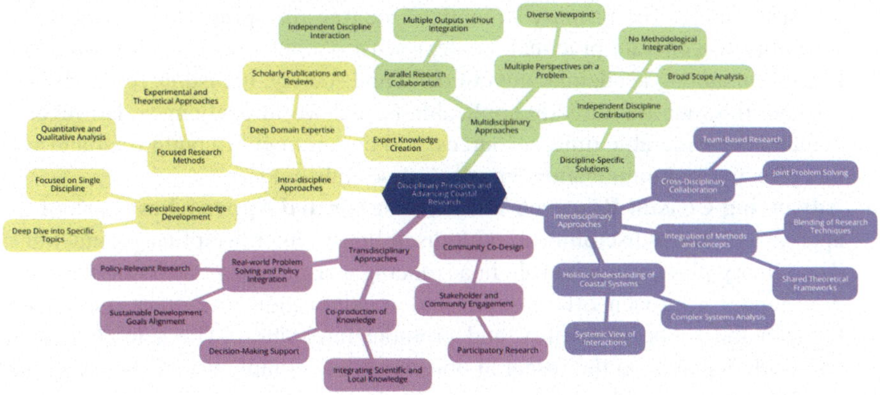

Fig. 17.1 Disciplinary principles and advancing coastal research: intra-discipline, multidisciplinary, interdisciplinary, and transdisciplinary approaches (Prepared by the authors)

ficulties in harmonizing findings across disciplines. The absence of methodological and conceptual integration may result in fragmented outcomes, limiting the full potential of collaborative expertise.

(c) **Interdisciplinary Approach:** The interdisciplinary approach moves beyond collaboration by integrating theories, methods, and perspectives from multiple disciplines to develop new frameworks for understanding coastal systems. It acknowledges that the dynamic interactions between natural and social processes in coastal areas cannot be fully captured through isolated disciplinary perspectives.

For example, interdisciplinary research on sea-level rise might combine physical oceanography, climate science, and urban planning to evaluate both the physical impacts and the socio-economic consequences of rising seas on coastal cities. By merging quantitative models with qualitative insights, such studies generate holistic solutions that are both scientifically rigorous and socially relevant. The strength of this approach lies in its ability to bridge knowledge gaps and foster innovative solutions. However, it demands effective communication, collaboration, and a commitment to transcending disciplinary boundaries. Institutional support and frameworks that facilitate integration are essential for the success of interdisciplinary initiatives.

(d) **Transdisciplinary Approach:** The transdisciplinary approach is the most integrative and participatory framework in coastal research, transcending academic boundaries by engaging non-academic stakeholders such as policymakers, community members, and industry representatives. It aims to co-create knowledge that is both academically rigorous and socially actionable, tailored to specific contexts.

For instance, addressing the climate vulnerability of fishing communities in the Sundarbans might involve scientists analysing ecosystem changes, local communities sharing indigenous knowledge, and policymakers offering insights on governance and resource allocation. This approach prioritizes participatory methods, ensuring diverse perspectives are included and solutions are equitable and sustainable. The primary strength of the transdisciplinary approach lies in its ability to generate practical, context-sensitive outcomes. By fostering collaboration among academia, government, and society, it ensures research findings are relevant and directly applicable to real-world challenges. However, it demands considerable time, resources, and commitment to build trust and sustain effective communication among diverse stakeholders.

(e) **Advancing Coastal Research Through Integrated Approaches:** Each of the approaches, intra-disciplinary, multidisciplinary, interdisciplinary, and transdisciplinary plays a crucial role in advancing coastal research. While the intra-disciplinary approach offers depth and specialization, the other approaches broaden the scope of inquiry and promote integration. The selection of an approach depends on the research objectives, the complexity of the issue, and the stakeholders involved.

In addressing contemporary challenges such as climate change, sea-level rise, habitat degradation, and coastal urbanization, integrated approaches are increasingly essential. For instance, managing coastal resilience in the face of rising seas requires expertise in climate modelling, engineering, policy development, and community engagement. Collaborative frameworks that leverage the strengths of different approaches can provide comprehensive solutions to these urgent issues.

In summary, advancing coastal research necessitates embracing the diversity of disciplinary frameworks and harnessing their collective strengths. By promoting collaboration, integration, and active participation, researchers can tackle the complexities of coastal systems and support their sustainable management. The future of coastal research depends on the ability to effectively navigate and integrate these approaches, ensuring that knowledge generation and application address the needs of both society and the environment.

17.1.3 Need for Transdisciplinary Approaches in Coastal Studies

Transdisciplinary approaches in coastal studies are essential due to the complex and interconnected nature of coastal systems, where natural processes, human activities, and socio-economic factors intersect. Coastal zones face urgent challenges such as climate change, sea-level rise, habitat loss, and socio-economic vulnerabilities, which cannot be effectively addressed through single-discipline perspectives. These approaches involve collaboration among scientists, policymakers, local communities, and other stakeholders to co-create practical, inclusive, and sustainable solutions. By integrating scientific knowledge with local and indigenous insights, transdisciplinary methods ensure strategies are context-specific and equitable, addressing both ecological and social dimensions.

For example, tackling coastal erosion in vulnerable areas like the Sundarbans requires not only technical expertise in geomorphology but also an understanding of community livelihoods, governance systems, and cultural practices. Transdisciplinary methods promote participatory decision-making and prioritize long-term resilience, making them indispensable for managing the multifaceted challenges of coastal zones in an era of rapid change.

17.1.4 Role of GIScience in Advancing Coastal Research and Transdisciplinary Decision-Making

GIScience is pivotal in advancing coastal research and supporting transdisciplinary decision-making through its robust tools for spatial analysis, data integration, and visualization. Coastal systems are inherently dynamic and complex, shaped by interactions among physical, ecological, and socio-economic processes. GIScience empowers researchers and stakeholders to map, monitor, and model these interactions with precision and efficiency. In coastal research, GIS tools are extensively applied to analyse shoreline changes, habitat distributions, flood risks, and climate impacts. The integration of remote sensing with GIS provides real-time data for monitoring coastal dynamics, while spatial modelling enables the projection of future scenarios under various environmental and anthropogenic pressures. Within transdisciplinary decision-making, GIS fosters collaboration by unifying diverse datasets, such as ecological studies, socio-economic surveys, and community knowledge into a cohesive spatial framework. This integration allows scientists, policymakers, and local communities to collectively visualize and analyse complex coastal issues, facilitating informed and equitable decision-making.

For instance, GIS-based participatory mapping allows local communities to share their knowledge and priorities, contributing directly to coastal management plans. By bridging science and practice, GIScience strengthens the capacity to address coastal challenges and promotes sustainable, inclusive solutions.

17.2 Transdisciplinary Integration in Coastal Research

Transdisciplinary integration in coastal research emphasizes the importance of bridging academic disciplines, policy frameworks, and community perspectives to tackle the complex challenges of coastal environments. Coastal zones are dynamic systems where natural processes, such as erosion and sea-level rise, intersect with human activities like urbanization, fishing, and tourism. These interconnections necessitate approaches that extend beyond traditional disciplinary boundaries. Transdisciplinary research fosters collaboration among scientists, policymakers, and local communities to co-create practical, inclusive, and context-specific knowledge and solutions. By integrating diverse expertise and stakeholder inputs, this approach promotes a comprehensive understanding and sustainable management of coastal systems, ensuring that ecological, social, and economic factors are all considered in the decision-making process (Fig. 17.2).

Fig. 17.2 Transdisciplinary Integration in coastal research (Prepared by the authors)

17.2.1 Bridging Disciplinary Silos: Science, Policy, and Practice

In coastal studies, where environmental, social, economic, and political factors intersect, this theme is vital for promoting collaboration among diverse academic disciplines, governmental bodies, local communities, and stakeholders. By dismantling traditional disciplinary boundaries, a more comprehensive approach to coastal research can be realized, effectively addressing the complex challenges faced by these ecosystems.

The Need for Bridging Disciplinary Silos
Coastal areas face a range of pressures, including overfishing, climate change, habitat degradation, and urbanization. Addressing these challenges requires a multidisciplinary understanding that encompasses ecology, geography, economics, policy, and social sciences. Traditionally, these disciplines operate in isolation, leading to fragmented knowledge that may fail to effectively inform policy or guide practical actions. Bridging these disciplinary silos is crucial for developing integrated solutions. In coastal research, this fragmented approach can result in several issues:

- **Incomplete Understanding**: Ecological studies may neglect socio-economic factors, while policy decisions might overlook scientific data or local insights.
- **Inefficient Resource Use**: Disciplinary silos can lead to duplicated research efforts or fail to incorporate the needs and perspectives of affected communities.
- **Limited Policy Implementation**: Policymakers may struggle to translate complex scientific findings into actionable policies without input from practitioners or stakeholders.

Transdisciplinary Integration
Transdisciplinary integration extends beyond interdisciplinary approaches, which generally involve collaboration among academic disciplines. It encompasses the active engagement of external stakeholders, such as local communities, industry experts, policymakers, and NGOs who contribute directly to the research process. This broader, more inclusive approach ensures that research is not only scientifically robust but also socially relevant and practically applicable. In the context of coastal research, transdisciplinary integration involves:

- **Collaborative Knowledge Generation**: Coastal scientists work alongside local communities to document traditional ecological knowledge (TEK) and combine it with modern scientific insights. This collaborative process ensures that policies and practices reflect the lived experiences of coastal populations.
- **Co-Designing Solutions**: Researchers, policymakers, and practitioners collaborate from the beginning to co-design research agendas and potential solutions, ensuring that both scientific and practical considerations are integrated into decision-making.

- **Feedback Loops**: Transdisciplinary research establishes continuous feedback between science, policy, and practice. Scientists provide up-to-date research to inform policies, while policymakers and practitioners identify knowledge gaps and propose areas for further exploration.
- **Adaptation to Local Contexts**: A transdisciplinary approach fosters the development of context-specific solutions, tailored to the unique ecological, cultural, and economic conditions of each coastal region.

Benefits of Bridging Silos in Coastal Research

- **Enhanced Problem-Solving:** A transdisciplinary approach provides a more holistic understanding of coastal issues, leading to more effective solutions that integrate ecological, social, and economic perspectives.
- **Informed Policy:** Involving scientists, policymakers, and local stakeholders ensures that policies are scientifically sound, socially acceptable, and practically feasible.
- **Sustainable Development:** Combining scientific, policy, and practical knowledge facilitates sustainable coastal management, balancing resource demands with ecosystem preservation.
- **Strengthened Community Involvement:** Engaging local communities in the research process fosters ownership and responsibility, improving the success of conservation and management efforts.
- **Increased Resilience:** Leveraging diverse knowledge systems enhances the development of resilient strategies for coastal communities to adapt to climate change and other environmental stressors.

Challenges in Bridging Disciplinary Silos

- **Cultural and Epistemological Differences:** Different disciplines and stakeholders offer distinct ways of knowing and understanding, requiring open communication and mutual respect to bridge these differences.
- **Institutional Barriers:** Rigid disciplinary structures in academic institutions and government bodies often hinder transdisciplinary research. Additionally, funding mechanisms and incentives tend to favor traditional, discipline-specific research.
- **Power Imbalances:** Involving local communities or non-scientific stakeholders may lead to power imbalances, with some voices marginalized or excluded from decision-making processes.
- **Time and Resource Intensive:** Collaborative transdisciplinary research demands significant time, effort, and resources for coordination among various parties, presenting a challenge, particularly in resource-constrained settings.

Therefore, the transdisciplinary integration is essential for addressing the complexities of coastal research. Engaging scientists, policymakers, practitioners, and local communities fosters an inclusive and holistic approach to coastal management. This collaborative process produces solutions that are scientifically robust, socially relevant, and practically viable, ensuring the sustainability of coastal ecosystems and the resilience of dependent communities. By bridging disciplinary divides, coastal research can effectively confront current challenges and build a foundation for long-term resilience.

17.2.2 Multi-scale Approaches to Coastal Management: Global to Local Perspectives

Multi-scale approaches to coastal management are pivotal to transdisciplinary integration in coastal research, as they highlight the importance of addressing coastal challenges across various spatial, temporal, and governance scales. Effective coastal management necessitates the integration of knowledge and practices at global, national, and local levels, ensuring that strategies are scientifically grounded and contextually appropriate (Fig. 17.3).

- **Global to Local Integration:** Coastal challenges, such as climate change, pollution, and biodiversity loss, carry global ramifications but are most acutely experienced at the local level. Addressing these issues requires a multi-scale approach that accounts for global drivers of change while responding to the unique realities of local contexts. This ensures that policies designed at international or national levels are effectively adapted and implemented on the ground.
- **Global Perspectives:** At the global level, research and policy frameworks provide overarching guidance on climate change mitigation, sustainable development, and transboundary environmental management. While these frameworks are essential for setting broad priorities, their practical applicability often requires customization to address the specific conditions of individual coastal areas, highlighting the need for local engagement.
- **Local Perspectives:** Local communities possess deep, context-specific knowledge of their environments and the challenges they face. Incorporating this

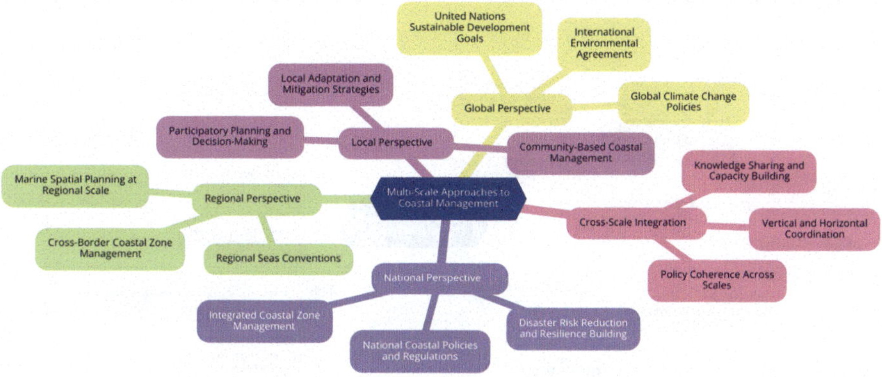

Fig. 17.3 Multi-scale approaches to coastal management: global to local perspectives (Prepared by the authors)

knowledge into coastal management strategies ensures solutions are grounded in the socio-economic, cultural, and ecological contexts of each area. This localized focus enhances the practicality and sustainability of interventions.

- **Transdisciplinary Integration in Multi-scale Approaches:** A transdisciplinary approach to multi-scale coastal management fosters collaboration among scientists, policymakers, practitioners, and local stakeholders. This approach bridges the gap between high-level global frameworks and the specific needs of local communities, ensuring coherent and effective governance.

  (a) **Feedback Loops:** Local experiences generate insights that inform global and national policies, while global research provides tools and frameworks to support localized efforts. This iterative exchange creates a dynamic and adaptive management process.
  (b) **Cross-Scale Collaboration:** Effective governance requires alignment and coordination across scales. Transdisciplinary efforts facilitate cohesive policies that resonate from the global level down to local implementation.

Therefore, the integrating global and local perspectives through a multi-scale, transdisciplinary approach is critical for addressing coastal issues. By combining scientific expertise with local knowledge and engaging diverse stakeholders, this approach enables the development of scalable, adaptable, and sustainable solutions for coastal governance.

17.2.3 Role of Stakeholders in Coastal Research

Stakeholders play a pivotal role in transdisciplinary coastal research, contributing to the co-creation and implementation of knowledge essential for sustainable coastal management. These stakeholders include scientists, government agencies, local communities, non-governmental organizations (NGOs), the private sector, and policymakers (Fig. 17.4). Each group brings unique expertise and perspectives, enriching the research and enhancing its applicability.

Fig. 17.4 Role of stakeholders in coastal research (Prepared by the authors)

Stakeholder Contributions

- **Scientists** contribute empirical data, theoretical models, and technical expertise to address ecological, environmental, and socio-economic challenges in coastal areas.
- **Local Communities** provide traditional ecological knowledge (TEK) and first-hand experiences in managing coastal resources, ensuring research and policies are tailored to local contexts and cultural nuances.
- **Government Agencies** play a central role in establishing policies, regulations, and funding mechanisms, bridging research findings with actionable strategies at regional and national levels.
- **NGOs** serve as intermediaries, facilitating communication between communities and policymakers. They also drive advocacy, mobilize resources, and enhance community capacity.
- **Private Sector** stakeholders, including industries such as tourism, fisheries, and infrastructure development, contribute by adopting sustainable practices, offering funding, or sharing technical resources and data.

Importance in Transdisciplinary Integration
The active involvement of diverse stakeholders in transdisciplinary coastal research ensures that outcomes are both scientifically rigorous and socially relevant. This collaboration leads to:

- **Inclusive Decision-Making:** Incorporating diverse viewpoints ensures research outcomes address the needs and values of all stakeholders, increasing acceptance and likelihood of implementation.
- **Effective Solutions:** Stakeholder engagement ensures management strategies are practical, context-sensitive, and aligned with broader environmental and socio-economic objectives.
- **Sustainability:** Stakeholder-driven approaches foster long-term solutions by balancing ecological conservation with the economic and social needs of coastal communities.

The integration of diverse stakeholders in transdisciplinary coastal research is indispensable for developing holistic, effective, and sustainable management strategies. By leveraging the unique contributions of each group, this approach addresses the multifaceted challenges facing coastal ecosystems and communities, ensuring resilient and equitable outcomes.

17.3 GIScience in Coastal Research: Tools and Techniques

Geographic Information Science (GIScience) plays a crucial role in fostering transdisciplinary integration within coastal research by offering advanced spatial analysis and visualization tools. These capabilities support decision-making across environmental science, policy development, and community engagement, addressing the complex challenges of coastal areas.

Role of GIScience in Coastal Research
GIScience provides researchers and stakeholders with essential tools and techniques for collecting, analysing, and visualizing spatial data. This enhances understanding of the complex interactions between human activities and coastal ecosystems at various scales, offering valuable insights for sustainable management.

Key Tools and Techniques

- **Remote Sensing:** Satellite and drone imagery enable the monitoring of coastal changes, such as land-use modifications, vegetation health, and shoreline dynamics. These data reveal long-term trends, such as erosion and the impacts of climate change.
- **Spatial Analysis:** GIS integrates diverse datasets, including population demographics, environmental variables, and infrastructure. Analytical techniques like buffer analysis, overlay analysis, and spatial interpolation identify hotspots, assess vulnerabilities, and guide conservation or development efforts.
- **Modeling and Simulation:** GIS tools model coastal processes, including tidal patterns, storm surges, and flood risks. These predictive models are vital for adaptive management, particularly in addressing climate change impacts.
- **Geospatial Databases:** GIScience supports the development of comprehensive geospatial databases that store and manage vast amounts of coastal data. These databases bridge scientific research with local knowledge, ensuring accessibility and usability for stakeholders across sectors.

Contribution to Transdisciplinary Integration

- **Bridging Knowledge Gaps:** GIS provides a common platform for collaboration among scientists, policymakers, local communities, and other stakeholders, enabling joint analysis and interpretation of spatial data.
- **Inclusive Decision-Making:** The visual and interactive nature of GIS outputs, such as maps and dashboards, simplifies complex environmental and social dynamics, empowering non-experts, including local communities, to actively participate in decision-making.
- **Evidence-Based Policy:** GIScience delivers spatially explicit data that informs regulatory frameworks and management practices, ensuring policies are both scientifically grounded and locally relevant.

GIScience is a transformative tool in coastal research, driving transdisciplinary integration through its capacity for spatial analysis, visualization, and data management. By offering a shared framework for collaboration among scientists, policymakers, and communities, GIScience facilitates data-driven, inclusive decision-making processes essential for the sustainable management of coastal ecosystems.

17.4 Innovations and Emerging Trends in GIScience for Coastal Research

Advances in GIScience are transforming the study and management of coastal environments by integrating cutting-edge technologies and fostering transdisciplinary collaboration. These innovations enhance adaptive, inclusive, and efficient strategies for addressing complex coastal challenges by bridging gaps between disciplines, stakeholders, and communities.

(a) **Role of Artificial Intelligence (AI) and Machine Learning (ML):** AI and ML are revolutionizing coastal research by automating data analysis and improving predictive modelling.

- **Enhanced Predictive Models:** AI-driven tools forecast coastal changes like erosion and habitat loss, aiding long-term sustainability planning.
- **Automated Data Interpretation:** Machine learning processes large, complex datasets quickly, supporting timely and informed decision-making.

(b) **Big Data and Cloud-Based GIS:** Big data and cloud-based platforms streamline the collection, processing, and sharing of spatial information.

- **Real-Time Monitoring:** These systems enable dynamic responses to coastal issues such as pollution or storms.
- **Collaborative Decision-Making:** Cloud platforms enhance cross-sector collaboration by providing shared access to up-to-date data.

(c) **Advancements in Remote Sensing Technology:** Technologies like LiDAR and UAVs (drones) offer high-resolution data for detailed mapping and monitoring.

- **LiDAR:** Provides precise 3D topographic data critical for assessing erosion, flooding, and land-use changes.
- **UAVs:** Affordable drones capture fine-scale imagery, enabling real-time habitat and impact assessments.

(d) **Citizen Science and Mobile Applications:** Citizen science initiatives and mobile apps empower local communities to actively participate in coastal monitoring.

- **Mobile Apps:** Tools allow citizens to contribute geospatial data on conditions like water quality and erosion.
- **Community Engagement:** Local involvement integrates cultural insights and fosters shared responsibility for sustainable management.

Innovations in GIScience encompassing AI, big data, remote sensing, and citizen science are reshaping coastal research and governance. By fostering transdisciplinary integration and inclusive decision-making, these advancements ensure more data-driven, collaborative, and adaptive approaches to the sustainable management of coastal ecosystems.

17.5 Strategic Decision-Making for Coastal Sustainability

Strategic decision-making is vital for maintaining resilient and sustainable coastal ecosystems amid challenges such as climate change, population growth, and economic development. Effective decisions require transdisciplinary integration, where collaboration among scientists, policymakers, and communities ensures that diverse knowledge systems shape sustainable management strategies.

(a) **Integrating GIScience into Coastal Policy:** GIScience provides spatially explicit, data-driven insights that guide policy development by visualizing and analysing complex coastal data.

- **Spatial Planning:** Identifies priority areas for conservation, urban development, or disaster preparedness, ensuring policies are context-specific.
- **Informed Decisions:** Offers real-time data and predictive models to craft responsive strategies for dynamic coastal conditions.

(b) **Adaptive Management for Dynamic Systems:** Adaptive management allows coastal strategies to evolve with continuous monitoring and feedback, addressing the unpredictable nature of coastal environments.

- **Continuous Learning:** Policies are adjusted based on real-time evaluation to remain effective amid uncertainty.
- **Stakeholder Engagement:** Ensures diverse perspectives, including local and scientific inputs, shape decision-making processes.

(c) **Multi-criteria Decision Analysis (MCDA):** MCDA supports policy prioritization by evaluating options against environmental, social, economic, and political factors.

- **Holistic Evaluation:** Balances diverse stakeholder priorities for comprehensive decision-making.
- **Trade-Off Analysis:** Identifies sustainable solutions that balance conservation with development.

(d) **Monitoring and Evaluation of Policies:** Robust monitoring systems ensure that policies meet sustainability goals and identify areas for improvement.

- **Indicators and Metrics:** Track biodiversity, water quality, and socio-economic impacts, providing actionable feedback.
- **Evaluation Frameworks:** Assess policy outcomes to inform future strategies and improve effectiveness.

Therefore, the strategic coastal sustainability decision-making relies on integrating GIScience, adaptive management, MCDA, and monitoring tools. These approaches ensure data-driven, flexible, and inclusive policies capable of addressing the complex and dynamic challenges of coastal ecosystems, fostering their long-term resilience.

17.6 Challenges in Advancing Coastal Research Through GIScience

The use of GIS technologies in addressing coastal management challenges is complex, particularly within the framework of transdisciplinary integration. This approach requires collaboration among diverse stakeholders, including scientists, policymakers, and local communities, each contributing unique perspectives and expertise. Although GIScience holds considerable potential to advance coastal research and inform sustainable management strategies, several critical challenges must be addressed to ensure its effective and equitable application.

(a) **Data Gaps and Accessibility:** A lack of comprehensive, high-quality, and accessible data is a critical challenge in coastal research, particularly in remote or under-resourced regions.

- **Inconsistent Data:** Incomplete, outdated, or incompatible datasets hinder effective integration and analysis.
- **Data Accessibility:** Restricted access to spatial data due to proprietary, legal, or logistical constraints limits research progress.

(b) **Technological Barriers and Capacity Needs:** Technological limitations and insufficient training restrict the effective use of GIS tools, particularly in resource-limited regions.

- **Infrastructure Deficits:** Limited access to GIS hardware, software, and connectivity affects data utilization.
- **Capacity Building:** Training stakeholders at all levels is crucial to ensure GIS technologies are effectively applied.

(c) **Institutional and Governance Challenges:** Fragmented governance and disciplinary silos impede the integration of GIS into coastal decision-making.

- **Fragmented Governance:** Overlapping jurisdictions and uncoordinated efforts weaken the application of GIS-based insights.
- **Limited Collaboration:** Institutional silos hinder the interdisciplinary integration needed for effective coastal management.

(d) **Ethical Considerations:** The participatory use of GIS raises ethical concerns related to data privacy and community equity.

- **Data Privacy:** Clear guidelines are needed to safeguard sensitive information about communities and ecosystems.
- **Empowerment vs. Exploitation:** Participatory GIS must prioritize equitable benefits for communities, avoiding exploitative practices.

Addressing data gaps, technological limitations, institutional barriers, and ethical concerns is essential for realizing the full potential of GIScience in coastal management. By fostering collaboration, building capacity, and upholding ethical research practices, GIS technologies can support sustainable coastal management, benefiting ecosystems and communities alike.

17.7 Conclusion and Endnotes

The integration of GIScience into coastal research presents transformative opportunities for advancing the understanding and management of coastal systems. By offering spatially explicit data and robust analytical capabilities, GIS technologies enhance decision-making to address complex coastal challenges, including climate change, erosion, and habitat degradation. A transdisciplinary approach ensures the inclusion of diverse perspectives scientific, policy-oriented, and community-based leading to more inclusive and sustainable solutions. Key components of this integration include:

- **Adaptability:** Advanced tools such as AI, remote sensing, and cloud-based GIS facilitate adaptive management, enabling continuous monitoring and dynamic adjustments to evolving coastal conditions.
- **Collaborative Decision-Making:** GIS fosters effective collaboration among stakeholders across sectors, ensuring management strategies address both ecological and socio-economic considerations.
- **Long-Term Sustainability:** By embedding GIS tools within policy frameworks, coastal research promotes sustainability through data-driven policies grounded in accurate and up-to-date spatial information.

Endnotes

- The successful integration of GIScience into coastal research hinges on addressing challenges like data gaps, technological barriers, and the need for capacity building at both local and institutional levels.
- Effective transdisciplinary coastal management demands active stakeholder participation, prioritizing ethics, data privacy, and community empowerment.
- Advancements in GIS technologies and enhanced cross-disciplinary collaboration offer significant potential for tackling emerging coastal sustainability challenges.

Further Readings

1. Agnew, S., Kopke, K., Power, O. P., Troya, M. D. C., & Dozier, A. (2022). Transdisciplinary research: Can citizen science support effective decision-making for coastal infrastructure management? *Frontiers in Marine Science, 9*, 809284. https://doi.org/10.3389/fmars.2022.809284
2. Bartlett, D., & Celliers, L. (2017). *Geoinformatics for marine and coastal management*. CRC Press ISBN: 9780367873684.

3. Bartlett, D., & Smith, J. (Eds.). (2019). *GIS for coastal zone management*. CRC Press ISBN 9780367393977.
4. Batchi, M., & Moumane, A. (2024). *Applying remote sensing and GIS for spatial analysis and decision-making*. IGI Global Scientific Publication ISBN: 9798369364529.
5. Benham, C. F., & Daniell, K. A. (2016). Putting transdisciplinary research into practice: A participatory approach to understanding change in coastal social-ecological systems. *Ocean & Coastal Management, 128*, 29–39. https://doi.org/10.1016/j.ocecoaman.2016.04.005
6. Bill, R., Blankenbach, J., Breunig, M., et al. (2022). Geospatial information research: State of the art, case studies and future perspectives. *Journal of Photogrammetry, Remote Sensing and Geoinformation Science, 90*, 349–389. https://doi.org/10.1007/s41064-022-00217-9
7. Celliers, L., Scott, D., Ngcoya, M., et al. (2021). Negotiation of knowledge for coastal management? Reflections from a transdisciplinary experiment in South Africa. *Humanities and Social Sciences Communications, 8*, 207. https://doi.org/10.1057/s41599-021-00887-7
8. Dronkers, J., & de Vries, I. (1999). Integrated coastal management: The challenge of transdisciplinarity. *Journal of Coastal Conservation, 5*, 97–102. https://doi.org/10.1007/BF02802745
9. Foale, S. (2006). Challenging coasts: Transdisciplinary excursions into integrated coastal zone development. *Development, 49*, 138–139. https://doi.org/10.1057/palgrave.development.1100281
10. Green, D. R. (Ed.). (2009). *Coastal and marine geospatial technologies*. Springer . ISBN: 978-1-4020-9719-5. https://doi.org/10.1007/978-1-4020-9720-1
11. Jonathan, M. P., Krishnamurthy, R. R., Srinivasalu, S., & Glaeser, B. (Eds.). (2018). *Coastal management: Global challenges and innovations*. Elsevier Science ISBN: 9780128104750.
12. Malone, T.C., Malej, A., and Faganeli, J. (Eds) (2020). Coastal ecosystems in transition: A comparative analysis of the Northern Adriatic and Chesapeake Bay. . ISBN: 978-1-119-54358-9.
13. Ruggerio, C. A., Lindig-Cisneros, R., & Méndez-Lemus, Y. (2024). A complex transdisciplinary approach to achieve water sustainability: Lessons from a case study in Morelia, Mexico. *Sustainability Science, 19*, 921–933. https://doi.org/10.1007/s11625-024-01503-8
14. Siew, T. F., Aenis, T., Spangenberg, J. H., et al. (2016). Transdisciplinary research in support of land and water management in China and Southeast Asia: Evaluation of four research projects. *Sustainability Science, 11*, 813–829. https://doi.org/10.1007/s11625-016-0378-0
15. Strand, M., Ortega-Cisneros, K., Niner, H. J., Wahome, M., Bell, J., Currie, J. C., et al. (2022). Transdisciplinarity in transformative ocean governance research—Reflections of early career researchers. *ICES Journal of Marine Science, 79*(8), 2163–2177. https://doi.org/10.1093/icesjms/fsac165

Author Contributions

Monograph conceptualization: K.D.M.; project proposal development, correspondence, and management: K.D.M.; lead author and supervision: K.D.M.; co-author: S.R. and M.K.; field investigation and data calculation: K.D.M. and S.R.; field narrative thoughts: K.D.M.; methodologies and software: K.D.M., M.K., and S.R.; tables, figures, and mapping: K.D.M. and S.R.; resources, research, and writing: K.D.M., S.R., and M.K.; reviewing roles: K.D.M., S.R., M.K., and other anonymous reviewers; editing: K.D.M.; original draft design and preparation: K.D.M.; proofing and others: K.D.M., S.R., and M.K.

K.D.M.: Kousik Das Malakar; S.R.: Supriya Roy; and M.K.: Manish Kumar

K. D. Malakar et al., *Geospatial Technologies in Coastal Ecologies Monitoring and Management*, Advances in Geographic Information Science,
https://doi.org/10.1007/978-3-031-92017-2

Glossary

Accuracy The degree of closeness between a derived position and the true location.

Aerial Photography The process of capturing images of the surface from an elevated position, typically from an aircraft.

Air Photo Interpretation The process of extracting information from aerial photographs to understand the landscape.

Attribute Query A search or selection of data based on specific attribute criteria.

Backup and Recovery Procedures for safeguarding and restoring data in case of loss or damage.

Buffer Analysis Creating zones or areas around geographic features based on a specified distance or criteria.

Buffer A zone or area around a point, line, or polygon feature defined by a specified distance.

Capacity Building Strengthening the skills and knowledge of community members in GIS and related areas.

Coastal Ecology The study of ecosystems and habitats in coastal areas.

Coastal Monitoring The ongoing observation and assessment of environmental conditions in coastal regions.

Coastal Resilience The capacity of coastal ecosystems to endure and adapt to disturbances and environmental shifts while maintaining their ecological functions and resilience.

Coastal Wetlands Low-lying, water-saturated areas along coastlines that support unique ecosystems.

Coastal Zone Management Regulations and strategies for the sustainable development and conservation of coastal regions.

Community Mapping The participatory approach to map-making that incorporates input from community members.

Conservation Planning Strategies and actions to protect and manage coastal ecosystems and biodiversity.

K. D. Malakar et al., *Geospatial Technologies in Coastal Ecologies Monitoring and Management*, Advances in Geographic Information Science,
https://doi.org/10.1007/978-3-031-92017-2

Data Dictionary A repository of metadata that describes the structure and properties of data in a database.

Data Validation The process of verifying the accuracy and quality of collected geographic information.

Data Visualization The graphical representation of data to communicate information effectively.

Database Management System (DBMS) Software that manages and controls access to a database.

Database A structured collection of data organized for efficient retrieval and management.

Digital Image A visual representation of an object or scene in digital form, composed of discrete pixels.

Elevation The vertical distance above or below a reference point, often expressed in meters or feet.

Ellipsoid A mathematical representation of the Earth's shape, typically an oblate spheroid.

Geodesy The science of measuring and understanding the Earth's shape, size, and gravitational field.

Geodetic Control Points Surveyed and precisely located points on the Earth's surface used for geospatial reference.

Geodetic Datum A reference system that defines the Earth's shape and position.

Geometric Transformation Altering the spatial relationships and geometry of an image.

Geospatial Analysis Analyzing and interpreting images and data in the context of geographic information.

Geospatial Technologies Tools and methods that capture, analyze, and visualize geographic data.

Geographic Information System (GIS) A system for capturing, storing, analyzing, and visualizing geographic data.

Global Positioning System (GPS) A satellite-based navigation system for determining precise locations on Earth.

Image Classification The process of categorizing pixels in a remote sensing image into different land cover or land-use classes.

Image Fusion Combining information from multiple images or sensors to create a single, more informative image.

Image Registration Aligning multiple images or datasets to a common coordinate system.

Land-Use Change The transformation of coastal land from one use to another, often driven by human activities.

Local Knowledge Information and expertise about a specific area or community held by residents.

Participatory Mapping The collaborative creation of maps that reflect community perspectives and needs.

Pixel The smallest unit of a digital image, representing a single point with color or intensity values.

Relational Database A type of database that organizes data into tables with defined relationships.

Remote Sensing Platforms Vehicles or platforms (e.g., airplanes, drones) used to carry remote sensing instruments.

Satellite An artificial or natural object in orbit around a celestial body, often used for data collection.

Scale The ratio between a distance on an aerial photograph and the actual distance on the ground.

Signal Acquisition The initial phase in which a GPS receiver locks onto satellite signals.

Spatial Data Collection Gathering geographic information using various methods, including GPS and mobile apps.

Spatial Data Information that describes the location and shape of geographic features.

Spectral Signature The unique pattern of reflectance or emission of an object in different spectral bands.

Index

K. D. Malakar et al., *Geospatial Technologies in Coastal Ecologies Monitoring and Management*, Advances in Geographic Information Science, https://doi.org/10.1007/978-3-031-92017-2

The manufacturer's authorised representative in the EU is Springer Nature Customer Service Centre GmbH, Europaplatz 3, 69115 Heidelberg, Germany. If you have any concerns regarding our products, please contact ProductSafety@springernature.com

Printed and bound by CPI Group (UK) Ltd, Croydon, CR0 4YY
06/10/2026
02228482-0004